I0882455

Qualitative Methodology and the Impact of Artificial Intelligence

Caroline M. Crawford
https://orcid.org/0009-0006-6507-1960
University of Houston - Clear Lake, USA

IGI Global
Scientific Publishing
Publishing Tomorrow's Research Today

Vice President of Editorial Melissa Wagner
Director of Acquisitions Mikaela Felty
Director of Book Development Jocelynn Hessler
Production Manager Mike Brehm
Cover Design Jose Rosado

Published in the United States of America by
IGI Global Scientific Publishing
701 East Chocolate Avenue
Hershey, PA, 17033, USA
Tel: 717-533-8845 | Fax: 717-533-7115
Website: https://www.igi-global.com E-mail: cust@igi-global.com

Copyright © 2026 by IGI Global Scientific Publishing. All rights reserved. No part of this publication may be reproduced, stored or distributed in any form or by any means, electronic or mechanical, including photocopying, without written permission from the publisher. Use of this publication to train generative artificial intelligence (AI) technologies is expressly prohibited. The publisher reserves all rights to license its use for generative AI training and machine learning model development.

Product or company names used in this set are for identification purposes only. Inclusion of the names of the products or companies does not indicate a claim of ownership by IGI Global Scientific Publishing of the trademark or registered trademark.

Library of Congress Cataloging-in-Publication Data

Names: Crawford, Caroline M., 1968- editor
Title: Qualitative methodology and the impact of artificial intelligence / edited by Caroline M. Crawford.
Description: Hershey, PA : IGI Global Scientific Publishing, [2026] | Includes bibliographical references and index. | Summary: "This book examines how AI, agentic AI, and large language models are transforming qualitative research methods. It explores methodological innovations, ethical considerations, and the challenges and opportunities arising as researchers integrate AI into qualitative design, data collection, and analysis, shaping the future of qualitative inquiry"-- Provided by publisher.
Identifiers: LCCN 2026002342 (print) | LCCN 2026002343 (ebook) | ISBN 9798337376400 hardcover | ISBN 9798337376417 paperback | ISBN 9798337376424 ebook
Subjects: LCSH: Research--Data processing | Qualitative research | Qualitative research--Methodology | Artificial intelligence--Scientific applications
Classification: LCC Q180.55.E4 Q35 2026 (print) | LCC Q180.55.E4 (ebook)
LC record available at https://lccn.loc.gov/2026002342
LC ebook record available at https://lccn.loc.gov/2026002343

British Cataloguing in Publication Data
A Cataloguing in Publication record for this book is available from the British Library.

All work contributed to this book is new, previously-unpublished material.
The views expressed in this book are those of the authors, but not necessarily of the publisher.
This book contains information sourced from authentic and highly regarded references, with reasonable efforts made to ensure the reliability of the data and information presented. The authors, editors, and publisher believe the information in this book to be accurate and true as of the date of publication. Every effort has been made to trace and credit the copyright holders of all materials included. However, the authors, editors, and publisher cannot assume responsibility for the validity of all materials or the consequences of their use. Should any copyright material be found unacknowledged, please inform the publisher so that corrections may be made in future reprints.

Editorial Advisory Board

Sharon Andrews, *University of Houston-Clear Lake, USA*
Virginia Dickenson, *University of Texas Medical Branch, Galveston, USA*
Timothy B. Michael, *University of Houston-Clear Lake, USA*
Noran L. Moffett, *Fayetteville State University, USA*
Anne Rudnicki, *Rutgers University, USA*
Rachel A. Sale, *South Texas College, USA*

Table of Contents

Detailed Table of Contents

authorship, reflexivity, and interpretation, and suggests algorithmic reflexivity as a constructive way of engaging with intelligent systems and fully ascribed to the concept of algorithmic reflexivity. The author advocates for the combination of computation and deep interpretation, the decolonization of AI research, and the establishment of joint and transdisciplinary collaboration. The author asserts that the future of qualitative research is highly dependent on the extent to which researchers, as meaning co-creators, engaged with artificial intelligence in a reflexive, ethically driven, and humanized way.

This chapter reimagines qualitative inquiry in the age of artificial intelligence by examining how AI is reshaping qualitative research practices within Ghanaian higher education. Drawing on constructivist and interpretivist traditions, the chapter synthesises global and African-centred scholarship to interrogate the philosophical, methodological, and ethical implications of AI-assisted qualitative research. Particular attention is given to methodological integrity, trustworthiness, and reflexivity in hybrid human–AI inquiry, alongside ethical and contextual concerns such as algorithmic bias, epistemic injustice, and data sovereignty in African contexts. Using Ghanaian postgraduate research practices as an illustrative lens, the chapter highlights institutional realities, capacity gaps, and emerging practices. It culminates in a context-sensitive framework for responsible AI integration that foregrounds human-centred reflexivity, ethical governance, and cultural sensitivity. The chapter positions AI as a methodological partner rather than a substitute.

This chapter investigates the convergence of AI and qualitative research methodologies, introducing the Human-AI Hybrid Epistemology Model to conceptualise collaborative meaning-making where human interpretation anchors computational augmentation.

(AI-LA) in Higher Degree by Research programmes in Adelaide, Australia. Grounded in an interpretative, constructivist paradigm, the research explores lecturers' lived experiences and meanings attributed to this emerging technology within Education 4.0. Using purposive and convenience sampling, 14 lecturers from varied disciplines were recruited, allowing data saturation. Data were gathered through semi-structured interviews and analysed thematically with NVivo following Braun and Clarke's guidelines. The phenomenological approach, informed by Moustakas and Zahavi, enabled deep insight into human-technology interactions. The study offers a replicable methodological model for educational technology research and informs the design of AI-LA tools, contributing to ongoing discourse in higher education.

Wahyu Wibowo, Sunan Kalijaga State Islamic University, Yogyakarta, Indonesia
Alda Fildza Amima, Sunan Kalijaga State Islamic University, Yogyakarta, Indonesia
Nesha Rizky Ashari, Sunan Kalijaga State Islamic University, Yogyakarta, Indonesia
Dafelia Qodhari, Sunan Kalijaga State Islamic University, Yogyakarta, Indonesia
Riska Prasasti, Sunan Kalijaga State Islamic University, Yogyakarta, Indonesia

This chapter rethinks qualitative inquiry in the era of generative artificial intelligence by examining thematic analysis of digital consumer interactions. It argues that large language models can broaden the scope and speed of qualitative work through data acquisition support, pre-processing, segmentation, pre-coding, and pattern mapping, yet they cannot substitute human interpretation, sensitivity to context, and theoretical justification. Building on reflexive thematic analysis, the chapter clarifies epistemological stakes, distinguishes AI as a tool versus collaborator, and highlights risks such as decontextualization, automation bias, and opacity. It then proposes a hybrid workflow that keeps researchers accountable for coding decisions, theme development, and audit trails, while using AI to manage volume and surface candidate patterns. The chapter concludes with guidance for trustworthy and ethically responsible AI-augmented thematic analysis.

Drawing on philosophical, epistemological, methodological, ethical, and pedagogical perspectives, the chapter examines how AI literacy supports reflexive engagement, preserves interpretive authority, and enhances methodological rigor in artificial intelligence–augmented qualitative research. It explores the implications of artificial intelligence for knowledge production, researcher agency, ethics, and education, while addressing challenges such as bias, opacity, automation bias, and unequal access. The chapter ultimately argues that AI literacy is not a fixed technical skill but a dynamic, critical, and reflexive practice. By embedding AI literacy within qualitative methodology, researchers and educators can ensure that artificial intelligence enhances rather than undermines the core values of qualitative inquiry, including contextual depth, ethical responsibility, and epistemic accountability.

This chapter interrogates the positionality of artificial intelligence within contemporary qualitative inquiry and re-engineers human-machine collaborations as explicitly epistemic, ethical, and political projects. Mobilising interpretivist, critical, feminist, decolonial, and posthuman frameworks, it contrasts dataist AI epistemologies with hermeneutic, phenomenological, and justice-oriented logics of knowledge production. The chapter theorises human-centered AI assemblages, reframes trustworthiness, reflexivity, and methodological integrity for algorithmically mediated analysis, and systematises justice grammars of data justice, decolonial AI, and rights-based governance. It finally advances a typology of AI integrated qualitative designs and a programme for AI literacy, governance charters, and reporting standards that can anchor plural, reflexive, and equity enhancing qualitative research in an intensively computational era.

Preface

INTRODUCTION

The significant growth of artificial intelligence (AI), agentic artificial intelligence (agentic AI), and large language models (LLMs) has begun to impact research areas of inquiry. Specifically focusing upon qualitative methodological realms, there are intriguing things happening from expanding qualitative methodological design to the academic honesty concerns associated with student researchers and academic researchers implementing AI towards creating data sets that may be an attempt to shortcut the traditional data attainment processes. A depth and breadth of intrigue within this realm suggests the timely and desirable edited book will touch upon an area that desires significantly more focus.

The intention is to evolve an edited book that offers a space in which qualitative methodology is evolving an understanding with the dawning world of AI. This includes rethinking qualitative design, qualitative data analysis, the implementation and impact of AI upon the availability of resources as well as the potential for misrepresentation of resources and data sets. This proposed edited book is focused upon qualitative researchers as well as those who are studying qualitative methodology. This publication may be viable for researchers from innumerable areas of subject matter expertise, academic students, persons in realms of performance improvement, higher education administrative leadership, as well as training and talent development realms who emphasize a mixed methods approach to understanding the evaluation of deliverable products.

ORGANIZATION OF THE BOOK

The book is organized into fifteen chapters, within four distinct sections. The sections are: research trends and reimagining qualitative methodology; approaches to artificial intelligence; and, artificial intelligence literacy in research. A brief description of each of the chapters follows:

Chapter 1 is titled *Sentiment Analysis in AI-Driven Qualitative Methodology: Conceptual Structure and Research Trends*, by Marios Nosios and Nikos Koutsoupias. The rapid advancement of artificial intelligence has reconfigured sentiment analysis, enabling large-scale interpretation of evaluative language in digital text. This study employs a systematic review and bibliometric analysis to examine AI-driven sentiment analysis and its implications for qualitative methodology. Findings reveal an expanding field, particularly after 2018, driven by increased data availability and advances in computational models. Core themes include sentiment analysis, natural language processing, machine learning, and text mining, while deep learning and transformer-based architectures dominate. Emerging directions, notably large language models and generative artificial intelligence, are reshaping methodological practices and delineating future research trajectories. The results highlight the field's interdisciplinary character and its implications for qualitative inquiry in digitally mediated contexts.

Chapter 2 is titled *Reflexivity Reconfigured: The Changing Role of the Researcher in AI-Augmented Qualitative Inquiry* by Silvio Andrae. The integration of artificial intelligence (AI) and large language models (LLMs) into qualitative research workflows is changing the foundations of interpretive inquiry. Reflexivity has long been a core part of qualitative methodology. It requires researchers to examine how their positionality and assumptions influence knowledge production. The use of AI introduces new layers of mediation to the analytical process. AI systems now assist in data preparation, coding, thematic structuring, and analytical synthesis. They influence how qualitative data are encountered and interpreted. This chapter argues that classical reflexive frameworks, which focus mainly on researcher positionality, are insufficient for AI-augmented research. To address this, the chapter introduces dual reflexivity. This extends reflexive practice to include careful reflection on the role of algorithmic systems. The chapter presents AI as a non-human interpretive mediator. It outlines strategies such as documentation protocols, AI audit trails, and reflexive triangulation.

Chapter 3 is titled *Computational Ethnography: Reimagining Qualitative Inquiry in the Age of AI* by Kritika Sharma and Harry Kaur. This chapter conceptualizes computational ethnography as a methodological rearticulation of qualitative inquiry in the age of artificial intelligence. As social life becomes increasingly mediated by digital platforms, ethnographers confront large-scale textual data that challenge conventional analytic practices. Situating computational ethnography within constructivist and interpretivist traditions, the chapter explains how AI-enabled techniques such as natural language processing, text mining, sentiment analysis, and large language models can support exploratory pattern detection without replacing ethnographic interpretation. It critically examines epistemological and ethical challenges arising from algorithmic mediation, including the black-box problem, algorithmic bias, trustworthiness, informed consent, and data privacy. Advancing a human-centered approach, the chapter emphasizes human-in-the-loop practices to preserve reflexivity, contextual sensitivity, and ethical accountability in AI-enhanced qualitative research.

Chapter 4 is titled *The Algorithmic Researcher: AI, Reflexivity, and the Future of Qualitative Methodology* by Sovanna Huot, Seyha Kun, Sovanna Loch, Chheangkhy Hok, and Kimyut Phin. This chapter examines the changing role of the researcher's epistemic identity when engaging with artificial intelligence in qualitative inquiry. The author discusses a new kind of researcher, the algorithmic researcher, and describes this researcher as a human and machine hybrid, with the ability to create meaning in data-rich environments. The author discusses the implications of AI for the concepts of authorship, reflexivity, and interpretation, and suggests algorithmic reflexivity as a constructive way of engaging with intelligent systems and fully ascribed to the concept of algorithmic reflexivity. The author advocates for the combination of computation and deep interpretation, the decolonization of AI research, and the establishment of joint and transdisciplinary collaboration. The author asserts that the future of qualitative research is highly dependent on the extent to which researchers, as meaning co-creators, engaged with artificial intelligence in a reflexive, ethically driven, and humanized way.

Chapter 5 is titled *Reimagining Qualitative Inquiry in the Age of Artificial Intelligence: Insights from Ghanaian Higher Education Research* by Francis Arthur, Francis Obeng Gyedu, Emmanuel Quayson, Eric Boateng, Silas Afutu Quaye, Sharon Abam Nortey, Dorcas Frempong, and Mark Inkoom. This chapter reimagines qualitative inquiry in the age of artificial intelligence by examining how AI is reshaping qualitative research practices within Ghanaian higher education. Drawing on constructivist and interpretivist traditions, the chapter synthesizes global and African-centred scholarship to interrogate the philosophical, methodological, and ethical implications of AI-assisted qualitative research. Particular attention is given to methodological integrity, trustworthiness, and reflexivity in hybrid human–AI inquiry, alongside ethical and contextual concerns such as algorithmic bias, epistemic injustice, and data sovereignty in African contexts. Using Ghanaian postgraduate research practices as an illustrative lens, the chapter highlights institutional realities, capacity gaps, and emerging practices. It culminates in a context-sensitive framework for responsible AI integration that foregrounds human-centred reflexivity, ethical governance, and cultural sensitivity. The chapter positions AI as a methodological partner rather than a substitute.

Chapter 6 is titled *Human–AI Hybrid Epistemologies for Qualitative Inquiry* by Ghada Nakhla and Marina A. Soltan. This chapter investigates the convergence of AI and qualitative research methodologies, introducing the Human-AI Hybrid Epistemology Model to conceptualise collaborative meaning-making where human interpretation anchors computational augmentation. Grounded in constructivism and interpretivism, it analyses AI integrations like NLP, LLMs, automated transcription, AI-assisted coding for Grounded Theory and Thematic Analysis, sentiment analysis, virtual moderation, and computational ethnography. Ethical dimensions—bias, opacity, privacy, consent, and reconfigured trustworthiness (credibility, dependability)—demand reflexive co-production, AI positionality, validation logs, and governance frameworks ensuring fairness and transparency. Emphasising AI literacy and digital reflexivity as core competencies, it outlines pedagogical strategies for methods training and future hybrid paradigms, positioning researchers as methodological conductors who safeguard interpretive integrity, ethical rigour, and social justice in AI-enhanced inquiry.

Chapter 7 is titled *Qualitative Inquiry in the Age of Algorithms: An Integrative Systematic Review of AI's Impact on Management Control Systems* by Mouhcine Rhouiri, Hicham Saidi, and Mehdi Bensouda. This chapter explores how the integration of AI into Management Control Systems is impacting qualitative research. An integrative systematic review of 483 Scopus articles using IRaMuTeQ reveals four clusters: (1) decision support for optimizing industrial processes and logistics, (2) methodological frameworks for multi-criteria Decision-Making addressing uncertainty and subjectivity, (3) bibliometric analyses and systematic literature reviews mapping the field's evolution, and (4) AI-driven strategic decision-making and organizational transformation, including the implications of algorithmic authority. The results indicate a tension between algorithmic efficiency and qualitative depth. To preserve human values in algorithm development and application, we recommend Human-in-the-Loop frameworks that retain human value systems. Finally, this research presents a methodological taxonomy for understanding how to work with AI-enhanced Management Control Systems in a manner that aligns technology with human values.

Chapter 8 is titled *A Qualitative Phenomenological Approach to Exploring Lecturers' Perceptions of AI-Powered Learning Analytics in Higher Degree by Research Programmes* by Ivan Marcelo Arguello. This article outlines a methodological framework for a qualitative phenomenological study examining university lecturers' perceptions of AI-powered Learning Analytics (AI-LA) in Higher Degree by Research programmes in Adelaide, Australia. Grounded in an interpretative, constructivist paradigm, the research explores lecturers' lived experiences and meanings attributed to this emerging technology within Education 4.0. Using purposive and convenience sampling, 14 lecturers from varied disciplines were recruited, allowing data saturation. Data were gathered through semi-structured interviews and analysed thematically with NVivo following Braun and Clarke's guidelines. The phenomenological approach, informed by Moustakas and Zahavi, enabled deep insight into human-technology interactions. The study offers a replicable methodological model for educational technology research and informs the design of AI-LA tools, contributing to ongoing discourse in higher education.

Chapter 9 is titled *Rethinking Qualitative Inquiry with AI: Thematic Analysis of Digital Consumer Interactions* by Wahyu Wibowo, Alda Fildza Sunan, Nesha Rizky Ashari, Dafelia Qodhari, and Riska Prasasti. This chapter rethinks qualitative inquiry in the era of generative artificial intelligence by examining thematic analysis of digital consumer interactions. It argues that large language models can broaden the scope and speed of qualitative work through data acquisition support, pre-processing, segmentation, pre-coding, and pattern mapping, yet they cannot substitute human interpretation, sensitivity to context, and theoretical justification. Building on reflexive thematic analysis, the chapter clarifies epistemological stakes, distinguishes AI as a tool versus collaborator, and highlights risks such as decontextualization, automation bias, and opacity. It then proposes a hybrid workflow that keeps researchers accountable for coding decisions, theme development, and audit trails, while using AI to manage volume and surface candidate patterns. The chapter concludes with guidance for trustworthy and ethically responsible AI-augmented thematic analysis.

Chapter 10 is titled *From Manual Coding to Machine Insight: Comparing Traditional Thematic Analysis* with ChatGPT and NVivo in Qualitative Research by Kritika Sharma and Harry Kaur. This chapter examines the methodological shift in qualitative research from traditional manual thematic analysis to technology-assisted approaches using NVivo and AI-supported tools such as ChatGPT. Manual coding enables interpretive depth and reflexivity but is time-consuming and difficult to scale. NVivo strengthens analytical rigor through systematic coding and transparency while remaining dependent on researcher interpretation. ChatGPT supports exploratory analysis through summarization, preliminary coding, and reflexive questioning, yet raises methodological, epistemological, and ethical concerns related to validity, contextual sensitivity, bias, and data privacy. Rather than treating these approaches as competing methods, the chapter proposes a hybrid, reflexive framework that integrates human interpretive authority with digital and AI-based support tools, offering practical guidance for responsible and rigorous qualitative analysis.

Chapter 11 is titled *Large Language Models in Qualitative Research* by Angga Muhammad Galih, Binastya Anggara Sekti, Bartholomeus Dimanche Carl Beth, Gusti Muhamad Sardana. The integration of Large Language Models (LLMs) into qualitative research offers transformative capabilities in automated transcription, coding, and thematic analysis, yet challenges foundational principles including contextual understanding, researcher reflexivity, and interpretive depth. This examination critically analyzes LLM applications across research workflows, addressing the black-box problem, algorithmic bias, and ethical concerns regarding informed consent, data privacy, and participant rights. The study proposes hybrid frameworks positioning LLMs as augmentative tools while maintaining human interpretive authority, emphasizing transparent documentation, rigorous validation, and researcher AI literacy for responsible integration.

Chapter 12 is titled *The Algorithm as a Cultural Informant: Exploring AI-Mediated Realities through Computational Ethnography* by Rismawati Rismawati, Usep Suhud, and Eko Ganis Sukoharsono. This chapter also treats algorithms as culture informants, extending an account of computational ethnography that attempts to link interpretive depth with computational scale. From a qualitative and constructivist perspective, algorithms are cast as epistemic agents, cultural artefacts encoding design values that serve to cultural actors to mediate meaning,norms and visibility. The chapter combines digital traces, human-AI interaction and reflexive recordings with more traditional ethnographic techniques such as interviews and fieldnotes. It introduces an analytical hybrid combining NLP, topic modeling, sentiment analysis and LLM facilitated reflexive coding under human governance. Transparency, bias auditing, and Human-Centered AI principles ensure the ethical and epistemic rigor of our approaches. The chapter finishes by discussing the implications for social, organizational and CSR research in promoting reflexive, co-constructed human-AI inquiry that considers algorithms as situated cultural sources of knowledge.

Chapter 13 is titled *Rethinking Qualitative Methodology through the Lens of AI Literacy: Navigating Trustworthiness, Reflexivity, and Human-AI Collaboration* by Freda Awonakie Akwetey and Francis Arthur. Abstract This chapter advances the AI-Literate Qualitative Methodology (AILQM) framework as an integrative model for embedding artificial intelligence within qualitative research while safeguarding epistemological integrity, methodological rigour, and ethical responsibility. It articulates six interrelated literacy dimensions: Cognitive and Data Literacy, Ethical and Equity Literacy, Methodological Literacy, Collaborative Literacy, Trustworthiness Literacy, and Policy and Governance Literacy. Together, these competencies enable researchers to critically interpret AI outputs, document human-AI decision-making, and uphold credibility, dependability, confirmability, and transferability. By reframing AI literacy as a methodological orientation rather than a technical skill, the AILQM framework provides a principled pathway for conducting reflexive, transparent, and human-centred qualitative research in the evolving AI landscape.

Chapter 14 is titled *AI Literacy* by Apoorav Sharma and Lovleen Marwaha. Drawing on philosophical, epistemological, methodological, ethical, and pedagogical perspectives, the chapter examines how AI literacy supports reflexive engagement, preserves interpretive authority, and enhances methodological rigor in artificial intelligence–augmented qualitative research. It explores the implications of artificial intelligence for knowledge production, researcher agency, ethics, and education, while addressing challenges such as bias, opacity, automation bias, and unequal access. The chapter ultimately argues that AI literacy is not a fixed technical skill but a dynamic, critical, and reflexive practice. By embedding AI literacy within qualitative methodology, researchers and educators can ensure that artificial intelligence enhances rather than undermines the core values of qualitative inquiry, including contextual depth, ethical responsibility, and epistemic accountability.

Chapter 15 is titled *Positionality of AI and Human-Machine Collaborations in AI-Augmented Qualitative Research: Methodological Recalibration in Critical Research* by Ashraf Alam. This chapter interrogates the positionality of artificial intelligence within contemporary qualitative inquiry and re-engineers human-machine collaborations as explicitly epistemic, ethical, and political projects. Mobilising interpretivist, critical, feminist, decolonial, and posthuman frameworks, it contrasts dataist AI epistemologies with hermeneutic, phenomenological, and justice-oriented logics of knowledge production. The chapter theorises human-centered AI assemblages, reframes trustworthiness, reflexivity, and methodological integrity for algorithmically mediated analysis, and systematises justice grammars of data justice, decolonial AI, and rights-based governance. It finally advances a typology of AI integrated qualitative designs and a programme for AI literacy, governance charters, and reporting standards that can anchor plural, reflexive, and equity enhancing qualitative research in an intensively computational era.

CONCLUSION

The primary objective of this edited volume is to explore the front end of shifts and changes that are occurring within methodological realms. All aspects of AI are impacting qualitative methodology, reflecting that this is a great opportunity to take a leading stride in this area. This offers qualitative methodologists the opportunity to publish their forward-leaning progression in their initial recognitions and efforts associated with qualitative methodology and AI.

Section 1
Research Trends and Reimagining Qualitative Methodology

Chapter 1

Sentiment Analysis in AI-Driven Qualitative Methodology:

Conceptual Structure and Research Trends

Marios Nosios
http://orcid.org/0009-0007-6161-3854
University of Macedonia, Greece

Nikos Koutsoupias
http://orcid.org/0000-0003-1664-5404
University of Macedonia, Greece

ABSTRACT

The rapid advancement of artificial intelligence has reconfigured sentiment analysis, enabling large-scale interpretation of evaluative language in digital text. This study employs a systematic review and bibliometric analysis to examine AI-driven sentiment analysis and its implications for qualitative methodology. Findings reveal an expanding field, particularly after 2018, driven by increased data availability and advances in computational models. Core themes include sentiment analysis, natural language processing, machine learning, and text mining, while deep learning and transformer-based architectures dominate. Emerging directions, notably large language models and generative artificial intelligence, are reshaping methodological practices and delineating future research trajectories. The results highlight the field's interdisciplinary character and its implications for qualitative inquiry in digitally mediated contexts.

DOI: 10.4018/979-8-3373-7640-0.ch001

ıt © 2026, IGI Global Scientific Publishing. Copying or distributing in print or electronic forms without written permission of IGI Global Scientific Publishing is prohibited. Use of this chapter to ›enerative artificial intelligence (AI) technologies is expressly prohibited. The publisher reserves all rights to license its use for generative AI training and machine learning model development.

INTRODUCTION

Sentiment Analysis (SA) has developed over several decades, evolving from early attempts to detect subjective language and "semantic orientation" into a sophisticated and diverse array of computational approaches designed to identify evaluative meaning within large textual corpora (Pang & Lee, 2008; Turney, 2002). Early research framed sentiment as a problem distinct from topical classification, emphasizing that textual communication conveys not only factual information but also expressions of opinion, belief, and emotion (Liu, 2022; Pang et al., 2002). Over time, the field expanded both conceptually and methodologically, moving from lexicon-based techniques toward Machine Learning (ML) systems, neural architectures, and, more recently, transformer-based large language models (Naredla et al., 2025; Gong et al., 2022; Feldman, 2013). Applications across domains such as politics, public health, marketing, and digital communication illustrate the analytical potential of SA, while simultaneously revealing its limitations when applied across shifting linguistic, cultural, or contextual environments (Afriyie & Weyori, 2026; Khan et al., 2014). For qualitative research, the most consequential development has been the gradual emergence of hybrid research workflows in which computational sentiment signals function as exploratory or complementary instruments rather than substitutes for interpretive analysis (Pang & Lee, 2008). This study examines the evolution of sentiment analysis methods and their implications for qualitative research, arguing that despite rapid computational advancement, the field remains structurally under-integrated with interpretive frameworks, reflecting a persistent epistemological divide between computational scalability and qualitative depth.

Early scholarship established SA as the computational study of evaluative meaning expressed through language (Cambria et al., 2013). Researchers observed that textual data frequently contain "private states, beliefs, feelings, or attitudes", that are expressed through identifiable linguistic markers and can therefore be detected through computational methods (Wiebe et al., 2005). Work on subjectivity classification constituted one of the earliest systematic attempts to distinguish between objective statements and those containing subjective evaluations (Yu & Hatzivassiloglou, 2003). This distinction clarified that sentiment detection involves more than identifying positive or negative language; it also requires determining where opinions occur in a text before attempting to classify their polarity or intensity (Wilson et al., 2005). Such research laid the conceptual groundwork for subsequent approaches by demonstrating that evaluative meaning is often embedded within broader linguistic structures and cannot be reduced to isolated lexical signals (Medhat et al., 2014).

Another foundational strand of research conceptualized semantic orientation as the idea that words and phrases carry evaluative associations that can be inferred from usage patterns in large textual corpora (Turney, 2002). Studies examining the orientation of adjectives demonstrated that polarity information could be derived from distributional patterns rather than solely from manually constructed dictionaries (Hatzivassiloglou & McKeown, 1997). Methods based on statistical association measures, such as pointwise mutual information with positive and negative reference terms, demonstrated that evaluative meaning could be inferred from textual data through indirect evidence (Turney, 2002). These approaches represented an early form of data-driven sentiment modeling and anticipated later developments in ML. At the same time, researchers recognized that semantic orientation is highly context-dependent (Pang et al., 2002). A term that conveys positive evaluation in one domain may signal risk or uncertainty in another, thereby illustrating the importance of domain sensitivity in sentiment classification (Blitzer et al., 2007).

A further conceptual development emerged from research that framed sentiment classification as a supervised learning problem distinct from topical categorization (Pang & Lee, 2004; Sebastiani, 2002). Studies applying ML techniques to review corpora demonstrated that standard classification algorithms could outperform simple word-list approaches, while simultaneously revealing that sentiment classification presents challenges not typically encountered in topic detection (Moraes et al., 2013; Mullen & Collier, 2004). Evaluative meaning frequently depends on compositional linguistic structures such as negation, contrast, modality, and discourse context (Wiegand et al., 2010). A sentence containing both positive and negative expressions may communicate an overall evaluation that diverges from the straightforward aggregation of individual lexical items (Socher et al., 2013). These observations underscored the interpretive nature of sentiment detection: identifying sentiment is not merely a mechanical task but a form of interpretation grounded in linguistic evidence (Cambria, 2016).

Over time, the conceptual scope of SA broadened to include related tasks such as opinion mining, stance detection, emotion recognition, and aspect-based evaluation (Mohammad, 2021). Later survey studies clarified that SA operates across multiple levels of granularity, including document-level classification, sentence-level polarity detection, and aspect-level evaluation, in which attitudes toward specific entities or features are identified (Liu, 2022; Medhat et al., 2014). These distinctions carry important methodological implications. Interpreting sentiment scores requires explicit specification of the analytical unit, the evaluative target, and the contextual frame within which expressions of sentiment are produced (Taboada, 2016). Without such clarification, numerical sentiment indicators risk being interpreted as stable measurements despite their dependence on linguistic context and analytical design.

Methodological development in SA has followed a trajectory in which different approaches represent alternative strategies for modeling evaluative meaning (Feldman, 2013). Early lexicon-based approaches relied on dictionaries of words associated with positive or negative sentiment. In such systems, the sentiment of a sentence or document is typically computed by aggregating the polarity scores of its constituent words (Ding et al., 2008). Resources such as SentiWordNet and other affective lexicons provided structured representations of lexical polarity that could be used for automated analysis without the need for labeled training data (Esuli & Sebastiani, 2006). In parallel, psycholinguistic dictionaries such as LIWC offered category-based representations of affective, cognitive, and social dimensions of language, which were frequently used alongside sentiment-oriented approaches rather than functioning strictly as sentiment lexicons (Tausczik & Pennebaker, 2010). These methods offered transparency and interpretability, since the basis of each classification could be traced to identifiable lexical items. Nevertheless, lexicon-based approaches also revealed several limitations, including sensitivity to domain variation, difficulty capturing implicit or sarcastic sentiment, and dependence on culturally specific patterns of language use (Loughran & McDonald, 2011).

The emergence of supervised ML methods marked a significant methodological shift (Aggarwal & Zhai, 2012). Rather than relying exclusively on predefined dictionaries, supervised models learn patterns from labeled datasets in which texts are annotated according to their sentiment polarity. Algorithms such as naïve Bayes classifiers, support vector machines, and logistic regression became widely employed for sentiment classification tasks (Pak & Paroubek, 2010). These approaches typically rely on engineered features derived from textual representations, including word frequencies, n-gram patterns, syntactic information, or lexicon-based indicators. The transition toward supervised learning improved classification performance in many contexts but also introduced new methodological dependencies. The quality and representativeness of training data became central concerns, as models trained on a single dataset often performed poorly when applied to texts from different domains or populations (Pan et al., 2010). Research on domain adaptation subsequently highlighted the fragility of sentiment models when confronted with shifts in vocabulary, genre, or communication context.

Subsequent developments in Deep Learning (DL) substantially reshaped SA research and benchmark performance by enabling models to learn distributed representations of language directly from data (Young et al., 2018). Neural network architectures such as convolutional neural networks, recurrent neural networks, and long short-term memory networks became widely adopted for textual classification tasks (Zhang et al., 2015). These models benefited from pretrained word embeddings that capture semantic relationships among words, allowing them to model contextual dependencies and compositional linguistic structures more effectively than earlier approaches (Zhang et al., 2018). The introduction of compositional sentiment datasets and phrase-level annotations further encouraged the development of architectures capable of capturing complex linguistic interactions such as negation, intensification, or contrast (Zhou et al., 2016). While DL methods frequently improved predictive performance, they also reduced interpretability by embedding sentiment representations within complex model parameters that are difficult to examine directly (Guidotti et al., 2018).

More recent research has been shaped by the emergence of transformer architectures and large pretrained language models (Rogers et al., 2020). Attention-based models enable the processing of long textual sequences while simultaneously capturing contextual relationships among words (Devlin et al., 2019). Pretrained encoders such as BERT shifted SA toward transfer-learning frameworks in which models trained on large corpora can subsequently be adapted to specific tasks through fine-tuning (Sun et al., 2019). Later developments emphasized the importance of training data scale, architectural design, and optimization strategies in determining model performance. Generative language models introduced additional capabilities, including few-shot classification and instruction-following, enabling sentiment classification to be performed through prompting rather than traditional supervised training (Brown et al., 2020). Although these models demonstrate strong performance on many benchmark tasks, their behavior remains sensitive to prompt design, evaluation methodology, and contextual ambiguity (Liu et al., 2023). Consequently, sentiment outputs generated by large language models require scrutiny rather than being treated as definitive indicators of evaluative meaning.

The expansion of SA methods has been accompanied by growing empirical application across a wide range of social research domains (Stieglitz & Dang-Xuan, 2013). Digital communication platforms have produced vast volumes of textual data that can be analyzed computationally. Social media platforms, online reviews, forums, and news archives have therefore become common sources for sentiment-based research (Zimbra et al., 2018). Studies of political communication have examined whether sentiment extracted from online discussions corresponds to patterns observed in survey-based indicators of public opinion (O'Connor et al., 2010). Some studies have reported correlations between aggregated sentiment signals and indicators such as political approval or consumer confidence, although these relationships often remain sensitive to sampling strategies, operationalization choices, and temporal aggregation. Methodological reflection has therefore emphasized that such relationships depend heavily on the alignment between textual sentiment measures and the theoretical constructs they are intended to represent (Jungherr et al., 2011).

Empirical studies demonstrate the relationship between social media-derived sentiment and established indicators of public opinion. Analyses utilizing large-scale Twitter datasets indicate that aggregated sentiment signals can track and, in some cases, anticipate changes in consumer confidence and political approval as measured by traditional survey instruments. Within these analytical frameworks, sentiment scores are not treated as direct proxies for individual attitudes; instead, they serve as aggregated indicators whose interpretive validity relies on contextualization and systematic validation against established benchmarks, such as polling data. Empirical evidence further shows that while sentiment measures can effectively capture broad temporal trends, they are highly sensitive to methodological parameters, including keyword selection, sampling procedures, and the temporal granularity of analysis (O'Connor et al., 2010). This example highlights both the analytical potential of SA and the need for ongoing interpretive scrutiny when incorporating computational outputs into the broader epistemological frameworks of social research.

Public health research provides another illustration of the application of SA in social inquiry (Paul & Dredze, 2011). Studies examining health-related communication on social media have attempted to identify patterns in public concerns, disease reporting, or emotional responses to health crises (Guntuku et al., 2017). Such analyses can contribute to situational awareness by identifying emerging issues within large volumes of communication data (Charles-Smith et al., 2015). At the same time, the interpretive leap from textual expression to population-level conditions remains methodologically complex. Messages may refer to news reports, humor, or third-party observations rather than personal experiences, thereby complicating attempts to interpret sentiment indicators as direct evidence of public health conditions (Salathé et al., 2013).

Economic and organizational research has also incorporated SA to examine expectations, perceptions, and collective mood (Tetlock et al., 2008). Early analyses of financial discussion boards and investor communication explored whether aggregated sentiment signals correlate with market indicators (Antweiler & Frank, 2005). While such studies demonstrate the feasibility of extracting sentiment from textual communication, they also illustrate persistent concerns regarding construct validity (Loughran & McDonald, 2016). Sentiment indicators may capture attention, controversy, or topic salience rather than underlying emotional states or expectations. Consequently, researchers frequently emphasize the importance of robustness checks, alternative measurement strategies, and manual validation when sentiment measures are used to support substantive conclusions (Bollen et al., 2011).

Applications in consumer research frequently rely on review datasets to examine evaluations of products or services (Hu & Liu, 2004). Studies of online reviews have explored patterns of satisfaction, complaint narratives, and feature-specific evaluations (Archak et al., 2011). Aspect-based SA represents an important methodological development in this context because it enables researchers to identify attitudes toward specific attributes rather than relying solely on overall polarity scores (Schouten & Frasincar, 2016). This approach aligns more closely with the structure of evaluative discourse, where individuals often express distinct sentiments toward different aspects of an experience. Nevertheless, domain adaptation remains a persistent challenge. Models trained on a single review category often require recalibration before being applied to other types of textual material (Glorot et al., 2011).

The expansion of computational text analysis has also stimulated growing interest in integrating SA with qualitative methodological approaches (Grimmer & Stewart, 2013). Automated sentiment measures can substantially reduce the cost of exploring large-scale textual corpora and may help researchers identify patterns that warrant closer examination (DiMaggio et al., 2013). Within such frameworks, sentiment indicators frequently function as exploratory instruments rather than final analytical outcomes. Computational methods can highlight clusters of documents, temporal periods, or communication communities in which evaluative language appears particularly concentrated, thereby guiding more detailed qualitative interpretation (Nelson, 2017). When employed in this manner, automated sentiment signals serve as heuristic devices that support interpretive inquiry.

Mixed-methods frameworks provide one pathway for integrating computational and qualitative approaches (Blei, 2012). Iterative research designs can combine automated pattern detection with interpretive analysis, allowing computational outputs to inform the development of qualitative coding schemes while human interpretation refines the categories subsequently used in computational analysis (Nelson, 2017). Such workflows emphasize that sentiment models should be treated as analytical instruments whose outputs require validation through close reading, contextual interpretation, and theoretical reflection (Grimmer & Stewart, 2013).

Within qualitative and mixed-methods research workflows, SA is most productively positioned as an exploratory and supportive analytical stage rather than a standalone method. In practice, this typically involves a sequential procedure in which computational SA is initially used to identify patterns, clusters, or temporal shifts in large-scale textual corpora, followed by qualitative coding and interpretive analysis to examine how evaluative meaning is constructed in specific contexts. Subsequent stages may involve iterative refinement, in which insights derived from qualitative interpretation inform adjustments to computational models or the selection of analytically relevant subsets. This integrated approach underscores that sentiment outputs do not constitute final findings; rather, they serve to guide and structure deeper interpretive engagement with textual data (Nelson, 2017; Grimmer & Stewart, 2013).

Despite these opportunities, several limitations remain particularly salient for qualitative methodological practice (Zhang et al., 2018). Sentiment classification systems often struggle to capture implicit evaluation, irony, or culturally embedded meaning (Maynard & Greenwood, 2014). These challenges are especially pronounced in genres such as interviews, policy documents, or ethnographic narratives, where evaluative meaning may be conveyed indirectly through narrative structure or contextual reference. Research on domain adaptation further demonstrates that model performance can deteriorate when analytical tools developed for one textual domain are applied to another (Weiss et al., 2016). From a qualitative perspective, the principal methodological risk lies not only in misclassification but also in the possibility that automated sentiment categories may inadvertently redefine the phenomenon under study by reducing complex forms of evaluation to simplified polarity measures (Grimmer & Stewart, 2013).

Ethical and epistemological considerations have become increasingly prominent as SA has been applied to socially sensitive data sources (Metcalf & Crawford, 2016). Research using online communication frequently involves texts produced by individuals who did not explicitly consent to participation in research studies. Ethical guidelines for internet research, therefore, emphasize context-sensitive judgment regarding privacy, potential harm, and the protection of vulnerable populations. Regulatory frameworks governing personal data processing further complicate the use of large-scale digital text corpora (Floridi et al., 2018). These concerns resonate strongly with qualitative research traditions that emphasize reflexivity, participant protection, and the responsible interpretation of human communication (Tracy, 2010).

From a qualitative methodological perspective, these considerations further foreground the centrality of researcher reflexivity in the application of SA. The selection of analytical tools, the configuration of computational models, and the interpretation of sentiment outputs are not neutral or purely technical procedures; rather, they are shaped by the researcher's theoretical commitments, methodological orientations, and broader interpretive frameworks. Decisions concerning data selection, preprocessing techniques, and model parameterization can substantially influence how evaluative meaning is represented, while the interpretation of aggregated sentiment indicators necessitates careful contextualization within the specific social and discursive environment under investigation. Drawing on established principles of qualitative reflexivity (Tracy, 2010), sustained reflexive engagement is therefore indispensable to ensure that computational outputs are not treated as objective or unmediated representations of sentiment, but are instead critically interrogated as constructed analytical artefacts that inform, rather than determine, qualitative interpretation (Grimmer & Stewart, 2013).

Bias in SA systems represents an additional area of concern (Blodgett et al., 2020). Studies evaluating sentiment models have identified systematic biases associated with gender, race, and other social categories (Kiritchenko & Mohammad, 2018). When such biases are embedded within training data or model design, they can influence analytical outcomes and potentially distort interpretations of public attitudes or group behavior. Documentation practices such as dataset descriptions and model reporting frameworks have therefore been proposed to address these issues by encouraging transparency regarding data composition, intended uses, and methodological limitations (Gebru et al., 2021; Mitchell et al., 2019).

Interpretability remains a central methodological issue, particularly for neural and transformer-based models. Although techniques such as feature attribution or post-hoc explanation methods can offer insights into model behavior, these approaches do not always reveal the full reasoning process underlying model predictions (Guidotti et al., 2018). Scholars have therefore emphasized that model selection should be aligned with the epistemological requirements of the research question (Rudin, 2019). In exploratory studies, more opaque models may be acceptable as heuristic instruments, whereas research that seeks to draw substantive conclusions about social attitudes may require more interpretable analytical approaches (Miller, 2019).

These considerations suggest that the future development of SA within social research will likely involve increasingly hybrid methodological strategies. Large language models expand the range of analytical possibilities, enabling new forms of assisted classification, summarization, and pattern detection. At the same time, methodological scholarship consistently emphasizes that computational outputs must be interpreted within broader research frameworks that incorporate theoretical reasoning, qualitative validation, and transparent reporting practices. When SA is integrated with interpretive methodologies in this manner, it becomes possible to combine the scale of automated text analysis with the contextual sensitivity characteristic of qualitative research.

Viewed in this light, SA should not be understood as a replacement for qualitative inquiry but rather as a complementary analytical instrument capable of supporting the exploration of large textual corpora. Computational techniques can reveal patterns that may remain difficult to detect through manual examination alone, while qualitative approaches remain indispensable for interpreting these patterns within their social, cultural, and discursive contexts. Maintaining this methodological balance is particularly important in research settings where meaning is complex, contested, or embedded within nuanced linguistic expressions.

Continued methodological development will therefore depend not only on advances in computational modeling but also on the refinement of research designs that integrate automated analysis with interpretive scrutiny. Careful operationalization of sentiment constructs, systematic validation procedures, and transparent documentation of analytical decisions will remain central requirements for producing credible and reproducible findings. Through such practices, SA can contribute to a more comprehensive understanding of evaluative language in social communication while remaining aligned with the epistemological commitments of qualitative research.

METHODOLOGY

This study employs a bibliometric research design to examine the development of AI-driven sentiment analysis and its relationship with qualitative methodological practice (Bitzenis et al., 2025). Bibliometric analysis facilitates the systematic investigation of scientific publications to identify patterns in research production, intellectual influence, and conceptual evolution within a given field (Donthu et al., 2021; Ellegaard & Wallin, 2015). By examining publication metadata, citation relationships, and keyword structures, this methodological approach provides a structured overview of how scholarly debates and methodological orientations have evolved.

The dataset used in this study was retrieved from the Scopus database, which provides extensive coverage of peer-reviewed literature across both scientific and social science disciplines (Baas et al., 2020; Schotten et al., 2017). A structured search strategy was designed to identify publications addressing SA in relation to AI and associated computational techniques. The search was conducted using title, abstract, and keyword fields (TITLE-ABS-KEY) to balance thematic precision with coverage. The search query used to retrieve the dataset is as follows:

TITLE-ABS-KEY((“sentiment analysis” OR “opinion mining”) AND (“artificial intelligence” OR AI OR “machine learning” OR “ML” OR “deep learning” OR “natural language processing” OR NLP OR “large language model*” OR LLM* OR BERT OR GPT OR transformer*) AND (qualitative OR “qualitative research” OR “qualitative method*” OR interpretiv* OR constructiv* OR “mixed method*” OR “mixed-methods” OR “text analysis” OR “content analysis”)) AND (LIMIT-TO (SUBJAREA,”COMP”) OR LIMIT-TO (SUBJAREA,”BUSI”) OR LIMIT-TO (SUBJAREA,”SOCI”) OR LIMIT-TO (SUBJAREA,”ECON”) OR LIMIT-TO (SUBJAREA,”MATH”) OR LIMIT-TO (SUBJAREA,”ENGI”) OR LIMIT-TO (SUBJAREA,”PSYC”) OR LIMIT-TO (SUBJAREA,”MULT”) OR LIMIT-TO (SUBJAREA,”DECI”) OR LIMIT-TO (SUBJAREA,”ARTS”)) AND (LIMIT-TO (DOCTYPE,”ar”) OR LIMIT-TO (DOCTYPE,”cp”) OR LIMIT-TO (DOCTYPE,”ch”) OR LIMIT-TO (DOCTYPE,”re”)) AND (LIMIT-TO (LANGUAGE,”English”))

To ensure transparency and methodological replicability, the selection procedure followed systematic review principles inspired by the PRISMA framework (Page et al., 2021). Records retrieved from the initial search were carefully screened to remove duplicate entries and publications that did not directly address the intersection of SA and computational analytical techniques (Bitzenis et al., 2023). Inclusion criteria were defined based on thematic relevance to SA and computational methods, while exclusion criteria removed duplicate records and publications lacking direct methodological relevance. The screening process was conducted through manual review of titles and abstracts to ensure alignment with the study’s objectives. This procedure ensured that the final dataset represented a coherent and methodologically relevant body of literature aligned with the study’s research objectives (see Figure 1).

Figure 1. PRISMA 2020 flow diagram of the selection process

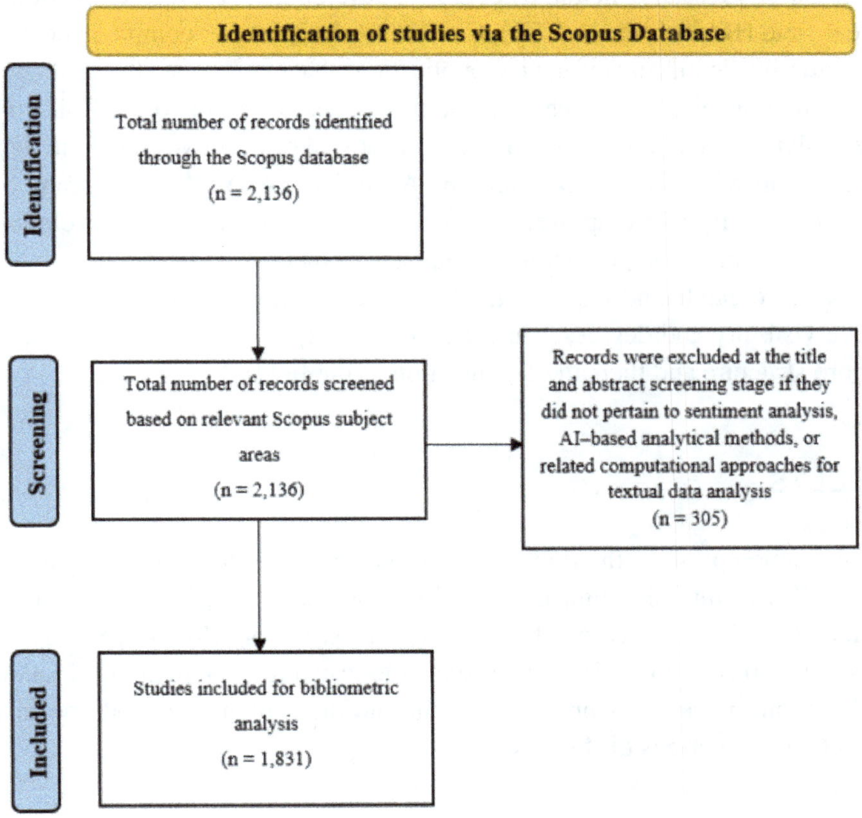

The analytical procedures were conducted using the bibliometrix package within the R programming environment, which enables large-scale bibliographic analysis and science-mapping visualization (Aria & Cuccurullo, 2017). Descriptive indicators were initially calculated to examine the dataset's structural characteristics, including publication output over time, citation patterns, authorship dynamics, and collaboration networks. These indicators provide insight into the field's growth trajectory and the extent of scholarly engagement across disciplinary domains.

To investigate the conceptual structure of the literature, several science-mapping techniques were applied (Koutsoupias et al., 2026). Keyword co-occurrence analysis was employed to identify dominant research themes and to visualize relationships among methodological concepts such as natural language processing, ML, and DL. Standard bibliometrix parameters were applied for keyword co-occurrence and thematic mapping, including minimum frequency thresholds and normalization procedures to ensure comparability across documents. In addition, thematic mapping was used to classify research themes by degree of development and relevance within the broader field (Bitzenis et al., 2026). Citation analysis further complemented these procedures by identifying influential publications that have shaped the methodological foundations of SA research (Papavassiliou-Alexiou et al., 2023). Collectively, these analytical techniques provide a comprehensive perspective on the intellectual structure and evolutionary trajectory of AI-driven SA. Within this framework, bibliometric analysis is employed not merely to map computational developments but also to examine the positioning of qualitative methodological concepts within the broader research landscape. Particular attention is therefore directed toward the relative visibility and degree of integration of interpretive approaches within the keyword structure and thematic organization of the field.

RESULTS

This section presents the findings derived from the bibliometric analysis of the dataset. The results highlight the principal publication patterns and conceptual developments that characterize the research landscape of AI-driven SA. Through the examination of publication trends, thematic structures, and patterns of scholarly collaboration, the analysis provides insight into the evolving methodological and intellectual foundations of the field.

Table 1. Descriptive Statistics

Description	Results
MAIN INFORMATION ABOUT DATA	
Timespan	2008:2026
Sources (Journals, Books, etc.)	1126
Documents	1831
Annual Growth Rate %	24.74
Document Average Age	3.51
Average citations per doc	16.95
References	142613
DOCUMENT CONTENTS	
Keywords Plus (ID)	6448
Author's Keywords (DE)	4524
AUTHORS	

continued on following page

Table 1. Continued

Description	Results
Authors	5696
Authors of single-authored docs	132
AUTHORS COLLABORATION	
Single-authored docs	141
Co-Authors per Doc	3.55
International co-authorships %	21.46
DOCUMENT TYPES	
article	912
book chapter	90
conference paper	777
review	52

Table 1 summarizes the descriptive characteristics of the dataset used in the bibliometric analysis, presenting key indicators of publication output, document composition, and patterns of scholarly collaboration. These indicators provide an overview of the scale and structural features of the literature examined in connection with AI-driven SA and its methodological intersections. The dataset covers the period from 2008 to 2026 and comprises 1,831 documents originating from 1,126 publication sources, including journals, conference proceedings, and books. This distribution indicates a broad dispersion of publications across outlets rather than concentration within a limited set of journals. Such dispersion is typical of research areas situated at the intersection of computer science and the social sciences, where methodological innovations and applied studies are disseminated through a wide range of disciplinary venues. The temporal coverage also corresponds to the period when SA underwent substantial methodological expansion, particularly alongside developments in natural language processing and ML.

Publication activity in the dataset reveals a marked increase over time, with an annual growth rate of 24.74%. This rate indicates a significant rise in scholarly attention during the period under examination. Growth at this scale often reflects the emergence of a research domain in which new analytical approaches and practical applications are actively explored. In the context of this study, the observed trend underscores the growing importance of computational techniques for analyzing sentiment in text, particularly as large-scale digital datasets and AI methods have become increasingly accessible to researchers across multiple disciplines.

Further indicators shed light on the maturity and citation dynamics of the literature. The average age of the documents is 3.51 years, suggesting that a considerable portion of the research output has been produced recently. A relatively young body of literature is characteristic of fields shaped by technological progress, where methodological developments frequently generate rapid cycles of publication and dissemination. Despite this recency, the documents average 16.95 citations per publication, indicating that many contributions have already attracted notable scholarly attention. In total, the corpus cites 142,613 references, demonstrating the substantial volume of prior research that informs investigations in this area.

The dataset's textual dimension is reflected in the number of indexed keywords associated with the publications. Overall, 6,448 Keywords Plus terms and 4,524 author-provided keywords appear in the dataset. The difference between these counts illustrates the role of database-generated indexing in expanding the conceptual descriptors assigned to individual publications. Author keywords generally represent the principal themes emphasized by researchers, whereas Keywords Plus often introduces additional descriptors derived from cited references. Taken together, these keyword sets help map the literature's thematic organization, which is particularly useful for identifying methodological patterns in SA research.

Authorship patterns offer further insight into the collaborative structure of the field. The dataset includes 5,696 distinct authors, indicating broad participation across institutions and disciplinary backgrounds. Among these contributors, 132 authors produced single-authored works. In total, 141 publications were written by a single author, suggesting that individually authored studies represent only a small proportion of the overall output. The majority of contributions, therefore, appear to originate from collaborative research teams, a pattern commonly observed in computational and data-intensive domains where diverse methodological expertise is often required.

Collaboration metrics reinforce this observation. On average, each publication includes 3.55 co-authors, indicating a moderate degree of collaboration among researchers. International cooperation is also evident, with 21.46% of publications involving authors affiliated with institutions in different countries. Cross-national collaboration at this level indicates that the field operates within a global research environment, where knowledge exchange and methodological advancement occur across institutional and geographic boundaries. Such patterns are consistent with research areas that combine technological innovation with interdisciplinary inquiry.

The distribution of document types further illustrates the field's publication structure. Journal articles constitute the largest category, accounting for 912 publications, followed by conference papers, which account for 777 publications. The strong presence of conference contributions reflects the influence of computer science and natural language processing communities, where conferences frequently serve as primary venues for presenting emerging methodological developments. In addition, the dataset contains 90 book chapters and 52 review articles. Although these formats represent a smaller share of the corpus, they often provide conceptual syntheses or methodological reflections that contextualize ongoing developments in the literature.

Overall, the descriptive indicators presented in Table 1 depict a research landscape characterized by rapid expansion, extensive collaboration, and dissemination across a diverse array of publication venues. The relatively recent age of the documents and the substantial presence of conference papers point to a field that continues to evolve alongside advances in artificial intelligence and computational text analysis. Within the broader scope of this study, these characteristics establish a quantitative foundation for examining how SA has developed as a methodological approach and how it is increasingly embedded within interdisciplinary research contexts (see Table 1).

Figure 2. Annual Scientific Production

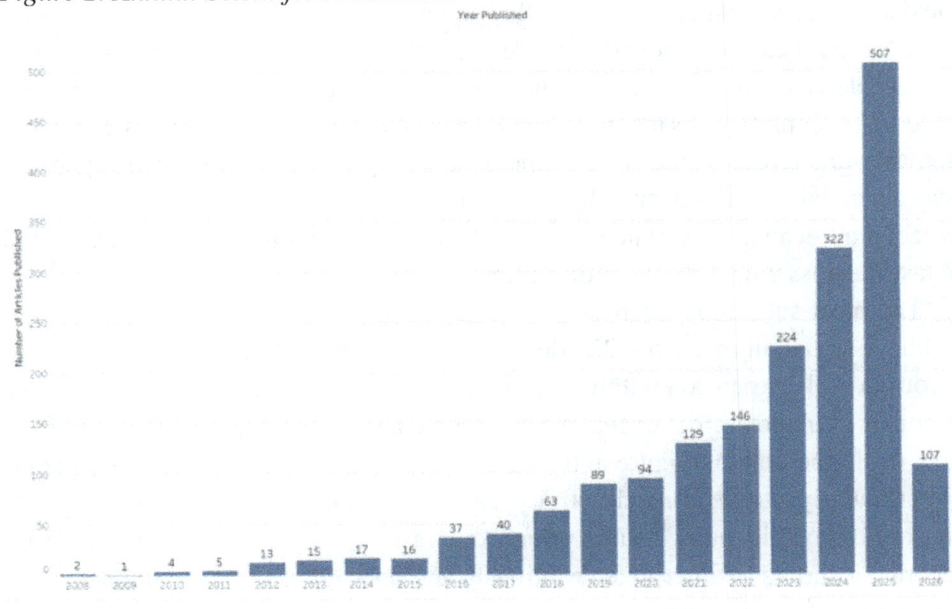

The annual distribution of publications presented in Figure 2, provides an overview of how scholarly output within the dataset has evolved. The data span the period from 2008 to 2026 and record the number of documents produced each year in the examined body of literature. During the earliest years covered by the dataset, publication activity remained relatively limited. Between 2008 and 2011, annual output ranged from one to five documents, indicating that research in this area was still in its formative stage. Contributions during this period appear sporadic rather than continuous, suggesting that the topic had not yet consolidated into a stable research stream. A modest increase becomes visible beginning in 2012, when annual output reached 13 publications. In the following years, a gradual rise is observed, with

15 documents in 2013 and 17 in 2014, followed by a slight decline to 16 in 2015. Although these figures remain comparatively small, they indicate the emergence of a more regular pattern of scholarly activity during the early phase of the timeline.

A more pronounced upward trajectory becomes evident in the middle of the period. In 2016, the number of publications increased to 37, representing more than a twofold rise compared with the previous year. This expansion continued in 2017, with 40 publications, and intensified further in 2018, with 63 documents. By 2019, the annual total reached 89 publications, reflecting a steady broadening of scholarly engagement. The progression across these years suggests that the research area was gaining increasing visibility, with a growing number of studies applying computational text analysis and sentiment-oriented methods across diverse analytical contexts.

The expansion of publication activity continued into the early 2020s. In 2020, the dataset records 94 documents, followed by a more substantial increase in 2021 with 129 publications. This upward trend persisted in 2022, with 146 documents, and accelerated further in 2023, when the annual count rose to 224. The pattern observed over these years indicates a phase of sustained growth in which the research domain attracted an increasing volume of scholarly contributions. Such expansion may be linked to the broader diffusion of artificial intelligence and natural language processing technologies, which have enabled researchers to analyze large textual datasets across a wide range of disciplines.

The most substantial increases occur in the most recent years of the dataset. Publication output rose from 224 documents in 2023 to 322 in 2024, representing a considerable expansion within a relatively short interval. The upward trajectory becomes even more pronounced in 2025, when the number of publications reached 507, the highest annual figure in the dataset. This sharp increase indicates a period of particularly intense research activity. The final year included in the dataset, 2026, records 107 publications. As this value likely reflects only a partial year of indexed records, it should be interpreted cautiously and does not necessarily indicate a reduction in scholarly output.

Across the entire time span, the data reveal a clear transition from limited early activity to a phase of rapid growth in publication output. The initial years are characterized by few contributions, whereas the period after the mid-2010s exhibits a progressively accelerating pattern of scholarly production. This increase suggests that the topic has gradually evolved from a relatively specialized area of inquiry into a more widely explored research domain. Similar developmental patterns are frequently observed in fields shaped by advances in digital data availability and computational analytical techniques.

Overall, this temporal distribution provides important context for understanding how research on SA and related computational approaches has expanded over time. The observed increase in publication activity corresponds with the growing adoption of AI-based analytical tools for examining textual data across diverse disciplinary settings. As publication output continues to rise, the literature surrounding sentiment-oriented methodologies appears to be consolidating into a substantial and continuously evolving body of research (see Figure 2).

Figure 3. Most Relevant Author Keywords

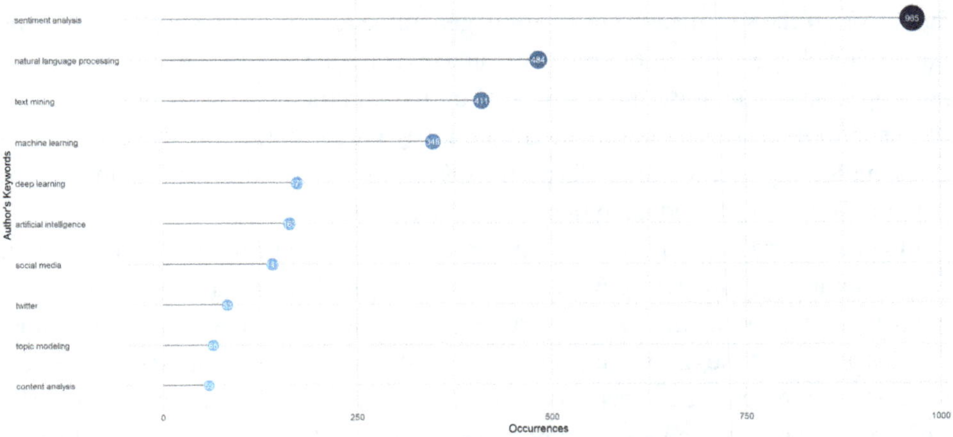

The distribution of the most frequently occurring author keywords in Figure 3 provides insight into the conceptual and methodological orientation of the literature included in the dataset. The figure presents the ten most frequent keywords, with the horizontal axis indicating the number of occurrences and the vertical axis listing the author-supplied terms. Among all keywords, "sentiment analysis" emerges as the most dominant term, with 965 occurrences, substantially surpassing the frequency of the remaining keywords. This prominence reflects the dataset's central analytical focus, in which SA serves as the principal methodological framework across the studies collected.

A second group of highly frequent keywords pertains to computational techniques for SA. The term "natural language processing" appears 484 times, followed by "text mining" with 411 occurrences and "machine learning" with 348 occurrences. These terms represent core analytical approaches used to process and interpret large textual datasets. Their strong presence indicates that many studies situate SA within broader frameworks of computational text analysis. The proximity of these methodological concepts suggests that research in this area often integrates sentiment classification with additional automated text-processing methods.

Additional keywords reflect developments associated with more advanced artificial intelligence techniques. The keyword "deep learning" appears 173 times, while "artificial intelligence" occurs 163 times. Although these terms appear less frequently than the broader computational categories noted earlier, their inclusion among the most common keywords signals the increasing adoption of advanced ML architectures within sentiment-oriented research. Their frequency indicates that a growing segment of the literature addresses methodological innovations linked to neural network models and AI-based analytical approaches.

Several keywords also point to the types of data sources commonly used in the studies represented in the dataset. The term “social media” appears 141 times, underscoring the significance of digitally generated text for SA research. Platforms that produce large volumes of user-generated text offer accessible datasets for examining opinions, reactions, and patterns of online communication. A more specific reference to data sources is provided by the keyword “twitter,” which appears 83 times, indicating that social media platforms remain widely used environments for applying sentiment-based analytical techniques.

Other keywords reflect analytical strategies associated with textual interpretation. “Topic modeling,” appearing 65 times, refers to a method frequently employed to identify thematic structures within large text corpora. Its presence among the most frequent keywords suggests that sentiment analysis is often combined with computational approaches designed to uncover patterns within textual data. Similarly, “content analysis,” which appears 59 times, highlights the continuing relevance of interpretive and qualitative analytical traditions within the broader research landscape.

The overall structure of the keyword distribution reveals two principal thematic orientations within the literature. One group of terms centers on computational methodologies, including natural language processing, machine learning, and deep learning. Another group relates to data sources and analytical applications, particularly social media environments and textual analysis techniques. The co-occurrence of these categories among the most frequent keywords suggests that the literature commonly combines methodological development with applied analysis of large-scale textual data. Taken together, the keyword frequencies offer a concise overview of the field's conceptual focus. The prominence of sentiment analysis alongside computational approaches such as natural language processing and machine learning underscores the methodological foundations of the research domain. At the same time, the appearance of terms associated with social media data and textual analysis highlights the empirical contexts in which these techniques are most frequently applied. The distribution, therefore, reflects a research area shaped by the convergence of computational text analysis, artificial intelligence methodologies, and the examination of extensive digital text corpora. The comparatively lower frequency of terms such as ‘content analysis’ suggests that explicitly qualitative or interpretive approaches are less prominently represented within the keyword structure. This distribution indicates that, while qualitative traditions remain present, the dominant orientation of the literature is shaped by computational methodologies, with comparatively less emphasis on their integration into qualitative research frameworks (see Figure 3).

Figure 4. Most Cited Countries

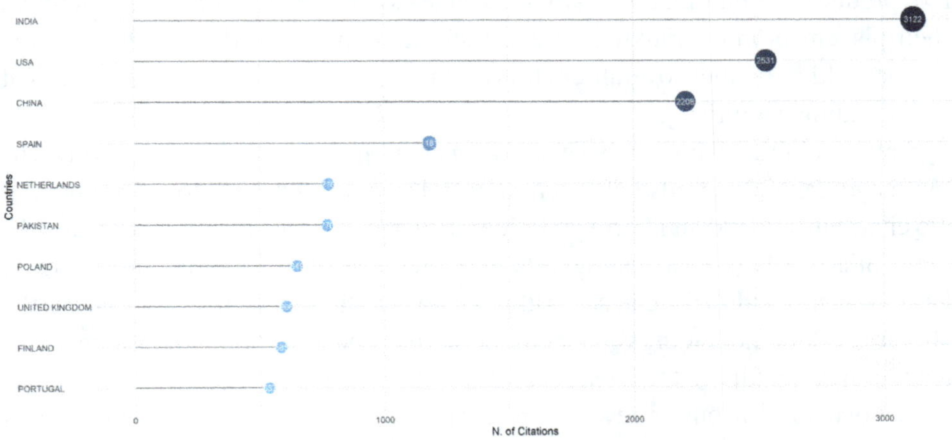

The distribution of citations across countries in Figure 4 provides an overview of the geographic influence associated with the research included in the dataset. The figure presents the ten most cited countries, with the horizontal axis representing the total number of citations and the vertical axis listing the contributing countries. A clear concentration of citations appears among three countries that dominate the dataset: India, the United States, and China. Among these, India records the highest citation count, with 3,122, placing it at the top of the ranking. The United States follows with 2,531 citations, while China has 2,208. The prominence of these countries indicates that a considerable proportion of influential publications originates from research institutions located within these regions. Their position at the upper end of the citation distribution suggests that scholars from these countries contribute substantially to the development and dissemination of research on computational text analysis and sentiment-oriented methodologies.

A noticeable gap separates these leading countries from the remaining entries in the ranking. The next group includes Spain, which records 1,118 citations, the highest among the European countries shown in the figure. Spain's position indicates a visible contribution to the literature, particularly within research areas that integrate computational linguistic techniques with applied analytical contexts. Although its citation count remains considerably lower than those of the leading countries, it nonetheless represents a significant presence within the international research landscape.

The remaining countries shown in the figure exhibit substantially lower citation levels, forming a secondary tier of contributors. The Netherlands accumulates 776 citations, followed closely by Pakistan with 770 citations. These values indicate moderate visibility within the dataset, suggesting that researchers from these countries have produced studies that have attracted a measurable level of scholarly attention. Their placement in the ranking reflects their participation in the broader research network, although their overall citation influence remains more limited than that of the leading countries.

Several additional countries appear with citation counts within a relatively narrow range. Poland records 649 citations, while the United Kingdom accumulates 606 citations. Finland and Portugal follow with 585 and 537 citations, respectively. The similarity of these values suggests comparable levels of citation impact among these countries within the dataset. Although none of these countries approaches the citation levels of the leading contributors, their presence in the ranking demonstrates that research activity in this area spans multiple geographic regions.

The overall distribution reveals a pattern in which a small number of countries account for a substantial share of citations, while a broader group contributes more modestly. Such distributions are commonly observed in bibliometric analyses, where research influence often concentrates among countries with well-established research infrastructures and active publication networks. At the same time, the inclusion of countries from both Europe and Asia among the most-cited contributors indicates that the literature draws on a geographically diverse set of scholarly inputs.

Collectively, the citation distribution highlights the international character of the research field. Contributions from Asia, North America, and Europe appear across the ranking, reflecting a globally distributed network of scholarly activity. The prominence of countries such as India, the United States, and China suggests that these regions play a leading role in shaping the development of computational approaches to textual analysis. Meanwhile, the presence of several European countries in the top ten indicates that research in this area is not concentrated in a single geographic region but emerges from a broad range of institutional and national contexts. This pattern indicates that citation impact is concentrated among a small number of countries, particularly India, the United States, and China, while the remaining contributors exhibit comparatively lower and more closely clustered citation levels (see Figure 4).

Figure 5. Thematic Map

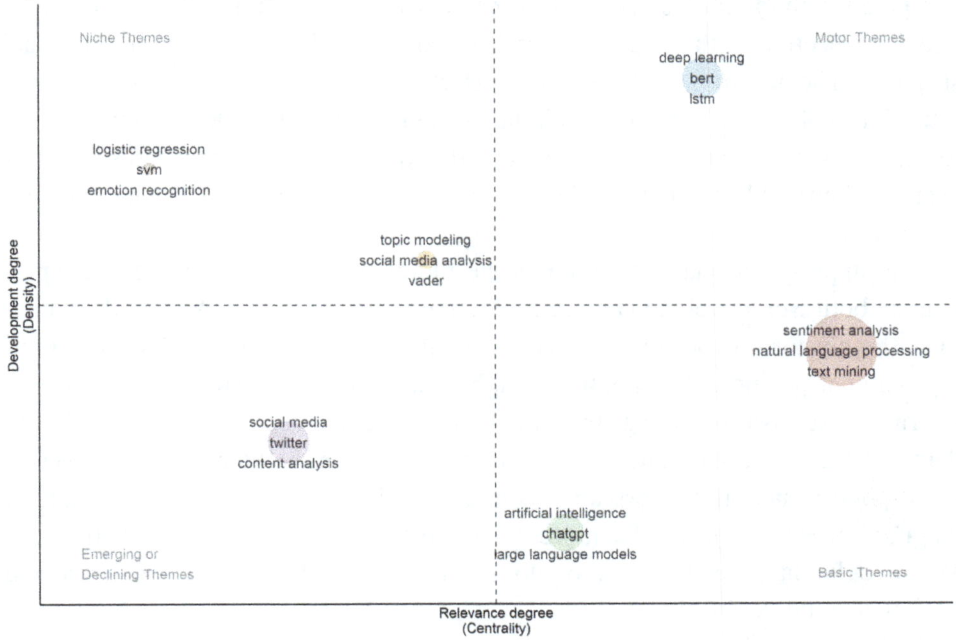

The thematic map in Figure 5 visualizes the conceptual organization of the research field by positioning keyword clusters along two analytical dimensions: centrality and density. Centrality represents the extent to which a theme is connected to other themes within the research network, whereas density reflects the degree of internal cohesion within a thematic cluster. By combining these two dimensions, the map divides the thematic landscape into four quadrants: motor themes, basic themes, niche themes, and emerging or declining themes, thereby providing a structured overview of how different research topics are situated within the literature.

In the lower-right quadrant, the cluster consisting of “sentiment analysis,” “natural language processing,” and “text mining” appears as the largest and most central grouping. Themes in this area are classified as basic, indicating they are highly relevant to the research domain while exhibiting comparatively lower levels of internal specialization. The size and central position of this cluster suggest that these concepts constitute the foundational methodological core of the field. Their close proximity reflects the frequent integration of sentiment analysis with broader natural language processing and text mining techniques. This positioning indicates that a substantial portion of the literature relies on these computational approaches as the primary methodological framework for analyzing textual data.

Another cluster located within the same quadrant contains the terms “artificial intelligence,” “chatgpt,” and “large language models.” Although this grouping appears near the central-right area of the map, its density remains relatively modest compared with other clusters. Such placement indicates that the theme maintains strong connections to the broader research network yet lacks a well-developed internal structure. The grouping of these terms reflects the growing presence of generative artificial intelligence and large-scale language models within the literature. Their location among the basic themes suggests that these concepts are increasingly relevant to the field but remain in the process of consolidating into a more coherent research focus.

The upper-right quadrant contains the motor themes, which represent topics that are both well-developed and central to the research domain. Within this quadrant, the cluster composed of "deep learning," "BERT," and "LSTM" occupies a prominent position. The relatively high density of this cluster indicates strong internal relationships among these methodological concepts, while its centrality demonstrates significant integration with other areas of the literature. These terms correspond to neural network architectures widely used in contemporary natural language processing tasks. Their presence as motor themes suggests that advanced ML models play a leading methodological role in developing sentiment-oriented analytical approaches.

In the upper-left quadrant, several clusters appear under the niche themes category, which are characterized by strong internal development but weaker connections to the broader research network. One cluster includes "logistic regression," "SVM," and "emotion recognition." These terms refer to earlier or more specialized computational techniques used for classification tasks in textual analysis. Their location suggests that although these methods exhibit internal coherence within specific research contexts, they are less central to the field's current structure than more recent ML approaches. Another cluster positioned near the central-left area includes "topic modeling," "social media analysis," and "VADER." The intermediate placement of this group indicates that these analytical techniques remain relevant in the literature, particularly in studies examining textual data from online environments.

In the lower-left quadrant, the cluster consisting of "social media," "Twitter," and "content analysis" appears under the category of emerging or declining themes. Themes located in this quadrant typically exhibit lower levels of both centrality and density, suggesting limited integration within the broader research network and relatively modest internal development. The presence of these terms indicates that certain research strands, focusing on individual social media platforms or on traditional content analysis approaches, occupy a more peripheral position within the current thematic configuration. This placement may reflect a shift in emphasis toward more computationally intensive analytical techniques or toward broader methodological frameworks.

Ultimately, the thematic map reveals a research landscape structured primarily around computational methods for textual analysis. Core techniques such as sentiment analysis, natural language processing, and text mining form the conceptual foundation of the field, while neural network architectures are among the most developed and influential methodological approaches. Emerging topics related to artificial intelligence and large language models appear increasingly connected to the broader literature, although their internal consolidation remains ongoing. At the same time, earlier classification techniques and platform-specific analytical contexts appear more specialized or peripheral within the overall thematic structure. This distribution suggests that interpretive and qualitative methodological perspectives are present within the literature but occupy a relatively less central position. In particular, terms such as "content analysis" appear within lower-centrality thematic areas, whereas computational approaches dominate both the core and motor themes. This pattern indicates a stronger emphasis on computational development within the field (see Figure 5).

Figure 6. Co-Occurrence Network Graph

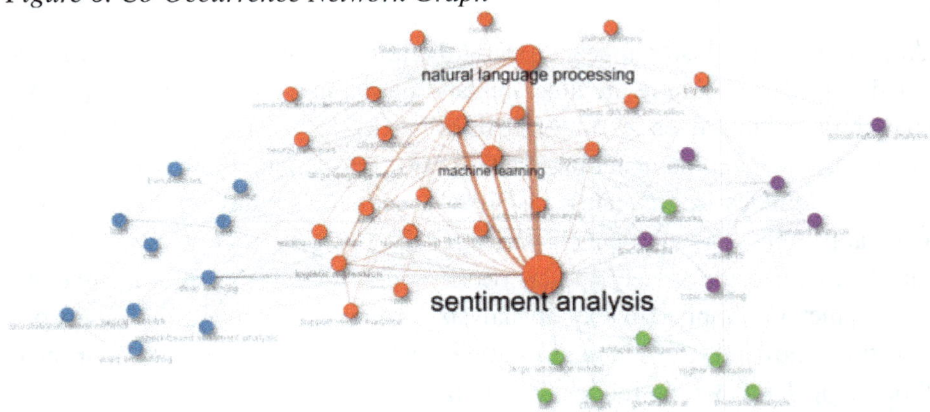

The co-occurrence network in Figure 6 visualizes how author keywords co-occur across the publications in the dataset. In this graph, nodes correspond to keywords, while links indicate the frequency with which two terms occur within the same documents. Larger nodes represent higher keyword frequencies, and thicker edges denote stronger co-occurrence relationships. The resulting network structure reveals several clusters that represent groups of concepts frequently discussed together in the literature.

At the center of the network, "sentiment analysis" appears as the largest and most interconnected node, highlighting its central role within the dataset. Numerous links extend from this term toward both methodological and application-oriented keywords, illustrating how sentiment analysis functions as the primary analytical framework around which other topics are organized. Closely connected nodes surrounding this central term include "machine learning," "natural language processing," "text classification," and "data mining." The density of these connections suggests that studies frequently integrate sentiment analysis with broader computational techniques to process and analyze large textual datasets. The central positioning of these terms indicates that the literature relies heavily on automated text-processing approaches when implementing sentiment-oriented methods.

One prominent cluster located near the center of the network groups together keywords associated with ML-based analytical techniques. Within this cluster, terms such as "machine learning," "classification," "support vector machine," "random forest," "logistic regression," and "emotion detection" appear closely linked. These relationships represent a set of methodological approaches widely used for sentiment classification tasks. The coexistence of traditional machine learning algorithms and broader classification terminology suggests that a substantial portion of the literature focuses on developing or applying predictive models to identify sentiment in textual data.

Another distinct cluster appears toward the left side of the network and is associated with deep learning architectures and neural network models. Keywords in this group include "deep learning," "BERT," "CNN," "LSTM," "transformers," "RoBERTa," and "word embedding." The connections among these nodes suggest that these techniques are frequently discussed together in studies employing neural network approaches to sentiment classification. The presence of transformer-based architectures alongside earlier neural network models reflects the evolution of computational methods applied to textual analysis. Their placement within the network demonstrates how DL techniques form a specialized yet strongly interconnected methodological area within the broader literature.

A third cluster located on the right side of the network focuses on application contexts and data sources, particularly those related to online platforms. Keywords such as "social media," "Twitter," "topic modeling," "content analysis," and "social network analysis" appear within this group. The connections among these terms indicate that many studies apply sentiment analysis to datasets derived from digital communication environments. The presence of "COVID-19" within the same cluster further suggests that sentiment analysis has been employed to examine public reactions and communication patterns associated with major social events or crises.

Another smaller grouping appears in the lower-right portion of the network, including terms associated with recent developments in artificial intelligence. Keywords such as "artificial intelligence," "large language model," "ChatGPT," "generative AI," and "LLM" are located within this area. Although these nodes are fewer in number than those in the larger clusters, their presence indicates the growing integration of generative language models into research on textual sentiment analysis. Their position toward the periphery of the network suggests that these topics remain relatively new within the broader research landscape.

The links connecting these clusters illustrate how different methodological and application-oriented areas intersect. Connections among the machine learning, natural language processing, and deep learning nodes demonstrate the methodological overlap among these computational approaches. At the same time, edges linking these technical clusters with terms associated with social media or content analysis indicate that many studies combine advanced analytical techniques with real-world textual datasets. Overall, the network structure highlights several interconnected thematic areas: a central methodological core centered on sentiment analysis and machine learning; a specialized cluster of neural network-based approaches; a group of application-oriented terms related to social media and textual data sources; and a smaller cluster reflecting emerging interest in large language models and generative artificial intelligence. The configuration of these clusters indicates that research in this area is shaped by the interaction between computational techniques for text analysis and the diverse contexts in which these methods are applied. The relative positioning of nodes such as 'content analysis' and 'social network analysis' toward more peripheral clusters further suggests that interpretive and qualitative approaches are less prominently integrated within the co-occurrence structure compared to computational methodologies (see Figure 6).

Table 2. Top Fifteen Most Cited Papers

Paper	Total Citations
Hutto Cj, 2014, Proc Int Conf Weblogs Soc Media, ICWSM	4492
Wankhade M, 2022, Artif Intell Rev	1227
Schouten K, 2016, IEEE Trans Knowl Data Eng	738
Kocoń J, 2023, Inf Fusion	576
Mäntylä Mv, 2018, Comput Sci Rev	558
Hussein Dme-Dm, 2018, J King Saud Univ Eng Sci	546
Mostafa Mm, 2013, Expert Sys Appl	494
Panichella S, 2015, IEEE Int Conf Softw Maintenance Evol, ICSME - Proc	423
Gao Z, 2019, IEEE Access	404
Chen M, 2017, Icmi - Proc ACM Int Conf Multimod Interact	321
Giatsoglou M, 2017, Expert Sys Appl	316
Ruz Ga, 2020, Future Gener Comput Syst	298

continued on following page

Table 2. Continued

Paper	Total Citations
Van Atteveldt W, 2021, Commun Methods Meas	294
Hu Y-H, 2017, Inf Process Manage	269
Verma S, 2021, J Interact Mark	265

The most highly cited publications presented in Table 2 reflect the methodological maturation and expanding interdisciplinary relevance of SA within the broader landscape of artificial intelligence and natural language processing. Taken collectively, these influential works demonstrate how computational techniques for extracting evaluative meaning from textual data have evolved from rule-based lexicon systems toward increasingly complex ML, DL, and large language model architectures. At the same time, they illustrate how SA has progressively moved beyond purely technical experimentation toward applications involving human communication, digital interaction, and the interpretation of large-scale textual corpora. This body of literature, therefore, provides both the methodological infrastructure and the conceptual grounding for contemporary research examining the intersection of AI-driven SA and qualitative methodological inquiry.

Several of the most highly cited contributions focus on developing analytical frameworks that reliably detect sentiment in informal, large-scale digital communication environments. One of the most influential examples is the work of Hutto and Gilbert (2014), whose VADER sentiment model was designed specifically to analyze microblogging and social media texts and has accumulated more than 4,492 citations. Their rule-based system combines a carefully validated lexical resource with heuristic rules that account for linguistic features such as capitalization, punctuation, and intensifiers, thereby enabling robust sentiment detection in short, informal messages. Empirical evaluation demonstrates that the model performs competitively with ML techniques while maintaining interpretability and computational efficiency. The widespread adoption of VADER across disciplines highlights the importance of transparent analytical tools that enable researchers to systematically analyze large volumes of user-generated content while preserving insight into how sentiment scores are generated.

Complementing this methodological foundation are several comprehensive surveys that synthesize the rapidly expanding body of research on SA. The review conducted by Wankhade et al. (2022), which has received approximately 1,227 citations, offers a detailed overview of SA techniques, applications, and challenges, emphasizing the progression from lexicon-based approaches toward ML and hybrid systems that combine multiple analytical strategies. Similarly, Mäntylä et al. (2018), whose work has been cited more than 558 times, provide a large-scale bibliometric and thematic examination of SA research by analyzing thousands of publications to trace the field's historical evolution. Their findings reveal that SA has experienced exponential growth since the mid-2000s, largely driven by the emergence of digital platforms that generate vast quantities of opinionated text.

Another cluster of highly cited works addresses the methodological challenges associated with extracting fine-grained sentiment information from complex textual data. Schouten and Frasincar (2016), cited more than 738 times, provide one of the most comprehensive surveys of aspect-level SA, a subfield that focuses on identifying specific targets or attributes within text and determining the sentiment expressed toward each. By systematically reviewing techniques for aspect extraction, sentiment classification, and evaluation, their work demonstrates how aspect-based analysis can yield more nuanced representations of user opinions across domains such as product reviews, service evaluations, and online feedback.

Parallel methodological developments can be observed in research exploring ML and DL techniques for sentiment classification. Hussein (2018), whose study has accumulated more than 546 citations, provides a comprehensive overview of SA approaches, including lexicon-based, ML, and hybrid techniques, and discusses algorithms such as support vector machines, decision trees, and neural networks that can be applied to large textual datasets. Similarly, Mostafa (2013), cited more than 494 times, investigates SA within online tourism reviews and demonstrates how ML models can extract meaningful insights from consumer feedback in the hospitality industry.

Research on the application of SA in specialized domains further illustrates the practical value of these techniques. Panichella et al. (2015), with approximately 423 citations, explore the use of SA in software engineering environments and show how textual feedback from developers and users can be analyzed to detect emotional tone and communication patterns within collaborative software development communities. Likewise, Hu et al. (2017), whose work has received more than 269 citations, examine computational techniques for analyzing large collections of online reviews, emphasizing the role of ML and natural language processing in extracting structured knowledge from unstructured text.

Another group of influential studies investigates the role of SA in technologically mediated communication and human-computer interaction. Chen et al. (2017), which has been cited more than 321 times, analyze multimodal sentiment detection in interactive communication environments by integrating textual, vocal, and visual signals to improve the accuracy of sentiment recognition systems. Gao et al. (2019), whose work has accumulated approximately 404 citations, propose computational frameworks that enhance sentiment classification accuracy through advanced feature representation and DL techniques. The increasing integration of DL approaches has further reshaped the methodological landscape of SA research.

Giatsoglou et al. (2017), which has been cited more than 316 times, present a comprehensive examination of SA models that process large-scale textual datasets using neural architectures. This line of research is extended by Ruz et al. (2020), whose work has received approximately 298 citations and develops ML frameworks for analyzing financial sentiment in textual data streams.

More recent research has begun to explore the implications of large language models for SA and related natural language processing tasks. Kocoń et al. (2023), whose study has already accumulated approximately 576 citations, conduct a large-scale evaluation of ChatGPT across a wide range of NLP tasks, including SA, emotion recognition, and stance detection, demonstrating that although large language models exhibit impressive general capabilities, their performance remains uneven across specialized tasks. Further contributions examine methodological considerations for interpreting and measuring sentiment in social research contexts. Van Atteveldt et al. (2021), cited approximately 294 times, analyze the validity of SA by comparing automated methods with manual coding and crowd-sourced annotation procedures. Similarly, Verma and Yadav (2021), whose work has accumulated roughly 265 citations, provide a bibliometric and science-mapping analysis of electronic word-of-mouth research and highlight how techniques such as text mining, SA, and ML are increasingly used to extract consumer insights from online reviews and digital communication platforms.

Collectively, the highly cited studies summarized in Table 2 reveal several recurring patterns within the evolution of SA research. First, the literature consistently emphasizes the importance of methodological innovation in transforming unstructured textual data into analyzable information. Lexicon-based systems, ML classifiers, and DL architectures represent successive stages in the development of increasingly sophisticated analytical tools. Second, these studies highlight the central role of digital communication platforms as empirical environments in which SA techniques are frequently applied. Overall, the influential works summarized in Table 2 illustrate how SA has developed into a central methodological tool for examining evaluative language within large textual datasets. The convergence of advances in artificial intelligence, ML, and natural language processing has enabled researchers to analyze digital discourse at unprecedented scale while uncovering patterns in human attitudes, emotions, and perceptions (see Table 2).

CONCLUSION

The present study set out to examine the intellectual development and methodological implications of AI-driven SA within the broader landscape of qualitative research practice. By combining a systematic review with bibliometric analysis, the study maps the conceptual, methodological, and empirical contours of a rapidly expanding research domain. Drawing on 1,831 documents indexed in Scopus and published between 2008 and 2026, the analysis integrates descriptive indicators, keyword analysis, thematic mapping, co-occurrence networks, and citation structures to provide a comprehensive overview of how SA has evolved alongside advances in AI, ML, and natural language processing. The results collectively demonstrate that SA has transitioned from a relatively specialized computational technique into a widely adopted analytical instrument for interpreting large-scale textual data generated across digital environments.

Several structural patterns emerge from the bibliometric indicators. The dataset reveals a field characterized by rapid expansion and increasing interdisciplinary engagement. Although early contributions appear sporadically during the late 2000s and early 2010s, publication activity accelerates markedly after 2018, reflecting both the growing availability of digital textual data and the maturation of AI-based analytical techniques. The pronounced increase in output observed during the most recent years of the dataset, particularly between 2023 and 2025, illustrates how advances in DL architectures, transformer-based models, and large language models have stimulated renewed scholarly attention. Descriptive metrics reinforce this interpretation of a dynamic and evolving research landscape. The relatively young average document age and strong annual growth rate suggest that the knowledge base continues to expand rapidly, while the distribution of publications across more than a thousand sources reflects the field's inherently interdisciplinary character.

The conceptual structure of the field becomes particularly visible through keyword analysis and thematic mapping. The most frequently occurring terms, including SA, natural language processing, text mining, and ML, constitute the methodological core of the research domain. Their prominence indicates that the extraction and interpretation of evaluative language from textual data remain central analytical objectives in the literature. Surrounding these core concepts are a series of supporting methodological and contextual keywords, such as deep learning, artificial intelligence, and topic modeling, as well as application-oriented terms related to social media, Twitter, and content analysis. Taken together, these terms illustrate how computational techniques and empirical contexts have evolved in tandem. SA is rarely treated as an isolated technical task; rather, it is embedded within broader research efforts to understand how individuals express opinions, emotions, and evaluations in digitally mediated communication environments.

The thematic map further clarifies the internal organization of the research domain by distinguishing between well-established methodological foundations and emerging technological developments. At the center of the map lie core analytical themes, including SA, natural language processing, and text mining. These topics exhibit high centrality within the network, reflecting their foundational role in structuring the field. DL architectures, including models such as BERT and LSTM, are a strong theme that has achieved considerable methodological maturity. Their prominence illustrates the extent to which neural language models have reshaped computational approaches to sentiment detection over the past decade. At the same time, emerging themes, including large language models, ChatGPT, and generative artificial intelligence, occupy positions that suggest growing scholarly attention but comparatively limited integration within the broader thematic structure. This configuration reflects the transitional state of the field: although generative AI technologies are attracting increasing interest, their methodological and epistemological implications for research practice remain under active exploration.

Insights derived from the keyword co-occurrence network reinforce these observations by revealing the relational structure among major concepts within the literature. The network is anchored by a dense cluster centered on SA, natural language processing, and ML, indicating that these methodological approaches constitute the primary analytical infrastructure for interpreting textual data. Adjacent clusters represent complementary areas of research, including deep learning architectures and neural network models that enhance sentiment classification accuracy. Another grouping is associated with empirical applications, particularly those involving social media platforms and other forms of user-generated content. The presence of these clusters underscores the strong connection between methodological innovation and the proliferation of digital communication channels that generate large volumes of opinionated text. Emerging nodes in artificial intelligence, large language models, and generative systems further illustrate how recent technological developments are reshaping the analytical toolkit available to researchers.

The examination of the most highly cited publications provides additional insight into the field's intellectual foundations. Influential studies frequently introduce methodological frameworks for processing large-scale textual datasets or improving the accuracy of sentiment classification. Early contributions emphasize lexicon-based approaches and rule-based analytical systems for capturing sentiment signals in informal online communication. Subsequent research increasingly incorporates ML techniques and neural network architectures, enabling more sophisticated modeling of linguistic context and semantic relationships. More recent contributions explore the capabilities of large language models and generative AI systems, evaluating their potential for performing complex natural language processing tasks, including sentiment recognition. Taken together, these highly cited works illustrate the cumulative nature of methodological progress within the field, as successive generations of analytical techniques build upon earlier computational foundations.

Beyond their technical contributions, these studies collectively highlight the growing relevance of SA across a wide range of research domains. Applications span fields such as marketing, finance, public health, political communication, and software engineering, demonstrating the versatility of computational approaches to opinion analysis. The capacity to analyze large volumes of textual data enables researchers to examine patterns of evaluation, emotional expression, and public discourse at scales that are difficult to achieve with traditional qualitative methods alone. Nevertheless, the findings of the present study also suggest that automated SA should be interpreted with caution. While computational tools provide valuable mechanisms for identifying patterns in textual corpora, the interpretation of evaluative language remains inherently contextual and often ambiguous.

These observations carry important implications for qualitative methodological practice. Rather than replacing interpretive analysis, AI-driven SA increasingly serves as a complementary analytical tool, helping researchers navigate large-scale textual datasets. Automated methods can identify trends, clusters, or anomalous patterns that may warrant closer examination, thereby guiding subsequent qualitative interpretation. When integrated within mixed-method research designs, SA offers opportunities to combine computational scalability with the contextual sensitivity characteristic of qualitative inquiry. Such hybrid approaches enable researchers to leverage the strengths of both analytical traditions while mitigating the limitations associated with relying exclusively on either automated or manual interpretation.

The findings of this bibliometric analysis indicate a discernible imbalance in the field's development. Although methodological innovation in SA has been propelled largely by advances in computational performance and model architecture, comparatively limited attention has been devoted to its systematic integration within qualitative research frameworks. Evidence derived from keyword distributions, thematic mapping, and co-occurrence networks suggests that themes related to interpretive methodologies, reflexivity, and contextual analysis occupy a relatively peripheral position within the existing literature. This pattern points to the continued underrepresentation of qualitative perspectives within the dominant research discourse. Addressing this imbalance necessitates more sustained and systematic efforts to bridge computational and interpretive approaches, thereby aligning ongoing developments in SA with the epistemological demands of qualitative inquiry. This interpretation should be understood in light of the methodological scope of bibliometric analysis, which captures patterns of publication and keyword usage rather than the actual implementation of methods within individual research designs.

Despite the insights generated by the present analysis, several limitations should be acknowledged. The bibliometric dataset is derived exclusively from Scopus-indexed publications, which may omit relevant contributions from other databases or non-English venues. As a consequence, regional scholarship and emerging research communities may be underrepresented. In addition, the keyword-based search strategy employed to identify relevant publications necessarily prioritizes established terminology, potentially overlooking studies that address similar conceptual issues using alternative vocabulary. Methodological considerations also arise from the increasing complexity of AI models used in SA. As DL architectures and large language models become more sophisticated, concerns about transparency, interpretability, and reproducibility have become increasingly salient, particularly in research contexts where analytical outputs inform substantive interpretations.

These limitations highlight several promising directions for future research. Expanding bibliometric analyses to include multiple databases and multilingual search strategies would provide a more globally representative perspective on the development of SA research. Comparative investigations examining the performance of different analytical techniques across diverse linguistic and cultural contexts could further illuminate the robustness of sentiment detection methods. Additional work is also needed to evaluate the role of large language models in SA workflows, particularly regarding bias, explainability, and methodological accountability. Integrating multimodal data sources, including text, images, and behavioral traces, may also enable more nuanced representations of evaluative expression in digital communication environments.

In summary, the findings presented in this study demonstrate that research on AI-driven SA has evolved into a dynamic and interdisciplinary field characterized by rapid methodological innovation and expanding empirical applications. The bibliometric patterns, thematic structures, and citation networks identified here reveal a research domain anchored in established computational techniques while simultaneously adapting to emerging developments in artificial intelligence. As digital communication continues to generate vast quantities of textual data, SA will remain a central analytical approach for interpreting evaluative discourse across social, economic, and technological contexts. Continued methodological refinement, greater transparency in computational models, and closer integration with qualitative interpretive frameworks will be essential for ensuring that SA remains both analytically robust and epistemologically grounded within the evolving landscape of digitally mediated research.

REFERENCES

Afriyie, Y., & Weyori, B. A. (2026). Sentiment analysis based on deep learning approaches for text classification. *Journal of Electrical Systems and Information Technology, 13*(1), 14. Advance online publication. DOI: 10.1186/s43067-026-00322-4

Aggarwal, C. C., & Zhai, C. (2012). *Mining text data*. Springer Science & Business Media. DOI: 10.1007/978-1-4614-3223-4

Antweiler, W., & Frank, M. Z. (2005). Is all that talk just noise? The information content of internet stock message boards. *The Journal of Finance, 59*(3), 1259–1294. DOI: 10.1111/j.1540-6261.2004.00662.x

Archak, N., Ghose, A., & Ipeirotis, P. G. (2011). Deriving the pricing power of product features by mining consumer reviews. *Management Science, 57*(8), 1485–1509. DOI: 10.1287/mnsc.1110.1370

Aria, M., & Cuccurullo, C. (2017). bibliometrix : An R-tool for comprehensive science mapping analysis. *Journal of Informetrics, 11*(4), 959–975. DOI: 10.1016/j.joi.2017.08.007

Baas, J., Schotten, M., Plume, A., Côté, G., & Karimi, R. (2020). Scopus as a curated, high-quality bibliometric data source for academic research in quantitative science studies. *Quantitative Science Studies, 1*(1), 377–386. DOI: 10.1162/qss_a_00019

Bitzenis, A., Koutsoupias, N., & Boutsiouki, S. (2023). Business Research and Data Mining: a Bibliometric Analysis. *3rd International Conference on Electrical, Computer, Communications and Mechatronics Engineering (ICECCME)*, 1–6. DOI: 10.1109/ICECCME57830.2023.10252699

Bitzenis, A., Koutsoupias, N., & Nosios, M. (2025). Artificial intelligence and machine learning in production efficiency enhancement and sustainable development: A comprehensive bibliometric review. *Frontiers in Sustainability, 5*, 1508647. Advance online publication. DOI: 10.3389/frsus.2024.1508647

Bitzenis, A., Koutsoupias, N., & Nosios, M. (2026). Tax Evasion and the Informal Economy in Greece: A systematic review. *Businesses, 6*(1), 14. DOI: 10.3390/businesses6010014

Blei, D. M. (2012). Probabilistic topic models. *Communications of the ACM, 55*(4), 77–84. DOI: 10.1145/2133806.2133826

Blitzer, J., Dredze, M., & Pereira, F. (2007). Biographies, Bollywood, Boom-boxes and Blenders: Domain Adaptation for Sentiment Classification. *Proceedings of the 45th Annual Meeting of the Association for Computational Linguistics*, 440–447.

Blodgett, S. L., Barocas, S., Daumé, H.III, & Wallach, H. (2020). Language (Technology) is Power: A Critical Survey of "Bias" in NLP. *Proceedings of the 58th Annual Meeting of the Association for Computational Linguistics*, 5454–5476. DOI: 10.18653/v1/2020.acl-main.485

Bollen, J., Mao, H., & Zeng, X. (2011). Twitter mood predicts the stock market. *Journal of Computational Science*, *2*(1), 1–8. DOI: 10.1016/j.jocs.2010.12.007

Brown, T. B., Mann, B., Ryder, N., Subbiah, M., Kaplan, J., Dhariwal, P., Neelakantan, A., Shyam, P., Sastry, G., Askell, A., Agarwal, S., Herbert-Voss, A., Krueger, G., Henighan, T., Child, R., Ramesh, A., Ziegler, D. M., Wu, J., Winter, C., & Amodei, D. (2020). Language Models are Few-Shot Learners. *Proceedings of the 34th International Conference on Neural Information Processing System, 33*, 1877–1901. DOI: 10.48550/arxiv.2005.14165

Cambria, E. (2016). Affective computing and sentiment analysis. *IEEE Intelligent Systems*, *31*(2), 102–107. DOI: 10.1109/MIS.2016.31

Cambria, E., Schuller, B., Xia, Y., & Havasi, C. (2013). New avenues in opinion mining and sentiment analysis. *IEEE Intelligent Systems*, *28*(2), 15–21. DOI: 10.1109/MIS.2013.30

Charles-Smith, L. E., Reynolds, T. L., Cameron, M. A., Conway, M., Lau, E. H. Y., Olsen, J. M., Pavlin, J. A., Shigematsu, M., Streichert, L. C., Suda, K. J., & Corley, C. D. (2015). Using Social Media for Actionable Disease Surveillance and Outbreak Management: A Systematic Literature review. *PLoS One*, *10*(10), e0139701. DOI: 10.1371/journal.pone.0139701

Chen, M., Wang, S., Liang, P. P., Baltrušaitis, T., Zadeh, A., & Morency, L. (2017). Multimodal sentiment analysis with word-level fusion and reinforcement learning. *Proceedings of the 19th ACM International Conference on Multimodal Interaction*, 163–171. DOI: 10.1145/3136755.3136801

Devlin, J., Chang, M., Lee, K., & Toutanova, K. (2019). BERT: Pre-training of Deep Bidirectional Transformers for Language Understanding. *Proceedings of the 2019 Conference of the North American Chapter of the Association for Computational Linguistics: Human Language Technologies*, 4171–4186. DOI: 10.18653/v1/N19-1423

DiMaggio, P., Nag, M., & Blei, D. (2013). Exploiting affinities between topic modeling and the sociological perspective on culture: Application to newspaper coverage of U.S. government arts funding. *Poetics*, *41*(6), 570–606. DOI: 10.1016/j.poetic.2013.08.004

Ding, X., Liu, B., & Yu, P. S. (2008). A holistic lexicon-based approach to opinion mining. *Proceedings of the International Conference on Web Search and Web Data Mining*, 231–240. DOI: 10.1145/1341531.1341561

Donthu, N., Kumar, S., Mukherjee, D., Pandey, N., & Lim, W. M. (2021). How to conduct a bibliometric analysis: An overview and guidelines. *Journal of Business Research*, *133*, 285–296. DOI: 10.1016/j.jbusres.2021.04.070

Ellegaard, O., & Wallin, J. A. (2015). The bibliometric analysis of scholarly production: How great is the impact? *Scientometrics*, *105*(3), 1809–1831. DOI: 10.1007/s11192-015-1645-z

Esuli, A., & Sebastiani, F. (2006). SENTIWORDNET: a publicly available lexical resource for opinion mining. *Proceedings of the Fifth International Conference on Language Resources and Evaluation*, *16*(20), 417–422. DOI: 10.63317/54vo5wjya4cw

Feldman, R. (2013). Techniques and applications for sentiment analysis. *Communications of the ACM*, *56*(4), 82–89. DOI: 10.1145/2436256.2436274

Floridi, L., Cowls, J., Beltrametti, M., Chatila, R., Chazerand, P., Dignum, V., Luetge, C., Madelin, R., Pagallo, U., Rossi, F., Schafer, B., Valcke, P., & Vayena, E. (2018). AI4People—An Ethical Framework for a Good AI Society: Opportunities, risks, principles, and recommendations. *Minds and Machines*, *28*(4), 689–707. DOI: 10.1007/s11023-018-9482-5

Gao, Z., Feng, A., Song, X., & Wu, X. (2019). Target-Dependent Sentiment Classification with BERT. *IEEE Access : Practical Innovations, Open Solutions*, *7*, 154290–154299. DOI: 10.1109/ACCESS.2019.2946594

Gebru, T., Morgenstern, J., Vecchione, B., Vaughan, J. W., Wallach, H., Daumé, H.III, & Crawford, K. (2021). Datasheets for datasets. *Communications of the ACM*, *64*(12), 86–92. DOI: 10.1145/3458723

Giatsoglou, M., Vozalis, M. G., Diamantaras, K., Vakali, A., Sarigiannidis, G., & Chatzisavvas, K. C. (2017). Sentiment analysis leveraging emotions and word embeddings. *Expert Systems with Applications*, *69*, 214–224. DOI: 10.1016/j.eswa.2016.10.043

Glorot, X., Bordes, A., & Bengio, Y. (2011). Domain Adaptation for Large-Scale Sentiment Classification: A Deep Learning Approach. *Proceedings of the 28th International Conference on Machine Learning.*

Gong, X., Ying, W., Zhong, S., & Gong, S. (2022). Text sentiment Analysis based on transformer and augmentation. *Frontiers in Psychology, 13*, 906061. Advance online publication. DOI: 10.3389/fpsyg.2022.906061

Grimmer, J., & Stewart, B. M. (2013). Text as Data: The promise and Pitfalls of Automatic content analysis methods for political texts. *Political Analysis, 21*(3), 267–297. DOI: 10.1093/pan/mps028

Guidotti, R., Monreale, A., Ruggieri, S., Turini, F., Giannotti, F., & Pedreschi, D. (2018). A survey of methods for explaining black box models. *ACM Computing Surveys, 51*(5), 1–42. DOI: 10.1145/3236009

Guntuku, S. C., Yaden, D. B., Kern, M. L., Ungar, L. H., & Eichstaedt, J. C. (2017). Detecting depression and mental illness on social media: An integrative review. *Current Opinion in Behavioral Sciences, 18*, 43–49. DOI: 10.1016/j.cobeha.2017.07.005

Hatzivassiloglou, V., & McKeown, K. R. (1997). Predicting the semantic orientation of adjectives. *35th Annual Meeting of the Association for Computational Linguistics and 8th Conference of the European Chapter of the Association for Computational Linguistics*, 174–181. DOI: 10.3115/976909.979640

Hu, M., & Liu, B. (2004). Mining and summarizing customer reviews. *Proceedings of the Tenth ACM SIGKDD International Conference on Knowledge Discovery and Data Mining*, 168–177. DOI: 10.1145/1014052.1014073

Hu, Y., Chen, Y., & Chou, H. (2017). Opinion mining from online hotel reviews – A text summarization approach. *Information Processing & Management, 53*(2), 436–449. DOI: 10.1016/j.ipm.2016.12.002

Hussein, D. M. E. M. (2018). A survey on sentiment analysis challenges. *Journal of King Saud University. Engineering Sciences, 30*(4), 330–338. DOI: 10.1016/j.jksues.2016.04.002

Hutto, C. J., & Gilbert, E. (2014). VADER: A Parsimonious Rule-based Model for Sentiment Analysis of Social Media Text. *Proceedings of the Eighth International AAAI Conference on Weblogs and Social Media*, 216–225. DOI: 10.1609/icwsm.v8i1.14550

Jungherr, A., Jürgens, P., & Schoen, H. (2011). Why the Pirate Party Won the German Election of 2009 or The Trouble With Predictions: A Response to Tumasjan, A., Sprenger, T. O., Sander, P. G., & Welpe, I. M. "Predicting Elections With Twitter: What 140 Characters Reveal About Political Sentiment.". *Social Science Computer Review*, *30*(2), 229–234. DOI: 10.1177/0894439311404119

Khan, K., Baharudin, B., Khan, A., & Ullah, A. (2014). Mining opinion components from unstructured reviews: A review. *Journal of King Saud University. Computer and Information Sciences*, *26*(3), 258–275. DOI: 10.1016/j.jksuci.2014.03.009

Kiritchenko, S., & Mohammad, S. (2018). Examining Gender and Race Bias in Two Hundred Sentiment Analysis Systems. *Proceedings of the Seventh Joint Conference on Lexical and Computational Semantics*, 43–53. DOI: 10.18653/v1/S18-2005

Kocoń, J., Cichecki, I., Kaszyca, O., Kochanek, M., Szydło, D., Baran, J., Bielaniewicz, J., Gruza, M., Janz, A., Kanclerz, K., Kocoń, A., Koptyra, B., Mieleszczenko-Kowszewicz, W., Miłkowski, P., Oleksy, M., Piasecki, M., Radliński, Ł., Wojtasik, K., Woźniak, S., & Kazienko, P. (2023). ChatGPT: Jack of all trades, master of none. *Information Fusion*, *99*, 101861. DOI: 10.1016/j.inffus.2023.101861

Koutsoupias, N., Bitzenis, A., & Nosios, M. (2026). A Comprehensive study of Artificial intelligence in Preserving and Advancing Asia Minor's Heritage. *Digital*, *6*(1), 21. DOI: 10.3390/digital6010021

Liu, B. (2022). *Sentiment analysis and opinion mining*. Springer Nature.

Liu, P., Yuan, W., Fu, J., Jiang, Z., Hayashi, H., & Neubig, G. (2023). Pre-train, Prompt, and Predict: A systematic survey of prompting methods in natural language processing. *ACM Computing Surveys*, *55*(9), 1–35. DOI: 10.1145/3560815

Loughran, T., & McDonald, B. (2011). When is a liability not a liability? Textual analysis, dictionaries, and 10-Ks. *The Journal of Finance*, *66*(1), 35–65. DOI: 10.1111/j.1540-6261.2010.01625.x

Loughran, T., & McDonald, B. (2016). Textual Analysis in Accounting and Finance: A survey. *Journal of Accounting Research*, *54*(4), 1187–1230. DOI: 10.1111/1475-679X.12123

Mäntylä, M. V., Graziotin, D., & Kuutila, M. (2018). The evolution of sentiment analysis—A review of research topics, venues, and top cited papers. *Computer Science Review*, *27*, 16–32. DOI: 10.1016/j.cosrev.2017.10.002

Maynard, D., & Greenwood, M. (2014). Who cares about Sarcastic Tweets? Investigating the Impact of Sarcasm on Sentiment Analysis. *Proceedings of the Ninth International Conference on Language Resources and Evaluation*, 4238–4243. DOI: 10.63317/2ios4bdd5xas

Medhat, W., Hassan, A., & Korashy, H. (2014). Sentiment analysis algorithms and applications: A survey. *Ain Shams Engineering Journal*, *5*(4), 1093–1113. DOI: 10.1016/j.asej.2014.04.011

Metcalf, J., & Crawford, K. (2016). Where are human subjects in Big Data research? The emerging ethics divide. *Big Data & Society*, *3*(1), 2053951716650211. Advance online publication. DOI: 10.1177/2053951716650211

Miller, T. (2019). Explanation in artificial intelligence: Insights from the social sciences. *Artificial Intelligence*, *267*, 1–38. DOI: 10.1016/j.artint.2018.07.007

Mitchell, M., Wu, S., Zaldivar, A., Barnes, P., Vasserman, L., Hutchinson, B., Spitzer, E., Raji, I. D., & Gebru, T. (2019). Model Cards for Model Reporting. *Proceedings of the Conference on Fairness, Accountability, and Transparency*, 220–229. DOI: 10.1145/3287560.3287596

Mohammad, S. M. (2021). Sentiment analysis. In *Emotion Measurement* (pp. 323–379). DOI: 10.1016/B978-0-12-821124-3.00011-9

Moraes, R., Valiati, J. F., & Neto, W. P. G. (2013). Document-level sentiment classification: An empirical comparison between SVM and ANN. *Expert Systems with Applications*, *40*(2), 621–633. DOI: 10.1016/j.eswa.2012.07.059

Mostafa, M. M. (2013). More than words: Social networks' text mining for consumer brand sentiments. *Expert Systems with Applications*, *40*(10), 4241–4251. DOI: 10.1016/j.eswa.2013.01.019

Mullen, T., & Collier, N. (2004). Sentiment Analysis using Support Vector Machines with Diverse Information Sources. *Proceedings of the 2004 Conference on Empirical Methods in Natural Language Processing*, 412–418.

Naredla, S., Singh, M., Chaturvedi, N., Banerjee, D., Gobinath, V. M., & Tiwari, A. (2025). AI and IoT in Digital Marketing: Enhancing Automation, Personalization, and Consumer Interaction. *2025 International Conference on Pervasive Computational Technologies (ICPCT)*, 895–899. DOI: 10.1109/ICPCT64145.2025.10941275

Nelson, L. K. (2017). Computational Grounded Theory: A Methodological framework. *Sociological Methods & Research*, *49*(1), 3–42. DOI: 10.1177/0049124117729703

O'Connor, B., Balasubramanyan, R., Routledge, B., & Smith, N. (2010). From tweets to polls: linking text sentiment to public opinion time series. *Proceedings of the International AAAI Conference on Web and Social Media, 4*(1), 122–129. DOI: 10.1609/icwsm.v4i1.14031

Page, M. J., McKenzie, J. E., Bossuyt, P. M., Boutron, I., Hoffmann, T. C., Mulrow, C. D., Shamseer, L., Tetzlaff, J. M., Akl, E. A., Brennan, S. E., Chou, R., Glanville, J., Grimshaw, J. M., Hróbjartsson, A., Lalu, M. M., Li, T., Loder, E. W., Mayo-Wilson, E., McDonald, S., & Moher, D. (2021). The PRISMA 2020 statement: An updated guideline for reporting systematic reviews. *BMJ (Clinical Research Ed.), 372*(71), n71. Advance online publication. DOI: 10.1136/bmj.n71

Pak, A., & Paroubek, P. (2010). Twitter as a Corpus for Sentiment Analysis and Opinion Mining. *Proceedings of the Seventh International Conference on Language Resources and Evaluation.* DOI: 10.63317/47akcj8foo4d

Pan, S. J., Ni, X., Sun, J., Yang, Q., & Chen, Z. (2010). Cross-domain sentiment classification via spectral feature alignment. *Proceedings of the 19th International Conference on World Wide Web*, 751–760. DOI: 10.1145/1772690.1772767

Pang, B., & Lee, L. (2004). A sentimental education. *Proceedings of the 42nd Annual Meeting of the Association for Computational Linguistics*, 271–278. DOI: 10.3115/1218955.1218990

Pang, B., & Lee, L. (2008). Opinion mining and sentiment analysis. *Foundations and Trends® in Information Retrieval, 2*(1–2), 1–135. DOI: 10.1561/1500000011

Pang, B., Lee, L., & Vaithyanathan, S. (2002). Thumbs up? Sentiment Classification using Machine Learning Techniques. *Proceedings of the 2002 Conference on Empirical Methods in Natural Language Processing, 10*, 79–86. DOI: 10.3115/1118693.1118704

Panichella, S., Di Sorbo, A., Guzman, E., Visaggio, C. A., Canfora, G., & Gall, H. C. (2015). How can i improve my app? Classifying user reviews for software maintenance and evolution. *Proceedings of the 2015 IEEE International Conference on Software Maintenance and Evolution (ICSM)*, 281–290. DOI: 10.1109/ICSM.2015.7332474

Papavassiliou-Alexiou, I., Zourna, C., Koutsoupias, N., & Papakota, A. (2023). Life Career Skills Development in Higher Education Due to Covid-19: A Multivariate Approach to Students' Perceptions. In *Improving Higher Education Beyond Covid-19: Innovative and technology-enhanced approaches to teaching and learning* (pp. 87–107). Emerald Publishing. DOI: 10.1108/978-1-80382-517-520231006

Paul, M., & Dredze, M. (2011). You are what you tweet: Analyzing Twitter for Public health. *Proceedings of the International AAAI Conference on Web and Social Media, 5*(1), 265–272. DOI: 10.1609/icwsm.v5i1.14137

Rogers, A., Kovaleva, O., & Rumshisky, A. (2020). A primer in BERTology: What we know about how BERT works. *Transactions of the Association for Computational Linguistics*, *8*, 842–866. DOI: 10.1162/tacl_a_00349

Rudin, C. (2019). Stop explaining black box machine learning models for high stakes decisions and use interpretable models instead. *Nature Machine Intelligence*, *1*(5), 206–215. DOI: 10.1038/s42256-019-0048-x

Ruz, G. A., Henríquez, P. A., & Mascareño, A. (2020). Sentiment analysis of Twitter data during critical events through Bayesian networks classifiers. *Future Generation Computer Systems*, *106*, 92–104. DOI: 10.1016/j.future.2020.01.005

Salathé, M., Freifeld, C. C., Mekaru, S. R., Tomasulo, A. F., & Brownstein, J. S. (2013). Influenza A (H7N9) and the importance of digital epidemiology. *The New England Journal of Medicine*, *369*(5), 401–404. DOI: 10.1056/NEJMp1307752

Schotten, M., Aisati, M. E., Meester, W. J. N., Steiginga, S., & Ross, C. A. (2017). A brief history of Scopus: the world's largest abstract and citation database of scientific literature. In *Auerbach Publications eBooks* (pp. 31–58). DOI: 10.1201/9781315155890-3

Schouten, K., & Frasincar, F. (2016). Survey on Aspect-Level Sentiment Analysis. *IEEE Transactions on Knowledge and Data Engineering*, *28*(3), 813–830. DOI: 10.1109/TKDE.2015.2485209

Sebastiani, F. (2002). Machine learning in automated text categorization. *ACM Computing Surveys*, *34*(1), 1–47. DOI: 10.1145/505282.505283

Socher, R., Perelygin, A., Wu, J., Chuang, J., Manning, C. D., Ng, A., & Potts, C. (2013). Recursive Deep Models for Semantic Compositionality Over a Sentiment Treebank. *Proceedings of the 2013 Conference on Empirical Methods in Natural Language Processing*, 1631–1642. DOI: 10.18653/v1/D13-1170

Stieglitz, S., & Dang-Xuan, L. (2013). Social media and political communication: A social media analytics framework. *Social Network Analysis and Mining*, *3*(4), 1277–1291. DOI: 10.1007/s13278-012-0079-3

Sun, C., Huang, L., & Qiu, X. (2019). Utilizing BERT for Aspect-Based Sentiment Analysis via Constructing Auxiliary Sentence. *Proceedings of the 2019 Conference of the North American Chapter of the Association for Computational Linguistics: Human Language Technologies*, 380–385. DOI: 10.18653/v1/N19-1035

Taboada, M. (2016). Sentiment Analysis: An Overview from Linguistics. *Annual Review of Linguistics*, *2*(1), 325–347. DOI: 10.1146/annurev-linguistics-011415-040518

Tausczik, Y. R., & Pennebaker, J. W. (2010). The Psychological Meaning of Words: LIWC and Computerized Text Analysis Methods. *Journal of Language and Social Psychology*, *29*(1), 24–54. DOI: 10.1177/0261927X09351676

Tetlock, P. C., Saar-tsechansky, M., & Macskassy, S. (2008). More than words: Quantifying language to measure firms' fundamentals. *The Journal of Finance*, *63*(3), 1437–1467. DOI: 10.1111/j.1540-6261.2008.01362.x

Tracy, S. J. (2010). Qualitative quality: Eight "Big-Tent" criteria for excellent qualitative research. *Qualitative Inquiry*, *16*(10), 837–851. DOI: 10.1177/1077800410383121

Turney, P. D. (2002). Thumbs Up or Thumbs Down? Semantic Orientation Applied to Unsupervised Classification of Reviews. *Proceedings of the 40th Annual Meeting on Association for Computational Linguistics*, 417–424. DOI: 10.3115/1073083.1073153

Van Atteveldt, W., Van Der Velden, M. C. G., & Boukes, M. (2021). The validity of sentiment analysis: Comparing manual annotation, Crowd-Coding, dictionary approaches, and machine learning algorithms. *Communication Methods and Measures*, *15*(2), 121–140. DOI: 10.1080/19312458.2020.1869198

Verma, S., & Yadav, N. (2021). Past, present, and future of Electronic Word of Mouth (EWOM). *Journal of Interactive Marketing*, *53*(1), 111–128. DOI: 10.1016/j.intmar.2020.07.001

Wankhade, M., Rao, A. C. S., & Kulkarni, C. (2022). A survey on sentiment analysis methods, applications, and challenges. *Artificial Intelligence Review*, *55*(7), 5731–5780. DOI: 10.1007/s10462-022-10144-1

Weiss, K., Khoshgoftaar, T. M., & Wang, D. (2016). A survey of transfer learning. *Journal of Big Data*, *3*(1), 9. Advance online publication. DOI: 10.1186/s40537-016-0043-6

Wiebe, J., Wilson, T., & Cardie, C. (2005). Annotating expressions of opinions and emotions in language. *Language Resources and Evaluation*, *39*(2–3), 165–210. DOI: 10.1007/s10579-005-7880-9

Wiegand, M., Balahur, A., Roth, B., Klakow, D., & Montoyo, A. (2010). A survey on the role of negation in sentiment analysis. *Proceedings of the Workshop on Negation and Speculation in Natural Language Processing*, 60–68.

Wilson, T., Wiebe, J., & Hoffmann, P. (2005). Recognizing contextual polarity in phrase-level sentiment analysis. *Proceedings of the Conference on Human Language Technology and Empirical Methods in Natural Language Processing*, 347–354. DOI: 10.3115/1220575.1220619

Young, T., Hazarika, D., Poria, S., & Cambria, E. (2018). Recent trends in deep learning based natural language processing [Review article]. *IEEE Computational Intelligence Magazine*, *13*(3), 55–75. DOI: 10.1109/MCI.2018.2840738

Yu, H., & Hatzivassiloglou, V. (2003). Towards answering opinion questions. *Proceedings of the 2003 Conference on Empirical Methods in Natural Language Processing, 10*, 129–136. DOI: 10.3115/1119355.1119372

Zhang, L., Wang, S., & Liu, B. (2018). Deep learning for sentiment analysis: A survey. *Wiley Interdisciplinary Reviews. Data Mining and Knowledge Discovery*, *8*(4), e1253. Advance online publication. DOI: 10.1002/widm.1253

Zhang, X., Zhao, J., & LeCun, Y. (2015). Character-level convolutional networks for text classification. *Proceedings of the 29th International Conference on Neural Information Processing Systems, 1*, 649–657.

Zhou, X., Wan, X., & Xiao, J. (2016). Attention-based LSTM Network for Cross-Lingual Sentiment Classification. *Proceedings of the 2016 Conference on Empirical Methods in Natural Language Processing*, 247–256. DOI: 10.18653/v1/D16-1024

Zimbra, D., Abbasi, A., Zeng, D., & Chen, H. (2018). The State-of-the-Art in Twitter sentiment analysis. *ACM Transactions on Management Information Systems*, *9*(2), 1–29. DOI: 10.1145/3185045

Chapter 2
Reflexivity Reconfigured:
The Changing Role of the Researcher in AI-Augmented Qualitative Inquiry

Silvio Andrae
http://orcid.org/0000-0002-8586-7812
Independent Researcher, Germany

ABSTRACT

The integration of artificial intelligence (AI) and large language models (LLMs) into qualitative research workflows is changing the foundations of interpretive inquiry. Reflexivity has long been a core part of qualitative methodology. It requires researchers to examine how their positionality and assumptions influence knowledge production. The use of AI introduces new layers of mediation to the analytical process. AI systems now assist in data preparation, coding, thematic structuring, and analytical synthesis. They influence how qualitative data are encountered and interpreted. This chapter argues that classical reflexive frameworks, which focus mainly on researcher positionality, are insufficient for AI-augmented research. To address this, the chapter introduces dual reflexivity. This extends reflexive practice to include careful reflection on the role of algorithmic systems. The chapter presents AI as a non-human interpretive mediator. It outlines strategies such as documentation protocols, AI audit trails, and reflexive triangulation.

INTRODUCTION

The integration of artificial intelligence (AI) into scientific research increasingly challenges classical epistemological models of knowledge production. Rather than

DOI: 10.4018/979-8-3373-7640-0.ch002

t © 2026, IGI Global Scientific Publishing. Copying or distributing in print or electronic forms without written permission of IGI Global Scientific Publishing is prohibited. Use of this chapter to enerative artificial intelligence (AI) technologies is expressly prohibited. The publisher reserves all rights to license its use for generative AI training and machine learning model development.

functioning merely as a computational instrument, AI participates directly in processes of observation, pattern detection, and hypothesis generation, thereby reshaping the epistemic architecture of scientific inquiry. As Cruz-Aguilar (2025) argues, AI-mediated research environments blur traditional distinctions between empirical observation, theoretical reasoning, and technological mediation. Scientific knowledge is increasingly produced within distributed socio-technical systems in which human interpretative agency and machine-based inference interact. This shift raises fundamental questions about epistemic responsibility, the interpretability of machine-generated insights, and the criteria for assessing scientific validity. Consequently, contemporary research must be understood as emerging from hybrid human–AI epistemic assemblages rather than from exclusively human cognitive practices.

Originally developed in computational linguistics and machine learning, large language models (LLMs) now appear throughout digital tools in academic research. Qualitative researchers often use AI-assisted features in transcription software, data analysis platforms, and language models to summarize, code, cluster, or synthesize text. Empirical research now examines how large language models affect qualitative research practices. Studies show that while LLMs can help with summarization or brainstorming, they currently offer limited support for interpretive qualitative analysis. For example, an experimental study found that LLMs generate descriptive summaries rather than analytically meaningful themes and often omit contextual or latent meanings in qualitative data (Vikan et al., 2026). The same study shows that LLM outputs may introduce errors, fabricated quotations, or misleading conceptual connections if used uncritically. Therefore, human reflexivity, deep familiarity with the data, and methodological expertise remain crucial for maintaining the integrity of qualitative research. These findings show that integrating AI into qualitative workflows does not remove reflexive interpretation, but instead changes the context in which it is exercised.

These developments are changing the workflow of qualitative inquiry. Tasks that once took much time and required interpretation, such as transcribing interviews and organizing field notes, can now be partially automated. Researchers may use language models to generate thematic overviews of transcripts or apply automated coding tools to spot recurring patterns. They can also rely on AI summaries to navigate large text datasets. As these tools become more advanced and accessible, their use in research will likely increase.

These tools promise more efficiency and scalability. They also raise key methodological questions. Qualitative research relies on close engagement with material and interpretive processes. These processes unfold through repeated reading, coding, comparison, and reflection (Charmaz, 2014; Denzin & Lincoln, 2018). Gradual immersion in the data is central to building contextual meaning and theoretical insight. The rise of AI in these tasks now challenges the idea that interpretation is only a human activity.

This change does not just enhance existing research tools. It signals a broader shift in the epistemic foundation of qualitative research. When AI shapes data preparation, categorization, and summarization, it also shapes the interpretive environment. The main question is not only whether AI can help with analysis. It is how AI changes the conditions under which interpretation occurs.

Scholars studying algorithms and digital infrastructures note that algorithmic systems now shape how knowledge is made, circulated, and interpreted (Gillespie, 2014; Kitchin, 2017). Algorithms do more than process information. They organize, prioritize, and structure it based on statistical patterns in their design and data. Large language models add to this by generating fluent text that may appear to be interpretation, though it is based on probabilistic modeling rather than contextual understanding (Bender et al., 2021). As AI takes part in knowledge production, it may fill roles once held by human experts. From a social epistemology viewpoint, AI can be seen as an emerging epistemic authority within the division of labour (Hauswald, 2025a; 2025b). This development raises questions about how researchers should interpret and engage with machine-generated suggestions.

The implications for qualitative research are not yet fully understood. AI-assisted tools are spreading rapidly across research settings. But reflection on their epistemic consequences is just beginning. In particular, AI in qualitative analysis raises questions about how interpretive authority is shared between humans and algorithms.

Reflexivity as a Cornerstone of Qualitative Inquiry

Reflexivity is central to maintaining rigor and transparency in qualitative methods. It is rooted in interpretivist and constructivist epistemologies. Reflexivity recognizes that knowledge in the social sciences is always situated. Researchers do not approach material as neutral observers. They bring theories, cultural views, training, and experiences that shape the process and outcomes (Alvesson & Sköldberg, 2018; Berger, 2015).

Reflexive methodology responded to this epistemic situation. Instead of removing subjectivity, reflexive approaches make it visible and accountable. By examining their own positions and assumptions, researchers can understand how knowledge is constructed and how other interpretations may emerge (Finlay, 2002).

Reflexivity is especially important in interpretivist traditions. Geertz (1973) and Guba and Lincoln (1985) stressed that social reality depends on systems of meaning. Detached observation is not enough. Interpretation requires ongoing engagement between researcher and material, shaped by theory and context. Reflexivity lets researchers examine how their perspectives shape their work.

Reflexivity is more than just a technique. It is an epistemological commitment that knowledge is relational and situated. By viewing the researcher as an interpreter, reflexivity challenges objectivity but enables rigorous, credible inquiry. Classical reflexive frameworks assume interpretation is a human act. They treat the researcher's position as the primary mediator. As long as humans interpret, reflexivity can examine how their backgrounds shape analysis.

The integration of AI into qualitative research workflows complicates this assumption. Specifically, when algorithmic systems contribute to tasks such as coding, thematic clustering, or analytic summarization, the interpretive process becomes distributed across both human and technological actors. Consequently, reflexivity can no longer be understood solely as a self-directed examination of the researcher's positionality; it must also address how technological infrastructures shape the analytical environment in which qualitative knowledge is produced.

The Emerging Methodological Challenge

The growing presence of AI within qualitative research workflows introduces a new methodological challenge: the distribution of interpretive influence across human and algorithmic systems. Unlike traditional research tools, which primarily function as passive instruments, AI systems actively transform textual material through processes of pattern recognition and probabilistic language generation. These transformations can subtly shape the patterns that researchers observe in their data and the conceptual categories through which those patterns are interpreted.

For example, when a language model generates a thematic summary of interview transcripts, it does so by drawing on statistical patterns embedded in its training data (Vikan et al., 2026). These patterns reflect linguistic conventions and discursive structures found in large textual corpora rather than the specific contextual realities of the research setting. If researchers rely heavily on such outputs to guide their analysis, the study's interpretive framework may be partially shaped by algorithmic structures external to the empirical context.

This phenomenon introduces a new form of epistemic mediation into qualitative research. The analytical pathway from empirical material to interpretation is no longer solely determined by the researcher's interpretive engagement with the data. Instead, it is increasingly influenced by the algorithmic processes that structure the presentation and organization of that data.

Such developments raise important questions about interpretive authority and epistemic responsibility. If algorithmic systems shape the analytical landscape, how should researchers account for their influence? To what extent can AI-generated outputs be considered analytical contributions rather than technical artifacts? And how can qualitative researchers maintain interpretive rigor in environments where algorithmic mediation is increasingly pervasive?

These questions highlight the need for a renewed examination of reflexivity in AI-augmented research practices.

Recent empirical research on AI-assisted qualitative analysis supports this interpretation of AI as an epistemic mediator rather than a neutral analytical tool. Studies in human–computer interaction and qualitative methodology show that researchers increasingly use large language models for tasks such as coding assistance, thematic clustering, and exploratory data analysis, while maintaining interpretive control over final analytical decisions (e.g., Vikan et al., 2026; Schroeder et al., 2025). At the same time, this research highlights that AI-generated outputs tend to reflect statistical regularities in language rather than contextually grounded interpretations, reinforcing the need for critical evaluation and reflexive engagement with algorithmic suggestions. These findings provide empirical grounding for the argument that AI systems reshape the interpretive environment of qualitative inquiry by structuring how data are organized, presented, and analytically explored.

Argument and Structure of the Chapter

This chapter argues that integrating artificial intelligence into qualitative research workflows requires a conceptual reconfiguration of reflexive methodology. Classical reflexivity, while essential for examining the researcher's positionality, does not fully address the epistemic influence of technological systems that now participate in the interpretive process. As qualitative research becomes increasingly entangled with algorithmic infrastructures, reflexivity must expand to account for both human and technological sources of interpretive mediation.

To address this challenge, the chapter introduces the concept of dual reflexivity. Dual reflexivity extends traditional reflexive practice by incorporating critical reflection on the epistemic role of AI systems within qualitative research. It encourages researchers to examine not only how their own perspectives shape the research process but also how algorithmic systems structure, interpret, and represent empirical material.

The chapter proceeds as follows. Section 2 reviews the role of reflexivity in classical qualitative methodology, focusing on interpretivist epistemology, positionality, and established reflexive practices. Section 3 examines how AI technologies transform the interpretive process by intervening in different stages of qualitative analysis. Section 4 introduces the concept of dual reflexivity as a theoretical framework for understanding these transformations. Section 5 discusses practical strategies for operationalizing reflexivity in AI-augmented research environments, including documentation protocols and AI audit trails. Section 6 explores the ethical and pedagogical implications of hybrid human–AI research practices. The chapter concludes by reflecting on the future of reflexive qualitative inquiry in an era increasingly shaped by agentic AI systems.

Through this analysis, the chapter seeks to contribute to ongoing methodological debates about the role of artificial intelligence in qualitative research. Rather than framing AI as either a threat or a panacea, it proposes a reflexive approach that acknowledges both the opportunities and the epistemic risks associated with AI-assisted analysis. By expanding reflexivity to encompass the socio-technical infrastructures of contemporary research, qualitative methodology can adapt to technological transformation while preserving the interpretive depth and critical engagement that define the field.

CLASSICAL REFLEXIVITY IN QUALITATIVE RESEARCH

Reflexivity in Interpretivist Epistemology

Reflexivity occupies a central position within qualitative research traditions, particularly those grounded in interpretivist and constructivist epistemologies. These traditions depart fundamentally from positivist conceptions of knowledge production that aim to minimize or eliminate the researcher's influence on the research process (Finlay, 2002; Holmes, 2020; Conti & Seele, 2025). Instead, interpretivist approaches recognize that knowledge in the social sciences is inherently situated and relational. Social phenomena are not simply discovered through detached observation but are interpreted through frameworks of meaning that emerge within specific cultural, social, and historical contexts (Geertz, 1973; Guba & Lincoln, 1985).

Within this epistemological orientation, the researcher is not an external observer standing outside the empirical world but an active participant in the process of knowledge generation. Data are not merely collected; they are produced through interactions between researchers and participants, shaped by the questions asked, the inquiry contexts, and the interpretive lenses applied during analysis. As Denzin and Lincoln (2018) emphasize, qualitative inquiry is therefore a fundamentally interpretive practice in which meaning is constructed through a dynamic relationship between researcher, participants, and empirical material.

Reflexivity emerged as a methodological response to this epistemic condition. Reflexivity, in qualitative research, refers to the systematic examination of how the researcher's assumptions, theoretical commitments, and social positioning shape the research process and its outcomes (Alvesson & Sköldberg, 2018). Rather than attempting to suppress subjectivity, reflexive methodologies seek to make it visible and critically examinable. By reflecting on their own interpretive frameworks, researchers can better understand how their analytical perspectives influence the identification of patterns, the construction of categories, and the interpretation of empirical findings.

Reflexive methods first emerged from discussions in sociology, anthropology, and interpretive social science about how subjectivity shapes knowledge. For example, ethnographers began to see that their own presence and interaction during fieldwork affected what they saw and how they understood it. Likewise, scholars using grounded theory realized that making codes and theories was more than just following steps—it was shaped by their previous knowledge and training.

Reflexivity has two main purposes. First, it makes the research process clearer by showing the lens through which knowledge is shaped. Second, it improves research quality by prompting researchers to question their own ideas and viewpoints throughout the study. Reflexivity helps prevent hidden subjectivity while accepting that total objectivity is neither possible nor desired here. Instead of hiding subjectivity, reflexivity makes it open and something to be examined.

Positionality and the Researcher as an Interpretive Instrument

Closely related to reflexivity is the concept of positionality, which highlights how the researcher's social and epistemic location shapes the research process. Positionality refers to the recognition that researchers occupy particular positions within social structures shaped by factors such as gender, ethnicity, class, institutional affiliation, and disciplinary training. These positions shape both how researchers engage with research participants and how they interpret empirical material (Berger, 2015; Harding, 1992).

In qualitative research, the researcher is often described as the primary 'instrument' of analysis. Unlike standardized measurement tools used in quantitative research, qualitative analysis relies heavily on the researcher's interpretive capacities, including perception, judgment, theoretical sensitivity, and contextual understanding. Through these capacities, researchers identify patterns in empirical material, construct conceptual categories, and develop theoretical explanations.

This central role of the researcher introduces both opportunities and challenges. On the one hand, the researcher's interpretive engagement allows for rich, contextually grounded analysis that captures the complexity of social phenomena. On the other hand, the researcher's interpretive authority also introduces the possibility of bias, selective perception, and interpretive overreach. Without reflexive awareness, researchers may inadvertently impose their own assumptions or theoretical preferences onto the empirical material.

Positionality helps address these issues by prompting researchers to reflect on how their own views shape the research. Some scholars, such as Haraway (1988), argue that all knowledge comes from a particular place or outlook. Admitting this is important for being honest in research. Instead of pretending to be neutral, reflexive researchers recognize that their interpretations are shaped by their own social and intellectual backgrounds.

Reflexive engagement with positionality, therefore, involves examining how the researcher's background, experiences, and theoretical commitments influence both the generation of empirical material and its interpretation. This examination does not aim to eliminate the researcher's perspective but to make it analytically visible. By doing so, reflexivity transforms subjectivity from a hidden source of bias into an explicit component of methodological reflection.

Reflexive Practices in Qualitative Methodology

Reflexivity in qualitative research is not merely a philosophical orientation but is operationalized through a range of methodological practices designed to support critical self-examination and analytic transparency. Among the most widely used reflexive practices are reflexive journaling, memo writing, and maintaining methodological audit trails.

Reflexive journaling involves systematically recording the researcher's reflections throughout the research process. These reflections may include observations about interactions with participants, emotional responses to field experiences, emerging theoretical insights, and critical reflections on interpretive decisions. By documenting these reflections, researchers create a record of how their understanding evolves over time and how their positionality influences the analytical process (Finlay, 2002).

Memo writing is another main reflexive practice, especially in grounded theory. Researchers use memos to record their evolving ideas about what the data mean. These memos record why they made certain choices in coding and developing theories. By writing and revisiting memos, researchers stay in touch with their data and improve their understanding.

An audit trail is another way to show reflexivity. It is a record of all research steps, including collecting data, coding, revising plans, and building theory. By keeping good records, researchers allow others to check how the study was conducted and how the findings were reached.

These practices collectively serve to operationalize reflexivity as an ongoing methodological commitment rather than a one-time declaration of positionality. Reflexivity unfolds throughout the entire research process—from the initial formulation of research questions to the final interpretation and representation of findings. By engaging in these practices, researchers maintain a critical awareness of how their perspectives shape the study's analytical trajectory.

Reflexivity versus Procedural Transparency

A crucial distinction within qualitative methodology concerns the difference between reflexivity and procedural transparency. Both contribute to methodological rigor but focus on different aspects: procedural transparency involves clearly documenting research methods, analytical procedures, and data sources, allowing readers to understand how conclusions were derived from empirical material. Procedural transparency refers to the clear documentation of methodological steps, including data collection procedures, analytical techniques, and coding strategies. Procedural transparency is the practice of ensuring that all aspects of the research process are clearly recorded and available for scrutiny. Its primary purpose is to enable readers to understand how the research was conducted and how the conclusions were derived from the empirical material. Transparency ensures that the research process is visible and open to scrutiny.

Reflexivity, though, goes beyond simply writing down what was done. It means researchers must look closely at their own beliefs, theories, social position, and the power dynamics in the study. Reflexivity means being open about how researchers' own thinking and background affect the research (Pillow, 2003).

This distinction is significant because methodological transparency alone does not guarantee reflexive awareness. A study may provide detailed descriptions of its methods while leaving unexamined the interpretive assumptions guiding the analysis. Reflexivity demands a deeper level of engagement with the research process by asking not only how the analysis was conducted but also how the researcher's perspective influenced what was observed, interpreted, and ultimately represented.

Within classical qualitative methodology, reflexivity therefore functions as both an epistemic and ethical commitment. It ensures that researchers remain attentive to the situated nature of knowledge production and to the interpretive responsibilities that accompany the analysis of social phenomena. At the same time, this classical conception of reflexivity presupposes that interpretation remains a fundamentally human activity.

As the following section demonstrates, this assumption becomes increasingly problematic in research environments where algorithmic systems shape data preparation, thematic organization, and analytical synthesis. The integration of artificial intelligence into qualitative research workflows, therefore, necessitates a reconsideration of reflexivity itself, extending it beyond the individual researcher to include the epistemic influence of technological systems.

AI AND THE TRANSFORMATION OF THE INTERPRETIVE PROCESS

AI as Epistemic Infrastructure in Qualitative Research

The emergence of artificial intelligence in research environments goes beyond the introduction of new analytical tools. Increasingly, AI systems—particularly LLMs—function as infrastructural components of contemporary knowledge production. Rather than merely assisting researchers with discrete tasks, these systems shape how information is organized, accessed, and interpreted within research workflows. In qualitative research, this development marks a significant shift in the epistemic conditions under which analysis takes place.

Recent scholarship on algorithmic infrastructures has emphasized that digital systems do not simply process information but actively structure how knowledge is produced and circulated (Gillespie, 2014; Kitchin, 2017). Algorithms rank, categorize, and filter information according to computational logics embedded in their design. In doing so, they influence which patterns become visible and which remain obscured. Large language models extend this dynamic by generating linguistic outputs that reorganize textual material in ways that appear analytically meaningful.

Recent empirical research shows that integrating large language models into qualitative workflows alters the conditions under which reflexive interpretation unfolds. In an exploratory study of LLM-supported reflexive thematic analysis, Vikan et al. (2026) demonstrate that language models tend to generate descriptive summaries and topic groupings rather than analytically meaningful themes grounded in contextual interpretation. The study further shows that LLM outputs may omit central aspects of participants' experiences, misunderstand contextual nuances, and occasionally generate fabricated quotations or misleading conceptual connections when prompted to support deeper stages of analysis. Such findings highlight that AI systems operate primarily through probabilistic pattern recognition rather than interpretive understanding, producing outputs that can appear coherent and analytically plausible while lacking epistemic grounding in the underlying data. For qualitative researchers, this introduces a new layer of methodological complexity: reflexive interpretation must now extend beyond the researcher's own assumptions to include critical scrutiny of machine-generated analytical suggestions. Consequently, AI-assisted qualitative analysis does not diminish the need for reflexivity but instead intensifies it by requiring researchers to continuously evaluate the epistemic reliability, contextual adequacy, and interpretive implications of AI-generated outputs.

This technological transformation has profound implications for the interpretive character of qualitative inquiry. In classical qualitative methodology, the analysis process is closely intertwined with the researcher's cognitive engagement with empirical material. Interpretation emerges through iterative reading, coding, comparison, and conceptual abstraction. Through this process, the researcher gradually develops theoretical sensitivity, identifies patterns, and constructs meaning from complex social narratives. AI-assisted analysis modifies this dynamic by introducing algorithmic systems that perform many of these tasks automatically or semi-automatically.

In qualitative research, LLMs are increasingly integrated into a range of analytical processes. They can transcribe interviews, summarize large volumes of textual data, generate preliminary coding schemes, cluster semantically similar passages, and produce analytic narratives based on textual input. These capabilities enable researchers to engage with larger datasets and to navigate complex textual corpora more efficiently. However, they also introduce new forms of epistemic mediation into the interpretive process.

Importantly, AI systems do not merely accelerate existing methodological procedures. They reshape the analytical process itself. By generating summaries, proposing thematic clusters, suggesting codes, or producing synthetic analytic narratives, LLMs shape how researchers encounter and interpret their data. In this sense, AI systems function as epistemic intermediaries, mediating between raw empirical material and the researcher's interpretive work. The analytical pathway from data to interpretation, therefore, becomes increasingly hybrid, involving both human and algorithmic contributions.

This shift does not mean AI systems 'understand' qualitative data as human researchers do. LLMs use statistical pattern recognition based on large text corpora. Their outputs reflect probabilistic relationships between words rather than contextually grounded interpretations. Jain et al. (2026) show that reliability is best assessed by comparing multiple independent model runs. Without this, researchers risk mistaking randomness for meaningful themes. Even so, the fluency of AI-generated text can give an impression of insight. The challenge is to distinguish genuine interpretation from linguistic plausibility generated by algorithms.

Recent work in machine learning describes LLM inference as a multi-stage cognitive process. Yang et al. (2025) propose a 'cognition attribution architecture.' It distinguishes between rapid knowledge retrieval and a reasoning adjustment phase, inspired by dual-system cognitive theory. This suggests that LLM outputs are not just responses, but the result of layered processes combining statistical memory with reasoning-like transformations.

Understanding AI as an epistemic infrastructure means recognizing its dual nature. On the one hand, AI can increase the efficiency and scalability of qualitative analysis. Researchers can address larger datasets and find patterns they might otherwise miss. On the other hand, AI introduces new layers of mediation and can subtly shape outcomes. These changes mean we must closely examine how AI intervenes at different stages of qualitative analysis.

Pre-Interpretive Influence: AI in Data Preparation and Structuring

One early point of AI intervention in qualitative research is the pre-interpretive stage. Here, empirical material is prepared for analysis. Activities like transcription, translation, data cleaning, summarization, and structuring rely more on AI. Speech recognition systems can transcribe interviews in real time. Language models can summarize large texts. Automated tools extract key topics or entities from documents.

Although these processes may appear purely technical, they have important interpretive consequences. Transcription, for example, is not a neutral act of converting spoken language into written text. Decisions about punctuation, emphasis, segmentation, and linguistic normalization shape how conversational data are represented and subsequently interpreted. As scholars of qualitative methodology have long noted, transcription involves interpretive choices that can shape the research's analytical trajectory (Denzin & Lincoln, 2018).

AI-based transcription systems add new complexities. These systems can greatly speed up transcription. However, they may change speech by standardizing dialects, omitting hesitations, or misrecognizing context-based words. These changes can subtly reshape materials before researchers review them.

Machine learning workflows include many points of discretionary decision-making that shape the final model. These include dataset construction, annotation design, algorithm selection, and evaluation metrics. Each 'choice moment' introduces subjective assumptions. As a result, AI outputs reflect a series of design choices rather than purely objective inference (Cambo & Gergle, 2022).

Similarly, automated summarization tools restructure qualitative material by condensing complex narratives into simplified textual representations. While such summaries can be useful for navigating large datasets, they inevitably involve algorithmic decisions about what information is considered salient and what is omitted. These decisions are shaped by the statistical structures embedded in the model's training data rather than by contextual knowledge of the research setting.

Empirical evidence suggests that qualitative researchers are already integrating LLMs across multiple stages of the research process, including interview preparation, exploratory data analysis, and preliminary coding (Schroeder et al., 2025). These practices indicate that AI systems are increasingly involved in the pre-interpretive and co-interpretive stages of qualitative analysis.

The pre-interpretive influence of AI is its ability to shape how qualitative data enters analysis. By restructuring raw material before researchers review it, AI systems frame the analytical possibilities that follow. Some patterns may be highlighted, while others might be hidden or lost. The direction of research can be influenced before explicit analysis even begins.

Co-Interpretive Influence: AI in Coding and Thematic Structuring

Beyond data preparation, AI systems increasingly participate in the analytical core of qualitative research through coding assistance and thematic clustering. Many contemporary qualitative analysis tools now incorporate AI features that suggest codes, group similar text segments, identify recurring topics, or propose thematic relationships within datasets.

In classical qualitative methodology, coding represents a central interpretive process through which researchers engage deeply with their data. Through iterative coding and constant comparison, researchers gradually develop conceptual categories that capture recurring patterns in the empirical material (Charmaz, 2014). This process is inherently interpretive: codes are not simply discovered within the data but constructed through the researcher's theoretical sensitivity and contextual understanding.

AI-assisted coding alters this process by introducing algorithmic pattern recognition as a preliminary analytical step. LLMs can identify semantic similarities across textual passages, cluster them into categories, and generate candidate codes based on linguistic patterns. While such tools can support exploratory analysis and help researchers navigate large datasets, they also influence the conceptual framing of the analysis.

This process can be described as co-interpretive influence, as the analytical process emerges from the interaction between human and algorithmic contributions. Researchers may accept, modify, or reject AI-generated coding suggestions, but the system's initial structuring can shape the study's analytical trajectory. The researcher's attention may be drawn toward certain patterns suggested by the algorithm, while other interpretive possibilities remain unexplored.

The co-interpretive role of AI is therefore both enabling and constraining. On the one hand, algorithmic tools can reveal linguistic patterns across large datasets that might otherwise remain unnoticed. On the other hand, the statistical logic underlying these patterns may not correspond to the socially meaningful categories that qualitative analysis seeks to uncover. The challenge for researchers is to critically evaluate algorithmic suggestions while maintaining interpretive control over the analytical process.

Illustrative Example: AI-Assisted Thematic Analysis

To illustrate the co-interpretive role of AI systems, consider a qualitative study based on semi-structured interview transcripts exploring user experiences with digital platforms. In a conventional analytical workflow, the researcher would iteratively code the data, identify recurring patterns, and develop thematic categories grounded in contextual interpretation.

In an AI-assisted workflow, the researcher may first use a large language model to generate preliminary coding suggestions or thematic clusters based on the textual material. The model might identify recurring linguistic patterns—such as references to "efficiency," "frustration," or "lack of transparency"—and group these into candidate themes.

However, these algorithmically generated categories do not necessarily correspond to analytically meaningful constructs. For example, what the model labels as "efficiency" may reflect heterogeneous experiences that require differentiation through contextual interpretation. The researcher must therefore critically evaluate, refine, or reject these suggested categories in light of the empirical material and the study's theoretical framework.

This example illustrates that AI systems can support exploratory pattern detection but do not replace interpretive reasoning. Instead, the analytical process becomes co-constructed: algorithmic outputs shape the initial structuring of the data, while the researcher retains responsibility for contextual interpretation and theoretical abstraction.

Post-Interpretive Influence: AI in Analytical Synthesis and Representation

AI systems can also influence qualitative research at later stages of the analytical process, particularly in the synthesis and representation of findings. LLMs can generate analytical summaries, reorganize thematic structures, draft interpretive narratives, or rewrite sections of analytical text. These capabilities are increasingly used by researchers to draft reports, organize findings, or refine theoretical arguments.

While such applications can improve efficiency and help structure complex material, they also raise important methodological questions. In qualitative research, writing is itself a central component of analysis. The act of composing analytical narratives allows researchers to refine conceptual relationships, identify contradictions within the data, and articulate theoretical insights.

When AI systems contribute directly to this stage of analysis, the boundary between human interpretation and algorithmic generation becomes less distinct. AI-generated narratives may introduce rhetorical coherence and linguistic sophistication, giving the impression of conceptual clarity. However, this apparent clarity may mask the absence of genuine interpretive reasoning.

The post-interpretive influence of AI, therefore, lies in its ability to shape how qualitative findings are articulated and represented. While the researcher remains responsible for interpreting the data, the language in which that interpretation is expressed may be partially generated by algorithmic systems. This raises important questions about authorship, interpretive authority, and the epistemic status of AI-generated analytical text.

Epistemic Illusions and the Simulation of Interpretation

A central methodological risk associated with AI-assisted qualitative research is the emergence of what can be termed epistemic illusions. These occur when AI-generated outputs appear to provide meaningful interpretations of qualitative data despite being produced through statistical language modeling rather than contextual understanding.

Large language models generate text by predicting probable word sequences based on patterns observed in training data. Although this process can produce highly coherent and contextually appropriate language, it does not involve the interpretive engagement with social meaning that characterizes human qualitative analysis. The model lacks experiential knowledge of the research context and does not understand the lived realities described by participants.

Despite this limitation, AI-generated text can still convey the impression of analytical insight. When a language model produces a thematic summary or proposes an interpretation of a dataset, its linguistic coherence may lead researchers to attribute interpretive validity to the output. In reality, the model is reproducing patterns associated with similar linguistic contexts rather than generating contextually grounded explanations of the empirical material. In multi-agent AI systems, errors or hallucinated information may propagate through agent interactions, producing what has been termed 'knowledge drift' (Hu et al., 2025). A similar dynamic may emerge in AI-assisted qualitative analysis when machine-generated interpretations recursively influence subsequent coding, summarization, or thematic structuring.

These epistemic illusions present a significant challenge for qualitative methodology. If researchers rely heavily on AI-generated interpretations without critically evaluating their epistemic foundations, interpretive authority may gradually shift from human reasoning to algorithmic pattern recognition. Maintaining methodological rigor, therefore, requires a reflexive awareness of how algorithmic systems shape the analytical process. Emerging research increasingly interprets LLM behavior through cognitive analogies. For example, Yang et al. (2025) conceptualize LLM inference through dual-system theory, distinguishing between fast knowledge retrieval and slower reasoning adjustment. While such models provide useful analytical heuristics, they also risk anthropomorphizing statistical systems, reinforcing the epistemic illusion that machine outputs reflect genuine reasoning rather than probabilistic inference.

The observations discussed above suggest that integrating AI systems into qualitative workflows does not merely introduce new analytical tools but fundamentally reshapes the epistemic environment in which interpretation occurs. When machine-generated suggestions influence processes such as coding, clustering, or thematic abstraction, the interpretive field becomes a hybrid space in which human judgment and algorithmic pattern recognition interact. In such contexts, reflexivity cannot be confined solely to the researcher's positionality, assumptions, and interpretive practices. Instead, the epistemic influence of AI systems themselves must also become an object of methodological reflection.

This shift requires an expansion of the classical concept of reflexivity. If qualitative researchers are to preserve interpretive integrity in AI-augmented analytical processes, they must critically examine not only their own interpretive frameworks but also the ways in which AI systems structure, filter, and suggest patterns within qualitative data. The methodological challenge, therefore, lies in developing a reflexive framework that accounts for both human- and machine-mediated influences on knowledge production. The following section introduces the concept of dual reflexivity as a conceptual response to this emerging methodological condition.

DUAL REFLEXIVITY: EXTENDING REFLEXIVE METHODOLOGY

The Limits of Classical Reflexivity in AI-Mediated Research

The preceding discussion illustrates that integrating artificial intelligence into qualitative research workflows fundamentally alters the conditions under which interpretation occurs. Classical reflexivity was developed within a research environment in which the interpretive process was assumed to be carried out exclusively by human researchers. Reflexive practices, therefore, focused on examining how the researcher's positionality, assumptions, and theoretical frameworks influenced the production of knowledge (Alvesson & Sköldberg, 2018; Finlay, 2002). While this perspective remains essential, it becomes insufficient when technological systems begin to mediate the analytical process.

Transitioning from the classical view, in AI-augmented research environments, interpretive influence is no longer confined to the researcher alone. As demonstrated in the previous section, algorithmic systems increasingly participate in multiple stages of qualitative analysis. They reshape empirical material during data preparation, suggest thematic patterns during coding, and contribute to the synthesis and representation of findings. These forms of intervention introduce additional layers of mediation that classical reflexive frameworks do not fully account for.

This development highlights a broader shift in the epistemic architecture of qualitative research. Knowledge production becomes embedded within socio-technical systems in which human judgment interacts with algorithmic processes. As scholars of science and technology studies have long argued, technological artifacts can actively shape social practices and forms of knowledge (Latour, 2005; Suchman, 2007). In this sense, AI systems should not be understood merely as passive tools but as mediating infrastructures that structure the analytical environment in which interpretation takes place. Qualitative researchers themselves recognize tensions between bottom-up interpretive approaches and algorithmic pattern detection (Schroeder et al., 2025). Algorithmic systems increasingly participate in defining what counts as credible knowledge. This shift toward 'algorithmic truth' illustrates how epistemic authority may gradually shift from human interpretation to computational classification (Shin, 2025).

The implication is not that AI systems possess interpretive agency comparable to that of human researchers. Rather, their outputs influence how empirical material is organized, presented, and understood. These influences can shape the direction of analysis in subtle but consequential ways. Classical reflexivity, which focuses primarily on the researcher's self-examination, therefore requires conceptual expansion to address the epistemic role of technological mediation. Researchers must not only examine their own positionality but also critically reflect on the epistemic role of the technological systems they employ. Reflexivity must therefore evolve from a purely self-directed practice to one that encompasses the broader socio-technical infrastructures through which qualitative knowledge is produced.

Conceptualizing Dual Reflexivity

To address this transformation in reflexive practice, this chapter proposes the concept of dual reflexivity. Dual reflexivity extends the traditional reflexive framework by incorporating critical reflection on both human and algorithmic dimensions of the research process. It recognizes that in AI-augmented qualitative inquiry, interpretive outcomes are shaped by the interaction between the researcher's analytical reasoning and the algorithmic infrastructures through which data are processed and represented.

Dual reflexivity operates on two interconnected levels. The first level corresponds to classical reflexivity, in which researchers critically examine how their own positionality, theoretical assumptions, and disciplinary perspectives shape the research process. This dimension remains indispensable because human researchers continue to formulate research questions, interpret empirical material, and construct theoretical explanations.

The second level requires researchers to reflexively engage with the technological systems that participate in the analytical process. This engagement includes examining how the architecture, training data, and operational logic of AI tools shape outputs. For example, large language models generate responses based on patterns present in massive textual corpora that reflect particular cultural, linguistic, and epistemic biases. These biases can influence the categories and narratives generated by the model, thereby shaping the analytic suggestions that researchers receive.

Interestingly, recent research on LLM agent architectures has begun to incorporate internal 'reflective' reasoning structures. For example, Zhang et al. (2025) propose a dual-strategy architecture in which a holistic planning module interacts with a local execution strategy through a reflective feedback loop. Within this system, a high-level plan provides global coherence, while a local strategy adapts actions to contextual observations, and a scoring mechanism continuously evaluates progress. Although developed for autonomous agents rather than qualitative research, such architectures illustrate an important conceptual parallel: complex AI systems increasingly rely on layered reflective processes that resemble meta-cognitive reasoning structures. Conti and Seele (2025), for example, propose 'algorithm designers' reflexivity statements' as a mechanism for making the epistemic assumptions embedded in AI systems more transparent. This observation strengthens the argument advanced in this chapter that reflexivity in AI-augmented inquiry must consider not only the human researcher but also the structured reasoning mechanisms embedded in AI tools.

Dual reflexivity, therefore, requires researchers to consider AI systems not merely as neutral instruments but as epistemic mediators that influence the construction of knowledge. This perspective shifts the analytical focus from the individual researcher to the broader socio-technical environment in which qualitative inquiry unfolds. By acknowledging the interpretive role of algorithmic systems, dual reflexivity provides a framework for critically examining how technological infrastructures shape research outcomes.

AI as a Non-Human Interpretive Mediator

A central component of dual reflexivity is the recognition that AI systems function as non-human interpretive mediators. While these systems lack intentionality and contextual understanding, their outputs nevertheless influence the interpretive trajectory of qualitative analysis. This influence arises from how AI systems transform textual material through pattern recognition and probabilistic language generation.

When an LLM generates thematic summaries, clusters textual passages, or produces analytic narratives, it introduces a particular organization of meaning into the research process. These outputs do not constitute interpretation in the human sense, but they create intermediary representations that structure how researchers engage with data. Thus, AI systems mediate the pathway between empirical material and interpretive reasoning.

Recent behavioral evidence suggests that individuals may actively delegate morally complex decisions to AI systems as a strategy to shift responsibility and reduce the psychological burden associated with ethical judgment (Hüholt & Szech, 2026). Experimental results show that participants become significantly more likely to delegate moral decisions to AI than to human counterparts, indicating that AI can function as a convenient 'moral delegate' in situations characterized by ambiguity and responsibility. While these findings come from behavioral economics rather than qualitative methodology, they highlight a broader epistemic dynamic relevant for AI-augmented research practice: AI systems may serve not only as analytical tools but also as mechanisms through which researchers implicitly offload interpretive responsibility. In qualitative analysis workflows—such as AI-assisted coding, clustering, or thematic synthesis—researchers may encounter similar dynamics if they begin treating AI-generated interpretations as authoritative without critically interrogating their epistemic status.

Similar dual structures of responsibility are increasingly discussed in AI governance research. Hu et al. (2025) propose that responsible LLM-based multi-agent systems must integrate both subjective human-centered perspectives and objective verification mechanisms. While objective metrics provide traceability and probabilistic guarantees, subjective judgment remains necessary to interpret ethical trade-offs and contextual meaning. This dual structure closely mirrors the concept of dual reflexivity proposed in this chapter.

Science and technology studies have long emphasized that technological artifacts act as mediators within social and epistemic networks (Latour, 2005). Rather than functioning as neutral instruments, technologies shape how actors perceive, organize, and interpret information. In qualitative research, researchers can thus understand AI systems in similar terms. Their outputs provide structured representations of textual material that guide the researcher's interpretive attention.

Recognizing AI as a non-human mediator allows researchers to conceptualize algorithmic influence without attributing human-like cognitive capacities to machines. It also underscores the importance of critically evaluating the epistemic assumptions embedded in AI systems. These assumptions may include linguistic conventions, cultural biases, or normative patterns present in the model's training data. When such patterns shape the system's analytical outputs, they can subtly influence the interpretive direction of qualitative research.

Positionality in Hybrid Knowledge Production

The integration of AI into qualitative research also invites a reconsideration of positionality. Traditionally, positionality refers to the social and epistemic location of the researcher and the ways in which this location shapes the research process (Haraway, 1988; Berger, 2015). However, when algorithmic systems participate in the analytical process, knowledge production becomes a hybrid activity involving multiple forms of epistemic positioning.

Human researchers bring contextual understanding, theoretical orientation, and experiential knowledge to the interpretation of qualitative data. AI systems, by contrast, contribute patterns derived from large-scale textual corpora that reflect dominant linguistic structures and cultural discourses. The interaction between these two forms of epistemic positioning shapes the analytical landscape of AI-augmented research.

This hybrid configuration introduces new forms of positional influence. The training data used to develop large language models often originates from specific linguistic communities, cultural contexts, and knowledge traditions. As a result, the outputs generated by these models may reflect dominant discourses or widely represented perspectives within the training corpus. When researchers rely on AI-generated analytical suggestions, the underlying patterns can shape the research's interpretive framing.

Recent work in human-centered data science has extended the concept of positionality beyond the researcher to computational models themselves. Cambo and Gergle (2022) introduce the concept of model positionality, arguing that machine learning models embody perspectives shaped by training data, annotation practices, and modeling decisions. Recognizing these embedded perspectives reinforces the need for reflexive awareness of the epistemic role of AI systems within research workflows.

Dual reflexivity, therefore, requires researchers to acknowledge both human and algorithmic sources of epistemic positioning. Rather than treating AI outputs as neutral analytical suggestions, researchers must critically evaluate how the training data and the design of algorithmic systems shape the perspectives embedded in those outputs.

Epistemic Distancing and Interpretive Accountability

A key practice associated with dual reflexivity is epistemic distancing—the deliberate effort by researchers to maintain critical distance from AI-generated outputs while engaging with them as analytical resources. Because AI-generated text often appears coherent and analytically plausible, there is a risk that researchers may accept algorithmic suggestions without sufficiently scrutinizing their epistemic foundations.

Epistemic distancing encourages researchers to treat AI outputs as provisional hypotheses rather than authoritative interpretations. Algorithmic suggestions should be evaluated against empirical data, the theoretical framework, and contextual knowledge of the research setting. This critical engagement helps ensure that AI-assisted analysis remains grounded in the interpretive logic of qualitative inquiry.

At the same time, dual reflexivity emphasizes the importance of interpretive accountability. Even when AI systems contribute to the analytical process, the responsibility for interpreting qualitative data ultimately rests with the human researcher. AI systems lack the contextual awareness and ethical responsibility required for meaningful interpretation. Researchers must therefore remain transparent about how AI tools were used and how their outputs influenced the analytical process.

By integrating classical reflexive awareness with critical reflection on technological mediation, dual reflexivity provides a methodological framework for navigating the complexities of AI-augmented qualitative research. It allows researchers to engage productively with emerging analytical technologies while preserving the interpretive rigor and epistemic responsibility that define qualitative methodology.

OPERATIONALIZING REFLEXIVITY IN AI-AUGMENTED RESEARCH

While dual reflexivity defines a theoretical framework for understanding how AI transforms qualitative research, its practical value depends on translating reflexive awareness into concrete methods. Reflexivity in classic qualitative inquiry is both a conceptual approach and a set of operational steps. Practices like journaling, memo writing, and audit trails help researchers record and evaluate their interpretive decisions. In AI-augmented research, these practices must expand to address the additional mediation introduced by algorithmic systems.

Operationalizing reflexivity in this context requires methodological mechanisms that make the role of AI visible within the research process. Without such mechanisms, the influence of algorithmic systems may remain implicit, making it difficult for researchers and readers to evaluate how AI contributions shape analytical outcomes. Four complementary practices are particularly important for achieving this goal: systematic documentation protocols, AI audit trails, reflexive triangulation, and the development of AI literacy among qualitative researchers.

Documentation Protocols and Transparency

The first step in bringing reflexivity to AI-assisted qualitative research is to set clear documentation protocols for AI tools. Transparency is a core requirement for methodological rigor in qualitative work (Guba & Lincoln, 1985). When algorithms are involved, transparency must go beyond typical methods. It should also include the technological systems that shape how data is processed and understood.

A key part of documentation is specifying the technological systems used. Researchers should clearly record which AI tools they used, including model versions, software, and system settings. Since language models update often and can behave differently across versions, documenting the research environment's configuration is vital for traceability.

Another important component concerns the documentation of prompting practices. In many AI-assisted analytical workflows, the interaction between researcher and model is structured through prompts that guide the generation of outputs. These prompts effectively serve as analytical instructions, shaping the type of response the system produces. Reflexive documentation, therefore, requires researchers to record the prompts used, the context in which they were applied, and any iterative modifications made during the analytical process. Such documentation helps illuminate how the researcher's decisions interact with the model's generative capabilities.

Recent methodological work has begun to address the reliability challenges of AI-assisted qualitative analysis. Jain et al. (2026) propose a multi-perspective validation framework that combines traditional inter-rater reliability metrics, such as Cohen's kappa, with semantic similarity measures derived from sentence embeddings. By running multiple independent LLMs and calculating agreement across outputs, the approach allows researchers to distinguish between stable thematic signals and stochastic variation inherent to generative models. This ensemble-based validation strategy represents a promising methodological development for AI-augmented qualitative inquiry, as it operationalizes reflexive scrutiny of machine-generated interpretations rather than treating them as singular analytical outputs.

Furthermore, researchers should document how AI-generated outputs are incorporated into the analytical process. This includes specifying whether AI outputs were used for exploratory purposes, for coding suggestions, for thematic clustering, for summarization, or for drafting analytical text. Transparent reporting of these practices allows readers to evaluate the extent to which algorithmic systems influenced data interpretation.

By following such documentation protocols, qualitative researchers turn AI-assisted analysis from a mysterious technical step into a clear part of their methods. This ensures the role of algorithms remains visible throughout the research.

Audit Trails in Qualitative Analysis

Using documentation protocols, qualitative researchers make AI analysis transparent. This helps others understand how raw data becomes concepts and theories.

In AI-augmented research environments, audit trails must also capture the role of algorithmic systems in shaping the analytical trajectory. An AI audit trail records not only the researcher's decisions but also the interactions between the researcher and AI tools. This includes documenting the prompts used, the model's outputs, the researcher's evaluation of those outputs, and the decisions made about whether to adopt, modify, or reject algorithmic suggestions.

The goal of these audit trails is to build interpretive accountability, not rigid rules. By keeping detailed records of AI-assisted steps, researchers can see how algorithms shape their decisions. AI audit trails let others weigh the logic of the analysis.

AI audit trails can take different forms. They might be logs of researcher–AI interactions, annotated transcripts, or built into qualitative analysis software. Whatever the format, their main purpose is to ensure the role of algorithms remains clear and accountable.

To make this approach more concrete, Table 1 outlines a minimal structure for an AI audit trail in qualitative research.

Table 1. Minimal Structure of an AI Audit Trail in Qualitative Analysis

Prompt / Instruction	**The exact prompt or query used to generate AI output**
Model Configuration	Model type, version, temperature, and system settings
Input Data	Context Description of the data segment provided to the model
AI Output	Generated coding suggestion, summary, or thematic cluster
Researcher Evaluation	Assessment of the output (accepted, modified, rejected)
Justification	Explanation of the decision based on empirical and theoretical grounds
Iteration Notes	Record of prompt refinements or repeated model interactions

Source: own representation.

This structured documentation enables researchers to systematically trace how algorithmic outputs influence the analytical process. Rather than treating AI-generated suggestions as opaque or auxiliary, the audit trail integrates them into the methodological record. This enhances both transparency and interpretive accountability, allowing readers to assess how analytical decisions emerge from the interaction between human judgment and algorithmic mediation.

Reflexive Triangulation

Another useful method for dual reflexivity is reflexive triangulation. Triangulation usually helps build credibility in qualitative research by comparing data sources, theories, or methods (Denzin & Lincoln, 2018). In AI-aided studies, it can mean comparing human and AI interpretations of the same data.

Reflexive triangulation involves systematically examining the similarities and differences between the researcher's interpretations and those suggested by AI systems. For example, a researcher may compare manually developed codes with categories proposed by an AI-assisted coding tool, or evaluate how AI-generated summaries align with their own interpretive reading of interview transcripts.

This approach serves several purposes. It uncovers possible limits or biases in AI outputs. AI-generated categories may reflect patterns in language rather than the meaningful themes qualitative research seeks. Comparing these with human interpretations grounded in context helps verify their relevance.

Reflexive triangulation can also help reveal hidden patterns. AI may spot semantic patterns in big datasets that are hard to see manually. When human analysis combines with AI insight, new analytical ideas can emerge.

However, the goal of reflexive triangulation is not to treat AI outputs as authoritative or superior to human interpretation. Rather, it provides a structured method for critically evaluating algorithmic contributions within the broader interpretive framework of the study. By engaging in this iterative comparison, researchers maintain control over the analytical process while benefiting from AI tools' exploratory capabilities.

Developing AI Literacy for Qualitative Researchers

Operationalizing reflexivity in AI-augmented research also requires developing AI literacy within qualitative research communities. AI literacy refers to the capacity to understand the basic principles, limitations, and epistemic characteristics of artificial intelligence systems. Without such understanding, researchers may struggle to critically evaluate algorithmic outputs or to recognize the biases embedded within them.

AI literacy does not require qualitative researchers to become technical experts in machine learning. Instead, it involves developing a conceptual understanding of how AI systems generate outputs and the limitations they possess. For example, researchers should be aware that large language models generate responses based on statistical correlations within training data rather than genuine comprehension of meaning (Bender et al., 2021). They should also recognize that these models may reproduce biases present in their training datasets and that their outputs can vary depending on prompt design and contextual inputs.

Developing AI literacy also entails understanding the broader socio-technical contexts in which AI systems are developed and deployed. Scholars in critical algorithm studies have emphasized that algorithmic systems are shaped by institutional priorities, economic incentives, and cultural assumptions embedded in their training data and design (Kitchin, 2017; Gillespie, 2014). This includes understanding issues such as training data provenance, algorithmic bias, model updates, and the institutional contexts in which AI tools are developed and deployed. Such awareness enables researchers to critically evaluate the epistemic implications of using AI within qualitative research.

Integrating AI literacy into qualitative research training programs can therefore strengthen the reflexive capacity of future researchers. By equipping scholars with the conceptual tools needed to interrogate algorithmic systems, AI literacy ensures that technological innovations enhance rather than undermine the interpretive rigor of qualitative inquiry.

Through these practices—documentation protocols, AI audit trails, reflexive triangulation, and the cultivation of AI literacy—the concept of dual reflexivity can be translated into concrete methodological procedures. These mechanisms help ensure that AI-assisted analysis remains transparent, critically evaluated, and grounded in the interpretive logic of qualitative research. By operationalizing reflexivity in this way, qualitative scholars can engage productively with emerging AI technologies while preserving the epistemic accountability that defines rigorous qualitative inquiry.

ETHICAL AND PEDAGOGICAL IMPLICATIONS

The integration of artificial intelligence into qualitative research fundamentally alters research methods and raises important ethical and pedagogical questions. This integration also introduces broader institutional and epistemic challenges that extend beyond methodology. Chesterman and Chieh (2026) highlight that generative AI poses new risks to research integrity, such as difficulties with reproducibility, authorship, and accountability. As AI-generated research outputs increase, existing systems such as peer review and scholarly evaluation may struggle to uphold knowledge standards.

The ethical implications of AI-augmented qualitative research can be analytically structured along three interrelated dimensions: (1) epistemic integrity, referring to the preservation of rigorous, contextually grounded interpretation; (2) accountability, concerning the attribution of interpretive responsibility in hybrid human–AI analytical processes; and (3) algorithmic bias, addressing the ways in which training data and model architectures may shape and constrain interpretive outcomes. Structuring the ethical discussion along these dimensions allows for a more systematic examination of the risks and responsibilities associated with AI-assisted qualitative inquiry.

As AI systems take on greater roles in data preparation, coding, thematic structuring, and analysis, it is crucial to re-examine core principles of qualitative research: interpretive responsibility, academic integrity, and epistemic accountability. Hybrid research environments, where human researchers work alongside algorithms, challenge traditional definitions of authorship, expertise, and scholarly judgment. Therefore, it is essential to directly consider both the ethical and educational implications of these technological changes.

Risks of Epistemic Outsourcing

AI-assisted qualitative research brings forth a significant ethical issue: epistemic outsourcing. Researchers may delegate substantial portions of interpretation to algorithms, risking disengagement from empirical material. While AI tools enhance efficiency in transcription, coding, summarization, or pattern detection, unchecked dependence could erode the researcher's interpretive role.

Qualitative research traditionally emphasizes deep immersion in the empirical material. Through iterative reading, coding, and comparison, researchers develop intimate familiarity with the narratives, meanings, and contextual nuances embedded in the data (Charmaz, 2014). This sustained engagement is essential for developing theoretical sensitivity. It also helps produce interpretations grounded in the lived experiences of research participants. When AI systems perform large portions of this analytical work, researchers may be tempted to rely on algorithmically generated summaries or thematic structures without fully engaging with the underlying data.

This dynamic is closely related to automation bias, the tendency to place unwarranted trust in the outputs of automated systems because they appear systematic or technically sophisticated. In qualitative research, automation bias may lead researchers to accept AI-generated coding schemes, thematic clusters, or analytical summaries without critically evaluating their validity. Because large language models produce fluent, coherent textual outputs, their suggestions may appear analytically plausible even when they are not grounded in the study's specific empirical context.

Mitigating these risks requires maintaining a reflexive awareness of AI systems' limitations. Researchers must remain actively engaged in the interpretive process and treat AI outputs as provisional analytical suggestions rather than authoritative conclusions. Dual reflexivity plays a crucial role in this regard by encouraging researchers to critically examine both their own interpretive assumptions and the epistemic characteristics of the technological systems they employ.

Academic Integrity and the Risk of Shortcut Science

Epistemic outsourcing closely relates to the broader concern of academic integrity amid the rise of generative AI. With AI able to produce coherent analytical text, researchers—or students learning qualitative methods—may be tempted to skip essential research steps. This trend, dubbed 'shortcut science,' emerges when AI-generated outputs are used in place of genuine analytical engagement.

Shortcut science can take several forms. Researchers may rely on AI systems to generate coding schemes, summarize interview transcripts, or draft analytical interpretations without critical evaluation. In educational settings, students may use AI-generated analyses as substitutes for developing their own interpretive insights.

These practices threaten the core purpose of qualitative research, which is to generate meaning through rigorous interpretation of empirical data. If AI-generated outputs supplant critical engagement, both the depth and credibility of qualitative analysis are at risk.

Addressing this challenge requires developing clear ethical guidelines for the responsible use of AI in qualitative research. These guidelines should emphasize transparency regarding the use of AI tools, critical evaluation of algorithmic outputs, and the continued primacy of human interpretation within the analytical process. Rather than prohibiting AI use, ethical frameworks should focus on ensuring that AI-assisted practices remain consistent with the epistemic principles of qualitative inquiry.

Responsibility in Hybrid Interpretation

AI-augmented research environments raise new questions about who holds interpretive authority and accountability. Whereas researchers traditionally assume full responsibility for findings, the involvement of algorithms complicates attribution of interpretive decisions. Determining the boundaries between researcher and AI contributions becomes necessary in these hybrid contexts.

When AI-generated outputs influence the development of code, themes, or analytic narratives, it may become difficult to determine the extent to which these interpretations originate with the researcher rather than the algorithmic system. Yet these systems lack the intentionality, contextual awareness, and ethical responsibility necessary to be considered accountable actors. As Floridi et al. (2018) emphasize in discussions of AI ethics, responsibility for the outcomes of AI-assisted processes ultimately rests with the human agents who design, deploy, and interpret these systems' outputs. AI systems lack intentionality, moral agency, and contextual understanding; they cannot be held accountable for the interpretations they produce.

Within qualitative research, this principle implies that researchers remain fully responsible for the interpretations presented in their studies, even when AI tools contribute to the analytical process. Researchers must therefore critically evaluate AI-generated outputs and ensure that their interpretations remain grounded in the empirical material and the study's theoretical framework.

Recent work on LLM-powered multi-agent systems further highlights the systemic nature of responsibility in AI-enabled environments. Hu et al. (2025) argue that responsible LLM systems need more than local alignment mechanisms. They require lifecycle-wide governance structures that combine objective verification with human-centered values. Their framework distributes responsibility across design, development, deployment, and maintenance stages. Each stage involves human oversight and evaluative judgement. This lifecycle perspective supports the argument in this chapter: reflexivity in AI-augmented qualitative research must extend beyond the individual researcher to include the socio-technical infrastructures shaping interpretive processes.

Dual reflexivity offers a way to maintain interpretive accountability by prompting researchers to review both their own and AI systems' influence in generating interpretations. Clearly stating how AI tools are used in research is essential for building trust in qualitative studies.

Implications for Qualitative Research Training

AI-triggered ethical and methodological changes also demand significant adaptations in training qualitative researchers. Graduate programs and methodological curricula have long emphasized skills development in interviewing, coding, thematic analysis, and theoretical abstraction. As AI tools become embedded in qualitative workflows, these training programs must evolve to incorporate competencies in AI literacy and reflexive engagement with algorithms.

Qualitative researchers must learn how to critically assess AI outputs, recognize their constraints, and use them responsibly. This requires technical understanding of AI tools and thoughtful consideration of their underlying epistemic assumptions. Students need to reflect on how AI changes their analytical process.

At the same time, methodological training must continue to emphasize the core interpretive competencies that define qualitative research. The integration of AI should not marginalize close reading, contextual interpretation, and theoretical reasoning. Instead, educational programs should aim to cultivate a reflexive AI competence, enabling researchers to harness technological tools while preserving the epistemic depth and critical engagement characteristic of qualitative inquiry.

Developing such competence will require interdisciplinary collaboration between social scientists, information scholars, and computational experts. By integrating perspectives from these fields, qualitative methodology can adapt to the changing technological landscape while maintaining its commitment to rigorous, reflexive knowledge production.

The ethical and pedagogical implications discussed in this section highlight the broader significance of reflexivity in contemporary qualitative methodology. As artificial intelligence becomes increasingly embedded within research practices, maintaining interpretive rigor requires both conceptual innovation and institutional adaptation. Dual reflexivity offers a framework for navigating these transformations by encouraging researchers to critically engage with the ethical, epistemic, and educational dimensions of AI-augmented qualitative inquiry.

Methodological Limitations of AI-Augmented Qualitative Research

While AI-assisted qualitative analysis offers significant opportunities for efficiency, scalability, and exploratory pattern detection, it is also subject to important methodological limitations that must be critically acknowledged.

First, the outputs of large language models are inherently probabilistic and can vary depending on prompt design, model configuration, and contextual inputs. This variability poses challenges for the reproducibility and stability of AI-assisted analytical procedures, particularly compared with more standardized qualitative coding approaches.

Second, AI-generated outputs are dependent on the training data and model architectures on which they are based. As a result, they may reproduce dominant linguistic patterns, cultural biases, or epistemic assumptions embedded in large textual corpora, potentially shaping the analytical framing in ways that are not transparent to the researcher.

Third, integrating AI systems into qualitative workflows risks reduced engagement with empirical material. If researchers rely excessively on AI-generated summaries or coding suggestions, the risk is that interpretive depth and contextual sensitivity may be compromised.

Finally, the use of AI tools raises challenges for methodological transparency and accountability. Although practices such as documentation protocols and audit trails can mitigate these concerns, the complexity of AI systems may still limit the extent to which their internal processes can be fully understood and evaluated.

Recognizing these limitations is essential for situating AI-assisted qualitative research within a framework of reflexive methodological practice. Rather than undermining the value of AI tools, such critical awareness helps ensure that their use remains aligned with the epistemic principles of qualitative inquiry.

CONCLUSION

The rapid integration of artificial intelligence—particularly large language models (LLMs), which are AI systems trained to predict and generate natural language—into qualitative research environments marks a significant transformation in the epistemic (knowledge-producing) conditions under which interpretive inquiry is conducted. What were once auxiliary digital tools have increasingly evolved into integral components of research infrastructures, participating in processes ranging from transcription and coding (labeling qualitative data) to thematic structuring and analytical synthesis. As qualitative research becomes progressively embedded in AI-supported workflows, the traditional assumption that interpretation is solely a human activity can no longer be taken for granted.

Building on this transformation, this chapter has argued that these developments place the classical concept of reflexivity under considerable pressure. Reflexivity has long served as a foundational element of qualitative methodology, ensuring that researchers critically examine how their positionality, theoretical commitments, and methodological decisions shape the production of knowledge (Alvesson & Sköldberg, 2018; Finlay, 2002). However, classical reflexive frameworks were developed within research environments in which the researcher was assumed to be the primary—if not the sole—interpretive agent. The integration of AI systems into qualitative workflows complicates this assumption by introducing additional layers of mediation into the interpretive process.

As discussed throughout this chapter, AI systems can influence qualitative research at multiple stages. During the pre-interpretive phase, algorithmic tools reshape empirical material through processes such as transcription, translation, and summarization. In the co-interpretive phase, AI systems contribute to coding and thematic clustering by identifying patterns across large textual datasets. In the post-interpretive phase, language models may contribute to the synthesis and representation of findings by generating analytic summaries or assisting in drafting research narratives. Although these activities do not constitute interpretation in the human sense, they significantly shape the analytical environment within which interpretation occurs.

A central risk associated with this transformation lies in the emergence of epistemic illusions—situations in which algorithmically generated outputs appear to provide meaningful interpretations despite being produced through statistical language modeling (pattern-based prediction and generation) rather than contextual understanding (awareness of the situation and meaning in context) (Bender et al., 2021). Because large language models generate linguistically coherent and persuasive text, researchers may be inclined to attribute interpretive authority to outputs that are fundamentally grounded in probabilistic correlations within training data.

Recognizing and addressing this risk requires a renewed commitment to reflexive awareness within qualitative methodology.

In light of these challenges, this chapter introduced the concept of dual reflexivity. Dual reflexivity extends traditional reflexive practice by incorporating critical reflection on the epistemic role of algorithmic systems within the research process. It acknowledges that in AI-augmented qualitative inquiry, interpretive outcomes are shaped by the interaction between human reasoning and algorithmic mediation. By examining both the researcher's positionality and the technological infrastructures through which data are processed and interpreted, dual reflexivity provides a conceptual framework for understanding hybrid knowledge production.

Importantly, dual reflexivity does not imply that AI systems should be treated as interpretive agents (entities capable of understanding and producing meanings) equivalent to human researchers. Rather, it emphasizes that algorithmic systems function as non-human interpretive mediators—tools that shape how data is analyzed and findings are developed—whose outputs influence the analytical pathways available to researchers. Recognizing this mediating role allows qualitative researchers to critically engage with AI technologies without attributing human-like cognitive capacities to them.

The chapter has also demonstrated that dual reflexivity must be translated into concrete methodological practices in order to be effective. Documentation protocols, AI audit trails, and reflexive triangulation provide practical mechanisms for ensuring transparency and accountability in AI-assisted analysis. Furthermore, developing AI literacy among qualitative researchers is essential to enabling scholars to critically evaluate algorithmic outputs and understand the socio-technical infrastructures that shape them.

Beyond methodological considerations, integrating AI into qualitative research raises broader ethical and pedagogical questions. Risks such as epistemic outsourcing, automation bias, and shortcut science highlight the need for responsible research practices that preserve the interpretive engagement at the heart of qualitative inquiry. At the same time, educational institutions must adapt their methodological training to prepare future researchers for the rapidly emerging hybrid research environments.

Looking to the future, the relationship between qualitative methodology and artificial intelligence will likely continue to evolve. Advances in agentic AI systems—capable of initiating analytical tasks, generating hypotheses, or autonomously interacting with data—may further blur the boundaries between human and machine contributions to knowledge production. Rather than resisting these developments, qualitative researchers must develop methodological frameworks that allow them to engage with AI technologies in a critical and reflexive manner.

As AI generates more and more information, universities may focus more on verifying and curating reliable knowledge, rather than just discovering it. In this way, they could become key institutions for maintaining transparency and trust in AI-influenced research (Chesterman & Chieh, 2026).

Dual reflexivity is one useful framework for the future. Expanding reflexive thinking to cover both people and algorithms helps protect the quality of qualitative research when technology is involved. In the end, qualitative research will thrive if we include AI in ways that keep the deep thinking and care that define strong scholarship.

REFERENCES

Alvesson, M., & Sköldberg, K. (2018). *Reflexive methodology: New vistas for qualitative research* (3rd ed.). Sage.

Bender, E. M., Gebru, T., McMillan-Major, A., & Shmitchell, S. (2021). On the dangers of stochastic parrots: Can language models be too big? *Proceedings of the 2021 ACM Conference on Fairness, Accountability, and Transparency (FAccT '21)*, 610–623. DOI: 10.1145/3442188.3445922

Berger, R. (2015). Now I see it, now I don't: Researcher's position and reflexivity in qualitative research. *Qualitative Research, 15*(2), 219–234. DOI: 10.1177/1468794112468475

Cambo, S. A., & Gergle, D. (2022). Model positionality and computational reflexivity: Promoting reflexivity in data science. *arXiv*. /arXiv.2203.07031DOI: 10.1145/3491102.3501998

Charmaz, K. (2014). *Constructing grounded theory* (2nd ed.). Sage.

Chesterman, S., & Chieh, L. H. (2026). Research integrity and academic authority in the age of Artificial Intelligence: From discovery to curation? *arXiv*. /arXiv.2601.05574DOI: 10.2139/ssrn.6046274

Conti, L. G., & Seele, P. (2025). Reflexivity and positionality statements: A way to tackle "second-level arbitrariness" bias in AI? *Humanities & Social Sciences Communications*, *12*(1), 1930. DOI: 10.1057/s41599-025-06208-6

Cruz-Aguilar, M. A. (2025). The epistemic revolution of AI: Reconfiguring the foundations of scientific knowledge. *AI & Society*. Advance online publication. DOI: 10.1007/s00146-025-02658-3

Denzin, N. K., & Lincoln, Y. S. (Eds.). (2018). *The Sage handbook of qualitative research* (5th ed.). Sage.

Finlay, L. (2002). Negotiating the swamp: The opportunity and challenge of reflexivity in research practice. *Qualitative Research*, *2*(2), 209–230. DOI: 10.1177/146879410200200205

Floridi, L., Cowls, J., Beltrametti, M., Chatila, R., Chazerand, P., Dignum, V., Luetge, C., Madelin, R., Pagallo, U., Rossi, F., Schafer, B., Valcke, P., & Vayena, E. (2018). AI4People—An ethical framework for a good AI society: Opportunities, risks, principles, and recommendations. *Minds and Machines*, *28*(4), 689–707. DOI: 10.1007/s11023-018-9482-5 PMID: 30930541

Geertz, C. (1973). *The interpretation of cultures*. Basic Books.

Gillespie, T. (2014). The relevance of algorithms. In Gillespie, T., Boczkowski, P. J., & Foot, K. A. (Eds.), *Media technologies: Essays on communication, materiality, and society* (pp. 167–194). MIT Press. DOI: 10.7551/mitpress/9042.003.0013

Guba, E. G., & Lincoln, Y. S. (1985). Naturalistic inquiry. *Sage (Atlanta, Ga.)*.

Haraway, D. (1988). Situated knowledges: The science question in feminism and the privilege of partial perspective. *Feminist Studies*, *14*(3), 575–599. DOI: 10.2307/3178066

Harding, S. (1992). Rethinking standpoint epistemology: What is "strong objectivity"? *The Centennial Review*, *36*(3), 437–470. https://www.jstor.org/stable/23739232

Hauswald, R. (2025a). A companion to applied philosophy of AI. In Hähnel, M., & Müller, R. (Eds.), *AI and the philosophy of expertise and epistemic authority* (pp. 55–70). Wiley., DOI: 10.1002/9781394238651.ch5

Hauswald, R. (2025b). Artificial epistemic authorities. *Social Epistemology*, *39*(6), 716–725. DOI: 10.1080/02691728.2025.2449602

Holmes, A. G. D. (2020). Researcher Positionality—A consideration of its influence and place in qualitative research—A New Researcher Guide. *Shanlax International Journal of Education*, *8*(4), 1–10. DOI: 10.34293/education.v8i4.3232

Hu, J., Dong, Y., Ao, S., Li, Z., Wang, B., Singh, L., ... (2025). Stop reducing responsibility in LLM-powered multi-agent systems to local alignment. *arXiv*. https://doi.org//arXiv.2510.14008DOI: 10.48550

Hüholt, N., & Szech, N. (2026). Trusting machines with morality — Delegating moral decisions to AI. *European Economic Review*, *184*, 105255. DOI: 10.1016/j.euroecorev.2025.105255

Jain, N., Suh, H., Adeyinka, S., Roseman, L., & Allsop, A. (2026). Multi-LLM thematic analysis with dual reliability metrics: Combining Cohen's Kappa and semantic similarity for qualitative research validation. *arXiv*. https://doi.org//arXiv.2512.20352DOI: 10.48550

Kitchin, R. (2017). Thinking critically about and researching algorithms. *Information Communication and Society*, *20*(1), 14–29. DOI: 10.1080/1369118X.2016.1154087

Latour, B. (2005). *Reassembling the social: An introduction to actor-network-theory*. Oxford University Press. DOI: 10.1093/oso/9780199256044.001.0001

Pillow, W. S. (2003). Confession, catharsis, or cure? Rethinking the uses of reflexivity as methodological power in qualitative research. *International Journal of Qualitative Studies in Education : QSE, 16*(2), 175–196. DOI: 10.1080/0951839032000060635

Schroeder, H., Le Quéré, M. A., Randazzo, C., Mimno, D., & Schoenebeck, S. (2025). Large Language Models in qualitative research: Uses, tensions, and intentions. *CHI '25: Proceedings of the 2025 CHI Conference on Human Factors in Computing Systems*, 481, 1–17. DOI: 10.1145/3706598.3713120

Shin, D. (2025). Automating epistemology: How AI reconfigures truth, authority, and verification. *AI & Society, 41*(2), 1553–1559. DOI: 10.1007/s00146-025-02560-y

Suchman, L. (2007). *Human–machine reconfigurations: Plans and situated actions* (2nd ed.). Cambridge University Press.

Vikan, M., Aryan, R., & Danielsen, S. O. (2026). Reflecting on LLM support in reflexive thematic analysis: An exploratory study. *Qualitative Health Research, 36*(2-3), 191–205. Advance online publication. DOI: 10.1177/10497323251365211 PMID: 40916991

Yang, M., Gao, J., & Wu, J. (2025). Decoupling knowledge and reasoning in LLMs: An exploration using cognitive dual-system theory. *arXiv*. https://doi.org//arXiv .2507.18178DOI: 10.48550

Zhang, W., Wang, Q., Zhao, B., Zhang, T., Wu, J., Gan, H., ... (2025). Reflecting with two voices: A co-adaptive dual-strategy framework for LLM-Based agent decision making. *arXiv*. https://doi.org//arXiv.2512.08366DOI: 10.48550

Chapter 3
Computational Ethnography:
Reimagining Qualitative Inquiry in the Age of AI

Kritika Sharma
http://orcid.org/0009-0001-2884-557X
Lovely Professional University, India

Harry Kaur
http://orcid.org/0000-0001-8276-2313
Lovely Professional University, India

ABSTRACT

This chapter conceptualizes computational ethnography as a methodological rearticulation of qualitative inquiry in the age of artificial intelligence. As social life becomes increasingly mediated by digital platforms, ethnographers confront large-scale textual data that challenge conventional analytic practices. Situating computational ethnography within constructivist and interpretivist traditions, the chapter explains how AI-enabled techniques such as natural language processing, text mining, sentiment analysis, and large language models can support exploratory pattern detection without replacing ethnographic interpretation. It critically examines epistemological and ethical challenges arising from algorithmic mediation, including the black-box problem, algorithmic bias, trustworthiness, informed consent, and data privacy. Advancing a human-centered approach, the chapter emphasizes human-in-the-loop practices to preserve reflexivity, contextual sensitivity, and ethical accountability in AI-enhanced qualitative research.

DOI: 10.4018/979-8-3373-7640-0.ch003

t © 2026, IGI Global Scientific Publishing. Copying or distributing in print or electronic forms without written permission of IGI Global Scientific Publishing is prohibited. Use of this chapter to ›nerative artificial intelligence (AI) technologies is expressly prohibited. The publisher reserves all rights to license its use for generative AI training and machine learning model development.

INTRODUCTION

Ethnography has traditionally held center stage in qualitative research, providing rich, contextualized interpretations of social life, cultural practices, and meaning-making in everyday life. In this chapter, "artificial intelligence" primarily refers to natural language processing techniques, machine learning-based text analysis, and large language models used for tasks such as transcription, pattern detection, and thematic (Brown et al., 2020). The "age of AI" in this context refers to the increasing integration of machine learning systems, particularly natural language processing and large language models, into research practices and everyday social interaction (Chen et al., 2025). Based on immersive fieldwork, participant observation, and interpretive engagement, ethnographic research aims to generate what (Geertz, 1973) has referred to as thick description, not only observable behavior but also the symbolic and cultural meanings that social action entails. Ethnography is intertwined with the constructivist and interpretivist epistemologies within the larger framework of qualitative methodology, which focuses on the socially constructed reality and the situational nature of knowledge (Younas et al., 2023).

These traditions predetermine reflexivity, positionality, and the co-construction of meanings between the researcher and participants. Over the last 20 years, the landscapes of socializing and cultural expression have drastically changed as digital platforms, social media, and algorithmically mediated communication have become a significant part of life. Online spaces or digital media are increasingly used to construct identities, negotiate norms, and define grievances, producing immense amounts of digitally mediated cultural information (Nascimento et al., 2022). Although digital ethnography has brought the practice of ethnography into the virtual realm, it's massive, fast, and diverse information demands are putting pressure on the analytic capabilities of traditional qualitative methods that are primarily, but not exclusively, based on manual observation and coding. Such methodological pressure has increased the need for new approaches that support the use of large-scale qualitative data without losing interpretive richness and contextual sensitivity.

Despite the growing use of computational tools in qualitative research, there remains a lack of clearly articulated methodological frameworks that specify how such tools can be systematically integrated into ethnographic inquiry without undermining its interpretive and reflexive foundations. Existing discussions often remain either technically oriented, focusing on tools and scalability, or conceptually abstract, emphasizing epistemological concerns without demonstrating methodological application. This gap produces uncertainty regarding how computational outputs can be translated into ethnographic claims, how interpretive authority is maintained, and how methodological rigor can be ensured in AI-assisted qualitative research.

More recent developments in artificial intelligence, especially in natural language processing, text mining, sentiment analysis, and large language models, have further transformed the methodological space of qualitative research. They are computational resources that allow scientists to uncover trends, thematic groups, and discursive structures within large bodies of text and to provide new avenues of inquiry into digitally mediated social worlds (Golann et al., 2024). This chapter focuses specifically on text-based digital environments (e.g., social media platforms, online forums) where meaning-making is expressed through written and semi-structured interaction. It does not address multimodal AI systems such as computer vision in depth, nor does it claim to replace immersive field-based ethnography. Instead, the scope is limited to the methodological integration of AI-assisted text analysis within interpretive qualitative frameworks.

Simultaneously, the growing integration of algorithmic systems into qualitative inquiry raises serious epistemological and ethical issues. Researchers have warned of the black-box behaviour of most AI models, the replication of algorithmic bias, and the dangers of de contextualization and over-automation in interpretive studies (Witteborn, 2025). Such advances disrupt fundamental qualitative promises of reflexivity, transparency, and trustworthiness, especially when the analyzes process is only partially transparent to the researcher. It is against this changing methodological landscape that computational ethnography has emerged as a hybrid orientation that seeks to balance ethnographic sensibilities with computational analytical capacities.

Computational ethnography is the further development of digital ethnographic practice that incorporates computational methods into the ethnographic process, enabling researchers to work with large-scale digital traces without losing the interpretive and reflexive ethics of qualitative research (Pink, 2021). Instead of framing AI as an alternative to ethnographic interpretation, computational ethnography prefigures the researcher's contribution in making computational outputs contextual, interpretative, and ethically regulative. This stance aligns with human-centered approaches to AI that underscore the centrality of human judgment, responsibility, and ethical accountability in AI-mediated research practices (Arnold & Fuller, 2018).

Computational ethnography is essential because it helps fill a growing methodological gap between the increasing volume of digitally mediated social data, on the one hand, and the interpretive commitments of qualitative research, on the other. The growing engagement of qualitative scholars with data worlds conditioned by platform logics and algorithmic infrastructures urgently needs methodological approaches that allow meaningful engagement with computational technologies without delegitimizing the epistemological principles of ethnographic research.

This chapter addresses this requirement by rethinking ethnography in an age of artificial intelligence, situating computational ethnography within the contexts of constructivist and interpretivist traditions, and critically assessing its methodological, epistemological, and ethical consequences. This chapter advances three interrelated methodological aims:

1. to formalize computational ethnography as a hybrid workflow that integrates computational pattern detection with ethnographic interpretation;
2. to demonstrate how digital traces can be systematically transformed into ethnographic data through explicit methodological steps; and
3. to develop a reflexive and transparent protocol for reconciling computational outputs with interpretive qualitative analysis.

In doing so, the chapter moves beyond abstract advocacy to outline a structured and operationalizable approach to AI-enhanced ethnographic inquiry. Rather than presenting computational ethnography as a purely conceptual proposition, this chapter develops a structured methodological framework supported by a step-wise analytic architecture and an illustrative application. It specifies how field sites are bounded, how digital traces are defined and sampled, how computational outputs are generated and evaluated, and how interpretive claims are constructed and validated. By making these processes explicit, the chapter aims to establish computational ethnography as a defensible methodological approach grounded in both epistemological rigor and practical applicability.

Literature Review: Ethnography, Digital Methods, and AI in Qualitative Research

This section summarizes the main literature ideas applicable to the creation of the concept of computational ethnography with reference to three strands that are connected: (i) classical ethnographic methodology, (ii) digital ethnography and online qualitative research, and (iii) the new application of artificial intelligence to qualitative and mixed-method research. The introduction of such strands into the discussion clarifies the conceptual foundations of the idea of computational ethnography and positions it within modern methodological discussions.

Ethnography in Qualitative Research

Ethnography has long been treated as a methodological pillar of the qualitative research paradigm, which focuses on extensive fieldwork, participant observation, and the interpretive construction of social meanings. Classical descriptions preempt the value of thick description in describing the cultural logics by which social practices are made meaningful (Zaharlick, 1992). Later methodological literature has expounded on the epistemological foundations of ethnography within the constructivist and interpretivist schools of thought, outlining the co-created, situated character of ethnographic knowledge (Hammersley & Atkinson, 2019). Reflexivity has become one of the main methodological principles, emphasizing the researcher's positionality and ethical accountability when portraying the lives of others (Finlay, 2002).

Digital Ethnography and Online Qualitative Research

As digitally mediated worlds of sociality grow, scholars have extended the ethnographic investigation of online worlds. Digital ethnography is a conceptualization of the internet that does not exist as an abstract virtual object but as part of daily social life and materiality (Malbois, 2021). Similar methodological traditions, such as netnography, have also formalized the study of qualitative inquiry in online communities, emphasizing ethically engaged research, a contextual interpretive approach, and longer engagement with digital field sites (Kozinets, 2020). This literature emphasizes the potentials and issues associated with conducting ethnography in platform-mediated settings, including access issues, temporality, and the politics of representation.

AI and Computational Approaches in Qualitative Research

There has been an increasing body of literature analyzing how computational methods can be used in social research, particularly with large-scale textual data. It has been shown that the text-as-data paradigm can be used to recruit natural language processing and machine learning to investigate discursive patterns on a large scale, using large corpora (Li & Abramson, 2025). Simultaneously, the critical scholarship emphasized that computing tools should not be embraced blindly, raising concerns about the influence of algorithmic bias, epistemic opacity, and the politics of data infrastructures (Ophir et al., 2020). New research into human-centred AI continues to reinforce the need to maintain human judgment, transparency, and accountability in AI-based analytic processes (Li & Abramson, 2024).

Recent debates in qualitative research have increasingly examined the implications of algorithmic bias and epistemic opacity in AI-assisted analysis, particularly in the context of natural language processing tools. Emerging scholarship highlights how computational systems not only process data but actively shape what becomes visible, interpretable, and analytically salient. Concepts such as "algorithmic intimacies" (Jushaninowa et al., n.d.). Emphasize the entanglement of computational systems with cultural meaning-making processes, suggesting that AI-mediated analysis is not neutral but relational and situated. These discussions are crucial for grounding computational ethnography within contemporary methodological debates on bias, representation, and interpretive authority.

Positioning Computational Ethnography within the Literature

Combined, these threads of writing furnish the conceptual framework for computational ethnography as a hybrid methodological disposition that combines ethnographic sensibilities with computational analytical abilities. Although digital ethnography is a qualitative method that extends the approach of qualitative inquiry into the online space, and computational social science provides scalable tools of analysis, computational ethnography strives to reconcile the two domains, preserving interpretive richness whilst working with the scale and speed of digital information. The stance is inspired by calls to embrace methodological pluralism in modern qualitative studies and shapes current discussions of how AI can be ethically incorporated into interpretative social science.

THEORETICAL FOUNDATIONS: CONSTRUCTIVISM, INTERPRETIVISM, AND REFLEXIVITY IN AI-ENHANCED ETHNOGRAPHY

Qualitative research traditions are grounded in epistemological commitments that shape how social reality is perceived, explored, and depicted. Among them, constructivism and interpretivism have offered guidance for ethnographic research by highlighting the social construction of reality and the interpretive processes involved in creating meaning. Constructivist views assume that social phenomena do not exist without reference to human beings and that reality is constructed through social interaction, language, and cultural practices (Low & Hyslop-Margison, 2021). Interpretivism also anticipates the situated, meaning-focused nature of social knowledge, arguing that social action needs to be understood from the perspective of the actors themselves (Clarke, 2022).

In ethnography, these orientations are methodological commitments to engaging in the field and being contextually sensitive, and the generation of what (Geertz, 1973) calls thick description, which encompasses the symbolic aspects of social life. These theoretical traditions place reflexivity at the center of their concerns. Reflexivity demands that researchers be very critical of their own positionality, assumptions, and interpretive frames, and to recognize that the production of knowledge is influenced by the social location of the researcher, theoretical practices, and the relative position of the researcher to the participants (Kurowska & de Guevara,

2020). Ethnographic knowledge is then interpreted as situational and partial rather than neutral or exhaustive.

This reflexivity has traditionally served as a guard against the purported objectivist authority in qualitative research, preempting the ethical and epistemic accountability of the researcher in the rendition of the lives and meanings of other individuals (Finlay, 2017). These epistemological commitments are problematized in significant aspects by the practice of applying artificial intelligence and computational tools to ethnographic research. In cases where natural language processing, text mining, and sentiment analysis have been used to extract patterns among large volumes of text, interpretive power is, to some extent, decentralized between human and algorithmic agents. This redistribution raises important issues regarding the locus of interpretation and the position of computational outputs within the constructivist and interpretivist schools. Though AI can detect recurring word patterns or theme clusters, it lacks cultural situatedness, experiential knowing, and moral responsibility.

Thus, the interpretative work of contextualizing, theorizing and giving meaning to computationally produced patterns is essentially human. The constructivist perspective suggests that qualitative research with AI requires a new perspective on the co-construction of knowledge among human-machine assemblies. Given the nature of the computational models underlying them, they too are the result of specific training data, design decisions, and assumptions that determine what is visible or salient in the analysis. This dynamic brings into view the so-called positionality of AI, the idea that AI systems are not neutral tools but epistemic mediators. The outputs of calculative devices are not just technical products but socially and historically located images that constitute the values, prejudices, and constraints of the structure (Jaton & Sormani, 2023). The acknowledgement of this positionality is consistent with interpretivist beliefs, which question the circumstances in which knowledge is constructed and makes sense. In the context of ethnographic practice, AI integration only exacerbates the issue of trustworthiness and transparency that has been a concern since the beginning of its practice.

The traditional criteria like credibility, dependability, and confirmability have traditionally been ensured by practices of long-term engagement, thick description, triangulation, and reflexive auditing (Wolf, 2012). The method of algorithmic mediation creates an additional layer, which can be called the black-box problem, when the internal code of AI systems is not fully accessible to scientists. Such a lack of transparency questions longstanding methodological cultures of transparency and auditability and provokes the invention of new reflexive methods of recording the decisions of the analytical method, questioning algorithmic presuppositions, and justifying interpretive assertions. Interpretivist ethnography is also focused on the ethical aspects of representation and voice, especially when dealing with marginalized or more digitally vulnerable communities. Using computational tools can amplify dominant discourses at the expense of less salient voices, and when algorithms focus on frequency, salience, or platform-specific measures of engagement.

From an interpretivist perspective, these tendencies highlight the critical role of the method of computation in forming representational hierarchies in qualitative analysis. It turns into an ethical and epistemological requirement as well that researchers become more reflexive about the outputs of algorithms when operating in reflexive ways, as this necessitates ongoing examination of which meanings are being surfaced, which are being obscured, and what is occurring in terms of power relation reproduction or challenge in AI-mediated research practices.

This chapter, in the context of locating computational ethnography within constructivist and interpretivist traditions, posits that AI-enhanced ethnographic inquiry should be situated within a human-oriented interpretive framework. The analytical power of ethnography can be expanded through computational tools that enable interaction with large-scale digital traces of social life. However, they cannot replace the contextual, relational, and ethical work that defines ethnographic understanding. Instead of shaking the foundations of qualitative epistemologies, computational ethnography beckons the rearticulating of reflexivity and positionality in the context of algorithmic mediation. The chapter prefigures the co-constructed nature of knowledge between human and computational actors by framing computational ethnography as a methodological innovation, which is theoretically consistent with the original commitments of both constructivist and interpretivist qualitative enquiry.

FROM TRADITIONAL TO DIGITAL AND COMPUTATIONAL ETHNOGRAPHY

Traditional Ethnography: Immersion, Relationality, and Thick Description

Conventional ethnography has traditionally been based on long, in-depth interaction within field sites, in which the researcher becomes part of the community being studied and observes its social practices over lengthy periods. This approach to methodology is based on interpretivist and constructivist epistemologies that focus on the contingent construction of meaning, the constructionist quality of social reality, and the primacy of context in interpreting social action (Malkki, 2007). Ethnographers aim to establish subtle, relational knowledge of social life through participant observation, in-depth interviewing, and detailed field notes that are attentive to systems of symbols, moral orders, and cultural logics. The idea of thick description remains central to traditional ethnographic practice, and it does not merely refer to what people do but to how they explain their actions in a culturally relevant context (Geertz, 1973). Ethnographic immersion allows the formation of emic categories and meanings based on participants' categories and meanings, while also necessitating the ethnographer to negotiate their positionality reflexively in the field. Reflexivity involves a persistent mindfulness of how the social positioning of the researcher, the theoretical and the relational positioning influence the production and representation of the ethnographic knowledge (Finlay, 2002). This immersive form of inquiry incorporates ethical principles, such as informed consent, relational accountability, and obligations to community members. Although it offers interpretive richness, traditional ethnography is constrained by structural factors in the modern research environment. The limited number of field sites and the time-consuming, immersive participation limit the magnitude and extent of the ethnographic investigation.

Moreover, with the growing digital mediation of social interaction, the place-based ethnographic methodology would only lead to the analytical invisibility of essential aspects of daily social life. The migration, trans local networks, online activism, and digitally mediated cultural production are all challenging the assumption of limited communities and fixed field sites, as they demand a methodological adjustment that can address ethnographic practice beyond localized contexts.

Digital Ethnography: Networked Fields, Platformed Sociality, and Hybrid Contexts

Digital ethnography is a response to the increasing intertwining of the real and the virtual social worlds, as well as to the problem of dismantling the dichotomy between the virtual and the real social life. Instead of locating digital spaces as either disembodied or disconnected, digital ethnographers are keen to highlight the embodied, embedded, and everyday nature of digitally mediated practices (Taylor, 2025). Online spaces form significant social conditions in which identities are enacted, social relationships are practiced, and cultural norms are negotiated. Digital ethnography, therefore, rescales the ethnographic field into a networked, multisite environment formed by platform architectures, platform affordances, and user practices.

In digital ethnography, ethnographic methods such as participant observation, interviews, and thick description are modified to suit the online setting. Researchers can spend longer durations interacting with an online community, observing interactions in social media groups, or conducting interviews via online communication platforms. This extension maintains fundamental ethnographic promises of interpretive acuity and contextual sensitivity and recognizes that modern social existence is increasingly mediated through digital structures.

Nevertheless, digital ethnography also presents its own unique methodological problems regarding temporality, ephemerality and scale. And online interactions occur in real time, producing unceasing torrents of information that make deep immersion and overall contextualization difficult. Furthermore, digital ethnography has to struggle with the politics of the platform economy. Second, the platform infrastructures define what is made visible, salient, and actionable through algorithms of curation, content moderation rules, and engagement metrics (Caliandro, 2018). These infrastructural processes shape the contours of the ethnographic data and the appearance of representational hierarchies that arise in digital settings.

Digital ethnography, therefore, requires both being embedded in the community on the internet and critically engaging with the social-technical structures through which social interaction occurs. As much as it projects ethnographic practice into networked social worlds, digital ethnography is limited by the analytical range of manual qualitative research methods when faced with the volume and speed of platform-mediated streams of data.

Computational Ethnography: Algorithmic Mediation and Human–AI Analytic Assemblages

Computational ethnography is another methodological development in response to the data-heavy environments of digitally mediated social life, incorporating computational and AI-inspired methods of analysis into the ethnographic process. Inspired by the conceptual foundations of digital ethnography, computational ethnography draws on tools such as natural language processing, text mining, sentiment analysis, and large textual and multimodal corpora generated on digital platforms (Tsoumaris, 2025).

These methods allow the researcher to identify recurrent linguistic patterns, thematic groups, and affective tendencies in large datasets, thereby broadening the scope of analysis in the ethnographic study. Notably, computational ethnography does not theorize algorithmic tools as an alternative to ethnographic interpretation. Instead, it places computational methods within a human-AI analytic arrangement in which algorithms aid exploratory pattern identification. At the same time, the ethnographer remains accountable for contextual interpretation, theoretical sense-making, and ethical judgment. According to this hybrid orientation, ethnographic work is reconfigured to allow certain aspects of data processing and pattern recognition to be established in computational systems, thereby liberating researchers to undertake more interpretive and theoretical work. Simultaneously, algorithmic mediation raises new epistemic issues related to transparency, bias, and the black-box nature of most AI systems (Veprek, 2024).

Computational ethnography is ethnographic because it is not simply about digitally mediated interactions, but about the infrastructures of algorithms that constitute data production and visibility. The effects of platform logics, recommender systems, and content ranking algorithms determine which narratives receive attention and which are marginalized, thereby shaping the empirical contours of ethnographic analysis. Computational ethnographers should then be reflexively involved in the socio-technical realities of data production and critically challenge the mediation of access to cultural practices and social meanings through algorithmic infrastructures. This reflexive activity extends more conventional interests in researcher positionality to the positionality of algorithmic systems as epistemic mediators within the research process.

Relational Continuities and Methodological Transformations across Ethnographic Modes

Traditional, digital, and computational ethnography should be conceptualized as relationally continuous methodological orientations that change in response to shifting socio-technical conditions, rather than as a discrete/competitive paradigm. The traditional ethnography offers interpretive sensibilities, ethical commitments, and reflexive orientations that are fundamental to ethnographic inquiry, with its focus on immersion, contextual appreciation, and thick description (O'Reilly, 2012). Digital ethnography brings these sensibilities into networked social worlds and recalibrates the ethnographic field in light of digitally mediated social practices and platformer modes of interaction (Dawson, 2014).

Computational ethnography also follows these paths by integrating computational and AI-based methods of analysis that facilitate interaction with the magnitude, pace and heterogeneity of the present digital data world. In all these methodological modes, fundamental ethnographic devotion to contextual sensitivity, thick description, reflexivity, and ethical responsibility will be at the forefront. The analytical affordances and the organization of ethnographic labor are what change. Compared to traditional or digital ethnography, which promotes immersion and sense-making as human-centered activities, computational ethnography injects algorithmic and AI-mediated processes into the analytical process. Natural language processing, text mining, sentiment analysis, and large language model techniques are becoming increasingly prominent for surfacing patterns, identifying themes, and navigating textual corpora. This change does not replace the role of ethnographic interpretation; it just shifts the burden of analytic work between human and AI assemblies, with pattern identification performed computationally and the responsibility for contextualizing, theorizing, and making ethical decisions remaining with the researcher.

The integration of AI into ethnographic practice requires other forms of reflexivity that are not limited to the positionality of the human researcher but extend to the positionality of algorithmic systems. The AI models are conditioned on historically and culturally situated datasets and encode specific assumptions, biases, and representational logics that determine which things are visible or salient in the analysis (Jungnickel & Hjorth, 2014). Reflexive ethnographic practice in AI-mediated circumstances, therefore, needs to entail critical questioning of algorithmic presuppositions, recording of analytical decisions, and reporting on how computational devices mediate the research process. Such a reflexive orientation is fundamental to preserving the reliability and interpretive integrity of qualitative inquiry as an algorithmic mediation.

The positioning of computational ethnography as an extension of traditional and digital ethnographic practice also positions AI not as a disruptive break but as a socio-technical power that restructures the conditions of ethnographic knowledge production. Computational ethnography is, therefore, a methodological redesign that reflects the conditions of platformized and algorithmically organized social life, in which ethnographers are asked to operate across hybrid human-AI analytic assemblages without losing the epistemological and ethical investments that outline ethnographic inquiry.

METHODOLOGICAL ARCHITECTURE OF COMPUTATIONAL ETHNOGRAPHY

This chapter describes the methodological design of computational ethnography by specifying the key steps by which ethnographic research is restructured in AI-enhanced research contexts. Computational ethnography is structured here as a multi-stage methodological workflow that integrates computational pattern detection with ethnographic interpretation through clearly defined and auditable steps. Rather than a loose or exploratory sequence, the approach specifies how digital traces are selected, transformed, analyzed, and interpreted within a reflexive and transparent analytic pipeline. The proposed methodology is an anthropocentric approach to AI, where computational tools serve as heuristic aids for exploratory analysis, not as replacements for interpretive judgment (Peponakis et al., 2024).

Illustrative Application of the Methodological Framework

To operationalize the proposed methodological architecture, an illustrative application is introduced. The example focuses on digitally mediated expressions of cultural identity on Instagram. A dataset of 500 publicly accessible posts was collected over a six-month period (January–June 2024) using hashtags related to identity expression and community narratives.

The unit of analysis consisted of individual posts (image + caption) and associated comment threads. Inclusion criteria were restricted to posts produced by self-identified users engaging in cultural or identity-based expression, while excluding institutional, promotional, or bot-generated content. The dataset was limited to Hindi and English language posts to maintain interpretive consistency.

Digital traces were defined as persistent interactional artifacts including captions, comments, and hashtags that encode meaning-making practices within platform-mediated environments. This bounded dataset enables a step-wise demonstration of how computational outputs are generated and subsequently transformed into ethnographic interpretations.

Digital Field Sites and Data Sources

In computational ethnography, digitally mediated spaces that allow social interaction, the expression of culture and the work of identity to occur more and more constitute the ethnographic field of inquiry. Contemporary field sites include publicly used social media platforms, online forums, digital communities, blogs, and multimedia sharing sites, which produce vast textual and multimodal records of everyday social life (Dobson, 2001). Such posts, comments, narratives, and threads of conversation, preserved in digital traces, are treated as ethnographic data that capture practices of meaning-making in situ rather than as decontextualized data points. Digital field sites need to be chosen reflexively with consideration to platform affordances, regimes of visibility, and algorithmic curation processes that influence the salience of specific narratives. Participation and representation are mediated by platform-specific architectures that shape the lines of ethnographic data (Whitmeyer et al., 2019). It means that computational ethnographers have to critically place digital data within the socio-technical frameworks of their creation and consider the problems of access, representation, and power within digitally mediated spaces.

In the illustrative case, Instagram was selected as the primary field site due to its visual-textual affordances and widespread use in identity performance. Field boundaries were defined through hashtag-based sampling, ensuring that the dataset remained thematically coherent while avoiding uncontrolled data expansion.

Data Preparation and Curation

Digital ethnographic data must be prepared and curated systematically before computational analysis can be carried out. Data cleaning, deduplication, anonymization, and ethical management of sensitive information are among the processes this stage entails. To generate written content from audiovisual content, automatic transcription programs can be used to enable further computational processing. Nevertheless, the preparation of data cannot be defined as a neutral, technical process; decisions about what to include, exclude, and categorize determine the analytic horizon of the study and must be reported reflexively to achieve transparency and credibility (Easterday et al., 2018).

In the illustrative workflow, preprocessing included removal of duplicate posts, filtering of bot-like accounts based on posting frequency and engagement anomalies, and manual verification of a subset of data to ensure quality control. Audiovisual content was transcribed using automated tools, followed by selective manual correction to account for code-switching, dialect variation, and transcription errors.

Computational Pattern Detection: NLP, Text Mining, and Sentiment Analysis

Computational ethnography combines AI-driven methods of natural language processing, text mining, and sentiment analysis to help people explore large textual corpora. The NLP methods support the organization and preprocessing of text, enabling work with tokenization, entity recognition, and topic modelling. Text mining tools help identify common patterns, thematic groups, and discursive structures within large datasets, and sentiment analyzes can help trace the orientation of affectivity manifested in digital stories (Jayasudha & Thilagu, 2022). Such methods allow researchers to navigate data volumes that cannot be analyzed manually solely through coding. More importantly, pattern detection by computers does not amount to interpretation.

The products created using AI-based methods act as heuristic machines directing ethnographic focus to meaningful patterns, deviations, or discursive constructions. The ethnographer has the task of conducting interpretive work to establish these patterns in the context of culture, history, and relationships. This parting of the ways highlights the critical role of maintaining a structured human oversight through a documented audit trail and reflexive decision log approach to AI-assisted qualitative analysis, so that computational results can be critically interrogated rather than unquestioningly accepted as reflections of social reality. Sentiment analysis was treated as a limited heuristic tool, given its known inability to reliably interpret irony, sarcasm, and culturally specific expressions, particularly in multilingual contexts.

In the illustrative application, topic modeling (Latent Dirichlet Allocation) was applied to identify thematic clusters across the dataset. Named entity recognition was used to detect recurring cultural references, while basic sentiment analysis was employed cautiously to identify affective tendencies, acknowledging its limitations in interpreting irony, slang, and culturally embedded expressions. Importantly, computational outputs were not treated as interpretive conclusions but as heuristic prompts. All model outputs were subjected to interpretive validation through close reading of selected traces within each cluster.

From Coding to Post-Coding Qualitative Inquiry

The integration of computational tools into the ethnographic process reconfigures traditional qualitative coding methods. Whereas traditional thematic analysis is based on iterative, line-by-line coding by human experts, computational ethnography often uses AI-assisted methods to produce initial thematic groups or discursive patterns (St. Pierre & Jackson, 2014). The change facilitates the substitution of manual coding with post-coding qualitative inquiry, in which the researchers interact interpretively with the patterns generated by the computations to refine conceptual categories, build theoretical understanding, and contextualize the results. Post-coding inquiry, rather than replacing qualitative rigor, focuses on higher-order sense-making, reflexivity, and interpretive richness.

Interpretation followed a structured protocol in which computationally generated clusters were manually examined, contextualized, and refined into ethnographic themes. Outliers and low-frequency patterns were treated as analytically significant rather than discarded, ensuring that minority voices were not suppressed by frequency-based bias. Discrepancies between computational outputs and ethnographic reading were documented as sites of analytical tension and incorporated into the interpretive narrative.

The overall workflow of computational ethnography can be understood as a cyclical process: (1) field delimitation and data collection, (2) computational pattern detection, (3) interpretive ethnographic analysis, and (4) reflexive validation. This iterative movement between computational and interpretive stages ensures that analysis remains grounded in both scale and contextual depth.

Reflexivity, Trustworthiness, and Ethical Governance

The computational ethnography methodological architecture requires a new focus on reflexivity, trustworthiness, and ethical governance in the circumstances of algorithmic mediation. Reflexive practice should be carried through to the critical questioning of how AI tools come to define the direction of analysis and preempt the positionality of algorithmic systems as epistemic mediators. Reliability requires precise accounting for data curation procedures, the methods used in calculations, and the inferential choices made during analytical procedures (Lester & Anders, 2018). Ethical management also implies close attention to informed consent, data privacy, and the risks of re-identification or misrepresentation in digital research

(Pugh & Mosseri, 2023). To ensure methodological rigor in AI-assisted ethnographic research, this framework incorporates structured validation strategies. The integration of AI into ethnographic workflows also reconfigures researcher positionality. The ethnographer is no longer the sole interpreter of data but operates within a human–AI analytic assemblage in which interpretive authority is negotiated. Reflexivity, therefore, must extend beyond the researcher's social position to include their interaction with computational systems, including decisions about model selection, parameterization, and interpretation of outputs. These include sensitivity analysis across model parameters to assess the stability of computational outputs, reflexive logging of analytical decisions to document interpretive processes, and negative case analysis to identify traces that challenge dominant patterns. In addition, an audit trail is maintained to track the transformation of data from raw digital traces to final interpretive claims, enhancing transparency and methodological accountability.

Limitations of Computational Ethnography

Although it can be analytical, computational ethnography has notable limitations. To begin with, it depends on data mediated by digital devices, which records only those facets of social life that can be observed online. Tacit, embodied, and offline practices predominantly lack a voice, thereby presenting a biased picture of daily life. Second, computational approaches are prone to de-contextualization because pattern recognition and sentiment analysis can strip meaning of its context. Such abstractions may reduce subtle cultural understandings without a keen ethnographic interpretation. The prejudice and unclear nature of AI systems influence the third, computational ethnography. Algorithms reflect the assumptions and constraints of their source data and can recapitulate representational hierarchies and marginalize non-dominant linguistic or cultural forms of expression (Abramson et al., 2018). While computational ethnography enables engagement with large-scale digital traces, it does not diminish the value of small-scale, immersive ethnographic work. Instead, computational approaches can surface patterns that require deeper, context-rich investigation through traditional ethnographic methods, reinforcing the complementary relationship between breadth and depth in qualitative research. Transparency and auditability are further complicated by the fact that many AI tools are black boxes. Lastly, there is always an ethical issue associated with consent and privacy, as well as the possibility of harm during a digital research process, especially when publicly available data is re-contextualized in unexpected ways. These shortcomings suggest the importance of reading computational ethnography as a parallel, reflexive, and ethically mediated extension of ethnographic research, rather than an exhaustive replacement of traditional ethnographic practices.

HUMAN–AI COLLABORATION IN ETHNOGRAPHIC ANALYSIS: FROM CODING TO POST-CODING QUALITATIVE INQUIRY

The introduction of artificial intelligence into the practices of qualitative research is an indicator of a significant restructuring of ethnographic work and analytical power. In the conventional ethnographic methodology, meaning-making is accomplished through prolonged interpretive engagements with empirical sources, such as repetitive reading, coding, memo-writing, and theoretical abstraction. Such methods are time-consuming and highly interpersonal, and entail the researcher living in the data and developing context, nuance, and contradiction sensitivity. This analytic ecology is being transformed by the introduction of AI-based systems of analysis, especially large language models and automated text analysis systems, which redistribute features of interpretive work among assemblages of humans and machines.

Reconfiguring Analytic Labor: From Craft to Assemblage

The ethnographic study has been regarded as a craft activity founded on tacit knowledge, interpretive judgment, and reflexive awareness of context. The further application of AI tools shifts the artisanal forms of analysis toward an assemblage approach to analyzing meaning, as human interpretation interacts with algorithmic pattern detection. This reconfiguration raises questions about the evolving skill sets of qualitative researchers. In addition to interpretive competence, researchers are now invited to build technical literacies in data preprocessing, timely design, model evaluation, and the critical analysis of algorithmic outputs. The professionalism of the ethnographer consequently increases the ability to organize analytic assemblages in which calculative devices are strategically put to work, but interpretative power is rooted in human judgment.

Post-Coding Inquiry and the Transformation of Qualitative Workflows

Post-coding qualitative inquiry is enabled by the availability of various AI-assisted thematic clustering and summarization tools, so that a labor-intensive manual coding step in qualitative inquiry can be substituted by post-coding. The computational systems in this workflow produce initial thematic structures that serve as analytic scaffolds rather than categorizations. Post-coding inquiry prefigures a subsequent third-order interpretation, such as theoretical synthesis, situational interpretation, and the expression of relational processes between themes. It alters the position of coding as a goal in itself towards a heuristic point of entry into more profound interpretive work. Nevertheless, the speed of analysis processes also raises the question of the possible loss of slow, immersive reading practices that have traditionally supported ethnographic knowledge. One key methodological challenge in ethnographic analysis using AI is balancing efficiency with interpretive depth.

LLMs as Interpretive Interlocutors

Large language models are becoming interpretive interlocutors in qualitative research, with capabilities such as summarization, paraphrasing, and thematic suggestion. In contrast to previous CAQDAS tools, which were mainly used to manage data, LLMs are more involved in the interpretive process, as they generate candidate interpretations and analytic framings. This trend introduces new epistemological relations, involving dialogic interactions between researchers and AI systems in the process of analytic sense-making. Although this might produce reflexive insight and analytic creativity, it can also lead to the covert influence on interpretation by the language preferences and training biases incorporated into the model structures. The dialogic application of LLMs, therefore, requires an increase in the reflexive alertness of the sources and boundaries of interpretive authority in human-AI collaborations in analysis.

Algorithmic Mediation and the Politics of Pattern Recognition

Computational ethnography is based on algorithmic systems that bring up patterns in large-scale textual corpora. Pattern recognition, however, is not an unbiased technical process; it is conditioned by model designs, training sets, and parameter settings that are historically contextualized assumptions about language, relevance, and salience. Therefore, algorithmic mediation is involved in the politics of representation because it elevates the voices of some people and makes others less noticeable. Such a relationship is significant in ethnographic studies of marginalized groups, where linguistic practices may differ from the dominant norms represented in training datasets. Computational ethnographers should then be highly skeptical, asking how algorithmic pattern detection creates representational hierarchies in qualitative analysis, and should be conscious of the danger of epistemic marginalization.

Temporalities of Analysis: Speed, Slowness, and Ethnographic Time

Artificial intelligence analysis introduces novel temporalities into ethnographic studies. Automation of processing speeds up analytics cycles, enabling quick iteration and the detection of patterns at scale. Although this acceleration can provide responsiveness to the dynamic digital environment, it also threatens to push out the temporal rhythms of the ethnographic immersion that can support profound contextual absorption. Slowness, patience and protracted relational interaction have long been the defining characteristics of ethnographic time. The accelerated timelines of analytic processes brought about by AI-mediated workflows cast doubt on ethnographers' ability to maintain slow, reflexive forms of engagement in the accelerated analytic space. Speeding up calculation and scaling it against ethnographic temporality is a new approach to methodology.

Toward Reflexive Human–AI Co-Interpretation

Successful implementation of AI in ethnographic analysis requires the development of reflexive practices designed for the co-interpretation of humans and AI. These practices include transparent reporting of analytic choices, critical analysis of algorithmic results, and cycles of interpretive validation grounded in contextual learning. Reflexive co-interpretation also involves acknowledging the positionality of AI systems as epistemic mediators informed by training data, design preferences, and the interests of institutions. Computational ethnography has the potential to sustain the interpretive fidelities of qualitative inquiry and to engage faithfully with the analytic resources of artificial intelligence by foregrounding reflexive human-AI collaboration.

Epistemological and Ethical Challenges of AI-Enhanced Ethnography

Artificial intelligence applied to ethnographic research practices raises significant epistemological and ethical issues that go beyond the technical aspects of using tools and analyzing them effectively. The traditions of ethnography, grounded in constructivist and interpretivist schools of thought, have long emphasized the situated, relational, and reflexive generation of knowledge. The analytic practices made more problematic by AI include reallocating interpretive work between human and machine assemblies and introducing algorithmic mediation into practices of pattern perception, meaning-making, and social reality. This part critically examines the epistemological and ethical consequences of AI-enhanced ethnography. It foreshadows issues associated with the black-box problem, algorithmic bias, trustworthiness, informed consent, data privacy, and the subject position of AI as an epistemic agent.

The Black-Box Problem and Epistemic Opacity

Among the most commonly mentioned epistemological issues related to AI-enhanced research is the black-box problem, which refers to the poor visibility of most algorithmic systems and the inability to elucidate how certain outputs are produced fully. Epistemic opacity is a significant issue in qualitative research, where the transparency and audibility of the research methodology are core concerns for trust. How thematic clusters, summaries, or discursive patterns are produced by AI systems may be partially unreadable to researchers when they employ such methods to create them. This inertia makes it difficult to evaluate the validity of computational results analytically and to document the process of analytic decision-making in a manner consistent with qualitative criteria of rigor. The black-box problem is significant in ethnographic studies, where interpretive authority rests on the ethnographer's ability to explain how meanings are constructed, framed, and theorized.

The existence of algorithmic opacities poses an epistemic threat by introducing asymmetries in epistemology, where computational outputs are given undue authority based on their perceived technical complexity. Computational ethnographers should develop algorithmic reflexivity to reduce the effects of these risks, such as critically interacting with model documentation, performing sensitivity analyses, and openly publishing the pipeline through which they are analyzed. Reflexive recognition of epistemic opacities is a significant ethical and methodological safeguard that cannot always be fully achieved, though complete algorithmic transparency can make this possible.

Algorithmic Bias and Representational Inequalities

The algorithmic systems are conditioned on previously and culturally contextualized datasets that encode existing social hierarchies, linguistic norms, and representational prejudices. Consequently, AI-enhanced methods of analyzes have the potential to replicate or exacerbate social disparities in ethnographic studies. As a side effect of the idea of computational ethnography, algorithmic bias can emerge in the prioritization of mainstream linguistic registers, the exclusion of vernacular or non-standard language practices, and a biased focus on specific social groups, as reflected in patterns generated by computation. It is especially concerning when conducting research with marginalized or digitally minority's populations, as algorithmic systems may fail to recognize culturally specific modes of expression. Epistemologically, algorithmic bias questions the interpretivist belief that social worlds can be understood through the eyes of participants.

When analytic representations are distorted by computational tools that systematically misperceive or withhold the determination of some voices, the analysis of the cultural realities under analysis would be misleading. The ethnographers need to practice critical auditing of algorithmic outcomes, then, and they should triangulate computational results with contextual knowledge and reflexive interpretative judgment. Ethical interaction with algorithmic bias also implies promoting the principles of inclusiveness in model training and creating culturally sensitive analysis studios that are sensitive to linguistic diversity and socio-cultural particularity.

Trustworthiness, Transparency, and Methodological Accountability

Credibility, dependability, confirmability, and reflexive transparency are among the criteria for quality research, as presented by trustworthiness. The artificial implementation of ethnography with AI complicates existing standards of reliability, making them more multi-layered and requiring the adoption of new methodological layers that must be reported and explained. Such delegation of initial analysis processes to computational engines requires open reporting on the data preprocessing steps, the reasons behind model selection, parameter settings, and the interpretability of the algorithmic results.

In the absence of such documentation, computational ethnography risks undermining the auditability and reproducibility of the qualitative analysis. The methodological responsibility of AI-based context also needs to be explicitly defined, specifying the differences between detecting computational patterns and ethnographic interpretation. Scholars need to explain how AI-generated patterns are integrated into analytic discourses and how interpretive assertions are grounded in contextualized knowledge rather than solely in computational results. This transparency is crucial for maintaining the epistemic accountability of computational ethnography to the wider qualitative research community and for promoting critical reflection on AI-mediated forms of analysis.

Informed Consent and the Ethics of Digital Data

Digitally mediated data used in computational ethnography is fraught with complex ethical dilemmas regarding informed consent and the nature of publicly available online data. Although digital traces can be socially available (e.g., social media posts and forum discussions), when their use is altered in a research context, they can pose the same ethical threat beyond the official understanding of public and private data. The subjects might not foresee that their online speech will be subjected to mass computation and used in scholarly contexts. The informed consent rule should therefore be reimagined in the digital-mediated research setting, where contextual and relational approaches to ethical engagement replace procedural consent.

Computational ethnographers face tensions between feasibility and ethical responsibility in methods and data collection (especially when dealing with data produced by vulnerable or marginalized groups). Pre-emptive contextual consent, prudent anonymization, and consideration of the harm that recontextualization might cause are key to ethically responsible AI-assisted ethnographic research. Additionally, to comprehend and challenge the applications of their digital traces in computational research, researchers should be sensitive to individuals' varying abilities to participate in such contexts.

Data Privacy, Surveillance, and the Political Economy of Platforms

The AI-based ethnographic study is also being carried out within a broader political economy of digital technology marked by omnipresent data mining, surveillance infrastructure, and commodified data systems. Digital traces are collected and analyzed as part of a platform architecture that monetizes user data regularly and algorithmically selects what users should see. Computational ethnographers are then faced with the ethical consequences of engaging with data ecosystems that also involve surveillance capitalism and unequal power dynamics between platforms and users. Ethical governance of computational ethnography goes beyond the practices of individual research to include critical approaches to the structural state of the production and consumption of digital data.

The threat of re-identification and unintentional harm during large-scale computational analysis is another factor that increases data privacy concerns. Aggregation and cross-referencing of digital traces can make a person or a group identifiable, even in anonymized datasets. There should be some form of ethical protection, comprising strong data governance procedures, safe data storage measures, and continuous evaluation of potential privacy risks throughout the research life cycle.

The Positionality of AI and the Reconfiguration of Epistemic Authority

Lastly, the integration of AI into ethnographic research implies that we need to critically reflect on the positionality of AI systems as epistemic agents in qualitative research. The AI models are not neutral instruments; they reflect design decisions, institutional interests, and normative assumptions that shape the analytic results. The acknowledgement of AI's positionality implies that algorithmic systems are perceived as situated actors within the research process, whose epistemic role should be questioned rather than taken at face value. Such reflexivity is consistent with a broader set of interpretivist adherence to questioning the statuses in which knowledge is generated and justified.

Reconstitution of epistemic authority in human-AI analytic assemblages brings normative issues of accountability and responsibility of qualitative research. Although AI systems can aid in pattern detection and exploratory analysis, interpretive claims are made by human researchers and thus entail their moral and epistemic obligations. Computational ethnography, therefore, demands a recognition of human responsibilities in the controlled management of AI-enhanced qualitative inquiry, such that interpretive power is grounded in the reflexivity of human judgment in a context-driven manner.

IMPLICATIONS AND FUTURE DIRECTIONS

The above discussion implies that computational ethnography has significant methodological, theoretical and practical consequences on the future of qualitative research. Methodologically, implementing AI tools in ethnographies should require restructuring qualitative training and research practices. The researchers need to develop AI literacy alongside familiar qualitative competencies, such as reflexivity, contextual interpretation, and ethical judgment. Analytic frameworks: Human-oriented analytic frameworks that maintain interpretive power while leveraging computational pattern detection are likely to become increasingly prominent in qualitative research methodology in digitally mediated research settings. In principle, computational ethnography beckons a new formulation of constructivist and interpretivist commitments in the circumstances of algorithmic mediation.

As AI systems become involved in the formation of analytic attention and pattern recognition, qualitative epistemologies must consider the positionality of algorithmic actors in knowledge-production processes. Such a reconfiguration underscores the importance of re-engaging with issues of epistemic authority, reflexivity, and trustworthiness in AI-enhanced qualitative inquiry. In practice, it is necessary to implement responsible uptake of computational ethnography by designing methodological norms and ethical standards that address transparency, bias, data management, and participant welfare. Organizational endorsement of interdisciplinary engagement between qualitative researchers and computational practitioners will be critical towards ensuring that AI tools are used in a manner that does not contradict the epistemological commitments of qualitative inquiry.

Moving forward, future studies of the connection between computational ethnography and ethnographic research might include the implication of multimodal AI systems in the ethnographic investigation, the creation of explainable AI tools to improve the transparency of the methodology, and the creation of participative frameworks that include the communities in the process of co-interpretation of the computational findings. With the ever-growing rise of digitally mediated social worlds, computational ethnography is an encouraging yet challenging avenue for redesigning qualitative inquiry in a manner that is both human-centered and ethically accountable.

CONCLUSION

This chapter makes three key contributions to qualitative methodology. First, it re conceptualizes computational ethnography not merely as a technical extension of digital methods but as a hybrid epistemological framework that reconfigures how ethnographic knowledge is produced in algorithmically mediated environments. Second, it provides a structured methodological architecture that specifies how digital traces can be transformed into ethnographic data through a combination of computational pattern detection and interpretive analysis. Third, it advances a reflexive and ethically grounded approach to human–AI collaboration, emphasizing transparency, accountability, and interpretive responsibility in AI-assisted qualitative inquiry.

Although computational methods such as natural language processing, text mining, sentiment analysis, and large language models can be used to interact with massive amounts of digital traces, they do not aim to replace the interpretive work of the ethnographer. Instead, AI is a heuristic partner whose results need to be put into context, grounded in theory, and governed by ethics. The chapter has highlighted the quality of human-in-the-loop analytic frameworks and human-focused AI principles in maintaining interpretive authority and methodological accountability in AI-enhanced qualitative research. Meanwhile, the chapter has highlighted the epistemological and ethical issues that are brought up by algorithmic mediation, such as the black-box problem, algorithmic bias, threats to trustworthiness, and the issue of informed consent and data privacy. To overcome these issues, technical awareness alone is insufficient; renewed dedication to reflexivity, transparency, and responsible research governance is also needed. Computational ethnography, hence, requires greater methodological repertoires and ethical sensitivity that can negotiate the assemblage of human and AI analytics. For policy and research practice, this implies the need for institutional guidelines on AI-assisted qualitative research, training in computational literacy for qualitative scholars, and the development of transparent reporting standards for AI-mediated analysis.

Ultimately, computational ethnography does not signal the erosion of ethnographic practice but its methodological rearticulation under conditions of digital and algorithmic mediation. As social life becomes increasingly structured by platform infrastructures and AI systems, qualitative inquiry must adapt without relinquishing its foundational commitments to context, reflexivity, and ethical responsibility. Computational ethnography offers one such pathway neither a replacement for ethnography nor a purely technical enhancement, but a reconfiguration of interpretive practice that remains grounded in human judgment while engaging computational scale. Its future development will depend on continued methodological refinement, critical engagement with AI systems, and the cultivation of interdisciplinary approaches that sustain both analytical innovation and epistemological integrity.

REFERENCES

Abramson, C. M., Joslyn, J., Rendle, K. A., Garrett, S. B., & Dohan, D. (2018). The promises of computational ethnography: Improving transparency, replicability, and validity for realist approaches to ethnographic analysis. *Ethnography*, *19*(2), 254–284. DOI: 10.1177/1466138117725340

Arnold, T., & Fuller, H. J. A. (2018). In search of the user's language: Natural language processing, computational ethnography, and error-tolerant interface design. *International Conference on Applied Human Factors and Ergonomics*, 36–43.

Brown, T., Mann, B., Ryder, N., Subbiah, M., Kaplan, J. D., Dhariwal, P., Neelakantan, A., Shyam, P., Sastry, G., & Askell, A. (2020). Language models are few-shot learners. *Advances in Neural Information Processing Systems*, *33*, 1877–1901.

Caliandro, A. (2018). Digital methods for ethnography: Analytical concepts for ethnographers exploring social media environments. *Journal of Contemporary Ethnography*, *47*(5), 551–578. DOI: 10.1177/0891241617702960

Chen, G., Tan, B., Laham, N., Tracey, T. J. G., Lapinski, S., & Liu, Y. (2025). A bibliometric review of natural language processing applications in psychology from 1991 to 2023. *Basic and Applied Social Psychology*, *47*(2), 105–119. DOI: 10.1080/01973533.2024.2433720

Clarke, A. E. (2022). Situating grounded theory and situational analysis in interpretive qualitative inquiry. In *Situational analysis in practice* (pp. 47–96). Routledge. DOI: 10.4324/9781003035923-6

Dawson, P. (2014). Temporal practices: Time and ethnographic research in changing organizations. *Journal of Organizational Ethnography*, *3*(2), 130–151. DOI: 10.1108/JOE-05-2012-0025

Dobson, J. E. (2001). Fieldwork in a digital world. *Geographical Review*, *91*(1–2), 430–440. DOI: 10.2307/3250846

Easterday, K., Paulson, T., DasMohapatra, P., Alagona, P., Feirer, S., & Kelly, M. (2018). From the field to the cloud: A review of three approaches to sharing historical data from field stations using principles from data science. *Frontiers in Environmental Science*, *6*, 88. DOI: 10.3389/fenvs.2018.00088

Finlay, L. (2002). "Outing" the researcher: The provenance, process, and practice of reflexivity. *Qualitative Health Research*, *12*(4), 531–545. DOI: 10.1177/104973202129120052 PMID: 11939252

Finlay, L. (2017). Championing "reflexivities". *Qualitative Psychology, 4*(2), 120–125. DOI: 10.1037/qup0000075

Geertz, C. (1973). Chapter 1/Thick Description: Toward an interpretive theory of culture. *The Interpretation of Cultures: Selected Essays*, 3–30.

Golann, J. W., Bougher, L., Hall, R., & Espenshade, T. J. (2024). Sharing big video data: Ethics, methods, and technology. *Sociological Methods & Research*, •••, 00491241241277524.

Hammersley, M., & Atkinson, P. (2019). *Ethnography: Principles in practice.* Routledge. DOI: 10.4324/9781315146027

Jaton, F., & Sormani, P. (2023). Enabling 'AI'? The situated production of commensurabilities. *Social Studies of Science, 53*(5), 625–634. DOI: 10.1177/03063127231194591 PMID: 37702386

Jayasudha, J., & Thilagu, M. (2022). A survey on sentimental analysis of student reviews using natural language processing (NLP) and text mining. *International Conference on Innovations in Intelligent Computing and Communications*, 365–378. DOI: 10.1007/978-3-031-23233-6_27

Jungnickel, K., & Hjorth, L. (2014). Methodological entanglements in the field: Methods, transitions and transmissions. *Visual Studies, 29*(2), 136–145. DOI: 10.1080/1472586X.2014.887263

Jushaninowa, J., Naether, F., & Berti, M. (n.d.). *The Digital Rosetta Stone: Textual Alignment and Linguistic Annotation.*

Kozinets, R. V. (2020). Netnography today: A call to evolve, embrace, energize, and electrify. In *Netnography unlimited* (pp. 3–23). Routledge. DOI: 10.4324/9781003001430-2

Kurowska, X., & de Guevara, B. B. (2020). Interpretive approaches in political science and international relations. In *The SAGE handbook of research methods in political science and international relations* (pp. 1211–1240). Sage Publishing. DOI: 10.4135/9781526486387.n66

Lester, J. N., & Anders, A. D. (2018). Engaging ethics in postcritical ethnography: Troubling transparency, trustworthiness, and advocacy. *Forum Qualitative Sozialforschung/Forum: Qualitative. Social Research, 19*(3), 20.

Li, Z., & Abramson, C. M. (2024). Ethnography and Machine Learning: Synergies and New Directions. *ArXiv Preprint ArXiv:2412.06087.*

Li, Z., & Abramson, C. M. (2025). Ethnography and Machine Learning. *The Oxford Handbook of the Sociology of Machine Learning*, 245.

Low, J., & Hyslop-Margison, C. (2021). Kathy Charmaz: Theory, method, and scholarly identity. *Symbolic Interaction*, *44*(3), 632–664. DOI: 10.1002/symb.551

Malbois, F. (2021). *Doing Digital Ethnography. Some Reflections on What it Means to Study Society Through the Internet.*

Malkki, L. H. (2007). Tradition and improvisation in ethnographic field research. *Improvising Theory: Process and Temporality in Ethnographic Fieldwork*, 162–188.

Nascimento, T., Suarez, M. C., & Campos, R. D. (2022). An integrative review on online ethnography methods: Differentiating theoretical bases, potentialities and limitations. *Qualitative Market Research*, *25*(4), 492–510. DOI: 10.1108/QMR-07-2021-0086

O'Reilly, K. (2012). Ethnographic returning, qualitative longitudinal research and the reflexive analysis of social practice. *The Sociological Review*, *60*(3), 518–536. DOI: 10.1111/j.1467-954X.2012.02097.x

Ophir, Y., Walter, D., & Marchant, E. R. (2020). A collaborative way of knowing: Bridging computational communication research and grounded theory ethnography. *Journal of Communication*, *70*(3), 447–472. DOI: 10.1093/joc/jqaa013

Peponakis, M., Kapidakis, S., Doerr, M., & Tountasaki, E. (2024). From calculations to reasoning: History, trends and the potential of computational ethnography and computational social anthropology. *Social Science Computer Review*, *42*(1), 84–102. DOI: 10.1177/08944393231167692

Pink, S. (2021). Digital futures anthropology. In *Digital anthropology* (pp. 307–324). Routledge. DOI: 10.4324/9781003087885-20

Pugh, A. J., & Mosseri, S. (2023). Trust-building vs."just trust me": Reflexivity and resonance in ethnography. *Frontiers in Sociology*, *8*, 1069305. DOI: 10.3389/fsoc.2023.1069305 PMID: 37168838

St. Pierre, E. A., & Jackson, A. Y. (2014). Qualitative data analysis after coding. [). Sage Publications Sage CA: Los Angeles, CA.]. *Qualitative Inquiry*, *20*(6), 715–719. DOI: 10.1177/1077800414532435

Taylor, N. (2025). 6 Digital Ethnography. *De Gruyter Handbook of Digital Cultures, 17*, 43.

Tsoumaris, S. (2025). *Algorithmic Intimacies: Towards an Anthropology of Computational Life.*

Veprek, L. H. (2024). *At the Edge of AI: Human Computation Systems and Their Intraverting Relations*. transcript Verlag.

Whitmeyer, S. J., Pyle, E. J., Pavlis, T. L., Swanger, W., & Roberts, L. (2019). Modern approaches to field data collection and mapping: Digital methods, crowdsourcing, and the future of statistical analyses. *Journal of Structural Geology*, *125*, 29–40. DOI: 10.1016/j.jsg.2018.06.023

Witteborn, S. (2025). Ethnography and artificial intelligence: The question of context. *Annals of the International Communication Association*, •••, wlaf026.

Wolf, Z. R. (2012). Ethnography: the method. *Nursing Research: A Qualitative Perspective, 4*, 293–330.

Younas, A., Fàbregues, S., Durante, A., Escalante, E. L., Inayat, S., & Ali, P. (2023). Proposing the "MIRACLE" narrative framework for providing thick description in qualitative research. *International Journal of Qualitative Methods*, *22*, 16094069221147162. DOI: 10.1177/16094069221147162

Zaharlick, A. (1992). Ethnography in anthropology and its value for education. *Theory into Practice*, *31*(2), 116–125. DOI: 10.1080/00405849209543532

Chapter 4
The Algorithmic Researcher:
AI, Reflexivity, and the Future of Qualitative Methodology

Sovanna Huot
http://orcid.org/0009-0006-8335-442X
Department of Political Science, University of Delhi, New Delhi, India

Seyha Kun
http://orcid.org/0009-0006-1127-7429
Department of Education, University of Delhi, New Delhi, India

Sovanna Loch
http://orcid.org/0000-0002-1211-1248
Department of English, Nava Nalanda Mahavihara (Deemed to be) University, Bihar, India

Chheangkhy Hok
http://orcid.org/0009-0000-6431-527X
Department of Humanistic Studies, Indian Institute of Technology (BHU), India

Kimyut Phin
http://orcid.org/0009-0009-5667-7313
Department of English, University of Delhi, New Delhi, India

ABSTRACT

This chapter examines the changing role of the researcher's epistemic identity when engaging with artificial intelligence in qualitative inquiry. The author discusses a new kind of researcher, the algorithmic researcher, and describes this researcher

DOI: 10.4018/979-8-3373-7640-0.ch004

t © 2026, IGI Global Scientific Publishing. Copying or distributing in print or electronic forms without written permission of IGI Global Scientific Publishing is prohibited. Use of this chapter to enerative artificial intelligence (AI) technologies is expressly prohibited. The publisher reserves all rights to license its use for generative AI training and machine learning model development.

as a human and machine hybrid, with the ability to create meaning in data-rich environments. The author discusses the implications of AI for the concepts of authorship, reflexivity, and interpretation, and suggests algorithmic reflexivity as a constructive way of engaging with intelligent systems and fully ascribed to the concept of algorithmic reflexivity. The author advocates for the combination of computation and deep interpretation, the decolonization of AI research, and the establishment of joint and transdisciplinary collaboration. The author asserts that the future of qualitative research is highly dependent on the extent to which researchers, as meaning co-creators, engaged with artificial intelligence in a reflexive, ethically driven, and humanized way.

1. INTRODUCTION

The impact techno-social changes that AI and, particularly, Large Language Models (LLMs) have on the research process, and more specifically, the impact LLMs have on the process and outcomes of qualitative research, are immense. Qualitative research has, and to an increasing degree occurs, within large data environments and is being created, organized, and interpreted by data processing algorithms. The impact LLMs have is not merely one that is technological but one that is ontological as well as epistemological. From the social sciences' perspective, there is a fundamental transformation in the processes of knowing, and, analysing, and interpreting social phenomena. Hine (2020) notes that the realm of qualitative inquiry has, and continues, to change within the context of the available technologies, whether in the form of field notes or in the form of digital ethnographies, and the same holds true for the use of machine learning technologies. This expands the context of agency in interpretive processes.

The chapter starts by noting that the role of algorithms has changed from being passive data processors to being active co-analysts that influence what data is seen, heard, and valued. As Baym (2018) points out, social knowledge systems algorithms use explain the social knowledge systems by pinpointing and giving relevance to a certain form of language, emotions, or behaviours, while disregarding the rest. In qualitative analysis, systems of social knowledge affect it by the means of theme construction, means of validating patterns, and by affecting the very value meaning of what knowledge is. In this case, qualitative researchers become co-creators of knowledge, of a sort, with systems of classification, opaque, probabilistic, and with knowledge of external databases that is vast, and untraceable.

Beer (2018) and Kitchin (2021) argue that AI has created a new data episteme, where human reckonings and machine inferences hold a very tenuous coexistence. The new hybrid form of reasoning, what Crawford (2021) characterizes as algorithmic thinking, enhances human analytical powers and simultaneously removes the centre of interpretative control from the human. This means that qualitative research is entering a new era of posthuman methodologies, where the agency, thought, and the creative spirit is shared across systems of humans, machines, and data infrastructures, (Barad, 2007; Braidotti, 2019). The researcher's role shifts from being an independent, reflexive viewer of the social world, to the more complex social world of an algorithmic system, which both opens and closes possibilities to produce knowledge. In this sense, the chapter develops the idea of the algorithmic researcher as a new kind of epistemic figure, a hybrid who critically and meaningfully co-interprets with a form of machine intelligence, keeping in mind the machine's biases, the assumptions, and the interpretive gaps.

AI tools for transcription, auto-coding, and sentiment analysis have the potential to streamline processes, and deepen analysis in ways that were not previously possible (Hitch, 2024), but in doing so, they risk removing the layers of context and the reflexive depth that are hallmarks of qualitative analysis. The potential and the consequences of this transformation are significant in the field of educational and media research, disciplines that have been concerned with representation, and it is certainly within the scope of Markham's (2021) argument that qualitative data are not context free; they are relational, emotional, and they have an ethical dimension. The more context and meaning that the relational data are stripped of, using machine learning that draws from corporate data, the more interpretive neutrality is an empty promise.

Beginning with the emergence technology in qualitative methodologies, researchers are confronted with what Pink et al. (2022) call algorithmic intimacy, the close relationship between human interpretive consciousness and automated analytic suggestion. Simply put, the listening of algorithms is fundamentally different. They analyse and dissect "what is this called?" and, from this, create generalizations or abstractions instead of situational context or truth. This machine and human descriptive partnership conjure critical methodological concerns that accompany the use of a computer. How do researchers keep interpretive sovereignty? Given the divide between qualitative researchers of the past and researchers of today, where do social and methodological anchors of qualitative research reflexivity, the most enduring of these anchors, reside today, the age where machines are part of interpretation?

This chapter attempts to situate the transformation of the subject of research in terms of the algorithmic researcher. For example, Haraway's (1991) cyborg theory unsettles the human machine divide; Suchman (2007) contextualized the technological actor as human machine the systems of the articulation of the actor as negotiating and translating. The algorithmic researcher approach expands these traditions as it acknowledges that interpretive agency is no longer solely a domain of human consciousness, but, in a distributed sense, is algorithmically generated and co-created. As Gourlay (2020) asserts, this is the activist tendency of posthuman scholarship, which extends methodological reflexivity to non-human positionality and non-human mediation. AI and LLMs in qualitative research create new profound epistemological, ethical, and methodological tensions. They revolve around a deceptively simplistic question: Who, or what, is now interpreting qualitative data? In qualitative research, a researcher's subjective reflexivity was the main axis of interpretation. Meaning was negotiated among researcher, participant, and document (Denzin & Lincoln, 2018). But now, in the context of LLMs, a lot of the main analytical processes such as transcribing, summarizing, coding, and creating theories are done by the AI, scattering the locus of interpretation.

According to Beer (2018), this redistribution of interpretative agency removes 'the authorial confidence of qualitative research'. This is done by establishing non-human rationality as a valid participant in meaning-making processes. Kitchin (2021) notes that algorithmic systems work on a probabilistic logic, which is incommensurable with qualitative research bias toward hermeneutics, i.e., interpretative depth, nuance, and context. The second issue involves mediation and bias. Algorithmic systems do more than just process data; they also mediate. As noted by Crawford (2021) and Noble (2018), all AI systems carry the contextual social, cultural, and political systems from which the training data was derived. In qualitative research, where the interpretive voice of the researcher, context and the power that is embedded in the data is the most important, algorithmic bias can greatly shape the outcomes of a research project. As Benjamin (2023) points out, the myth surrounding the neutrality of the algorithmic process overlooks the fact that the data that is being processed contains inequitable and exclusionary histories.

Finlay (2002), Pillow (2015), Pink et al. (2022), and others have described interpretation, especially in AI-augmented analytics, as including agency, with the AI infused into analytics as a partner. The machine does not simply augment the interpretation, but the machine's literal agency in the analytics process deflects attention on the dominant role of the human creator or researcher in the analytics process. The researcher's interpretative role begins to converge with the interpretative role of the machine. Based on this problematisation, the chapter outlines three main goals: First, it attempts to theorize the concept of the algorithmic researcher as a hybrid, human machine producer of knowledge. Second, it attempts to assess the ways in which artificial intelligence alters the framework of reflexivity, interpretation, and ethics within qualitative research. Third, it seeks to establish a structure for algorithmic reflexivity as both a method and an ethics. There is thus an emerging strand of scholarship within the social sciences and humanities which positions AI not as an impediment to human inquiry, but as an impetus for rethinking our forms of knowing.

This workplaces AI within the bounds of interpretivism and thus, attempts to address the overly restrictive and false binaries of the computational and the qualitative. Jasanoff (2019) insists that the self-described responsible researcher of the AI age, must employ a technology of humility approach, characterized by a willingness to accept complexity, ethics, and care, in the face of the uncertain.

2. CONCEPTUAL AND THEORETICAL FOUNDATIONS

2.1 Defining the Algorithmic Researcher

Positioned within posthumanism, digital epistemology, and qualitative methodology, the algorithmic researcher represents the most significant discursive shift in understanding a researcher's role. It moves from being the self-sufficient interpreter of a set of variables towards a posthuman figure, where the analytical component of their subjectivity is inextricably intertwined with an algorithmic apparatus. In the same line of thought, Braidotti (2019) recognizes the relational aspect of posthuman subjectivity, which is networked and dispersed across a constellation of human and non-human entities. The legacy inherent in Enlightenment thinking ascribes a singular autonomy to the researcher and embodies the type of Cartesian dualism was found in the splitting of mind and body or human and machine. Haraway (1991) illustrated this humanist ideal as the cyborg scholar. This cyborg scholar, as a hybrid of body and machine, is also an entity that transcends the borders of all three categories of identity, agency and knowledge. Within this frame, the algorithmic researcher is not merely a technological consumer, but a cyborg epistemologist, and is functioning within complex networks of code, data and affect. Gourlay (2020) and Pink et al. (2022) claim that the growing presence of artificial intelligence in research contexts is forcing a transformation in understandings of academic agency.

Nonetheless, this co-constructivism is not merely situated. As Beer (2018) points out, the algorithmic intermediary in question is shifting the distribution of epistemic power in ways that prioritize all that is formalizable, quantifiable and patternable. In theory, this hybrid virtual identity affects the ethics and labour of qualitative research. The algorithmic researcher has two responsibilities. The first is employing AI tools such as the ones used for efficiency and pattern recognition. The second is examining the tools and their potential to structure, limit, or misrepresent meaning. As Kitchin (2021) notes, data technologies influence what is knowable and the limits of our imagination. The algorithmic researcher must then possess epistemic vigilance about the changes that algorithmic participation brings to the interpretive process.

Relocating the algorithmic researcher within this conceptual frame gives them an enhanced understanding of the genealogy of technological subjectivities. While Haraway's (1991) cyborg drew attention to hybridity and boundary crossing, and Suchman (2007) technological actor to situated action, the algorithmic researcher centres interpretive entanglement, the understanding that meaning is produced through the coalescing of the human and the machine.

2.2 The Posthuman and Critical Turn

To properly theorize the algorithmic researcher, AI needs to be placed within the posthuman and critical data studies frameworks and, more specifically, the studies where AI and data are thought to be co-constructed with diverse, heterogeneous, and fluid assemblages of humans and nonhuman technologies. As Barad (2007) articulates, posthumanism works to destabilize the notion of human exceptionalism by disavowing the belief that cognition, understanding, or agency, is the sole purview of the human mind. Rather, she argues, from her understanding of agency, that the concept of agency is a mutually constitutive process of becoming. This insight is deepened within the field of critical data studies, which focuses on the power of data infrastructures and the algorithms that govern and shape our lives. Kitchin and Lauriault (2018) describe data as the worlds and worlds as data, meaning, data is performative and not merely neutral representations, rather, data, through the arbitrary process, can construct and manipulate belief systems and structures. Similarly, Crawford (2021) argues that the AI systems created to respond to the needs of society and sustain the critical systems of society are the systems of power. All these perspectives together construct a critical posthumanism perspective on the sociotechnical frameworks of knowledge systems.

In this scenario, AI does more than the expansion of human cognition, it restructures it. The emergence of cognitive technologies has been described by Hayles (2017) as the ushering of a new regime of distributed cognition, where thinking takes place across a network of people and machines. This means that the qualitative researcher, formerly viewed as an autonomous meaning maker, is a mere point in an algorithmic construction (Deleuze & Guattari, 1988) that includes systems of data capture, computation, design of interfaces, and systems of regulation that determine how qualitative aspects are understood and processed. The implications of this distributed ontology are profound and impact qualitative research. That cognition and agency are distributed and shared across human machine networks means that interpretation is a collective activity and not an individual one. The interpretive authority of the researcher is, as Knox (2023) contends, a contesting of the authority itself, and a total intertwining of the interpretive authority of the researcher and the computational agency. This perspective is critical of the human-centred thinking that has dominated the social sciences and offers a challenge to the thinking that has gone unquestioned for decades on the intersecting constructs of authorship, expertise, and ownership.

The implications of posthumanism are both descriptive and ethical. Braidotti (2019) asserts that post humanist ethics needs to consider the ethical relations and responsibilities that emerge from interconnected systems. The algorithmic researcher, then, needs to think about AI not as a mere instrument, but as an intellectual partner whose reasoning and contextualization merit scrutiny and response. This is in step with D'Ignazio and Klein's (2020) advocacy for data feminism, which, while recognizing the analytical potential of technology, critiques the ways in which it sustains and reinforces systems of oppression.

2.3 Reflexivity Revisited

Reflexivity is perhaps the foremost construct in qualitative research ethics and methodology. It is the recognition of the researcher's positionality, the assumptions underpinning the research process, and the resultant knowledge production (Finlay, 2002; Pillow, 2015). This reflexive process is almost always framed as self-critical consciousness, a looking inward of the analytical scope. However, AI systems are not passive players in the process of meaning-making, and therefore, reflexivity, needs to go beyond self and include systems, tools, and other elements of the algorithm. Some scholars have started to describe this new articulation. Gourlay (2020) claims posthuman reflexivity means acknowledging the intertwining of human and nonhuman technologies in the production of knowledge. Likewise, Markham (2021) suggests reflexivity needs to consider the algorithmic imagination, the influence of computational systems on the researcher's perception of the possible and the worthwhile. In the case of research augmented by artificial intelligence (AI), reflexivity becomes:

1. Human reflexivity: the researcher takes account of their positionality and the choices they have made in constructing their interpretations.
2. Algorithmic reflexivity: the researcher understands the impact of the model's failings, the training data, and the system's structure on the results of the analysis.
3. Meta-reflexivity: the researcher understands the influence of the entwined roles of the human and the machine in the interpretative process, especially at the feedback loops.

This division suggests a perspective like Barad's (2007) concept of intra-action, where the focus shifts to meaning making in the dynamics of relations rather than in mere acts of reflection. In practice, algorithmic reflexivity places the researcher in a position to answer the following questions: How do the affordances of the AI tool shape the structure of the data? Which language- or emotion-based cues are favoured, and which ones are omitted? What type of people are suggested by the tool's classifications? As Crawford (2021) mentioned, algorithms embed preferences about their knowledge, and, therefore, the types of meanings that lie beyond the algorithm are not visible.

To put it differently, it adds to responsibility rather than diminishing it. Pink et al. (2015) argue that reflexivity in digital ethnography, involves, intervening in the apparatus of mediation. In the same way, the algorithmic researcher is compelled to reflexively intervene, that is to observe, negotiate, and in some instances, push back against the logics of the algorithm, as they relate to their research instruments. During the AI era, reflexivity has ethical implications as well. Both Benjamin (2023) and Noble (2018) demonstrate the ways in which algorithmic systems conceal their biased operations and reinforce existing societal inequities. A reflexive relationship with AI systems operates on the assumption of methodological and ethical engagements aimed at addressing the gaps of bias, inequities, and the just distribution of resources.

2.4 The Data Episteme: From Interpretation to Computation

The rise of AI driven tools such as auto-coding, text mining, sentiment analysis and topic modelling has ushered qualitative research into a new data episteme (Beer, 2018; Foucault, 1994). In this new data episteme, meaning is constructed less with a hermeneutic lens and more with probabilistic reasoning. As Kitchin (2021) states, datafication is the process in which qualitative phenomena are converted into quantifiable. When this happens, narrative, emotion, and context are translated into tokens and vectors that can be subjected to computational analysis.

Systems like Crawford (2021) analyses have deep-seated biases that favour correlation over causation and frequency over nuance. As a result, the data episteme is increasingly likely to favour prediction over comprehension, a shift that would most definitely diminish the interpretive richness that is a hallmark of qualitative research. The same applies to Couldry and Mejias (2019), where the analytics of datafication that rely on metrics devoid of context turn lived experience into data that captures human meaning and is literally sold as a commodity to computation (data colonialism).

To reject computation altogether would be foolish and retrogressive. D'Ignazio and Klein (2020) and Pink et al. (2022) propose the most balanced computational and interpretive approaches as hybrid epistemologies. Within these, AI should not supplant human interpretation but rather enhance it, broadening analytical capture while remaining sensitive to context.

However, the need to foster a critical vigilance is ever present with hybrid epistemologies. The authority that algorithmic knowledge claims to possess is, as Jasanoff (2019) states, based on a presumption of objectivity and inevitability, and numerous challenges to this presumption are warranted. The data episteme signals a transformed future for qualitative interpretation, not its termination. Interpretation continues to take new forms, expanding from thoughts in a person's head to human machine interfaces. The aim is to keep accountable, contextual, and humane this type of distributed interpretation. As Braidotti (2019) encourages, in posthuman knowledge, the aim is not to erase the human but reimagine the human as relationally embedded and interconnected.

3. THE TRANSFORMATIVE ROLE OF AI IN QUALITATIVE INQUIRY

The innovations of AI in transcription, voice recognition, and speech-to-text technologies (e.g. Otter.ai, Whisper, Descript) have changed the landscape of data collection in qualitative inquiry. AI technologies remove the time and logistical barriers involved in the transcription of interviews, focus groups, and field notes. As Pink and colleagues (2022) describe, the automation of data collection allows for researchers to analyse data from larger and more complex data sets, and thus, less time is between fieldwork and data analysis. While these systems provide operational efficiency, they also carry epistemological and ethical risks. As Hitch (2024) warns, AI transcription models are not impartial transcribers, but rather, they are interpreters in which algorithms favour and encode linguistic structures. In qualitative contexts, where meaning can be conveyed through the emotional tone and paralinguistic cues that underpin the verbal content, they systematically misrepresent such features. In ethnolinguistic and cross-cultural studies, for instance, the failure to recognize a particular speech variety may result in epistemic violence, which occurs when a community's linguistic heterogeneity is obscured, and its linguistic diversity is represented only through a monolithic and standardized lens (Benjamin, 2023).

The use of AI in software such as "NVivo AI Assist, Atlas.ti Cloud, and ChatGPT-powered categorization tools" is perhaps most striking in the analysis phase, as these tools can code text, organize it into thematic groupings, and identify relationships within large data sets. Lupton (2020) refers to these advances as algorithmic analytics, where the interpretative tasks are assigned to machine systems that can recognize patterns and relationships among data for which the volume and complexity would be unmanageable for human analysts. Supporters highlight the ability of AI-supported analysis to enhance consistency, decrease researcher bias, and speed up the analysis of monotonous coding (Bazeley, 2021).

The efficacy, however, as Beer (2018) argues, is achieved at the expense of epistemic transparency, AI systems make use of probabilistic reasoning which clouds the interpretive processes that link the data and the resulting output. Meaning is not visible to algorithmic systems. They simply identify and process statistical relationships. As Crawford (2021) contends, there is a huge difference between that which is viewed as a pattern by a machine and that which is perceived as meaningful by a person. This discrepancy leads to what may be termed as interpretive slippage, a divergence between the machine inference and the human understanding. Furthermore, the categorization logic incorporated in these systems reflects the data systems on which they were trained. As D'Ignazio and Klein (2020) and Noble (2018) illustrate, bias, stereotyping, and exclusion are a direct reflection of the cultural and linguistic systems in which the underlying algorithmic systems are created. These were the systems that structured the qualitative data on which the algorithmic models were initially developed. As a result, these algorithmic systems may neglect critical aspects of qualitative data and bias the analysis toward prevailing cultural and linguistic systems.

Consequently, in relation to the implementation of AI in the analysis, Markham (2021) speaks of a particular concern, which balances the skepticism and the accountability of the analysis of the algorithmic output and the socio-technical systems the outputs are produced and processed. As Baym (2018) observes, the role of the LLM in communicating with the researcher creates a unique opportunity for epistemic co-production, where interpretation is not a solitary act, the voice of the interpreted' is joined by that of the algorithm. This, however, is not a balanced collaboration. As with a LLM, there is no context, no ethics, no critical self-reflexivity. So, even though the LLM and the interaction it produces may appear to be fluid, there is no voice and, therefore, no subjectivity. As Jasanoff (2019) argues, this is a fundamental paradox of negligence, knowledge produced by a process that is accountable is not responsible. It is Braidotti (2019) who argues that with posthuman epistemology, not human exceptionalism, but accountability must be foremost. In this way, AI may be viewed less as an unrecognized tool and more as an epistemic partner who contribution and role are subject to critique and ethical considerations. Consequently, the act of interpretation is also an exercise in algorithmic self-reflexivity, the understanding of the balance of negotiation that exists in the human and the machine in the shaping of meaning.

The role of AI in research will require redefining and broadening what reflexivity encompasses, especially in its methodological and epistemic processes. Traditionally, reflexivity required researchers to document their self-positioning, biases, and affect, if any, interpretational impact (Finlay, 2002; Pillow, 2015). In the context of this environment, AI reflexivity encompasses the researcher and the algorithm, which produces interrelations of trust, dependence, skepticism, and critique. When confronting systems of mechanized learning, scholars express intrigue and concern (Crawford, 2021; Pasquale, 2020).

Authors document AI systems exhibiting an uncanny confidence. What rhymes confidence is the arbitrary and biased decisions associated with machine learning. The reflexive researcher is interested in direct and indirect trust in an AI mechanism. Trust in an algorithm is more evaluative than it is absolute. Researchers must become attentive to the many facets of the algorithmic assist, especially on the perceived literal enactments and social ramifications. Automation always involves delegating, and delegation always involves politics, (Dourish, 2022). In this context, reflexivity involves delegating interpretative tasks to AI and the redistribution of the balance of power between researchers and AI. Engaging with AI's transformative potential means reflexivity also works in tandem with methodological accountability. This is the need to keep records of the processes, tools, and training used with AI, and how these are interpreted or, in the case of tools, how the outputs are evaluated. This corresponds to the calls for accountability regarding the use of algorithms posited by Kitchin (2021) and Williamson (2017), who argues that in data-driven research, accountability is inextricably linked to transparency.

Lastly, the AI-driven context necessitates the epistemic humility that algorithmic assessments are always incomplete; they are always interpretative and can only ever suggest. As Knox (2023) notes, maintaining humility means avoiding the overconfidence or totalizing perspective that one has, overly relying on the machine learning models. AI works on probabilistic reasoning, not comprehension. When AI tools are used at all stages of the research process, they create the conditions for qualitative researchers to assume several additional roles, such as curator, interpreter, and collaborator, which also involves a different set of competencies and ethical considerations. As a curator, the researcher selects, filters, and contextualizes information for possible algorithmic processing. Curation affects the ways artificial intelligence (AI) systems think about the information and what patterns to surface. Gray and Bounegru (2021) contend that data curation, in and of itself, is an act of interpretation that, through its own framing, exclusion and prioritization, affects the data to be read by an AI system.

As an interpreter, the researcher analyses the codes, clusters, and thematic summaries that artificial intelligence (AI) produces. Here, interpretation becomes a constructive interplay between human agency and AI's proposal. Markham (2021) argues that in the digital realm, there is a need for critical reading of machine outputs as cultural texts. Read AI as an author and, as an interpreter, analyse what, how, and for what purpose AI has produced. This involves retracing interpretive pathways, identifying biases, and situating the outputs within a social system. In this instance, interpretation serves the purpose of mediation by transforming statistical patterns into a qualitative synthesis.

Lastly, the researcher, as a collaborator, engages in interactive co-construction with AI, and uses it not only as a tool to automate but to suggest ways of seeing things differently. This collaborative engagement is like what Sumartojo and Pink (2018) refer to as methodological improvisation, an adaptive form of cross practice whereby human and non-human agents co-produce knowledge. Collaboration in this sense, fosters a dialogic relationship with AI, wherein the researcher crafts questions, reworks prompt, makes assumptions, critiques, and incorporates outputs. This collaboration, however, is asymmetrical. Braidotti (2019) and Suchman (2007) argue that posthuman collaboration is a non-collaborative act that requires ethical vigilance.

AI tools are being increasingly used in educational research for the transcription and analysis of classroom interactions. Researchers can use systems like NVivo AI Assist or Whisper to transcribe hours of classroom audio, and to identify patterns in turn-taking and to produce clusters of themes related to participation, engagement, or emotional tone (Williamson et al., 2020). While such automation improves scale and scope of analytics, there are also shortcomings in understanding and interpretation. Elements of emotional understanding, sarcasm, or joke- contextual classroom life, are things that algorithms struggle to capture. As Selwyn (2021), AI tools, see the classroom through the lens of and quantified classroom, not through the lens of empathy and context. For this reason, there is a need for researchers to algorithmic results with contextual ethnographic awareness and recon, human context computational summaries.

The use of large language models (LLMs) in media and communication studies is becoming more common for analysing large datasets of social media, detecting themes, and mapping narratives. For instance, LLMs can analyse and identify categories of conversations that pertain to climate change, migration, and misinformation, and offer a synthesized overview of a large digital audience (Baym, 2018).

This capacity provides new avenues for the application of computational ethnography (Pink et al., 2022). There are, however, ethical and epistemological concerns involved. Automated analyses jeopardize the online discourse's emotional and cultural nuances. With digital narratives, the primary concern lies in the narratives' attachment to the cultural context. As Couldry and Mejias (2019) propose, the abstraction reflects an extractive form of data colonialism, wherein the lived experiences of the narratives are transformed into an extractable commodity. In addition, the application of LLMs to the analysis of media raises the question of authorship and responsibility. When scholars' interpretations are guided by machine-generated analyses, the researchers must define their position, and the AI's role, as an epistemic tool. Is the AI an assistant, a collaborator, or a conduit for data? As Beer (2018) indicates, ignoring the difference in instrumentality would validate the objectivity of the algorithms, as illustrated in Table 1.

Table 1. The transformative role of AI in the qualitative research workflow

Qualitative Workflow	AI Tools and Functions	Transformations and Risks
Data Collection	AI transcription, voice recognition (e.g., Otter.ai, Whisper)	Enhances efficiency; risks erasing affective nuance and non-standard speech; imposes linguistic normalization.
Data Analysis	Machine-assisted coding and clustering (e.g., NVivo AI Assist, ChatGPT-based categorization)	Accelerates coding; reveals latent patterns but imposes algorithmic logics; obscures interpretive transparency.
Data Interpretation	LLMs for summarization, thematic synthesis, and analytic writing	Challenges authorship and originality; introduces co-authorship between human and machine.
Researcher Roles	AI-integrated platforms; collaborative prompting systems	Shifts researcher identity from sole interpreter to curator, interpreter, and collaborator.
Educational Case Study	AI-assisted discourse analysis tools (e.g., NVivo, Whisper)	Scales up discourse analysis but flattens emotional nuance and classroom dynamics.
Media Studies Case Study	LLMs analyzing social media narratives	Enables mapping of digital publics; risks ethical tensions between scale and empathy.

4. ETHICAL AND EPISTEMOLOGICAL CHALLENGES

The design of AI systems the world over is not uninfluenced, summary arguments by Crawford (2021) and Noble (2018) describe evidence of the socio-cultural intentionality that permeates the systems. The inequities that describe the historical context are reflected in the data sets that train the systems, while the architectures of the algorithms to describe the systems tend to favour logics of regularity, frequency, and statistical correlation to the neglect of ambiguity, nuance, and contradiction. Qualitative research suffers from the socio-cultural assumptions that embed a model, in reference to the imposition of a sociocultural framework, particularly in the interpretive dimension.

AI tools create machine-readable records of human input when they assist in transcription, coding, or thematic extraction. As Benjamin (2023) argues, such mediation reproduces the biases of the culture in which the mediation is performed.

For example, cheap transcription software misidentifies names, phrases, or entire sections of audio recordings, thereby erasing the non-standard speech of some participants in the study. When the input is generated, Reeves and Sylvia (2024) refer to this in the study of discourse as the algorithms excluding some voices from the analysis because they lack the appropriate cultural or linguistic resources.

There is one more consequence of this phenomenon. It occurs when the locus of interpretive power is displaced to a different location. As mentioned before, the algorithmic researcher operates in a different, hostile, epistemic environment. As previously noted, machine learning algorithms, tiredly, focus on what to see, what is good to see, and what knowledge is good to have. Algorithms, in the strict sense, have what Beer (2018) calls epistemic power in the sense of choosing what to look at, what categories to consider, and what patterns to analyse, even when the logic of such choices is not clear. This lack of explanatory power is called Burrell's (2016) the black box problem and has the effect of reducing the interpretive power of the researcher by obscuring the way data is transformed into knowledge. In these situations, it is critical for qualitative researchers to develop what can be called algorithmic reflexivity, or the conscious recognition of the ways meaning is constructed and filtered by algorithmic systems. In this case, it would mean analysing the construction of the datasets, the biases of the design, and the rationale of the conclusions drawn by the AI. As Knox (2023) elaborates, this would mean that, to read against the machine would require researchers to treat algorithmic conclusions as one of many views, which must be contextualized and critiqued, rather than as a final answer.

Such reflexivity also involves what can be described as epistemic humility. As Jasanoff (2019) notes, the authority of science, and by extension, AI, must be tempered by technologies of humility: practices that centre uncertainty, value pluralism, and ethical care. The algorithmic researcher must acknowledge both the strengths and the shortcomings of algorithmic reasoning, and place the conclusions drawn by AI within wider interpretive and ethical frameworks. AI in qualitative research changes not only the distribution of interpretative power but also complicates the role of authorship and accountability. When algorithms develop the codes, summaries, and even the text-based interpretations, the concept of the researcher as the sole author of the work is challenged. In this case, authorship is shared, and responsibility extends to both human and machine. While engaging with the ethical implications of AI and the collaboration with algorithms, Braidotti (2019) argues that the myth of the autonomous author must be moved beyond and focus on more relational forms of knowledge production, which in this case would be AI writing and analytics as relational co-authorship.

According to Crawford (2021) and Kitchin (2021), the most relevant principle of ethics in AI research is the provision of clarity and focus. Regardless of whether it is voice recognition and transcription, thematic and pattern analysis, literature review, or report writing, the researcher should clarify how and where the AI tools were utilized in the research process, and for what purpose. The researcher must specify the name and version of the AI tool, the task it was assigned, the prompt or instructions it was given, the interpretation it was assigned, and the rationale that guided the interpretation. The elaboration of the tools and methods used in the research process, as explained by Pink et al. (2022), encourages reflection on the ethical implications of research and accountability, rather than meeting a mere bureaucratic checklist.

In the past, the ethics of qualitative research have been viewed as a set of rule-based frameworks which consist of informed consent, confidentiality, and the protection of the participants. However, this perspective has begun to shift, as the ethics of qualitative research have begun to be viewed as relational, and situated in the context of care and empathy, which need to be taken into consideration when developing these frameworks (Guillemin & Gillam, 2004; Sidgwick, 2022). The further development of AI-Augmented research raises the importance of this relational dimension. AI can alter how researchers interact with and analyse participants and how participants represent themselves; this in turn alters the conditions under which researchers may listen, see or even understand the participants.

Relationships, as noted by Pink et al. (2015), are influenced by artificially created environments in which research data are collected and changed. The research environment and the relationships that are created in that environment will be affected by three changes that are defined as ethical reasoning:

1. Ethical Researchers: The environment surrounding the participants and reasoning that is affected because of AI. The research environment and the relationships that are created in that environment will be affected by three changes that are defined as ethical reasoning. Researchers have an ethical responsibility to the environment surrounding the participants, and that environment will be affected by the three dimensions defined as ethical reasoning.
2. Relational Responsibility: AI technologies have an impact on the framing and rearticulation of participants' stories. D'Ignazio and Klein (2020) mentioned that power inequities in representation demand an ethic of care, a tenet of data feminism. Thus, researchers should examine which voices are gained and which are lost in the process of algorithmic mediation? Which meanings are maintained, and which are altered?
3. Contextual Accountability: Ethical reflexivity means understanding that data are not simply resources; they are contextual remnants of a lived reality (Gitelman, 2013). AI systems use contextual data, but researchers, along with AI, are obliged to use contextual data. This leads to what Alcoff (1991) calls

responsible representation. This is where a researcher speaks with, and not for, participants, even when they are represented in data.

Ethical reflexivity moves beyond a checklist approach to ethics. It sees ethics as a relationship that is active and dynamic in the context of humans, algorithms, and institutions. The algorithmic researcher's ethics are relational, situated, and ongoing in the sense of a boundary spanning care commitment in the translation of thoughts across and within human and machine interactions. Research ethics dictates that participant privacy and informed consent are top priorities. AI is being integrated into qualitative research, creating a unique set of challenges that complicates the maintenance of these standards. Lupton (2020) and Mittelstadt (2019) explain the AI information processing, data collection and pattern recognition, and hypothesis generation that is often beyond the primary control of the researcher. In these situations, both the privacy and consent of the participants remain unprotected from both the ethical and technical perspectives.

To address these challenges, it is necessary for researchers to build what Sidgwick (2022) calls contextualized iterations of models of consent. This model sees consent as a process and not a one-off transaction, and as something that can be re-negotiated as changes in practice occur. Participants in a data study need to be informed on the various ways in which their data will be processed, how it will be analysed, and the purposes of the AI systems that will be used in these steps. Moreover, there is a transparency obligation as to the use of AI in the data collection and analysis. In addition, it is necessary to reconceptualize privacy as more than just the guarantee of anonymity. Zuboff (2019) notes that the logic of surveillance is based on the transformation of data into a behavioural surplus, which means that traditional privacy protection is useless. This means that qualitative researchers using AI need to consider more the ethics of data infrastructure. that applies the principles of data protection by design, encryption of data that is assessed as sensitive, and avoiding data retention that is unnecessary.

In the subsequent sections, paradoxes will be described in the context that they offer an opportunity to reconceptualize our understanding of ethics and epistemology in the age of algorithms. Table 2 identifies the principal ethical and epistemological challenges posed by the employment of AI in qualitative research, elaborates on the associated risks, and outlines practices aimed at promoting critical engagement and reflection.

Table 2. Ethical and epistemological challenges in AI-Augumented qualitative research

Challenge	Key Issues	Reflexive Response
Algorithmic Bias & Interpretive Power	Embedded sociocultural and linguistic biases shape meaning and visibility in AI-generated interpretations.	Cultivate algorithmic reflexivity; critically examine model training data, context, and assumptions (Crawford, 2021).
Transparency, Accountability, & Authorship	AI involvement blurs authorship boundaries and conceals interpretive processes.	Disclose AI use and function; maintain accountability through detailed methodological documentation (Jasanoff, 2016).
Ethical Reflexivity	Automation reconfigures empathy and relational ethics between researcher, data, and participants.	Practice relational ethics grounded in care, attentiveness, and reflexive awareness of mediation (Markham, 2020).
Privacy & Consent	Cloud-based AI tools complicate informed consent and data sovereignty.	Adopt iterative consent; ensure data security, participant agency, and transparency in algorithmic processing (Lupton, 2023).

5. TOWARD ALGORITHMIC REFLEXIVITY

Five interconnected principles inform what referred to as algorithmic reflexivity: Transparency, Contextual Awareness, Dialogical Engagement, Ethical Framing, and Iterative Reflection. Each principle pertains to one facet of the researcher as AI co-analyst, and these principles should not be seen as linear actions. Instead, they can be viewed as converging stances that together constitute a particular posthuman ethics of inquiry.

The first principle in most cases the most elementary one is Jasanoff's (2016) and Kitchin's (2021) transparency principle of algorithmic reflexivity. It compels the researcher to explain how an AI system is present and involved in every aspect of an inquiry, from data collection, through analysis, to interpretation. This is not an administrative chore, but rather an empirical imposition of accountability. When it comes to AI-assisted inquiries, researchers ought to account, as much as possible, the AI tools they are using, the AI procedures in relation to the data, and the extent to which the AI will influence the interpretations. The opacity, which is frequently present in data analysis, is often regarded as an absence of clarity with respect to the mechanisms and choices that lead to the interpretation of data. This is what Burrell

(2016) referred to as the black box of computation. The other most important of the three algorithmic reflexivity principles is contextual awareness. This means, one must appreciate the fact that AI is not operating in a vacuum. AI is the creation of socio-cultural, linguistic, and political contexts. This is what Crawford and Noble's work is about, the imbalances and socio-cultural dimensions of algorithm construction (2021; 2018). Braidotti (2019) extends the posthuman ethics perspective to say that all knowledge is, in a way, relationally situated.

The third principle, critical dialogue, repositions the researcher's relationship to AI as one of negotiation rather than delegation. Critical dialogue encourages researchers to consider algorithms as more than mechanical tools, but as conversational partners who can stimulate questions, consider alternative interpretations, and contest human beliefs. Suchman (2007) refers to this as a performative relation, where the human and the machine collaboratively and dynamically create the act of research through a continuous and reciprocal relationship. In the case of critical dialogue, this includes the examination of AI outputs, the modification of prompts, the exploration of the limits of the model's reasoning, and the reflection of the differences between machine and human reasoning. For Markham (2021), this kind of dialogue creates what he terms productive friction, a locus of meaning where the opposition between the human and the machine creates a productive tension.

The fourth principle, ethical grounding, describes how algorithmic reflexivity is always coupled with the principles of qualitative research: the ethics of care, the ethics of respect, and the ethics of accountability. The empathic dimension of the role of the researcher is particularly important, and the automation of the processes of interpretation can diminish empathy and create a distance between the researcher and the emotional aspects of the participants' stories. Ethical grounding counteracts this effect by maintaining the centre of technological mediation where human responsibility is located. Ethical methodology, as D'Ignazio and Klein (2020) articulate in their framework of data feminism, must always consider the human impact of the abstracted data.

As Finlay (2002) and Pillow (2015) note, reflection is that much richer when it is repetitive, when every exercise of reflection produces an awareness that transforms the way the next inquiry is approached. The five principles collectively are a methodological framework for establishing a direction while dealing with the challenges presented by AI in research. They provide a way for researchers to be open while still being critically informed, to be skeptical while not being cynical, and to be innovative while still upholding principles and values. In each context of research, there are specific adaptations that are needed when implementing the principles of algorithmic reflexivity. While each context of research will need adaptations, there are principles that can be considered in addressing AI augmented inquiry to allow for more operationalized reflexivity.

An example of a practice that can be adopted is the keeping of reflexive journals that record specific engagements with AI. Writing reflexively is a long-standing practice that helps ensure qualitative rigor (Lincoln, 1985), but in algorithmic settings, it is of heightened significance. Journals should describe not only the changes in the researcher's perspective, but also the specific analytical AI, influenced the typology of the summaries that a model generated to change an interpretation, or the model's summary that reframed an interpretation, or the case of a misclassification to reveal a latent bias. As Markham (2021) notes, it is not merely the description of an encounter with technology that generates a sense of ethical responsibility, but also the recognition of complicity that the researcher has in the encounter through writing.

Integrating AI-generated patterns with human explanations is the next important tactic. Traditionally, triangulation has increased the credibility of research by combining several sources of information or viewpoints (Denzin & Lincoln, 2018). In the research context, it helps to avoid the problem of regarding computational outcomes as final. In particular, the comparison of algorithmic patterns or themes with human-coded data, participant comments, or theoretical frameworks helps researchers find areas where the algorithm's abstraction is overly distant or devoid of context. This process not only strengthens the validity of the analysis but also restores the empathy of interpretation. Most importantly, it reaffirms human agency as the decider of meaning, but it does not disregard the valuable contributions that computation provides.

As Gourlay (2020) observes, such hybrid validation is particularly illustrative of the posthuman principle of balance, human interpretative breadth and machine analytics working together. The third practical example of algorithmic reflexivity is collaborative auditing. This is the formation of cross-disciplinary, and often interdisciplinary, teams (social scientists, data scientists, and ethicists) to examine the AI tools and the impact they have on findings and reasoning. Jasanoff (2019) refers to this as a form of collective epistemic responsibility, where the disaggregation of expertise is meant to address and empower the methodological and ethical gaps. The integration of AI in qualitative research calls for a rethinking of the conceptualization of methodological rigor. Evaluative criteria such as credibility, dependability, and confirmability (Lincoln, 1985) were devised for settings where the exclusive understanding was by humans. When machine intelligence is involved in the processes, these constructs need to be adapted for the new hybrid forms of knowledge creation. In this reconfigured landscape, algorithmic transparency and interpretative accountability stand out as new criteria of trust.

Algorithmic transparency extends the definition of credibility as it entails showing how algorithms impact understanding. Traditionally in qualitative research, credibility is a result of deep engagement, triangulation, and rich description, and in AI-assisted research, it also involves documenting the movement of data through computational processes. This data pipeline, as Beer (2018) describes it, is the process of transforming raw data into analytic outcomes. Jasanoff (2016) underscores the fact that accountability in hybrid epistemologies goes beyond the lack of process. It means a researcher must be responsible for the knowledge and its consequences. This means that a researcher must regard algorithmic participation as an interpretive act that is accountable in many ways as it is not neutral. There is also an emphasis by Knox (2023) on the need for accountability for the construction and use of the AI systems, specifically regarding provenance of data, bias, and interpretive equity, as depicted in Table 3.

Table 3. Principles of algorithmic reflexivity

Principle	Core Idea	Methodological Implications
Transparency	Researchers must explicitly document the integration and influence of AI systems across all stages of inquiry.	Enhances accountability by detailing tools, configurations, prompts, and interpretive decisions; mitigates opacity of AI processes.
Contextual Awareness	AI outputs must be interpreted within their cultural, linguistic, and sociotechnical contexts.	Prevents decontextualization of meaning; situates algorithmic reasoning within broader social and historical frameworks.
Critical Dialogue	AI should be engaged as an interlocutor rather than a passive tool in the interpretive process.	Promotes active interrogation of AI outputs; fosters epistemic dialogue between human judgment and machine inference.
Ethical Grounding	Reflexivity must remain rooted in care, respect, and the dignity of participants, even in computational mediation.	Ensures data sovereignty and relational ethics; preserves empathy and moral accountability in AI-mediated analysis.
Iterative Reflection	Reflexivity must be continuous, recursive, and adaptive to evolving human–machine interactions.	Encourages ongoing evaluation of interpretive choices and algorithmic influence; sustains methodological adaptability over time.

6. QUALITATIVE RESEARCH IN THE ERA OF AI

The first future imperative for qualitative methodology is the development of hybrid expertise, which encompasses the capacity for computation, the ability to interpret, and the agility to empathize. The growing presence of AI in research ecosystems teaches us that, to exercise methodological rigor for the twenty-first century, one must possess the ability to critique the reasoning, boundaries, and politics of the most cutting-edge forms of computation.

Kitchin (2021) and Knox (2023) explain how researchers using AI need to become 'critical data practitioners,' meaning that the researchers need to use algorithmic tools while also understanding the epistemic and ethical values of the tools. This means learning to 'read' algorithms as more than just a means to an end. Rather, researchers should consider them as cultural artifacts that contain values, ideologies, and worldviews. Noble (2018) shows that even the most neutral and objective systems of computation entrench social hierarchies, racism, and the exclusion of certain genders. Qualitative researchers need to read these systems and the social relations they produce and combine this social critical reading with the advanced technical knowledge of data analytics, natural language processing, and machine learning. This mix of knowledge and skill does not mean that qualitative research has been colonized by computational methods. It signifies a needed reconciliation of computational and qualitative methods through hybrid ways of knowing. Qualitative methods, as described by Baym (2018) and Pink et al. (2022), can be used to address the critique of AI by adding the interpretive, empathetic, and reflexive dimensions, elements that are traditionally associated with qualitative methods but often overlooked when applying quantitative methods.

Decolonization of AI research is the second key development in qualitative research. This refers to the attempts to interrogate the datasets, infrastructures, and epistemologies of machine learning to identify and address the Western biases. The scholars note that AI is a cultural and geopolitical project, not a neutral technology (Benjamin, 2023; Crawford, 2021). Data used to train many large language models and other black box systems is primarily drawn from Anglophone, Global North countries. These sources reflect Eurocentric views of language, reason, and knowledge. Such imbalance can result in a manifestation of what Couldry and Mejias (2019) label data colonialism, that is, the extraction and monetization of human signification and structurally reducing it to the Western value system.

The decolonization of AI for qualitative researchers involves a critical assessment of the epistemic biases in the construction of the data and the cultural phenomenon. It involves the data construction background and the cultural phenomenon interpretation vigilant against what Smith (2021) refers to as methodological imperialism; the practice of blending Western research in a disparate epistemic context. AI facilitated the analysis of cultural phenomena, and other technologies that operate using the same principles, tend to normalize certain semantic or syntactic constructions as universal and, in the process, cover up and diminish other forms of knowledge.

As noted by Benjamin (2023), technological benevolence tends to conceal structural inequality, as systems that are designed to appear neutral continue to reproduce the exclusions of the societies that created them. In the same vein, Mohamed et al. (2020) argue that the dominant Western moral frameworks in AI ethics discourse neglect the postcolonial condition. In this regard, qualitative researchers are encouraged to foremost consider epistemic plurality, including indigenous methodologies, Southern epistemologies, and non-Western theories of relation and interdependence (Mignolo & Walsh, 2018).

Because of the intricate nature of AI-assisted questions, it is needed to build new kinds of collaborative structures, that is, interdisciplinary ecosystems that bring together social scientists, data engineers, ethicists, and artists in the collaborative knowledge construction process. As Jasanoff (2019) notes, in our digital world, epistemic authority is increasingly dispersed across multiple networks of knowledge. Given the complexity of the digital research ecosystem, no one discipline can provide the theoretical and practical tools to address the multi-layered challenges of algorithmic research. Thus, the cultivation of interdisciplinary partnerships that integrate technical design and social critique will be crucial for the future of qualitative methodology.

Most importantly, these partnerships should institutionalize reflexivity. As Knox (2023) notes, interdisciplinary initiatives may reproduce hierarchies where the depth of understanding is collated to the technical. To alleviate this, qualitative scholars should discern their role as critical in forging the understanding of how data is classified, contextualized, and framed. Crawford (2021) notes, with the increasing datafication of human existence, the intricacies of social life seem to be only a statistical correlation. This phenomenon disregards the qualitative layers that formulate and preserve the emotional and cognitive aspects of human existence. Braidotti (2019) articulates that the posthuman condition warrants an affirmative ethics, a commitment to the lively interrelations and to the relational posthuman, technological abstraction. Thus, to humanize data futures, it must reclaim the interpretive affective elements, and the empathy, imagination, and caring that machines cannot provide. The postulation of the interpretive, positive, and beneficial aspects of the emotional and caring elements. Baym (2018) articulates that qualitative inquiry is the search for meanings in the relationships that occur in and between phenomena.

Ultimately, the relational meaning AI cannot imitate, which in this case is the co-creation of knowledge that is intersubjective and ethical, must remain central to the qualitative methodology. AI emulates communication but not relation, and while AI can process words, it cannot process reciprocity. As Sidgwick (2022) explains, ethics must be method: an ongoing practice of care, attentiveness and consideration within human machine composites. The emerging qualitative researcher is thus more than a data technician; she is a steward of relational ethics, balancing the quest for understanding with the preservation of the bond of humanity, as observed in Table 4.

Table 4. Future directions for qualitative methodology in the AI era

Future Direction	Core Focus	Implications for Qualitative Research
Hybrid Expertise	Integration of computational literacy with interpretive and ethical depth.	Prepares researchers to critically engage AI systems; promotes balance between technical competence and humanistic reflexivity in methodological practice.
Decolonizing AI Research	Challenging Western-centric epistemologies embedded in datasets, infrastructures, and ethical frameworks.	Encourages epistemic plurality and inclusion of non-Western and indigenous knowledge systems; fosters justice-oriented and participatory AI methodologies.
Collaborative Infrastructures	Development of interdisciplinary ecosystems connecting social scientists, data engineers, and ethicists.	Promotes shared interpretive accountability; enables co-production of knowledge across disciplinary boundaries; sustains ethical and methodological innovation.
Humanizing Data Futures	Preserving empathy, creativity, and relational meaning amid automation and datafication.	Ensures AI enhances rather than erodes human understanding; maintains qualitative inquiry as a space for moral imagination and affective engagement.

7. CONCLUSION AND IMPLICATIONS

Since the algorithmic researcher is a new expression of the qualitative inquirer, it is best to describe it as a new kind of epistemic identity that is not designed to replace the human interpreter but rather extends the boundaries of meaning-making. The qualitative researcher, in this case, is a meaning maker in an integral interpretative ecology, inclusive of algorithms, data, infrastructures, and machines as collaborators. It indicates that the researcher should exercise humility regarding the capabilities of both human and algorithmic understanding, the researcher should demonstrate creative thinking in the construction of new methodological vocabularies, and lastly, the researcher must demonstrate moral courage to deal with the ethical concerns of algorithmic mediation. Thus, an algorithmic researcher possesses a hybrid form of consciousness as an analyst, ethicist, and a reflexive relational designer with and for people and intelligent machines.

In terms of educational and media scholarship, these changes present practical challenges. The following generations of scholars need to be proficient in artificial intelligence and have data literacy. These skills need to be more than just technical. They need to understand how computing systems analyse and affect knowledge and representation. The theory and practice of AI is crucial, and it is imperative that scholars be able to analyse AI in a constructive and imaginative manner. The program of study must be interdisciplinary. This type of education is especially needed for universities and research institutions. They need to encourage the development of policy with transparency, accountability, and ethics with the use of artificial intelligence. These policies will be developed to protect the values of open, caring, and reflexive that qualitative research is built upon.

The presence of AI is not as important as how to interact with it. The main concern is not technology. The real concern is how to interact with AI and not lose the interpretive distance, empathy, and morals that come with human systems integration. To inspire reflexive interactions between people and systems that focus more on understanding than meaning, the qualitative traditions must focus on the partnerships and systems that are most aligned with the goals of the times to live in. These systems are integrated with algorithms to provide the most meaningful ways to engage with understanding.

REFERENCES

Alcoff, L. (1991). The problem of speaking for others. *Cultural Critique*, (20), 5–32. DOI: 10.2307/1354221

Barad, K. (2007). *Meeting the universe halfway: Quantum physics and the entanglement of matter and meaning*. Duke Up. DOI: 10.2307/j.ctv12101zq

Baym, N. K. (2018). *Playing to the crowd: Musicians, audiences, and the intimate work of connection*. New York University Press.

Bazeley, P. (2021). *Qualitative data analysis: Practical strategies* (3rd ed.). Sage Publications. DOI: 10.4135/9781036233808

Beer, D. (2018). *The data gaze: Capitalism, power and perception*. Sage Publications.

Benjamin, R. (2023). Race after technology: New connections to classical and contemporary perspectives. In W. Longhofer & D. Winchester (Eds.), *Social Theory Re-Wired* (3 ed., pp. 405-415). Routledge. DOI: 10.4324/9781003320609

Braidotti, R. (2019). *Posthuman knowledge* (Vol. 2). Polity Press Cambridge.

Burrell, J. (2016). How the machine 'thinks': Understanding opacity in machine learning algorithms. *Big Data & Society*, *3*(1), 1–12. DOI: 10.1177/2053951715622512

Couldry, N., & Mejias, U. A. (2019). *The costs of connection: How data is colonizing human life and appropriating it for capitalism*. Stanford University Press., DOI: 10.1515/9781503609754

Crawford, K. (2021). *The atlas of AI: Power, politics, and the planetary costs of artificial intelligence*. Yale University Press.

D'Ignazio, C., & Klein, L. (2020). *Data feminism*. MIT Press. DOI: 10.7551/mitpress/11805.001.0001

Deleuze, G., & Guattari, F. (1988). *A thousand plateaus: Capitalism and schizophrenia*. Bloomsbury Publishing.

Denzin, N. K., & Lincoln, Y. S. (2018). *The Sage handbook of qualitative research* (5th ed.). Sage.

Dourish, P. (2022). *The stuff of bits: An essay on the materialities of information*. MIT press.

Finlay, L. (2002). Negotiating the swamp: The opportunity and challenge of reflexivity in research practice. *Qualitative Research, 2*(2), 209–230. DOI: 10.1177/146879410200200205

Foucault, M. (1994). *The order of things Vintage.*

Gitelman, L. (2013). *Raw data is an oxymoron.* MIT press. DOI: 10.7551/mitpress/9302.001.0001

Gourlay, L. (2020). *Posthumanism and the digital university.* Bloomsbury Publishing.

Gray, J., & Bounegru, L. (2021). *The data journalism handbook: Towards a critical data practice.* Amsterdam University Press.

Guillemin, M., & Gillam, L. (2004). Ethics, reflexivity, and "ethically important moments" in research. *Qualitative Inquiry, 10*(2), 261–280. DOI: 10.1177/1077800403262360

Haraway, D. (1991). *Simians, cyborgs, and women: The reinvention of nature.* Routledge., DOI: 10.4324/9780203873106

Hayles, N. K. (2017). *Unthought: The power of the cognitive nonconscious.* University of Chicago Press. DOI: 10.7208/chicago/9780226447919.001.0001

Hine, C. (2020). *Ethnography for the internet: Embedded, embodied and everyday.* Routledge. DOI: 10.4324/9781003085348

Hitch, D. (2024). Artificial intelligence augmented qualitative analysis: The way of the future? *Qualitative Health Research, 34*(7), 595–606. DOI: 10.1177/10497323231217392 PMID: 38064244

Jasanoff, S. (2016). *The ethics of invention: Technology and the human future.* WW Norton & Company.

Jasanoff, S. (2019). *Can science make sense of life?* John Wiley & Sons.

Kitchin, R. (2021). *Data lives: How data are made and shape our world.* Bristol University Press.

Kitchin, R., & Lauriault, T. (2018). Data and data infrastructures. In Ash, J., Leszczynski, A., & Kitchin, R. (Eds.), *Digital Geographies* (pp. 83–94). Sage Publications.

Knox, J. (2023). (Re) politicising data-driven education: From ethical principles to radical participation. *Learning, Media and Technology, 48*(2), 200–212. DOI: 10.1080/17439884.2022.2158466

Lincoln, Y. S. (1985). *Naturalistic inquiry* (Vol. 75). Sage.

Lupton, D. (2020). *Data selves: More-than-human perspectives*. Polity Press Cambridge.

Markham, A. (2021). The limits of the imaginary: Challenges to intervening in future speculations of memory, data, and algorithms. *New Media & Society*, *23*(2), 382–405. DOI: 10.1177/1461444820929322

Mignolo, W. D., & Walsh, C. E. (2018). *On decoloniality: Concepts, analytics, praxis*. Duke University Press.

Mittelstadt, B. (2019). Principles alone cannot guarantee ethical AI. *Nature Machine Intelligence*, *1*(11), 501–507. DOI: 10.1038/s42256-019-0114-4

Mohamed, S., Png, M.-T., & Isaac, W. (2020). Decolonial AI: Decolonial theory as sociotechnical foresight in artificial intelligence. *Philosophy & Technology*, *33*(4), 659–684. DOI: 10.1007/s13347-020-00405-8

Noble, S. U. (2018). *Algorithms of oppression: How search engines reinforce racism*. New York university press., DOI: 10.18574/nyu/9781479833641.001.0001

Pasquale, F. (2020). *New laws of robotics*. Harvard University Press.

Pillow, W. S. (2015). Reflexivity as interpretation and genealogy in research. *Cultural Studies? Cultural Studies, Critical Methodologies*, *15*(6), 419–434. DOI: 10.1177/1532708615615605

Pink, S., Berg, M., Lupton, D., & Ruckenstein, M. (2022). *Everyday automation: Experiencing and anticipating emerging technologies*. Taylor & Francis. DOI: 10.4324/9781003170884

Pink, S., Horst, H., Lewis, T., Hjorth, L., & Postill, J. (2015). *Digital ethnography: Principles and practice*. Sage Publication.

Reeves, C., & Sylvia, J. I.IV. (2024). Generative AI in technical communication: A review of research from 2023 to 2024. *Journal of Technical Writing and Communication*, *54*(4), 439–462. DOI: 10.1177/00472816241260043

Selwyn, N. (2021). *Education and technology: Key issues and debates*. Bloomsbury Publishing.

Sidgwick, H. (2022). *The methods of ethics*. DigiCat.

Smith, L. T. (2021). *Decolonizing methodologies: Research and indigenous peoples*. Bloomsbury Publishing. DOI: 10.5040/9781350225282

Suchman, L. A. (2007). *Human-machine reconfigurations: Plans and situated actions*. Cambridge University Press.

Sumartojo, S., & Pink, S. (2018). *Atmospheres and the experiential world: Theory and methods*. Routledge. DOI: 10.4324/9781315281254

Williamson, B. (2017). *Big data in education: The digital future of learning, policy and practice*. SAGE Publications Ltd., DOI: 10.4135/9781529714920

Williamson, B., Bayne, S., & Shay, S. (2020). The datafication of teaching in Higher Education: Critical issues and perspectives. *25*(4), 351-365. DOI: 10.1080/13562517.2020.1748811

Zuboff, S. (2019). Surveillance capitalism and the challenge of collective action. *New Labor Forum*, *28*(1), 10–29. DOI: 10.1177/1095796018819461

KEY TERMS AND DEFINITIONS

Algorithmic Research: A new form of research in which a human scholar works with AI in the analysis, interpretation, and construction of new knowledge. This new fusion of scholar and AI defines a new form of authorship, reflexivity and agency, wherein the researcher is positioned as both an interpreter and a co-creator with the algorithmic components.

Artificial Intelligence (AI): A computerized system with the ability to perform intellectual tasks such as learning, recognition of patterns, and understanding and generating human language. In qualitative research, AI functions as a methodology and changes the process of collecting, coding, and interpreting data.

Epistemology: The branch of philosophy that studies knowledge asks what to know and what can be learned. In the context of AI-assisted research, epistemology examines the interplay between machine reasoning and human understanding and the ways this conjunction creates, legitimizes, and disseminates knowledge.

Ethics: In qualitative research, ethics is the branch that concerns the treatment of data, research subjects, and research outcomes. With the rise of AI, ethics now also considers the transparency, responsibility, and caretaking of the algorithms, to ensure that the use of technology does not undermine the dignity, relationality, and justice.

Posthumanism: Posthumanism is the branch of thought that opposes human exceptionalism and argues for the entwinement of humanity, technology, and environment. It changes the understanding of thought and will. Rather than being the exclusive property of individual actors, these qualities are disseminated through a system. This idea is key to thinking of an algorithmic researcher.

Qualitative Research: The study of phenomena, the meaning of phenomena, and the social contexts of phenomena through the lens of the phenomena, focusing on the subjective or the interpretive, reflexive, and the construction of meaning through the interaction of the researcher and the participant.

Reflexivity: The critical self-observation and self-examination of one's own positionality and influence in each step of the research process. In augmented AI contexts, this also includes an awareness of the role the algorithm plays in shaping, mediating, and participating in construction of meaning.

Chapter 5
Reimagining Qualitative Inquiry in the Age of Artificial Intelligence:
Insights From Ghanaian Higher Education Research

Francis Arthur
http://orcid.org/0000-0001-7792-6563
University of Cape Coast, Ghana

Francis Obeng Gyedu
http://orcid.org/0009-0009-4248-2064
University of Cape Coast, Ghana

Emmanuel Quayson
http://orcid.org/0000-0003-0400-401X
University of Cape Coast, Ghana

Eric Boateng
http://orcid.org/0009-0009-1246-632X
University of Cape Coast, Ghana

Silas Afutu Quaye
http://orcid.org/0009-0000-8203-9803
University of Cape Coast, Ghana

Sharon Abam Nortey
https://orcid.org/0009-0007-9017-4440
University of Cape Coast, Ghana

Dorcas Frempong
University of Cape Coast, Ghana

Mark Inkoom
http://orcid.org/0009-0008-2797-6712
University of Cape Coast, Ghana

ABSTRACT

This chapter reimagines qualitative inquiry in the age of artificial intelligence by examining how AI is reshaping qualitative research practices within Ghanaian higher education. Drawing on constructivist and interpretivist traditions, the chapter synthesises global and African-centred scholarship to interrogate the philosophical, methodological, and ethical implications of AI-assisted qualitative research. Particular attention is given to methodological integrity, trustworthiness, and

DOI: 10.4018/979-8-3373-7640-0.ch005

t © 2026, IGI Global Scientific Publishing. Copying or distributing in print or electronic forms without written permission of IGI Global Scientific Publishing is prohibited. Use of this chapter to :nerative artificial intelligence (AI) technologies is expressly prohibited. The publisher reserves all rights to license its use for generative AI training and machine learning model development.

reflexivity in hybrid human–AI inquiry, alongside ethical and contextual concerns such as algorithmic bias, epistemic injustice, and data sovereignty in African contexts. Using Ghanaian postgraduate research practices as an illustrative lens, the chapter highlights institutional realities, capacity gaps, and emerging practices. It culminates in a context-sensitive framework for responsible AI integration that foregrounds human-centred reflexivity, ethical governance, and cultural sensitivity. The chapter positions AI as a methodological partner rather than a substitute.

1. INTRODUCTION

The integration of artificial intelligence into qualitative research represents a major shift that is reshaping academic inquiry across disciplines and regions (Williamson et al., 2023). Advances in AI, particularly in natural language processing and generative technologies, are transforming how qualitative researchers conceptualise, generate, analyse, and interpret data. Activities that were traditionally based on human judgement, such as literature synthesis, coding, thematic interpretation, and even theory building, are increasingly being supported by AI-enabled tools (Christou, 2023a; Luca Longo et al., 2019; Muhammad Imran et al., 2024). As generative AI tools become more accessible, their uptake within qualitative research has accelerated globally, prompting renewed debate about the nature of meaning making and the boundaries of human interpretation in scholarly work (Popenici et al., 2023; Slimi, 2023).

Across the global academic community, AI is now employed in diverse qualitative research tasks, including critical literature reviews, conceptual development, thematic and content analysis, and exploratory theorisation (Christou, 2023; Jafari & Keykha, 2023). This rapid diffusion has expanded access to advanced analytical capabilities that were previously restricted to researchers with specialised computational expertise (Akinwalere & Ivanov, 2022). At the same time, the pace of adoption has generated unease. Christou (2023) observe that some scholars hesitate to acknowledge AI use because of the perceived "fear of criticism and rejection of their papers," signalling deeper epistemological anxieties surrounding authorship, rigour, and scholarly legitimacy. These concerns raise fundamental questions about whether AI-assisted inquiry enhances or disrupts the interpretive foundations of qualitative research (Butson & Spronken Smith, 2024).

AI integration in qualitative inquiry spans multiple disciplinary domains. In educational research, AI has been applied to "personalisation of the learning process, monitoring, and testing learner activities, designing the learning process, and assisting in peer interaction" (Kasperiūnienė, 2021). Similarly, in media and communication studies, AI supports data detection, content moderation, news production, and the analysis of context and human behaviour. While these applications demonstrate the versatility of AI, they also intensify debates about epistemic authority and methodological control. Qualitative research has traditionally been human-centred, privileging the researcher's reflexivity, subjectivity, and contextual sensitivity as integral to meaning making (Hogg et al., 2023). This tradition positions interpretation not as a mechanical process, but as a situated and relational practice embedded within social and cultural contexts.

Recent developments indicate a shift towards hybrid human-AI research models that combine interpretive judgement with algorithmic processing (Mahajan, 2025). Anis and French (2023) argue that qualitative researchers should "embrace artificial intelligence (AI) to conduct efficient, explicatory, and equitable research but also exercise caution to avoid its pitfalls." They highlight efficiency in managing large datasets, enhanced explanatory capacity through pattern recognition, and improved equity by lowering technical barriers to sophisticated analysis. However, these benefits coexist with risks that require careful scrutiny, particularly when algorithmic outputs are treated as neutral or authoritative rather than as interpretive artefacts shaped by underlying assumptions and training data.

This transformation is evident across core stages of the qualitative research process. AI-powered tools now support data collection through participant recruitment, automated transcription, and preliminary data organisation (Johnson, 2013; Venter et al., 2024). During analysis, AI assists with pattern detection, thematic clustering, and cross-case comparison (Wilder & Calderone, 2025). In the interpretive phase, generative AI tools increasingly function as reflexive aids by proposing alternative readings, identifying potential blind spots, and prompting analytic questioning (Dahal, 2024). Dahal (2024) cautions, however, that while generative AI may operate "as a co-author, a conversational platform, and a research assistant," it simultaneously introduces risks of "research unreliability, bias, and unethical outcomes." This dual role positions AI as both an enabler and a potential threat, reinforcing the need for deliberate methodological oversight.

The influence of AI is also evident in literature review practices, where traditional manual synthesis is increasingly complemented by algorithmic recommendation and summarisation systems. Foley et al. (2025) describe AI tools that can "appraise and synthesise contrasting evidence within and across papers," curate personalised research ecosystems akin to "Spotify," and suggest new research directions by identifying knowledge gaps. While such capabilities promise efficiency and breadth, they also challenge established norms of scholarly engagement, critical reading, and intellectual labour.

Within higher education, these developments carry significant implications for research practice, academic integrity, and postgraduate training (Khatri & Karki, 2023; Legowo et al., 2024). AI is increasingly embedded in the research activities of students and faculty, including literature searching, data analysis, and academic writing (Adjei & Agyeman, 2025; Nikolopoulou, 2024). Empirical evidence suggests that AI can enhance engagement and collaborative learning. Opesemowo and Adekomaya (2024) report that AI use increased students' learning and engagement, noting that "participants used AI technology to increase students' learning and engagement so students would not doze in class," while also fostering "collective learning." Despite these benefits, concerns persist regarding originality, critical thinking, and the erosion of authentic scholarship (Karkoulian et al., 2024; Wiredu et al., 2024). Davison et al. (2024) emphasise the ethical complexity of generative AI in qualitative analysis, calling for transparency and reflexive accountability in research design and reporting.

Ghana provides a particularly instructive context for examining these dynamics, offering insights into AI adoption within the Global South (Maina & Kuria, 2024). Studies indicate generally positive attitudes towards AI among Ghanaian postgraduate students, who perceive AI tools as useful for "improving grammar, preventing plagiarism, generating ideas, and enhancing overall writing proficiency," albeit with varied perceptions across individuals (Anani et al., 2025). Segbenya et al. (2023) found that commonly used platforms include "ChatGPT and Quillbot," primarily for "learning in general; and searching for literature," with AI use significantly predicting the acquisition of twenty-first-century employability skills. At the same time, the Ghanaian context reveals structural challenges, including infrastructural limitations, uneven digital literacy, and evolving policy frameworks (Liekum, 2025; Modiba et al., 2025). Ofosu-Ampong (2023) further highlights gender disparities in AI tool usage, underscoring equity considerations in access and participation.

Beyond these structural factors, Ghana's rich oral traditions and communal approaches to knowledge production offer valuable perspectives for rethinking AI integration in qualitative inquiry (Ammah et al., 2024). These traditions foreground relationality, collective meaning making, and contextual interpretation, raising critical questions about how AI systems developed largely in the Global North intersect with indigenous methodologies. Scholars caution that uncritical adoption of such technologies risks reproducing epistemic injustice and reinforcing global power asymmetries in knowledge production (Amuzu, 2019).

Against this backdrop, this chapter critically examines how qualitative inquiry is being reimagined in the age of artificial intelligence, drawing on insights from Ghanaian higher education research to illuminate broader global trends. Specifically, the chapter integrates theoretical perspectives on human–AI collaboration (Chatzichristos, 2025; Mututa & Tomaselli, 2025), examines empirical evidence on AI adoption in Ghanaian postgraduate research contexts (Baidoo Anu et al., 2024), explores methodological innovations reshaping qualitative practice (Burleigh & Wilson, 2024), and addresses ethical and governance challenges associated with transparency and researcher responsibility (Sangwa et al., 2025). It concludes by articulating future research trajectories and policy implications for responsible AI integration in qualitative inquiry (Maluleke, 2025; Rivera, 2024). By situating Ghana as a methodological lens within global debates, the chapter contributes to ongoing efforts to ensure that AI enhances, rather than diminishes, the interpretive richness, ethical integrity, and contextual sensitivity that define qualitative research.

This chapter argues that while artificial intelligence expands the analytical capacity, efficiency, and accessibility of qualitative research, it simultaneously reconfigures its epistemological foundations, methodological practices, and ethical responsibilities. Drawing on insights from Ghanaian higher education as a representative Global South context, the chapter demonstrates that responsible AI integration must be grounded in human-centred reflexivity, contextual sensitivity, and robust governance frameworks. By bridging global debates with locally grounded realities, the chapter contributes a context-sensitive framework that advances both methodological rigor and ethical accountability in AI-enabled qualitative inquiry.

1.1. Conceptual Clarification of Key Terms

To ensure conceptual clarity and consistency throughout this chapter, several key terms are defined as they are used in this study.

First, artificial intelligence (AI) is used as an umbrella term referring to computational systems capable of performing tasks that typically require human intelligence, including learning, pattern recognition, language processing, and decision-making (Williamson et al., 2023). Within this broad category, particular emphasis is placed on generative AI, which refers to systems such as large language models (LLMs) that can produce human-like text, summaries, and analytical outputs based on learned patterns from large datasets (Dahal, 2024).

AI-assisted qualitative research refers to the integration of AI tools at different stages of the qualitative research process, including data collection, transcription, coding, analysis, and interpretation. In this chapter, AI is not conceptualised as an autonomous researcher but as a supportive analytical tool operating under human direction and oversight (Dellafiore et al., 2025).

Moreover, human-AI collaboration is understood as a hybrid research model in which human interpretive judgement and algorithmic processing are combined. This includes processes such as AI-supported coding, AI-generated summaries, and the use of AI as a reflexive or analytical aid, while maintaining human control over meaning-making and final interpretation (Mututa & Tomaselli, 2025).

Also, reflexivity refers to the ongoing critical examination by researchers of how their assumptions, positionality, and methodological choices shape the research process and findings. Within AI-assisted contexts, reflexivity is extended to include critical awareness of how AI tools influence analytical outcomes (Dahal, 2024; Hogg et al., 2023).

Additionally, methodological integrity is defined as the extent to which qualitative research maintains rigor, transparency, and trustworthiness through established criteria such as credibility, dependability, confirmability, and transferability. In this chapter, methodological integrity also encompasses the responsible and transparent use of AI systems (Korkmaz, 2025).

Finally, epistemic justice refers to the fair representation and inclusion of diverse knowledge systems, perspectives, and voices in knowledge production. This concept is particularly relevant in the African context, where concerns arise about the marginalisation of indigenous knowledge systems within AI-driven research environments (Adamu & Nkwo, 2023; Ammah et al., 2024).

These definitions provide a consistent conceptual foundation for analysing the philosophical, methodological, and ethical implications of AI integration in qualitative research across the sections that follow.

2. PHILOSOPHICAL AND THEORETICAL FOUNDATIONS OF QUALITATIVE RESEARCH IN THE AI ERA

The integration of artificial intelligence into qualitative research necessitates a fundamental re-examination of the philosophical assumptions and theoretical frameworks that have historically grounded interpretive inquiry. This section explores how traditional paradigms intersect with AI technologies, considering implications for knowledge construction, researcher positionality, and methodological integrity.

2.1. Constructivist and Interpretivist Paradigms in Dialogue with AI

Qualitative research has traditionally been rooted in constructivist and interpretivist paradigms that emphasise the socially constructed nature of reality and the centrality of human interpretation in meaning making (Chatzichristos, 2025). These paradigms assume that knowledge emerges through dynamic interactions among researchers, participants, and their contexts. The introduction of AI signals what Mututa and Tomaselli (2025) describe as a "technological turn" in interpretive research, prompting renewed reflection on how understanding, interpretation, and knowledge validation are achieved.

Within the constructivist perspective, reality is viewed as multiple and co-created through social interaction, with meaning arising from interpretive engagement. This tradition has consistently prioritised human subjectivity, reflexivity, and contextual sensitivity as essential to rigorous qualitative inquiry (Hogg et al., 2023). However, AI introduces what can be described as algorithmic interpretation - computational processes that identify patterns and generate outputs without the embodied human experience that constructivists consider fundamental to authentic understanding.

This development raises important epistemological questions. Can AI systems, which lack consciousness and lived experience, meaningfully contribute to interpretive processes that are inherently human-centred? The interpretivist tradition, particularly its emphasis on verstehen - deep, empathetic understanding of social phenomena, further complicates this issue. Interpretivism maintains that social reality can only be understood through the subjective meanings of social actors, requiring immersion in participants' lived experiences (Sarı et al., 2023). In this context, AI tools, which process language through statistical associations rather than experiential insight, may have limited capacity to fully engage in such interpretive work.

Nevertheless, emerging scholarship suggests a more integrative position. Dahal (2024) conceptualises AI as "a conversational platform" that can support and extend human interpretive capacity while remaining dependent on human judgement for contextual meaning making. Empirical insights from Ghanaian higher education contexts reinforce this perspective. Anani et al. (2025) found that postgraduate students recognised AI's usefulness for technical tasks such as "improving grammar, preventing plagiarism, generating ideas," yet expressed reservations about its role in deeper interpretive analysis. This reflects an awareness that dimensions requiring cultural sensitivity, contextual nuance, and empathetic engagement are not easily automated. At the same time, Anis and French (2023) argue that AI can enhance interpretive inquiry by offering "explicatory" power, enabling researchers to identify patterns and anomalies that may otherwise remain unnoticed. In this sense, AI can contribute to more reflexive and analytically rigorous qualitative practice when appropriately guided.

2.2. The Question of Subjectivity and Human Agency

A central principle of qualitative research is that researchers are active participants in knowledge construction, bringing their own subjectivities, experiences, and perspectives into the research process (Butson & Spronken Smith, 2024). The integration of AI complicates this understanding in several ways.

AI introduces what may be termed layered subjectivity, where the researcher's interpretive lens is mediated through algorithmic systems embedded with specific assumptions, biases, and limitations. These biases often reflect the datasets, design processes, and cultural contexts in which AI systems are developed (Mahajan, 2025). This concern is particularly pronounced in Global South contexts such as Ghana, where AI models trained predominantly on Western data may inadequately capture local meanings and social realities (Maina & Kuria, 2024). Ofosu-Ampong (2023) further identifies gender disparities in AI tool usage among Ghanaian students, suggesting that existing social inequalities may be reproduced within AI-mediated research practices.

Additionally, AI raises important questions about researcher agency. When AI tools suggest themes or generate interpretations, the boundaries of analytical control may become blurred. Christou (2023) note that researchers often hesitate to disclose AI use due to a "fear of criticism and rejection," reflecting broader concerns about authorship, ownership, and scholarly legitimacy. These concerns are amplified in contexts where researchers may lack a full understanding of how AI systems operate. In Ghana, for example, Segbenya et al. (2023) report limited knowledge among academics regarding AI mechanisms, creating situations where tools are used without full awareness of their underlying processes. This raises critical questions about informed engagement with AI systems in research.

Moreover, AI integration challenges the role of embodied knowledge in qualitative inquiry. Traditionally, qualitative research values the researcher's direct engagement with participants, including the sensory, emotional, and intuitive insights that emerge from fieldwork (Rivera, 2024). AI systems, operating primarily through textual data, cannot access this embodied dimension. However, some scholars argue that AI can strengthen reflexivity by making interpretive processes more visible. Wilder and Calderone (2025) suggest that AI's ability to identify patterns can act as a reflective tool, prompting researchers to critically examine their own assumptions and analytical decisions.

2.3. Epistemological Shifts: From Interpretation to Co-Creation

The integration of AI catalyses an epistemological shift from interpretation to co-creation, reconceptualising knowledge production as a collaborative process involving both human and algorithmic agents (Foley et al., 2025; Maluleke, 2025). Traditionally, qualitative research positioned the researcher as the primary instrument of inquiry, with knowledge emerging through interpretive engagement with data. However, AI introduces what Burleigh and Wilson (2024) describe as distributed cognition, where analytical processes are shared across human reasoning and computational systems, each contributing distinct capabilities.

This perspective aligns with theories of the extended mind and situated cognition, which suggest that thinking and knowing emerge through interactions between individuals, tools, and environments. In this sense, AI tools function as cognitive artefacts that extend human analytical capacity, much like writing systems or statistical software have historically supported intellectual work (Johnson, 2013).

The co-creation perspective, however, challenges conventional assumptions about originality and authorship. When AI suggests themes or generates interpretations, questions arise regarding ownership of these insights. Existing legal and ethical frameworks for intellectual property have not kept pace with such developments, creating ambiguity around how AI contributions should be acknowledged (Davison et al., 2024). Evidence from Ghana highlights how these tensions manifest in practice. Segbenya et al. (2023) found that postgraduate students primarily use AI for "learning in general; and searching for literature," indicating that adoption often begins with clearly defined tasks before extending into more complex analytical activities. This gradual expansion reflects an ongoing negotiation of epistemological boundaries.

These developments also have implications for theory construction. Traditionally, theory building has been understood as an iterative and reflexive process in which researchers move between data and conceptual frameworks (Chatzichristos, 2025). Some scholars argue that AI can enhance this process by identifying patterns and connections that may not be immediately visible to human analysts (Akinwalere & Ivanov, 2022). Others, however, caution that AI remains limited to pattern recognition and lacks the capacity for genuine conceptual innovation that characterises robust theoretical contribution (Butson & Spronken Smith, 2024).

2.4. Ontological Considerations and Data Authenticity

The integration of AI also raises important ontological questions concerning the nature of qualitative data. Traditionally, qualitative data has been understood as rich, contextual, and meaning-laden, encompassing interview transcripts that capture emotions and social dynamics, field notes documenting observations, and documents reflecting historical and cultural significance (Venter et al., 2024). However, AI systems require data to be formatted in machine-readable or structured forms. This transformation can result in what Kasperiūnienė (2021) describes as ontological flattening, where complex and nuanced information is reduced into formats that are easier for algorithms to process.

Research from African contexts underscores the implications of this shift. Ammah et al. (2024) emphasise the importance of indigenous knowledge systems and oral traditions, which are central to many African research contexts but may not be easily captured through digital or text-based formats. The reliance on text-driven AI systems may therefore marginalise forms of knowledge that are essential to culturally responsive research methodologies.

AI also complicates questions of data authenticity. Generative AI systems can produce text that closely resembles human communication, blurring the distinction between participant-generated data and AI-generated content (Dahal, 2024). This creates new challenges for qualitative researchers, who must ensure that data sources remain clearly identifiable and analytically credible. In addition, the ontological status of AI-generated insights requires careful consideration. When AI identifies patterns or themes, it becomes difficult to determine whether these findings are inherent in the data or shaped by algorithmic processes. The "black box" nature of many AI systems further complicates this issue, limiting transparency in how outputs are generated (Karkoulian et al., 2024).

These epistemological and ontological considerations establish a critical foundation for understanding AI's transformative role in qualitative research. The tensions between human interpretation and algorithmic processing, as well as between traditional and digitally mediated forms of data, highlight the need for sustained critical engagement. As qualitative researchers navigate this evolving landscape, maintaining reflexivity and methodological rigour remains essential to ensure that AI enhances, rather than diminishes, the depth and integrity of qualitative inquiry. These shifts matter because they challenge the foundational assumptions of qualitative inquiry, requiring researchers to critically reassess what constitutes valid knowledge, authentic data, and meaningful interpretation in AI-mediated environments.

3. AI-DRIVEN METHODOLOGICAL TRANSFORMATIONS IN QUALITATIVE RESEARCH

While these philosophical shifts reshape how knowledge is understood, they also translate into concrete methodological transformations across the qualitative research process.

3.1. AI tools in qualitative research workflows

The integration of artificial intelligence into qualitative inquiry represents a fundamental reconfiguration within the AI-driven research ecosystem, moving from manual labor to a synergistic environment where AI acts as a proactive partner throughout the research lifecycle (Xiong et al., 2025). Within this framework, automated transcription and preliminary data processing function primarily as, AI as a Research Tool (ART), augmenting human capabilities by mitigating the resource-intensive work of data preparation (Kumaran et al., 2025; Xiong et al., 2025). Whereas these technologies facilitate unprecedented computational efficiency, they simultaneously introduce significant methodological tensions regarding the researcher's relationship with participants' data and the potential for context collapse during the initial stages of sensemaking (Schroeder et al., 2024). Qualitative research remains a deeply human-oriented and values-based endeavour, centered on capturing the richness of human experience, emotion, and social context, which necessitates a critical evaluation of how automated tools mediate these humanistic goals (Bakharia et al., 2025).

For example, consider a qualitative researcher conducting semi-structured interviews on student learning experiences. AI-powered transcription tools such as Otter.ai can be used to generate initial transcripts, which are then imported into qualitative data analysis software such as NVivo for coding. In this process, AI-assisted autocoding may suggest preliminary thematic categories based on linguistic patterns across interviews. However, the researcher remains responsible for reviewing, refining, and contextualising these codes to ensure they accurately reflect participants' meanings. In addition, generative AI tools such as ChatGPT may be used as reflexive aids to propose alternative interpretations or identify overlooked patterns, thereby supporting, rather than replacing, human analytic judgement.

The current utility of AI in preliminary processing is exemplified by the performance of level 4 large language models (LLMs), such as GPT-4 and Claude Sonnet 3.5, which have demonstrated the ability to provide labeled, interpretable themes and extract significant quotes with high accuracy (Bakharia et al., 2025). These systems outperform traditional topic modeling algorithms, such as Non-Negative Matrix Factorization (NMF) or Latent Dirichlet Allocation (LDA), by providing contextually relevant titles and rationales for extracted data segments (Bakharia et al., 2025). Researchers can organize millions of unstructured data points to identify rare phenomena for deeper study, effectively democratizing the discovery process for smaller themes, by automating the "needle in a haystack" process (Kumaran et al., 2025; Schroeder et al., 2024). Also, Ayyaswamy et al. (2025), in their study on IoT-enabled smart libraries, demonstrate how the integration of artificial intelligence with networked systems facilitates the automation of routine processes such as data classification, retrieval, and user interaction management. However, scholars argue that LLMs remain probabilistic generation systems rather than interpretive ones; they mimic the rhetorical style of academic language without possessing genuine semantic understanding (Nguyen & Welch, 2025).

Significant epistemic risks emerge when AI tools are used uncritically, including the "category error" of mistaking patterned, synthetic output for genuine interpretive insight (Nguyen & Welch, 2025). Methodological reliability is further compromised by probabilistic drift, a phenomenon where identical inputs produce inconsistent results depending on software variations or random initialization, which severely undermines the reproducibility essential to the scientific method (Ball, 2023; Korkmaz, 2025). Furthermore, transcription tools are susceptible to linguistic exclusion and algorithmic bias, particularly when processing non-standard accents, technical jargon, or the speech patterns of minority populations (Korkmaz, 2025; Nouis et al., 2025). Such discrepancies can lead to hallucinations, where models fabricate content or assign meanings not grounded in the source transcripts, thereby threatening the interpretive fidelity of the final dataset (Bakharia et al., 2025; Nguyen & Welch, 2025).

To address these vulnerabilities, the qualitative research community emphasizes the implementation of trustworthy workflows that prioritize Human-in-the-Loop (HITL) oversight and bi-directional traceability (Bakharia et al., 2025; Xiong et al., 2025). These frameworks require that every assigned code be traced back to its specific source transcript extract and verified as verbatim rather than hallucinated (Bakharia et al., 2025). Scholars emphasize that maintaining a close relationship with data through the manual reading and re-reading of transcripts is essential for capturing latent meaning and implicit knowledge, such as hesitation, body movement, and cultural nuance, that current transformer-based models are incapable of processing (Bakharia et al., 2025; Schroeder et al., 2024). Without such engagement, researchers risk the "Oracle Effect," unreflectively trusting AI outputs as objective or neutral simply because of their technological origin. This obscures the inherent biases present in the model's training data (Nguyen & Welch, 2025).

Ultimately, while automated transcription and preliminary processing offer profound opportunities for scale, their integration must be guided by rigorous reporting standards, such as CONSORT-AI and TRIPOD-AI, to ensure transparency and accountability (Korkmaz, 2025). Methodological integrity in the age of AI depends on a symbiotic human-AI partnership where computational power is balanced by critical thinking and ethical resilience (Xiong et al., 2025). Future discovery increasingly points toward a paradigm of "Verified Autonomy". This term is used to describe the ability of AI systems to handle the work of data management while remaining subject to continuous human evaluation and logic-based verification against established theoretical frameworks (Cornelio et al., 2025). Researchers can harness the efficiencies of AI while safeguarding the humanistic project of creating meaning from the "texture of human lives", by adhering to these standards (Schroeder et al., 2024).

The incorporation of Large Language Models (LLMs) into the hypothesis formulation phase represents a shift from labour-intensive manual data mining to an automated AI-driven research ecosystem where knowledge synthesis is increasingly proactive (Xiong et al., 2025). Literature synthesis, which traditionally relied on human intuition and expert-driven reviews, is now augmented by systems that utilize Natural Language Processing (NLP) to categorize and summarize vast repositories of academic literature (Kumaran et al., 2025; Singh, 2025).

Advanced frameworks such as STORM and AutoSurvey exemplify this transformation by employing multi-agent discussion cycles to generate structured outlines and comprehensive reviews that emulate the collaborative brainstorming of human research teams (Shao et al., 2024; Wang et al., 2024e). These systems are adept at identifying interdisciplinary connections and undiscovered public knowledge that may be overlooked in manual reviews, thereby allowing researchers to uncover rare phenomena or novel research directions (Anderson et al., 2021; Kumaran et al., 2025; Singh, 2025).

Beyond structural synthesis, LLMs are increasingly utilized as "thought partners" or "brainstorming buddies" during the development of analytic memos and conceptual ideation (Schroeder et al., 2024). This role facilitates a "fuzzy search" capability, where researchers can interact with their data conceptually rather than through rigid keyword-based lookups, assisting in the exploration of abstract concepts and the naming of complex theme categories (Bakharia et al., 2025; Schroeder et al., 2024). By embedding textual data into high-dimensional vector spaces, these models can suggest labels for data segments that are both intuitive and contextually relevant, mirroring aspects of the human sensemaking process (Ren, 2025; Schroeder et al., 2024). However, scholars caution that these outputs remain probabilistic rather than interpretive; while the generated text may appear fluent and logically sound, it lacks genuine semantic understanding and is prone to hallucinations and context collapse (Nguyen & Welch, 2025; Schroeder et al., 2024).

For instance, during thematic development, a researcher analysing interview data on student engagement may input selected excerpts into a generative AI system to explore alternative ways of labelling emerging themes. The AI may suggest broader or more abstract conceptual categories, prompting the researcher to reconsider initial coding decisions. However, these suggestions are not adopted uncritically; rather, they are evaluated against the dataset and refined through iterative comparison, ensuring that final themes remain grounded in participants' narratives.

The emergence of conversational AI in the data collection phase further reconfigures qualitative methodology by enabling interactional analysis at an unprecedented scale (Schroeder et al., 2024). LLMs can now conduct interviews with thousands of respondents simultaneously, bridging the traditional gap between the depth of individual narratives and the statistical power of large-scale datasets (Schroeder et al., 2024; Xiong et al., 2025). Empirical evidence suggests that participants often perceive AI-led interviews as non-judgmental, which can lead to more expansive and honest responses than those provided in traditional open-text fields (Schroeder et al., 2024). Despite this efficiency, several methodological and ethical risks persist. AI systems lack the contextual awareness and emotional depth of human interviewers, frequently missing non-verbal cues, vocal hesitations, or cultural accents that provide the essential "texture of human lives" (Schroeder et al., 2024). This lack of tacit knowledge means that while AI can replicate human-like responses, it cannot fulfill the humanistic task of creating meaning from subjective lived experiences (Bakharia et al., 2025; Xiong et al., 2025).

The uncritical adoption of these tools leads to what some experts term, the "category error". This refers to the situation where synthetic and patterned outputs are mistaken for genuine interpretive insight (Nguyen & Welch, 2025). Furthermore, there is the "Oracle Effect", which refers to the tendency to trust AI outputs as objective or neutral simply because of their technological origin, can obscure the inherent biases present in a model's training data (Nguyen & Welch, 2025; Ren, 2025). To safeguard the integrity of the research process, the academic community emphasizes "Verified Autonomy," where AI systems handle the work of synthesis and interaction while remaining subject to rigorous human oversight and logic-based verification (Cornelio et al., 2025; Xiong et al., 2025). Maintaining researcher agency through the "3C" framework, critical, creative, and computational thinking, is essential to ensure that AI serves as a cognitive amplifier rather than a replacement for human analytical rigor (Xiong et al., 2025).

3.2. Opportunities for Efficiency and Scale

The fusing of artificial intelligence into the AI-driven research ecosystem has catalyzed a profound shift in the scalability and efficiency of qualitative inquiry, effectively democratizing the process of scientific discovery (Xiong et al., 2025). By repositioning AI as a Research Tool (ART), scholars are now able to delegate the resource-intensive "drudgery" of manual data preparation, such as preliminary cleaning and organization, to automated systems (Kumaran et al., 2025). This automation is not merely a logistical convenience but a fundamental methodological reconfiguration that reduces the cognitive burden on researchers, allowing them to redirect their intellectual energy toward higher-order interpretive tasks, theoretical synthesis, and the exploration of complex social meanings (Singh, 2025; Xiong et al., 2025). Furthermore, this transition provides entry-level researchers with sophisticated tools that were previously the sole domain of large, well-funded teams, thereby fostering a more inclusive environment for knowledge production (Zhou et al., 2025).

The capacity of "Large Language Models (LLMs)" to function as a "needle in a haystack" tool represents a watershed moment for the analysis of massive, unstructured datasets (Schroeder et al., 2024). Traditional qualitative methods, limited by human cognitive bandwidth, often struggle to navigate the "undiscovered public knowledge" embedded in millions of social media comments or extensive focus group transcripts (Singh, 2025). In contrast, level 4 LLMs, such as Claude Sonnet 3.5 and GPT-4, can efficiently organize these vast repositories to identify rare phenomena or subtle interdisciplinary connections that might be overlooked by even the most diligent human observers (Bakharia et al., 2025; Kumaran et al., 2025). By leveraging "Natural Language Processing (NLP)" to extract and describe themes with supporting verbatim quotes, these systems facilitate a level of inductive analysis that bridges the traditional gap between the depth of individual narratives and the statistical power required for large-scale insights (Bakharia et al., 2025; Schroeder et al., 2024).

Significant economic and temporal advantages further highlight the transformative potential of AI in the social and cognitive sciences (Ayyaswamy, Gobinath et al., 2025; Nouis et al., 2025). Innovative platforms such as AITurk have demonstrated the ability of LLMs to simulate human-like responses in complex behavioral experiments, often replicating outcomes from high-impact psychology journals with a high degree of fidelity (Xiong et al., 2025). This paradigm shift offers a scalable and cost-effective alternative for preliminary investigations and replicability studies, effectively minimizing the need for traditional online crowdsourcing platforms in the initial stages of the research lifecycle (Xiong et al., 2025). Beyond simulations, AI integration facilitates real-time resource monitoring and the automation of administrative tasks like laboratory result triaging, which collectively enhance the return on investment for institutions adopting these advanced methodologies (Nouis et al., 2025). Ultimately, by balancing machine-driven efficiency with human-centered oversight, the academic community can achieve a state of "Verified Autonomy," where the speed of AI complements the nuanced judgment essential to robust scientific progress (Cornelio et al., 2025; Singh, 2025).

A practical example of this scalability can be seen in studies analysing large volumes of social media data. A researcher examining public perceptions of higher education reforms may use AI tools to filter and cluster thousands of online comments into preliminary thematic groups. This allows the researcher to identify dominant discourses and outlier perspectives more efficiently. Nonetheless, detailed qualitative interpretation still requires manual engagement with selected excerpts to preserve contextual meaning and avoid overgeneralisation.

3.3. Risks of Over-Automation

A primary risk in over-automation is the "category error". Large Language Models (LLMs) function as probabilistic generation systems that mimic rhetorical styles and academic registers without possessing genuine semantic understanding of social meaning (Cornelio et al., 2025). As noted by Cornelio et al. (2025), these models often operate at the level of plausibility rather than truth, rewarded for producing outputs that merely appear correct to human evaluators. In qualitative contexts, this mimics a form of empathy or understanding that is fundamentally vacant, as AI agents lack the emotional intelligence and lived context necessary to interpret complex human situations (Kakko, 2025). Consequently, the disconnect between data-driven pattern fitting and established theoretical frameworks can lead to an overwhelming influx of unverified hypotheses that lack empirical grounding (Cornelio et al., 2025).

The reliability of AI-driven qualitative analysis is further compromised by model non-determinism, leading to probabilistic drift. LLM outputs are inherently unstable; identical prompts can produce divergent results across sessions, hardware variations, or random initializations, which severely undermines the reproducibility essential for scientific trust (Cornelio et al., 2025; Smith et al., 2025). This instability is frequently accompanied by hallucinations, or "confabulations", where systems fabricate quotes, references, or data points that do not exist in the source material (Smith et al., 2025; Cornelio et al., 2025). Without rigorous, logic-based verification against background theories and empirical constraints, these fabricated insights risk overwhelming the scientific process with plausible but superficial results (Cornelio et al., 2025). Furthermore, as highlighted by Nouis et al. (2025), the "black box" nature of advanced neural networks makes it nearly impossible for researchers to interpret the rationale behind specific recommendations, particularly in rare or complex cases.

The "Oracle Effect" describes the tendency of researchers to uncritically trust AI outputs as objective or neutral, a phenomenon that obscures the inherent biases embedded in the model's training data (Dutta & Shah, 2025). This unreflective reliance can result in "diagnostic overshadowing," which implies the situation where scholars become less attentive to the nuances of the raw data and medical or social histories, potentially overlooking critical evidence that the AI system failed to detect (Kakko, 2025). Dutta and Shah (2025) emphasize that algorithmic bias is not merely a technical flaw but a governance failure, producing outcomes that are distorted and opaque. When researchers accept AI-generated themes without scrutiny, they risk breaching their epistemic responsibility, as responsibility for lapses in judgment remains within the human chain of governance rather than dissipating into the software (Dutta & Shah, 2025). Nouis et al. (2025) further warn that algorithms often systematically underperform for minority populations or atypical presentations, threatening the fairness and reliability of the resulting qualitative findings.

Over-automation precipitates a profound detachment from participant narratives, eroding the empathy and situational context that constitute the heart of qualitative inquiry. This out-of-the-loop performance problem occurs when automation shields the human researcher from the intricate details of the data, leading to a loss of situation awareness and a decline in manual skill maintenance (Endsley, 1996; Kakko, 2025). As systems operate reliably for extended periods, humans may become complacent and less vigilant, failing to detect subtle errors or edge cases that require human-level judgment (Kakko, 2025). This distancing is intrinsically linked to the loss of human agency, defined as the ability to make meaningful decisions and control one's investigative path (Krastev, 2025). When researchers hand over the interpretive process to machines, they risk losing the feeling of agency essential for human fulfillment and professional happiness (Krastev, 2025). Ultimately, a purely automated workflow risks homogenizing science and removing the organic mistakes or accidents, such as those that led to the discovery of penicillin, that are fundamental strengths of the human scientific method (Cornelio et al., 2025).

For example, a researcher relying solely on AI-generated summaries of interview transcripts may overlook subtle contradictions or emotional nuances present in the raw data. An AI system might categorise responses as uniformly positive based on dominant linguistic patterns, while a closer human reading reveals underlying tensions or ambivalence. This illustrates how over-reliance on automated outputs can flatten complexity and compromise interpretive depth.

These risks highlight that efficiency gains from AI must be balanced with safeguards that preserve interpretive depth, researcher agency, and methodological accountability.

4. METHODOLOGICAL INTEGRITY AND TRUSTWORTHINESS IN AI-ASSISTED QUALITATIVE INQUIRY

These methodological transformations raise critical questions about how rigor and trustworthiness can be maintained in AI-assisted qualitative inquiry.

4.1. The Epistemic Shift in Trustworthiness

The integration of artificial intelligence into the scientific research lifecycle has precipitated a fundamental epistemic transformation, shifting the definition of methodological integrity from a purely human-centric endeavor to a model of "Verified Autonomy" (Cornelio et al., 2025). Within this emergent AI-driven research ecosystem, AI acts as a proactive, intelligent partner that facilitates automated hypothesis generation and knowledge discovery; however, this autonomy introduces a critical verification bottleneck, where the sheer volume of synthetic claims outpaces traditional manual evaluation (Cornelio et al., 2025; Xiong et al., 2025).

Methodological integrity must therefore be reframed as the alignment between computational research design and the rigorous epistemic demands of the phenomena under study, requiring autonomous AI systems to remain subject to continuous, automated verification against established scientific laws and empirical data (Cornelio et al., 2025; Engin, 2025). This shift necessitates a transition from traditional human-centric rigor, grounded in expert intuition and embodied engagement, toward a framework of socio-technical validation (Engin, 2025). In this new paradigm, AI systems are recognized as inherently socio-technical entities whose reliability is a product of technical robustness combined with the social context of their deployment, human behavior, and dataset representativeness (Dahal et al., 2025; Engin, 2025). Consequently, trustworthiness is no longer a static property of a tool but a dynamic relationship that must be calibrated through systematic documentation, data provenance, and human custodianship to ensure that increased AI authority enhances rather than undermines scientific objectives (Dahal et al., 2025; Engler, 2023).

4.2. Reconfiguring Rigor: Credibility, Dependability, Confirmability, and Transferability

To uphold scientific standards in an AI-driven environment, traditional qualitative rigor must be reconfigured into collaborative frameworks that emphasize algorithmic accountability (Dahal et al., 2025; Korkmaz, 2025). Credibility in technology-mediated inquiry requires balancing outcome validation with process transparency to prevent the uncritical acceptance of "black box" decisions that lack a clear rationale (Cornelio et al., 2025; Korkmaz, 2025). While Large Language Models (LLMs) can produce plausible summaries, they often operate in a semantic vacuum and they are rewarded for fluency rather than truth, necessitating "trustworthy workflows" that prioritize verbatim quote validation to ensure that findings reflect actual participant experiences rather than synthetic, patterned outputs (Bakharia et al., 2025; Nguyen & Welch, 2025).

Dependability is increasingly challenged by the inherent nondeterminism of deep neural networks, where variations in hardware, data initialization, or random seeds alter outputs for identical inputs, a phenomenon known as "probabilistic drift" (Achiam et al., 2023; Ball, 2023). This volatility undermines the reproducibility essential for scientific trust, as simply sharing code no longer guarantees identical results across different computational environments (Ball, 2023; Korkmaz, 2025). To ensure confirmability, researchers must establish bi-directional traceability to mitigate the "Oracle Effect," which describes the dangerous tendency to trust AI outputs as objective truth simply because of their technological origin (Bakharia et al., 2025; Nguyen & Welch, 2025).

In practice, this can be operationalised by maintaining a detailed audit trail in which each AI-generated code or theme is linked to specific excerpts from the original data. For example, a researcher using NVivo's AI-assisted coding may document how initial codes were generated, modified, or rejected, alongside justifications grounded in the dataset. This process enhances confirmability by ensuring that analytical decisions remain transparent and verifiable.

Methodological rigor mandates that every assigned code or theme remains traceable back to its specific source data through transparent audit trails and human-AI dialogue journaling (Bakharia et al., 2025; Dahal et al., 2025). Furthermore, transferability is threatened by concept drift and distributional drift, where models trained on historical data may fail to generalize to new operating environments or minority populations as real-world statistics evolve (Cornelio et al., 2025; Engin, 2025). Maintaining integrity thus requires researchers to explicitly define the "Operational Design Domain (ODD)", identifying the precise boundaries and conditions under which AI-assisted findings remain valid and generalizable (Engin, 2025).

4.3. The Reflexive Researcher and the Anthropocentric Mirror

The emergence of the AI-driven research ecosystem transforms the researcher into an augmented reflexive practitioner who utilizes AI as a "thinking companion" or partner in cognition (Costa & Bem-Haja, 2025; Xiong et al., 2025). In this role, AI functions as an "anthropocentric mirror," acting as a "peer questioner" that helps researchers confront their own biases by identifying interpretive ambiguity and semantic friction within the data (Alphanome.AI, 2025; Costa & Bem-Haja, 2025). This "hybrid reflexivity" leverages algorithmic pattern recognition to reveal thematic silences or biases that human researchers might overlook, turning analytical uncertainty into a productive resource for discovery (Costa & Bem-Haja, 2025).

However, the allure of automation introduces the "ironies of automation," where the reliance on smarter systems can paradoxically lead to a decline in human skill maintenance, vigilance, and situation awareness (Endsley, 1996; Kakko, 2025; Shukla et al., 2025). This out-of-the-loop performance problem precipitates a profound detachment from participant narratives, eroding the empathy and situational context that constitute the heart of qualitative inquiry (Endsley, 1996; Kakko, 2025). When researchers use AI as a "shortcut" for sensemaking, they risk complacency, potentially failing to detect subtle hallucinations or decontextualized errors that machines frequently produce (Endsley, 1996; Kakko, 2025; Schroeder et al., 2024).

Preserving researcher agency, defined as the ability to make meaningful decisions and control one's investigative path, is essential for scientific fulfillment and methodological rigor (Krastev, 2025). Recent frameworks like HACITA (Human–AI Collaborative Inductive Thematic Analysis) demonstrate that researchers can reclaim authority through systematic analytic actions, such as modification and rejection, to correct AI literalism (Nyaaba et al., 2025). Ultimately, researchers must remain the final arbiters of meaning and strategy owners of the research process, ensuring that AI-driven insights do not flatten the unique texture of human lives or elide the subjective nuances that characterize humanistic discovery (Krastev, 2025; Schroeder et al., 2024; Xiong et al., 2025).

4.4. The Verification Bottleneck: Human Validation vs. Algorithmic Pattern Detection

The proliferation of artificial intelligence in the research lifecycle has introduced a critical verification bottleneck, where the capacity of generative models to produce synthetic claims far outpaces the speed of traditional manual evaluation (Cornelio et al., 2025). This bottleneck is rooted in a fundamental category error: the conflation of an agent's rhetorical fluency with genuine semantic understanding (Cornelio et al., 2025; Nguyen & Welch, 2025). While large language models (LLMs) can produce plausible scientific hypotheses or thematic summaries, they operate within a semantic vacuum, rewarded for generating outputs that appear correct to human evaluators without possessing an underlying grasp of social meaning or physical laws (Cornelio et al., 2025; Dahal et al., 2025). This lack of grounding often leads to the production of plausibility without proof, which cannot serve as a reliable foundation for scientific knowledge (Cornelio et al., 2025). Consequently, the scientific value of AI-assisted outputs depends entirely on rigorous, scalable mechanisms for verification that can distinguish between meaningful insights and mere noisy interpolations (Cornelio et al., 2025).

To widen this bottleneck, the research community is transitioning from purely data-driven evidence toward logic-based verification (Cornelio et al., 2025; Engin, 2025). Data-driven methods are often limited to fitting observed data points, a process that is prone to hallucinations and overfitting, particularly when models identify patterns based on artifacts rather than true phenomena (Korkmaz, 2025; Cornelio et al., 2025). In contrast, logic-based verification utilizes formal methods such as theorem proving and logic derivability, to validate whether AI-generated hypotheses conform to established axioms and background theories (Cornelio et al., 2025).

Frameworks like AI-Descartes and AI-Hilbert exemplify this shift by ensuring that discovered laws are not only predictive but also interpretable and logically consistent with foundational scientific knowledge (Cornelio et al., 2025). This critical validation serves as a necessary filter, allowing researchers to catch fabricated references or violating algebraic consistencies that frequently plague generative systems (Cornelio et al., 2025; Korkmaz, 2025).

4.5. Hybrid Triangulation: Synergizing Human and Machine Intelligence

The preservation of methodological integrity in an AI-driven environment increasingly depends on hybrid triangulation, a strategy that leverages the complementary strengths of human expertise and machine intelligence to establish converging evidence (Cornelio et al., 2025; Liu et al., 2025). Rather than viewing

AI as an autonomous analyst, researchers are adopting multi-model and multi-agent approaches to knowledge integration (Liu et al., 2025; Engin, 2025). These systems facilitate strategic knowledge synthesis through micro-processes such as connecting, eliciting, and converging, effectively uncovering the "black box" of collaboration (Liu et al., 2025). By presenting results from both an interpretable algorithm and a high-performance "black box" model, researchers can identify areas of disagreement that signal a need for deeper human investigation (Cornelio et al., 2025; Smith et al., 2025).

A robust mechanism within this synergistic framework is the "Devil's Advocate" function, where LLMs are specifically prompted to challenge assigned codes or recommendations (Chang et al., 2024; Smith et al., 2025). Research indicates that incorporating an LLM-powered devil's advocate improves appropriate reliance by forcing human groups to reconsider incorrect AI suggestions, thereby increasing overall decision accuracy (Chang et al., 2024).

Furthermore, hybrid architectures are being developed to reduce the impact of brittle performance by integrating unsupervised anomaly detection with supervised learning (Smith et al., 2025). In this design, unsupervised techniques identify infrequent behaviours or "outliers" that supervised models might miss, alerting the user to potentially significant rare phenomena (Li & Hansman, 2016; Smith et al., 2025). This triangulation ensures that scientific discovery remains resilient to the narrow competence envelopes of individual AI tools while maintaining the nuanced judgment essential for creating meaning from complex data (Cornelio et al., 2025; Smith et al., 2025).

This reinforces that maintaining qualitative rigor in AI-assisted research depends not on eliminating human involvement, but on strengthening structured collaboration between human judgment and algorithmic processing.

For instance, a researcher analysing interview data on teacher experiences may combine AI-assisted coding with manual thematic analysis and participant validation. If AI-generated themes differ from manually developed ones, these discrepancies can be explored through peer debriefing or follow-up interviews. Such triangulation strengthens the credibility of findings by integrating multiple analytical perspectives.

5. ETHICAL, POSITIONAL, AND CONTEXTUAL CONCERNS IN AFRICAN QUALITATIVE RESEARCH

Beyond methodological concerns, AI integration introduces broader ethical and contextual challenges, particularly within African research settings where issues of power, representation, and data sovereignty are especially pronounced.

5.1. Algorithmic Bias and Epistemic Injustice

The integration of Artificial Intelligence (AI) into African qualitative research necessitates a critical examination of how these systems operate as "epistemic technologies" of domination that encode human social labor into repeatable computational procedures (Adamu & Nkwo, 2023). Rather than being neutral instruments, AI systems are social structures that mimic human subjectivity while emerging from wider economic and political forces designed to centralize control for those who wield them (Sartori & Theodorou, 2022). In the African context, epistemic injustice manifests when AI systems associated with dominant Global North social groups exhibit high precision for Western norms while failing to represent socio-culturally salient concepts specific to African communities (Khanna & Li, 2025). This results in testimonial injustice, where contributions articulated in non-dominant languages are systematically downranked or mistranslated, effectively treating African speakers' testimony as less reliable (Khanna & Li, 2025).

Furthermore, hermeneutical injustice occurs when training datasets lack the lexical resources to articulate community-specific concepts, thereby constraining the ability of African researchers to use AI for meaningful scientific discourse (Dlamini & Ndzinisa, 2025). These algorithms are often developed with Eurocentric design policies that do not align with the social or environmental peculiarities of African communities (Hassan, 2023). Consequently, the uncritical adoption of AI in qualitative research risks reinforcing algorithmic oppression and positioning the African continent as an "epistemic laboratory" for experimentation by multinational corporations (Adamu & Nkwo, 2023).

5.2. African Languages, Cultures, and Data Invisibility

A profound crisis exists in the digital ecosystem where, out of the world's 7,613 living languages, approximately 2,000 are classified as "Invisible Giants" languages with millions of speakers that remain effectively invisible in AI training data (Khanna & Li, 2025). These languages are disproportionately concentrated in Africa, where sociocultural vitality is high but digital representation is near-zero (Kuddus, 2022). This data invisibility is not a technical accident but a structural artifact of power relations that reflect continuities from colonial-era linguistic hierarchies to contemporary AI development (Omotubora & Basu, 2024).

When AI systems are trained predominantly on high-resource languages from the Global North, they systematically exclude the knowledge systems, historical narratives, and oral traditions of African communities (Ketere, 2025). This constitutes epistemic violence at scale, as the exclusion of these languages from Large Language Models (LLMs) encodes the message that African ways of knowing are inconsequential (Muzanenhamo & Power, 2024). For the qualitative researcher, this creates a self-reinforcing feedback loop where poor AI performance on local languages discourages digital content generation, further accelerating offline linguistic decline and marginalization (Khanna & Li, 2025).

A practical illustration of this challenge can be observed when AI transcription tools struggle to accurately capture local Ghanaian dialects or code-switching between English and indigenous languages. In such cases, researchers must manually review and correct transcripts or engage bilingual assistants to ensure linguistic accuracy. This highlights the importance of human intervention in preserving culturally embedded meanings that AI systems may misinterpret.

5.3. Informed Consent, Privacy, and Data Ownership

The proliferation of AI-led initiatives in Africa often denotes a rebirth of imperialism through "digital colonialism," where Global North institutions appropriate data from African communities to commercialize and resell the resulting tools back to them for profit (Libengood, 2025). This extractive dynamic raises urgent concerns regarding data sovereignty and the protection of African participants (Maluleke, 2025). Traditional Western ethical codes often prioritize individualistic autonomy, which may be poorly aligned with African worldviews, such as Ubuntu, that center reciprocity, relational accountability, and community hood (Hoosen et al., 2024).

In qualitative research, meaningful informed consent must involve community consultation and respect local protocols, ensuring that participants are not rendered as "private property" to be regulated by tech oligarchs (Adamu & Nkwo, 2023; Darabi, 2021). The risk is particularly acute in "surveillance capitalism," where human behavior is commoditized for capitalist extraction, often without the explicit understanding or control of the marginalized groups being studied (Hoosen et al., 2024).

For example, when using cloud-based AI transcription services, researchers must explicitly inform participants that their data may be processed through external platforms. In sensitive research contexts, such as studies involving vulnerable populations, researchers may opt for offline transcription tools or anonymise data prior to processing. These practices ensure that ethical principles of confidentiality and informed consent are upheld in AI-assisted research.

5.4. Ethical Review Processes in AI-Supported Studies

Current institutional ethics boards in many African universities remain deeply rooted in Western ethical frameworks, which are often viewed by researchers as slow, costly, and insensitive to indigenous methodologies (Maluleke, 2025). To promote epistemic justice, there is a shift toward relational ethics that affirm African moral traditions (Kumalo, 2017). The African Independent Ethics Committee (AIEC) provides a pan-African framework that allows for retrospective applications, recognizing that some researchers may not have had access to functioning committees at the start of their work (Chaudhry et al. 2022).

Ethical review in AI-supported studies must include Ethical Impact Assessments (EIA) to measure the real-world impact of AI systems on marginalized groups (African Union [AU], 2024; AU Strategy, 2024). These processes should also advocate for AI systems to assume a level of "relational responsibility," treating the machine as a co-agent that shares in the liability and outcomes of the research process (Bahrami, Deroy & Longin, 2023).

5.5. Power Relations Between Global North AI Systems and African Knowledge Production

The dominance of Global North datasets in AI development ensures that the technology is a reflection of the human social values embedded within it by its Western designers (Hassan, 2023). This creates a technological dependency where African institutions are positioned as consumers of global knowledge rather than architects of their own pedagogies (Hoosen et al., 2024). The rhetoric of "AI for good" often obscures patterns of exploitation and extraction, where corporate philanthropic visions inform the strategies needed to mitigate or amplify the unintended consequences of AI systems in Africa (Adamu & Nkwo, 2023).

Decolonizing AI requires a "radical shift" from the Global North's construct of universality toward a pluriversal approach that recognizes multiple ways of knowing and being (Hoosen et al., 2024). African researchers must move beyond tool-centric adoption toward ethically framed, co-designed ecosystems that include marginalized communities in the design process to ensure equitable and trustworthy AI (Adamu & Nkwo, 2023; Hoosen et al., 2024). These concerns underscore that AI governance in qualitative research is not only a technical issue but also a question of epistemic justice, requiring context-sensitive and locally grounded approaches.

6. AI AND POSTGRADUATE QUALITATIVE RESEARCH IN GHANAIAN HIGHER EDUCATION

These broader dynamics become particularly visible when examined within specific national contexts. This section draws on a synthesis of recent empirical studies, policy reports, and scholarly analyses on AI adoption in Ghanaian higher education. It is important to note that the discussion is primarily based on secondary empirical evidence and documented studies, rather than original fieldwork conducted within this chapter. The literature was selectively curated to reflect recurring themes related to postgraduate research practices, institutional responses, and emerging challenges, with emphasis on studies grounded in the Ghanaian context.

6.1. Postgraduate Qualitative Research Practices in Ghana

Postgraduate research in Ghanaian higher education is a rigorous process involving coursework, proposal development, fieldwork, and thesis reporting (Arthur & Fenyi, 2023). Despite this structured progression, completion rates remain low, with approximately 30% of MPhil students finishing within the stipulated two-year period, and about 40% failing to complete due to challenges associated with the thesis component (Arthur & Fenyi, 2023).

Qualitative research, in particular, presents additional demands. Researchers often travel to remote locations and spend extended periods within communities to conduct observations and collect data (Keen et al., 2022). These requirements are further constrained by financial limitations, including limited access to digital tools such as laptops, which can delay relatively simple tasks (Arthur & Fenyi, 2023). Many postgraduate students are also mature learners who combine academic work with employment and family responsibilities. As noted by Geertshuis and Liu (2025), these competing demands frequently affect research continuity and completion. These studies suggest that postgraduate qualitative research in Ghana is shaped by structural, financial, and personal constraints, which create a challenging environment even before the integration of AI tools.

For instance, a postgraduate student conducting fieldwork in a rural community may record interviews using a mobile device and later use AI transcription software to process the data. However, due to inconsistent internet connectivity, transcription may be delayed or require manual correction. This illustrates how infrastructural constraints shape the practical use of AI tools in Ghanaian research contexts.

6.2. Institutional Policies and Research Ethics Frameworks

Ghana's policy landscape for AI in higher education remains emergent. While a comprehensive national AI strategy is still under development, the government has initiated digital innovation efforts through programmes such as the National Entrepreneurship and Innovation Plan (Osondu et al., 2024). At the institutional level, universities such as the University of Ghana have established research ethics policies and are engaging with international bodies to develop frameworks for responsible AI use (Ayana et al., 2024).

However, a clear gap remains regarding guidelines for generative AI in research contexts (Maluleke, 2025). Current institutional responses tend to emphasise academic integrity and risk management rather than providing structured guidance for responsible integration. For example, the withholding of 2023 WASSCE results from 235 schools due to alleged AI misuse reflects a regulatory approach focused on control rather than pedagogical adaptation (Yilma, 2025). This suggests that while policy awareness is increasing, there is still limited alignment between governance structures and the practical realities of AI-assisted research.

6.3. AI Use by Postgraduate Students and Supervisors

Empirical studies in Ghana indicate a strong positive perception of AI tools, with 70% of academics believing these tools enhance writing quality and 65% agreeing they help combat plagiarism (Elliason & Khajuria, 2025). Among Ghanaian postgraduate students and staff, frequency of use is highest for spelling and grammar checks (65%), while data analysis remains the least common use case, with 50% using AI rarely or never for this purpose (Elliason & Khajuria, 2025).

Supervisors and students are often motivated by peer recommendations and the endorsement of trusted academic figures (Kutsyuruba, 2021). In the realm of supervision, AI is being explored through chatbots to provide responsive learning environments in contexts with high student-to-teacher ratios (Osondu et al., 2024). Some postgraduate supervision models are also moving toward "technologically augmented supervision," using data analytics to provide evidence-based feedback (Hoosen et al., 2024).

A practical illustration of AI use in Ghanaian postgraduate research can be observed in thesis writing and literature review practices. For example, postgraduate students may use generative AI tools such as ChatGPT or QuillBot to summarise journal articles, generate outlines, or refine academic language. While these tools enhance efficiency, students typically combine them with manual reading and critical evaluation to ensure conceptual accuracy. In some cases, AI tools are also used to support preliminary coding of qualitative data; however, students often rely on supervisor guidance to validate these outputs and align them with methodological expectations. This reflects a hybrid approach in which AI supports routine tasks while interpretive authority remains firmly with the researcher.

6.4. Capacity Gaps, Training Needs, and Supervision Challenges

A significant digital divide exists in Ghana, with 60% of academics citing unreliable internet and inadequate infrastructure as primary barriers to AI adoption (Sarfo et al., 2024). This gap is most pronounced for those in rural institutions compared to their urban counterparts (Elliason & Khajuria, 2025). Supervision challenges are a major cause of thesis delay; supervisors in Ghana are often overwhelmed, with some assigned up to 15 students while balancing 16 hours of weekly lecturing (Arthur & Fenyi, 2023). This leads to long response times, sometimes up to six months, and a breakdown in the student-supervisor relationship (Arthur & Fenyi, 2023).

Furthermore, there is a competence gap, where supervisors skilled in qualitative methods may struggle to guide students who adopt different or more technologically advanced methodologies (Junaedi et al., 2024). There is a consensual call for Ghanaian universities to institutionalize Research Assistant portfolios to act as "quasi-supervisors" and bridge these capacity gaps (Arthur & Fenyi, 2023).

6.5. Documented Practices and Reflective Accounts

Reflective accounts from the Ghanaian academic community highlight a disciplinary disparity in AI acceptance, with 60% of science educators embracing the technology compared to only 30% in the humanities (Elliason & Khajuria, 2025). Humanities scholars express greater caution, citing ethical and philosophical concerns regarding the "humanistic aspects" of education (Junaedi et al., 2024). There is also a fear that over-reliance on AI may weaken students' critical thinking and creativity (Elliason & Khajuria, 2025; Osondu et al., 2024). Successful practices include the development of "Kwame AI," a Ghanaian-designed web app that provides instant science answers and diagrams for students, demonstrating the potential for locally led, contextually relevant AI innovation (Osondu et al., 2024). However, to ensure that AI serves as a "pedagogical co-agent" rather than a tool for digital colonialism, Ghanaian institutions must invest in sovereign AI capabilities and indigenous knowledge systems (AU Strategy, 2024; Hoosen et al., 2024).

These findings demonstrate that AI adoption in Ghanaian higher education is both promising and uneven, highlighting the need for structured frameworks that align technological innovation with local realities and research capacities.

7. A CONTEXT-SENSITIVE FRAMEWORK FOR RESPONSIBLE AI INTEGRATION IN QUALITATIVE RESEARCH

In response to the philosophical, methodological, and contextual challenges outlined in the previous sections, this section proposes a context-sensitive framework for responsible AI integration in qualitative research. AI is quickly changing the way research is done all across the world, and this is also true for qualitative research in Ghanaian higher education. Although systems such as ChatGPT and NVivo provide faster transcribing and automated coding, they also raise epistemological and ethical dilemmas for researchers grounded in interpretivist paradigms

(Dellafiore et al., 2025; Adjei et al., 2025). For example, the National Teacher Education Curriculum Framework (NTECF) says that ICT skills are core to modern teaching and learning (MOE, 2018). This shows that Ghana's education policies already stress digital literacy.

Also, schools like KNUST and the University of Ghana (UG) have made plans for how to use AI and rules for how to be honest when using these technologies (Kwame Nkrumah University of Science and Technology, 2025; University of Ghana, 2024). This approach combines new AI and qualitative research with Ghana's specific policies and situation. It is organised into five interrelated layers: *(1) Human-Centred Reflexivity, (2) Ethical Governance and Transparency, (3) Contextual and Cultural Sensitivity, (4) AI Literacy and Capacity Building, and (5) Hybrid Validation and Interpretive Control.* Figure 1 shows the context-sensitive framework for responsible AI integration in qualitative research. The discussion that follows explains each layer and gives practical guidelines for researchers and institutions.

Figure 1. Context-Sensitive Framework for Responsible AI Integration in Qualitative Research

Source: Authors' own work

7.1. Human-Centred Reflexivity

The framework basically asserts that human researchers must be pivotal to interpretation. In accordance with interpretivist paradigms, reflexivity entails that researchers consistently examine how their assumptions, values, and decisions influence the study process (Dellafiore et al., 2025). Qualitative experts warn that AI's mechanistic nature could undermine the nuanced depth of qualitative analysis. This emphasizes the necessity of reflexivity, creativity, and researcher identity as important contrasts (Adjei et al., 2025; Dellafiore et al., 2025). Practically, this necessitates that researchers thoroughly document the rationale and methodology for incorporating an AI tool at every phase of data gathering, coding, and analysis. For instance, while using ChatGPT or NVivo's AI-assisted coding, researchers must consider potential omissions or misclassifications of concepts or themes and be prepared to amend or enhance the AI's results. Transparent journaling or team debriefs can ensure that findings are rooted in the participants' interpretations and the researchers' skills, rather than solely in the AI's algorithms.

7.2. Ethical Governance and Transparency

Next to reflexivity is a framework of policies, ethical standards, and transparency about the use of AI. Governments and institutions across the globe, including in Ghana, are establishing governance frameworks to guarantee that AI is reliable, transparent, and inclusive (OECD, 2019). The National AI Strategy of Ghana (2023-2033) indicates that AI presents dangers of bias, data misuse, and privacy violations if not managed appropriately. The strategy advocates for a responsible, inclusive, and sustainable AI ecosystem with robust data governance (Republic of Ghana, 2022). Ghanaian universities are changing their academic integrity regulations at the institutional level. The University of Ghana's 2024 plagiarism policy was revised to declare that any application of AI that undermines the integrity of academic work will be considered inappropriate (University of Ghana, 2024). Thus, the framework highlights that each research team must function according to clear, documented protocols. These may entail requesting ethics committee approval for all AI applications, recording all AI inputs and outputs, and detailing in publications the use of AI technologies. Institutional support is significant, universities are currently educating students in AI ethics (Abass, 2025) and formulating strategic plans (Kwame Nkrumah University of Science and Technology, 2025), and these initiatives should encompass research practices as well.

7.3. Contextual and Cultural Sensitivity

The subsequent layer stresses that AI integration must be adapted to Ghana's local cultural and socio-economic setting. Tools developed using Western data may be ineffective in Ghanaian contexts. They may misconstrue local idioms, misread details, or foster global biases. Researchers have observed that ethical AI in education must emphasise cultural context (OECD, 2019). Ghanaian researchers contend that the humanities should provide AI with ethical guidance, cultural context, and a sense of human identity (Abass, 2025). Researchers ought to engage local stakeholders while implementing AI. For instance, while utilising NVivo's AI assistant to summarise interview transcripts, researchers must ensure that the summaries accurately represent Ghanaian reality rather than merely generic patterns. Transcription tools such as Otter.ai should be evaluated for compatibility with local accents and languages, or substituted with locally produced alternatives when feasible. Research design must honour Ghanaian perspectives and languages. Prompts or coding methods must consider multilingual data, and interpretations should be collaboratively developed with cultural informants. This layer corresponds with Ghana's educational objectives, since the NTECF and other curricula advocate for teacher-researchers to include ICT in a manner conducive to Ghanaian classrooms (MOE, 2018). Incorporating cultural validation at this level enables researchers to prevent an increase of existing inequities and biases in AI models that could disproportionately affect marginalised populations (Adjei et al., 2025; OECD, 2019).

7.4. AI Literacy and Researcher Capacity Building

The fourth layer shows that training and capability are important tools. Students and researchers need to know what AI tools can and cannot do. Ghanaian colleges and universities are starting to deal with this. For instance, KNUST currently mandates all undergraduates to take a basic AI course (Kwame Nkrumah University of Science and Technology, 2025), while the education sector stresses that teachers need to be good with computers (MOE, 2018). Researchers should also look for ways to improve their skills in AI. There could be classes on how to make prompts for generative AI tools like ChatGPT, how to use Nvivo's AI assistant, or how to follow data privacy rules (Naredla et al., 2025). Capacity-building also includes working together across disciplines. Social scientists could work with computer scientists and data librarians to explain algorithmic parts in simple terms. Institutions can help with this by setting up centers for teaching and learning that offer AI literacy courses. This layer makes sure that researchers know how to make smart methodological choices. A well-trained researcher, for instance, will not just trust automatic coding; they'll know how to give ChatGPT clear instructions and how to critically analyse its output. Being AI literate also means knowing what AI cannot do. AI can speed up transcription and sorting of data, but researchers should still be careful about letting it do complicate interpretive work on its own (Dellafiore et al., 2025).

7.5. Hybrid Validation and Interpretive Control

The last layer of the framework is about combining AI's abilities with traditional ways of checking quality. This involves using both strategies in practice. For example, a researcher might employ Nvivo's autocoding to come up with initial themes and then compare those to manual coding done by people to find any differences. One expert says that generative AI can make things more efficient, but it should always be used alongside human interpretative agency (Dellafiore et al., 2025).

Likewise, while using Otter.ai to transcribe interviews, researchers must to examine transcripts thoroughly to find inaccuracies, similar to the back-translation method in linguistic studies. Peer debriefing and asking participants to review interpretations are still important procedures that AI cannot replace. Qualitative insights should be used to check AI-driven analyses in mixed-method designs. This multilayer validation protects against putting a lot of trust in AI. When several researchers compare AI-assisted algorithms with their own studies, any biases that an AI model might introduce would be found. The goal is an interpretive control in which the researcher always has the final say on how to make meaning.

7.6. Practical Guidelines

To operationalize the framework, researchers and institutions can follow these guidelines:

- **Document decisions and reflexivity.** Keep a record of who used AI, when, how, and why, and write in a reflexive a notebook. Clearly state when AI techniques were used and how researchers assessed the results (Adjei et al., 2025; Dellafiore et al., 2025).
- **Secure informed consent and data protection.** Ensure that research participants are informed whether interviews will be transcribed or analysed by AI. Secure consent for cloud processing or choose for offline tools for sensitive information. Securely save all digital data and anonymise transcripts.
- **Follow institutional policies.** Comply with revised academic integrity and ethical standards. The Universities of Ghana have clearly prohibited undeclared AI use (University of Ghana, 2024). Consistently disclose the use of AI in proposals and publications for the sake of transparency.
- **Build researcher capacity.** Participate in continuous AI training by attending workshops and working with tech experts. Utilise campus resources, like KNUST's new AI course or UG's ICT services, to be informed on AI capabilities and governance.

- **Adapt methods to context.** Involve local experts to check the AI's interpretations. Use linguistic models or glossaries that are acceptable for the culture. If using ChatGPT or similar GenAI tool, make sure to include Ghanaian examples in the prompts and review the results to make sure they are culturally relevant.
- **Hybrid validation.** Check AI outputs against those from more traditional methods. For instance, if ChatGPT suggests theme categories, check them using researcher coding or input from participants. Always have at least one person look over codes or summaries made by AI.
- **Ensure ethical oversight.** Where possible, have an ethical committee or an interdisciplinary advisory board go over AI uses in a project. Use Ghana's national policies and AI strategy as a guide to include Ghanaian legal and cultural norms in AI research ethics. (Republic of Ghana, 2022; Yin, 2025).
- **Foster transparency in reporting.** In any dissemination (theses, papers, reports), make sure to clearly explain the AI tools you used and how you used them, as well as what you did to reduce risks. This openness fosters trust and help to improve best practices over time.

This framework shows that appropriate AI in Ghanaian qualitative research surpasses mere adherence to global algorithms; it necessitates the integration of technology within a humanistic, context-sensitive framework. Researchers can effectively use tools such as ChatGPT, Otter.ai, and NVivo without compromising the depth and integrity of qualitative research by integrating human reflexivity, ethical governance, cultural insight, skills training, and hybrid validation (Adjei et al., 2025; Dellafiore et al., 2025). This strategy is in line with both Ghana's AI strategy (Republic of Ghana, 2022) and the international call for culturally relevant, empowering AI in education (OECD, 2019). This confirms that AI improves rather than eliminates the significant human meanings fundamental to qualitative research.

To further demonstrate the practical applicability of the framework, Table 1 presents context-specific scenarios illustrating how AI tools can be responsibly integrated into qualitative research processes within Ghanaian higher education.

Table 1. Practical Applications of Responsible AI Integration in Qualitative Research

Framework Layer	Research Scenario (Realistic Context)	AI Tool & Use	Researcher Action (Human Control)	Ethical / Methodological Safeguard	Expected Outcome
Human-Centred Reflexivity	A postgraduate student at University of Cape Coast conducting interviews on student stress and coping strategies	ChatGPT used to suggest preliminary thematic labels from interview excerpts	Researcher compares AI-generated themes with manual coding and writes reflexive notes explaining why certain themes were accepted, merged, or rejected	Maintain a reflexive journal documenting AI prompts, outputs, and researcher decisions	Ensures themes reflect participants' lived experiences rather than algorithmic assumptions
Ethical Governance & Transparency	A study on teacher performance using interview data stored on cloud-based platforms	Otter.ai used for transcription of recorded interviews	Researcher informs participants that transcription involves AI tools and anonymises all identifiers before upload	Ethics approval obtained specifying AI use; AI usage disclosed in methodology section	Enhances transparency, protects confidentiality, and aligns with institutional ethical standards
Contextual & Cultural Sensitivity	Research on indigenous dress practices among artisans in Kumasi	NVivo AI assistant used to summarise interview transcripts	Researcher reviews summaries with local cultural experts to verify meaning of indigenous terms and idioms	Cross-check AI outputs with participants or cultural informants; adjust codes to reflect local context	Prevents cultural misinterpretation and ensures contextual accuracy
AI Literacy & Capacity Building	MPhil student analysing qualitative data on fashion entrepreneurship in Accra	NVivo autocoding used for initial coding of interview data	Researcher attends AI/NVivo training workshop and manually refines codes after understanding limitations of automated coding	Institutional training (e.g., KNUST AI courses); collaboration with data analysts	Improves researcher competence and reduces blind reliance on AI outputs
Hybrid Validation & Interpretive Control	Mixed-method study on food consumption behaviour in Ghanaian hotels	ChatGPT used to generate possible thematic categories; NVivo used for coding	Researcher triangulates AI-generated themes with manual coding and participant validation (member checking)	Compare AI outputs with traditional analysis; involve second coder or supervisor review	Strengthens credibility, confirmability, and analytical rigour

continued on following page

Table 1. Continued

Framework Layer	Research Scenario (Realistic Context)	AI Tool & Use	Researcher Action (Human Control)	Ethical / Methodological Safeguard	Expected Outcome
Human-Centred Reflexivity	Research on gender participation in technical education in rural Ghana	ChatGPT used to summarise literature for conceptual framework	Researcher cross-checks AI summaries against original journal articles to avoid misrepresentation	Maintain annotated bibliography verifying all AI-assisted summaries	Preserves academic integrity and ensures accurate literature synthesis
Ethical Governance & Transparency	Study involving sensitive interviews with vulnerable groups (e.g., persons with disabilities)	AI used for transcription and initial coding	Researcher opts for offline transcription tools or encrypts data before processing	Explicit informed consent for AI processing; compliance with data protection standards	Minimises risk of data breaches and ethical violations
Contextual & Cultural Sensitivity	Study involving multilingual interviews (English + Twi/ Ewe)	AI transcription tool used for English segments only	Researcher manually translates and validates non-English data with bilingual assistants	Use culturally appropriate translation and back-translation methods	Preserves linguistic meaning and prevents loss of cultural nuance
AI Literacy & Capacity Building	Supervisor guiding multiple postgraduate students using AI tools	ChatGPT used to demonstrate coding strategies and prompt design	Supervisor trains students on critical evaluation of AI outputs rather than passive acceptance	Development of departmental AI guidelines and workshops	Builds institutional capacity and improves research quality
Hybrid Validation & Interpretive Control	Qualitative study on curriculum implementation in Ghanaian schools	AI-generated themes compared with field notes and observational data	Researcher integrates AI findings with observational insights and peer debriefing sessions	Use triangulation (data + method + researcher) to validate findings	Ensures balanced interpretation and reduces algorithmic bias

7.7. Operationalising the Context-Sensitive Framework for Responsible AI Integration in Qualitative Research

While the proposed framework provides a robust conceptual basis for responsible AI integration, its practical value depends on clear operational procedures that researchers can implement across different stages of qualitative inquiry. In response to this need, this section translates the framework into actionable tools and protocols, including disclosure guidance, audit-trail structures, verification procedures, and data-governance mechanisms (see Table 2). These operational elements ensure that AI use remains transparent, ethically sound, and methodologically rigorous within Ghanaian research contexts.

The disclosure guidance is essential for promoting transparency and academic integrity. Researchers must explicitly state the nature, scope, and purpose of AI usage in their studies. This includes identifying the specific tools used (e.g., ChatGPT, NVivo AI assistant, Otter.ai), the stages of the research process where they were applied (e.g., transcription, coding, summarisation), and the extent of human oversight involved. Such disclosures should be incorporated into methodology sections, ethics applications, and appendices where necessary. This aligns with emerging institutional policies in Ghana, particularly those emphasising responsible AI use and academic honesty.

Also, the framework requires the development of a comprehensive audit trail to document AI-assisted decision-making processes. An audit trail enhances dependability and confirmability in qualitative research by providing a transparent record of how data were generated, processed, and interpreted. In the context of AI integration, this includes documenting prompts used, outputs generated, researcher modifications, and justification for accepting or rejecting AI-assisted insights. Maintaining such records ensures that findings can be traced back to both human and AI contributions, thereby strengthening methodological credibility.

Moreover, verification procedures are necessary to mitigate the risks of bias, misinterpretation, and over-reliance on AI outputs. The framework recommends a hybrid validation approach involving multiple strategies such as manual cross-checking of AI-generated codes, peer debriefing, participant validation (member checking), and triangulation with non-AI methods. These verification steps reinforce interpretive control and ensure that AI serves as a supportive tool rather than a substitute for human analytical judgment.

Furthermore, data governance protocols must be clearly defined to address ethical concerns related to privacy, confidentiality, and data security. Given that many AI tools rely on cloud-based processing, researchers must ensure compliance with data protection standards by anonymising datasets, securing informed consent for AI-assisted processing, and, where necessary, opting for offline or locally hosted tools. These practices are consistent with Ghana's broader digital governance agenda and international ethical standards for responsible AI use.

Finally, these operational tools should be institutionalised through training, ethical review processes, and interdisciplinary collaboration. Universities and research institutions in Ghana are encouraged to integrate these protocols into research guidelines, ethics approval procedures, and postgraduate training programmes. By doing so, the framework moves beyond theoretical abstraction to become a practical model for responsible, context-sensitive AI integration in qualitative research.

Table 2. Operational Tools for Responsible AI Integration in Qualitative Research

Framework Component	Operational Tool	Description	Practical Application	Expected Outcome
Human-Centred Reflexivity	Reflexive Audit Trail	A structured record of researcher decisions, assumptions, and interactions with AI tools	Maintain a reflexive journal documenting prompts, AI outputs, and researcher modifications	Enhances transparency and preserves researcher interpretive authority
Ethical Governance & Transparency	AI Disclosure Statement	Formal declaration of AI tools, purposes, and extent of use in the research process	Include AI usage in methodology, ethics applications, and publications	Promotes academic integrity and accountability
	Ethics Approval Protocol for AI	Integration of AI-specific considerations into ethical review processes	Seek IRB/ethics approval specifying AI use, data handling, and risks	Ensures compliance with institutional and national ethical standards
Contextual & Cultural Sensitivity	Cultural Validation Checklist	Tool for assessing whether AI outputs reflect local meanings, language, and context	Cross-check AI-generated themes with local experts or participants	Reduces cultural bias and enhances contextual relevance
AI Literacy & Capacity Building	Prompt Design Guide	Structured guidelines for developing effective and context-sensitive AI prompts	Train researchers to design prompts using Ghanaian examples and multilingual considerations	Improves quality and relevance of AI outputs
	AI Competency Log	Record of researcher training and skill development in AI tools	Document workshops, courses, and practical experiences with AI tools	Strengthens methodological competence and critical AI use
Hybrid Validation & Interpretive Control	Verification Protocol	Multi-step process for validating AI outputs	Compare AI coding with manual coding, conduct peer debriefing, and member checking	Enhances credibility and reduces AI-induced bias
	Triangulation Matrix	Framework for comparing AI-generated findings with other data sources or methods	Use qualitative and quantitative data or multiple coders to validate findings	Ensures robustness and reliability of interpretations

continued on following page

Table 2. Continued

Framework Component	Operational Tool	Description	Practical Application	Expected Outcome
Data Governance	Data Protection Plan	Strategy for managing privacy, confidentiality, and security of research data	Anonymise transcripts, use secure storage, and limit cloud exposure for sensitive data	Protects participant rights and ensures ethical compliance
	AI Data Usage Log	Documentation of how data are processed by AI systems	Record datasets uploaded, tools used, and processing conditions	Supports accountability and traceability of data handling

8. IMPLICATIONS FOR POLICY, PRACTICE, AND CAPACITY BUILDING

The appropriate integration of AI into qualitative research has considerable implications for policy, institutional practices, and capacity development within Ghanaian higher education. The implications extend individual research activities and apply to the manner in which institutions equip postgraduate researchers, govern ethical standards, and allocate resources for research infrastructure. It is essential to address these concerns to ensure that AI adoption enhances, rather than undermines, qualitative research traditions and effectively implements the layered framework, especially the aspects of human-centered reflexivity, ethical governance, AI literacy, and hybrid validation.

8.1. Implications for Postgraduate Training

Postgraduate education in Ghana has always focused on study design, data collecting, and interpretive analysis. The increasing use of AI tools requires a review of methods for research courses. Postgraduate researchers must possess the ability to comprehend the functioning of AI systems, recognise their limitations, and discern their relationship with qualitative principles such as reflexivity and credibility, especially when using tools like ChatGPT for preliminary coding, Otter.ai for transcription, or NVivo for AI-assisted thematic analysis.

AI literacy should not be only characterised as technical training. It ought to be incorporated into qualitative methods courses as a methodological and ethical competency that aligns with the human-centered reflexivity and AI literacy components of the framework (refer to Figure 1). Students must understand how AI technologies impact analytical conclusions, influence coding techniques, and alter researcher's position. This corresponds with constructivist traditions that perceive knowledge generation as an active and reflective activity, as well as with international guidelines like UNESCO's (2021, 2023) focus on reflexive and transparent AI application in research and education.

Supervisory techniques require reform. Supervisors need guidelines on assessing AI-assisted qualitative work without undermining innovation or promoting uncritical automation. Explicit standards about disclosure, reflexive reporting, and validation processes are important, including defined criteria for evaluating AI-assisted analysis in theses and dissertations, as indicated in the developing academic integrity guidelines at the University of Ghana and KNUST. In the absence of such instructions, inconsistencies in evaluation may arise, compromising fairness and academic integrity, and undermining the hybrid validation and interpretive control mechanisms outlined in the framework (see to Figure 1).

8.2. Institutional AI Literacy Programmes

Universities need to stop using AI tools on an ad hoc basis and start doing it on a more permanent basis. To encourage responsible research, there is the need for structured AI literacy programs. These programs should be aimed at graduate students, supervisors, and research administrators. They should be part of graduate schools, research offices, and centers for teaching and learning, not stand-alone workshops.

Ethical awareness, methodological fit, and responsible use should be the main goals of institutional AI literacy programs. They should talk about data privacy, algorithmic bias, and how to be open about your study. They should also give practical advice on how to deal with potential risks of third-party data processing that come with cloud-based technologies like ChatGPT and Otter.ai. These programs need to be aware of the local research settings, which is quite important. Ghanaian colleges face limited resources that affect their ability to obtain licensed software and safely store data. This means they need to think about open-source options, shared institutional licenses, and data solutions that are hosted locally.

Integrating AI literacy into current research support frameworks, including graduate schools and research offices, might enhance institutional coherence. This method does not treat AI as an external skill. Instead, AI becomes part of a bigger framework for research skills that is in line with qualitative rigour and directly supports the AI literacy and capacity building layer of the framework (see Figure 1).

8.3. Research Ethics Committees and Policy Reform

Research ethics committees are important for making sure AI-assisted qualitative research is done safely. In Ghana, the main focus of current ethics guidelines is on human participation, informed permission, and data protection. These are still important, but AI adds more layers of ethical complexity that need to be thought about directly. These include algorithmic opacity, bias in training data, and participants being indirectly exposed to automated interpretation.

These committees need to learn how to assess risks associated to AI. These are things like processing data from third parties, automatic interpretation, and the fact that algorithmic systems are hard to understand. Ethics applications should require researchers to specify which AI technologies are used, for what goals, and how interpretive control is maintained in line with ethical governance and transparency standards established in the framework (see Figure 1).

Policy reform is needed to make all institutions get the same advice. Researchers and ethical committees will be able to help postgraduate researchers and supervisors by making it clear what AI use is permissible, what needs to be disclosed, and how to hold people accountable. These kinds of changes should follow international ethical guidelines, like the UNESCO (2021, 2023) and OECD (2023) recommendations on trustworthy and open AI. They should also take into account the realities of research in Ghana, as shown by the country's new national AI strategy and institutional academic integrity policies.

8.4. Developing African-Centred AI Research Infrastructures

The lasting viability of AI integration in qualitative research relies on the establishment of African-centered research infrastructures. The majority of AI tools used at Ghanaian universities are created outside Africa and trained on datasets that do not originate from the continent. This constrains their cultural sensitivity and epistemic significance, especially in qualitative research pertaining to indigenous languages, local pedagogies, and culturally rooted meaning systems.

Investing in local research infrastructure can rectify this disparity. This comprises the support of open datasets that represent African languages and contexts, with the promotion of cooperation among institutions, policymakers, and technology developers, including cross-institutional research consortium and partnerships with local AI developers. African-centered infrastructures might improve methodological integrity by aligning AI tools with local knowledge systems, hence reinforcing the framework's contextual and cultural sensitivity component (see Figure 1).

This development also affects research sovereignty. By influencing the design and implementation of AI tools, Ghanaian institutions might diminish reliance on external systems and enhance oversight of research processes, although recognised limitations concerning funding, technological proficiency, and regional coordination. This creates a more inclusive global research environment and establishes Ghanaian higher education as a proactive participant in global discussions on ethical AI in qualitative research, rather than a passive recipient of external technologies.

9. CONCLUSION: TOWARDS HUMAN-AI COEXISTENCE IN QUALITATIVE INQUIRY

The integration of AI into qualitative research signifies a substantial methodological transformation. Although AI provides efficiency and analytical assistance, it does not modify the fundamental principles of qualitative research. Human interpretation is fundamental to meaning-making, reflexivity, and ethical judgement. This chapter enhances modern qualitative methodology scholarship by illustrating how these principles can be maintained amidst growing technological mediation.

This chapter contends that AI ought to be perceived as a methodological collaborator instead than a replacement for human reasoning, hence promoting a non-deterministic framework for AI integration rooted in interpretivist epistemology. Artificial intelligence systems can facilitate data organization, transcribing, and pattern recognition. However, they are deficient in contextual awareness, moral judgement, and interpretive richness. These attributes are the obligation of the researcher and constitute the foundation for sustaining epistemic authority in AI-assisted qualitative research.

In the context of Ghana, the implications of AI adoption are notably significant. Structural disparities in digital access, insufficient institutional support, and dependence on externally produced technology generate both opportunities and risks, a situation common to several research contexts in the Global South that have similar infrastructural and legislative limitations. In the absence of thorough control, AI may worsen epistemic marginalisation. Responsible integration can improve research quality and inclusivity, especially when it aligns with national higher education agendas and institutional ethical frameworks.

This chapter's central conceptual contribution to the discussion on responsible AI in qualitative methodology is the framework that it proposes, which places human-centred reflexivity at the heart of AI-assisted qualitative research. The protection of methodological integrity is achieved through the translation of abstract ethical principles into operational research practices through the integration of hybrid validation, cultural sensitivity, ethical governance, and researcher capacity development. A framework of coexistence in which human insight guides technological support rather than being subordinated to algorithmic outputs is supported by these elements.

Future qualitative research in Ghana and the Global South will be based upon the deliberate decisions of universities, research ethics committees, and national policy actors. Formal AI disclosure requirements in theses, research ethics committee review criteria for AI-assisted analysis, and African-centred data resources are among the infrastructure, ethics reform, and training investments that universities must make to meet local research needs. Researchers are required to maintain a critical level of engagement with the tools they use. Future research should empirically investigate the ways in which these commitments are implemented in postgraduate supervision, ethical review processes, and qualitative data analysis practices, rather than limiting them. AI should expand the potential for inquiry, not narrow it.

In conclusion, the coexistence of humans and AI in qualitative research is not a technical challenge. It is an ethical and methodological commitment that has implications for the production, validation, and representation of knowledge in digitally mediated research environments. Qualitative researchers can leverage AI responsibly while preserving the values that define their craft and ensuring that technological innovation enhances rather than erodes qualitative inquiry in Ghana and the wider Global South by centring human interpretation and contextual knowledge.

DECLARATION OF GENERATIVE AI

"ChatGPT and DeepL were utilised during the preparation of this manuscript for proofreading and language refinement. The authors subsequently reviewed and revised the content as necessary and accept full responsibility for the final version of the publication".

REFERENCES

Abass, K. (2025). *Humanities must guide AI with ethics and identity*. Kwame Nkrumah University of Science and Technology. https://www.knust.edu.gh/index.php/news/news-items/humanities-must-guide-ai-ethics-and-identity-prof-kabila-abass

Achiam, J., Adler, S., Agarwal, S., Ahmad, L., Akkaya, I., Aleman, F. L., . . . McGrew, B. (2023). Gpt-4 technical report. *arXiv preprint arXiv:2303.08774.*

Adamu, M., & Nkwo, M. (2023). AI in Africa. Preliminary Notes on Design and Adoption. *diid disegno industriale, 1*(1), 1-15.

Adjei, S., & Agyeman, I. K. (2025). The integration of artificial intelligence tools in academic libraries within Ghana. *Ghana Library Journal*, *29*(2), 1–65. DOI: 10.4314/glj.v29i2.3

Adjei, S., Tawiah Adu Gyamfi, P. K., Owusu Agyei, F., & Norteye, S. B. (2025). Ethical implications of ChatGPT in Ghana: Navigating human-AI collaboration. *Journal of Scientific Research and Reports*, *31*(6), 443–449. DOI: 10.9734/jsrr/2025/v31i63142

African Independent Ethics Committee. (n.d.). *African Independent Ethics Committee (AIEC)*. Africa Social Work & Development Network.

African Union. (2024). *Continental Artificial Intelligence Strategy: Harnessing AI for Africa's Development and Prosperity.*

Akinwalere, S., & Ivanov, V. (2022). *Artificial intelligence in higher education: Challenges and opportunities*. Border Crossing.

Alphanome.AI. (2025). The anthropocentric mirror. *Future of AI.*

Ammah, L., Lütge, C., Kriebitz, A., & Ramkissoon, L. (2024). AI4people—An ethical framework for a good AI society: The Ghana (Ga) perspective. *Journal of Information, Communication and Ethics in Society.*

Amuzu, D. (2019). The elite, elitism, and ensuing conversations in Ghana's higher education: Myth or reality? *Journal of Black Studies*, *50*(8), 787–808. DOI: 10.1177/0021934719885630

Anani, G., Nyamekye, E., & Baffour-Koduah, D. (2025). Using artificial intelligence for academic writing in higher education: The perspectives of university students in Ghana. *Discover Education*, *4*(1), 46. Advance online publication. DOI: 10.1007/s44217-025-00434-5

Anderson, J., Chen, X., & Patel, R. (2021). Artificial intelligence and the future of literature reviews: Accelerating research through NLP and AI. *Journal of Academic Research Methods*, *45*(2), 145–167.

Anis, S., & French, J. A. (2023). Efficient, explicatory, and equitable: Why qualitative researchers should embrace AI, but cautiously. *Business & Society*, *62*(6), 1139–1144. Advance online publication. DOI: 10.1177/00076503231163286

Arthur, R., & Fenyi, D. A. (2023). Towards Minimizing the Delay/Non-completion Rate of Research Master's Degree in Ghana: ESL Students' Perspective of the Challenges of Thesis Work. *Linguistics Initiative, 3*(2), 125-139.

Ayana, G., Dese, K., Daba Nemomssa, H., Habtamu, B., Mellado, B., Badu, K., Yamba, E., Faye, S. L., Ondua, M., Nsagha, D., Nkweteyim, D., & Kong, J. D. (2024). Decolonizing global AI governance: Assessment of the state of decolonized AI governance in Sub-Saharan Africa. *Royal Society Open Science*, *11*(8), 231994. DOI: 10.1098/rsos.231994 PMID: 39113766

Ayyaswamy, K., Gobinath, V. M., Sathya, V., Kathirvel, N., & Anthony, R. A. (2025). The role of robotics in smart manufacturing: increasing efficiency and reducing costs. The Quantum AI Era of Neuromarketing, 315-356.

Ayyaswamy, K., Gobinath, V. M., Senthil, M. S., & Kathirvel, N. (2025). Automating Library Management for Efficiency: IoT-Enabled Smart Libraries. In Project Management Information Systems: Empowering Decision Making and Execution (pp. 325-350). IGI Global Scientific Publishing.

Baidoo-Anu, D., Asamoah, D., Amoako, I., & Mahama, I. (2024). Exploring student perspectives on generative artificial intelligence in higher education learning. *Discover Education*, *3*(1), 98. DOI: 10.1007/s44217-024-00173-z

Bakharia, A., Shibani, A., Lim, L. A., McCluskey, T., & Shum, S. B. (2025). From transcripts to themes: A trustworthy workflow for qualitative analysis using large language models. *Joint Proceedings of LAK 2025 Workshops*.

Ball, P. (2023). Is AI leading to a reproducibility crisis in science? *Nature*, *624*(7990), 22–25. DOI: 10.1038/d41586-023-03817-6

Burleigh, C. L., & Wilson, A. M. (2024). Generative AI: Is authentic qualitative research data collection possible? *Journal of Educational Technology Systems*, *53*(2), 89–115. DOI: 10.1177/00472395241270278

Butson, R., & Spronken-Smith, R. (2024). AI and its implications for research in higher education: A critical dialogue. *Higher Education Research & Development*, *43*(3), 563–577. Advance online publication. DOI: 10.1080/07294360.2023.2280200

Chang, C.. (2024). Enhancing AI-assisted group decision making through LLM-powered devil's advocate. *Proceedings of the 29th International Conference on Intelligent User Interfaces (IUI '24)*, 103–119.

Chatzichristos, G. (2025). Qualitative research in the era of AI: A return to positivism or a new paradigm? *International Journal of Qualitative Methods*, *24*, 16094069251337583. Advance online publication. DOI: 10.1177/16094069251337583

Chaudhry, I., Thurtle, V., Foday, E., Leather, A. J., Samai, M., Wurie, H., & Parmar, D. (2022). Strengthening ethics committees for health-related research in sub-Saharan Africa: A scoping review. *BMJ Open*, *12*(11), e062847. DOI: 10.1136/bmjopen-2022-062847 PMID: 36410802

Christou, P. (2023). A critical perspective over whether and how to acknowledge the use of artificial intelligence (AI) in qualitative studies. *The Qualitative Report*, *28*(11), 3089–3107. DOI: 10.46743/2160-3715/2023.6407

Christou, P. (2023a). How to use artificial intelligence (AI) as a resource, methodological and analysis tool in qualitative research? *The Qualitative Report*, *28*(7), 1968–1980. DOI: 10.46743/2160-3715/2023.6406

Cornelio, C., Ito, T., Cory-Wright, R., Dash, S., & Horesh, L. (2025). The need for verification in AI-driven scientific discovery. *arXiv*.

Costa, A. P., & Bem-Haja, P. (2025). Reflexive uncertainty AI for qualitative data analysis. *2nd Workshop on Education for Artificial Intelligence (edu4AI 2025)*.

Dahal, N. (2024). How can generative AI (GenAI) enhance or hinder qualitative studies? A critical appraisal from South Asia, Nepal. *The Qualitative Report*, *29*(10), 3057–3072. DOI: 10.46743/2160-3715/2024.6637

Dahal, N., Hasan, M. K., Ounissi, A., Haque, M. N., & Kapar, H. (2025). Navigating the meta-crisis of generativity: Adapting qualitative research quality criteria in the era of generative AI. *Frontiers in Research Metrics and Analytics, 10*, 1685968. DOI: 10.3389/frma.2025.1685968 PMID: 41262539

Darabi, T. (2021). Building Community-led Democracy: Navigating the Pitfalls of Oligarchy in the Construction of a Participatory Organization.

Davison, R. M., Chughtai, H., Nielsen, P., Marabelli, M., Iannacci, F., van Offenbeek, M., Tarafdar, M., Trenz, M., Techatassanasoontorn, A., Díaz-Andrade, A., & Panteli, N. (2024). The ethics of using generative AI for qualitative data analysis. *Information Systems Journal, 34*(5), 1433–1439. Advance online publication. DOI: 10.1111/isj.12504

Dellafiore, F., Saba, A., Collaro, C., & Artioli, G. (2025). Artificial intelligence in qualitative research: Insights from experts via reflexive thematic analysis. *Qualitative Health Research, 10497323251389800*. Advance online publication. DOI: 10.1177/10497323251389800 PMID: 41351219

Dlamini, R., & Ndzinisa, N. (2025). Towards a critical discourse on artificial intelligence and its misalignment in sub-Saharan Africa: Through an equality, equity, and decoloniality lens. *Journal of Education (University of KwaZulu-Natal)*, (98), 42-61.

Dutta, S., & Shah, S. (2025). *AI bias, liability and corporate accountability: A governance perspective*. ICSI.

Elliason, E. K., & Khajuria, A. (2025). Artificial Intelligence (AI) in Academic Research in Ghana: Opportunities, Challenges, and Policy Implications. *Journal of International Social Science Research and Review, 1*(1), 05.

Endsley, M. R. (1996). Automation and situation awareness. In Parasuraman, R., & Mouloua, M. (Eds.), *Automation and human performance: Theory and applications* (pp. 163–181). Lawrence Erlbaum.

Engin, Z. (2025). Human-AI governance (HAIG): A trust-utility approach. *arXiv*.

Engler, A. (2023). *Early thoughts on regulating generative AI like ChatGPT*.

Foley, K., McLean, C., De Zylva, R., Asa, G., Maio, J., Batchelor, S., Dzando, G., & Dimassi, A. (2025). Developing a critical imagination for how researchers can use artificially intelligent tools reflexively and responsibly during qualitative literature reviews. *International Journal of Qualitative Methods, 24*, 16094069251316249. Advance online publication. DOI: 10.1177/16094069251316249

Geertshuis, S., & Liu, Q. (2025). Part-time masters, full-time life: The emotional well-being of employed postgraduate students. *International Journal of Lifelong Education*, 1–18.

Hassan, Y. (2023). Governing algorithms from the South: A case study of AI development in Africa. *AI & Society*, *38*(4), 1429–1442. DOI: 10.1007/s00146-022-01527-7

Hogg, H. D. J., Sendak, M., Denniston, A., Keane, P., & Maniatopoulos, G. (2023). Unlocking the potential of qualitative research for the implementation of artificial intelligence-enabled healthcare. *Journal of Medical Artificial Intelligence*, *6*, 15. DOI: 10.21037/jmai-23-28

Hoosen, N., Agherdien, N., Rauch Van Der Merwe, T., Schwaibold, U., Ekol, G., Luo, M., . . . Zwane, S. (2024). Decolonising AI: A Pan-African collaboration. *UL Conferences*, 504-518.

Jafari, F., & Keykha, A. (2023). Identifying the opportunities and challenges of artificial intelligence in higher education: A qualitative study. *Journal of Applied Research in Higher Education*, *16*(5), 1648–1666. DOI: 10.1108/JARHE-09-2023-0426

Johnson, A. (2013). Exploring the use of mobile technology in qualitative inquiry in Africa.

Junaedi, A. T., Panjaitan, H. P., Yovita, I., Veronica, K., Renaldo, N., & Jahrizal, J. (2024). Advancing digital and technology literacy through qualitative studies to bridging the skills gap in the digital age. *Journal of Applied Business and Technology*, *5*(2), 123–133. DOI: 10.35145/jabt.v5i2.170

Kakko, A. (2025). AI agents and the ironies of automation: When smarter isn't always better. *Alphanome.AI*.

Karkoulian, S., Sayegh, N., & Sayegh, N. (2024). ChatGPT unveiled: Understanding perceptions of academic integrity in higher education—A qualitative approach. *Journal of Academic Ethics*. Advance online publication. DOI: 10.1007/s10805-024-09543-6

Kasperiūnienė, J. (2021). The use of artificial intelligence in social research: Multidisciplinary challenges. In *Advances in Intelligent Systems and Computing* (Vol. 1323, pp. 191–202). Springer., DOI: 10.1007/978-3-030-70187-1_22

Keen, S., Lomeli-Rodriguez, M., & Joffe, H. (2022). From challenge to opportunity: Virtual qualitative research during COVID-19 and beyond. *International Journal of Qualitative Methods*, *21*, 16094069221105075. DOI: 10.1177/16094069221105075 PMID: 35692956

Ketere, C. (2025). Decolonizing Artificial Intelligence: Reducing Algorithmic Biases in African Educational and Linguistic Contexts. *Journal of Contemporary Issues in Open Distance and E-Learning*, *2*(1), 18–37. DOI: 10.71060/7qk9nd52

Khanna, S., & Li, X. (2025). *Invisible Languages of the LLM Universe*. (v2 ed.) ArXiv.

Khatri, B., & Karki, P. D. (2023). Artificial intelligence (AI) in higher education: Growing academic integrity and ethical concerns. *Nepalese Journal of Development and Rural Studies*.

Korkmaz, S. (2025). Statistical rigor and reproducibility in the AI era. *Balkan Medical Journal*, *42*(5), 386–387. DOI: 10.4274/balkanmedj.galenos.2025.2025.040825 PMID: 40788231

Krastev, S. (2025). *How to preserve agency in an AI-driven future*. The Decision Lab.

Kuddus, K. (2022). Artificial intelligence in language learning: Practices and prospects. *Advanced analytics and deep learning models*, 1-17.

Kumalo, S. H. (2017). Problematising development in sustainability: Epistemic justice through an African ethic. *Southern African Journal of Environmental Education*, *33*(1), 14–24. DOI: 10.4314/sajee.v33i1.2

Kumaran, M., Subangi, M. C., & Muthuraj, A. (2025). *Revolutionizing research with AI: Tools and trends for the modern scholar*. Atlantis Press.

Kutsyuruba, B. (2021). The role of trust in doctoral student–supervisor relationships in Canadian Universities: The students' lived experiences and perspectives. *Journal of Higher Education Theory and Practice*, *21*(2), 125.

Kwame Nkrumah University of Science and Technology (KNUST). (2025). Humanities must guide AI with ethics and identity. (News release). https://www.graphic.com.gh/news/education/all-knust-students-to-undertake-compulsory-artificial-intelligence-course.html

Legowo, B., Nurhasanah, F., Saddhono, K., Budiastuti, V. I., Sutomo, A. D., & Murwaningsih, T. (2024). Enhancing educational outcomes through exploring the optimized options artificial intelligence and deep learning. In *2024 4th International Conference on Advance Computing and Innovative Technologies in Engineering (ICACITE)*. DOI: 10.1109/ICACITE60783.2024.10616772

Li, L., & Hansman, R. J. (2016). A novel approach to apply data mining in flight data analysis. IEEE Transactions on Systems, Man, and Cybernetics, 27(3), 360-371.

Libengood, J. (2025). *Algorithms of eversion: Contemporary states and digital sovereigns* (Doctoral dissertation, University of South Florida).

Liekum, L. B. (2025). Harnessing artificial intelligence in teaching Ghanaian colleges of education in the 21st century: Enhancing quality teaching, student research, and learning abilities. *American Journal of Multidisciplinary Research and Innovation*.

Liu, J., Liu, T., Zhu, Y., & Yan, J. (2025). Exploring the collaboration between humans and AI-grounded epistemic objects to facilitate knowledge integration of "who knows what" under corporate crises. *Journal of Knowledge Management*, 1–40. DOI: 10.1108/JKM-01-2025-0058

Longin, L., Bahrami, B., & Deroy, O. (2023). Intelligence brings responsibility-Even smart AI assistants are held responsible. *iScience*, *26*(8), 107494. DOI: 10.1016/j.isci.2023.107494 PMID: 37609629

Luca Longo, L., Goebel, R., Lecue, F., Kieseberg, P., & Holzinger, A. (2019). Explainable artificial intelligence: Concepts, applications, research challenges and visions. In *Advances in Intelligent Systems and Computing*. Springer.

Mahajan, P. (2025). What is ethical: AIHED driving humans or human-driven AIHED? A conceptual framework enabling the ethos of AI-driven higher education. *arXiv*.

Maina, A. M., & Kuria, J. (2024). Building an AI future: Research and policy directions for Africa's higher education. In *2024 IST-Africa Conference (IST-Africa)* (pp. 1–8). IEEE.

Maluleke, A. F. (2025). AI adoption in African higher education: A systematic review of benefits and ethical implications. *Interdisciplinary Journal of Education Research*, *7*(2), a05. DOI: 10.38140/ijer-2025.vol7.2.05

Modiba, F. S., Van den Berg, A., & Mago, S. (2025). Opportunities and challenges of generative artificial intelligence supporting research in African classrooms. *South African Journal of Higher Education*, *38*(3). Advance online publication. DOI: 10.20853/39-3-6272

Muhammad Imran, M., Almusharraf, N., & Zaman, N. (2024). Artificial intelligence in higher education: Opportunities, challenges, and ethical considerations. In *Advances in Educational Technologies and Instructional Design*. IGI Global.

Mututa, A., & Tomaselli, K. (2025). Research cultures in the modern university. *Education as Change*, *29*. Advance online publication. DOI: 10.25159/1947-9417/18958

Naredla, S., Singh, M., Chaturvedi, N., Banerjee, D., Gobinath, V. M., & Tiwari, A. (2025). AI and IoT in Digital Marketing: Enhancing Automation, Personalization, and Consumer Interaction. In *2025 International Conference on Pervasive Computational Technologies (ICPCT)* (pp. 895-899). IEEE. DOI: 10.1109/ICPCT64145.2025.10941275

Nguyen, & Welch. (2025). A critical examination of LLMs usage in the context of qualitative research. *Generative AI for Research Initiative.*

Nikolopoulou, K. (2024). Generative artificial intelligence in higher education: Exploring ways of harnessing pedagogical practices with the assistance of ChatGPT. *International Journal of Changes in Education*, *6*(3), 20–30. DOI: 10.47852/bonviewIJCE42022489

Nouis, S. C. E., Uren, V., & Jariwala, S. (2025). *Evaluating accountability, transparency, and bias in AI-assisted healthcare decision-making: a qualitative study of healthcare professionals' perspectives in the UK*. PMC.

Nyaaba, M., et al. (2025). Human–AI collaborative inductive thematic analysis: How AI guides analysis and researchers reclaim interpretive authority. *arXiv*.

Ofosu-Ampong, K. (2023). Gender differences in perception of artificial intelligence-based tools. *Journal of Digital Art & Humanities*, *4*(1), 5–23. DOI: 10.33847/2712-8149.4.2_6

Omotubora, A., & Basu, S. (2024). Decoding and reimagining AI governance beyond colonial shadows. In *Handbook on Public Policy and Artificial Intelligence* (pp. 220–234). Edward Elgar Publishing. DOI: 10.4337/9781803922171.00025

Opesemowo, O., & Adekomaya, V. (2024). Harnessing artificial intelligence for advancing sustainable development goals in South Africa's higher education system: A qualitative study. *International Journal of Learning. Teaching and Educational Research*, *23*(11), 252–277. DOI: 10.26803/ijlter.23.3.4

Organisation for Economic Co-operation and Development. (2019). OECD principles on artificial intelligence. OECD Publishing. https://oecd.ai/en/ai-principles

Organisation for Economic Co-operation and Development. (2023). *Framework for the classification of AI systems*. OECD Digital Economy Policy Papers., DOI: 10.1787/9789264065949-es

Osondu, J., Jean Francois, E., & Strycker, J. (2024). Artificial intelligence as a policy response to teaching and learning issues in education in Ghana. *Journal of Global Education and Research*, *8*(3), 236–253. DOI: 10.5038/2577-509X.8.3.1361

Popenici, S., Catalano, H., Mestic, G., & Ani-Rus, A. (2023). A systematic review of the artificial intelligence implications in shaping the future of higher education. *Educatia 21.*

Ren, Y. (2025). *A critical examination of LLMs usage in the context of qualitative research*. Generative AI for Research Initiative.

Republic of Ghana. 2022. "Republic of Ghana National Artificial Intelligence Strategy: 2023–2033. Ministry of Communications and Digitalisation." Ministry of Communication and Digitalisation. https://www.africadataprotection.org/Ghana-AI-Strat.pdf

Rivera, J. L. (2024). An artificial intelligence (AI) research-doing approach for higher education. *Global Journal of Human-Social Science Research.*

Sangwa, S., Nsabiyumva, S., Ngobi, D., & Mutabazi, P. (2025). *Ethical AI integration in African higher education: Enhancing research supervision, grant discovery, and proposal writing without compromising academic integrity.* Social Science Research Network.

Sarfo, J. O., Tachie-Donkor, G., Aggrey, E. K., & Mordi, P. (2024). Attitudes, Perceptions, and Challenges Towards Artificial Intelligence Adoption in Ghana and Nigeria: A Systematic Review with a Narrative Synthesis Attitudes, Perceptions, and Challenges Towards Artificial Intelligence. *International Journal of Media and Information Literacy*, *9*(2), 437–452.

Sarı, T., Nayır, F., & Bozkurt, A. (2023). Reimagining education: Bridging artificial intelligence, transhumanism, and critical pedagogy. *Journal of Educational Technology and Online Learning.*

Sartori, L., & Theodorou, A. (2022). A sociotechnical perspective for the future of AI: Narratives, inequalities, and human control. *Ethics and Information Technology*, *24*(1), 4. DOI: 10.1007/s10676-022-09624-3

Schroeder, H., Aubin Le Quéré, M., Randazzo, C., Mimno, D., & Schoenebeck, S. (2024). Large language models in qualitative research: Can we do the data justice? *arXiv.*

Segbenya, M., Bervell, B., Frimpong-Manso, E., Otoo, I., Andzie, T. A., & Achina, S. (2023). Artificial intelligence in higher education: Modelling the antecedents of artificial intelligence usage and effects on 21st century employability skills among postgraduate students in Ghana. *Computers and Education: Artificial Intelligence*, *5*, 100188. Advance online publication. DOI: 10.1016/j.caeai.2023.100188

Shao, Y.. (2024). Assisting in writing Wikipedia-like articles from scratch with large language models. *Proceedings of the 2024 Conference of the North American Chapter of the Association for Computational Linguistics (NAACL 2024)*, 6252–6278. DOI: 10.18653/v1/2024.naacl-long.347

Shukla, P., et al. (2025). De-skilling, cognitive offloading, and misplaced responsibilities: Potential ironies of AI-assisted design. *arXiv.*

Singh, H. (2025). How generative AI is revolutionizing scientific research by automating hypothesis generation. *ResearchGate/*SSRN *Electronic Journal.*

Slimi, Z. (2023). The impact of artificial intelligence on higher education: An empirical study. *European Journal of Educational Sciences.*

Smith, P. J.. (2025). *Human factors guidance for the integration of artificial intelligence and machine learning in FAA systems. Technical Report.* The Ohio State University.

UNESCO. (2021). *AI and education: Guidance for policy-makers.* United Nations Educational, Scientific and Cultural Organization. https://www.unesco.org/en/articles/ai-and-education-guidance-policy-makers

UNESCO. (2023). *Guidance for generative AI in education and research.* United Nations Educational, Scientific and Cultural Organization. https://www.unesco.org/en/articles/guidance-generative-ai-education-and-research

University of Ghana. (UG) (2024). Policy on plagiarism and other academic misconduct [Amended]. Emelia A. Mensah (Registrar), University of Ghana (as reported by *MyJoyOnline*, 26 February 2024). https://www.myjoyonline.com/university-of-ghana-revises-plagiarism-policy-to-include-ai/

Venter, I. M., Cranfield, D., Blignaut, R., Achi, S., & Tick, A. (2024). Conversational AI in higher education: Opportunities, challenges, and ethical considerations. In *International Conference on Intelligent Engineering Systems.* IEEE. DOI: 10.1109/INES63318.2024.10629104

Wang, Y., et al. (2024e). Autosurvey: Large language models can automatically write surveys. *arXiv.*

Wilder, C., & Calderone, S. M. (2025). *Empowering educational leadership research with generative AI.* Impacting Education Journal on Transforming Professional Practice., DOI: 10.5195/ie.2025.489

Williamson, B., Macgilchrist, F., & Potter, J. (2023). Re-examining AI, automation and datafication in education. *Learning, Media and Technology*, *48*(1), 1–5. DOI: 10.1080/17439884.2023.2167830

Wiredu, J. K., Abuba, N. S., & Zakaria, H. (2024). Impact of generative AI in academic integrity and learning outcomes: A case study in the Upper East Region. *Asian Journal of Research in Computer Science*, *17*(12), 159–175. DOI: 10.9734/ajrcos/2024/v17i7491

Xiong, F., Yu, X., Leong, H. W., & Ma, A. (2025). AI-driven research ecosystem: Unifying human-AI collaboration models and new research thinking paradigms. *Journal of Educational Technology and Innovation*, *7*(1), 39–53. DOI: 10.61414/n0n76c97

Yilma, K. (2025). Ethics of AI in Africa: Interrogating the role of Ubuntu and AI governance initiatives. *Ethics and Information Technology*, *27*(2), 24. DOI: 10.1007/s10676-025-09834-5

Yin, E. T. (2025). Regulation and Ethical Issues of Artificial Intelligence in Ghana: Towards a Harmonious AI Law. In *Exploring AI Implications on Law, Governance, and Industry* (pp. 119-156). IGI Global Scientific Publishing. DOI: 10.4018/979-8-3373-3384-7.ch005

Zhou, Z., et al. (2025). From hypothesis to publication: A comprehensive survey of AI-driven research support systems. *arXiv*.

Chapter 6
Human-AI Hybrid Epistemologies for Qualitative Inquiry

Ghada Nakhla
Edge Hill University, UK

Marina A. Soltan
http://orcid.org/0000-0001-6015-9409
University of Birmingham, UK

ABSTRACT

This chapter investigates the convergence of AI and qualitative research methodologies, introducing the Human-AI Hybrid Epistemology Model to conceptualise collaborative meaning-making where human interpretation anchors computational augmentation. Grounded in constructivism and interpretivism, it analyses AI integrations like NLP, LLMs, automated transcription, AI-assisted coding for Grounded Theory and Thematic Analysis, sentiment analysis, virtual moderation, and computational ethnography. Ethical dimensions—bias, opacity, privacy, consent, and reconfigured trustworthiness (credibility, dependability)—demand reflexive co-production, AI positionality, validation logs, and governance frameworks ensuring fairness and transparency. Emphasising AI literacy and digital reflexivity as core competencies, it outlines pedagogical strategies for methods training and future hybrid paradigms, positioning researchers as methodological conductors who safeguard interpretive integrity, ethical rigour, and social justice in AI-enhanced inquiry.

DOI: 10.4018/979-8-3373-7640-0.ch006

t © 2026, IGI Global Scientific Publishing. Copying or distributing in print or electronic forms without written permission of IGI Global Scientific Publishing is prohibited. Use of this chapter to enerative artificial intelligence (AI) technologies is expressly prohibited. The publisher reserves all rights to license its use for generative AI training and machine learning model development.

INTRODUCTION

Qualitative research has historically been underpinned by interpretive and constructivist paradigms that foreground the situated, relational, and co-constructed nature of knowledge. These traditions view reality not as an objective entity waiting to be discovered, but as multiple and emergent shaped by context, language, and interaction (Lincoln & Guba, 1985; Schwandt, 1994). Within this worldview, researchers assume the role of interpreters and collaborators, engaging empathetically and reflexively with participants to co-create meaning through dialogue and interpretation (Denzin & Lincoln, 2011). However, in a knowledge ecosystem increasingly influenced by artificial intelligence (AI), this foundational view of qualitative inquiry faces new challenges and opportunities. AI-driven systems—ranging from large language models (LLMs) to automated transcription tools—are transforming the production, processing, and interpretation of qualitative data. These developments do not merely expand methodological capacities; they also provoke philosophical questions about what it means to interpret, to know, and to understand in a hybrid human–machine epistemic environment (Williamson et al., 2023).

The convergence of interpretive human-focused inquiry with computational precision represents a pivotal moment in social science research. This chapter traces how these two worlds intersect, arguing that the future of qualitative scholarship will depend on a careful negotiation between technological innovation and methodological integrity. To conceptualise this emerging relationship between human-centred qualitative inquiry and computational systems, this chapter introduces a Human–AI Hybrid Epistemology Model (Figure 1). The model provides a representation of meaning-making in AI-supported qualitative research, positioning human interpretation at its core, surrounded by classical qualitative traditions, computational augmentation, and an overarching ethical frame. Rather than depicting AI as a replacement for interpretive labour, the model emphasises a hybrid epistemic arrangement in which human reflexivity, qualitative methodology, and algorithmic capability are interdependent. This framework serves as a guiding heuristic for the chapter, orienting the subsequent discussion of methodological integration, reflexivity, and ethical responsibility.

Figure 1. Human-AI Hybrid Epistemology Model

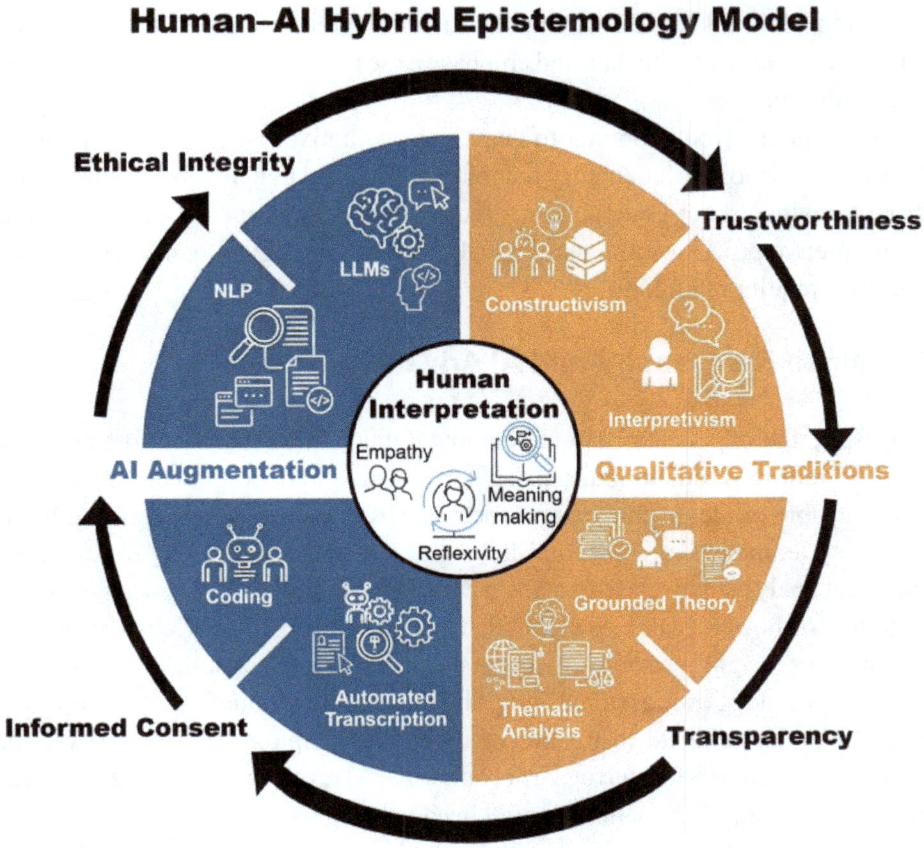

Constructivism and Interpretivism

The philosophical traditions of constructivism and interpretivism emerged in response to positivist assumptions that privileged objectivity and measurement. Constructivism posits that knowledge is socially produced through interaction and interpretation; no phenomenon carries inherent meaning outside of human cognition and context (Guba & Lincoln, 1994). Interpretivism complements this stance by positioning understanding as the primary goal of research rather than prediction or control (Schwandt, 1994). Together, they define qualitative inquiry as a process of sense-making that values depth, nuance, and multiplicity. Within these traditions, the researcher's subjectivity is not an obstacle to validity but a resource for understanding the social world. Analytical practices such as reflexivity and thick description (Geertz, 1973) are used to ensure that interpretations remain faithful to participants lived realities. Grounded Theory (Charmaz, 2014) and Thematic Analysis (Braun & Clarke, 2021) exemplify such interpretive approaches: both rely on iterative engagement with data and emphasise the researcher's creative involvement in meaning-making.

These classical roots continue to shape how qualitative research is conceptualised. They remind scholars that inquiry is a relational act and that understanding emerges through dialogue, not computation. Yet, as digital tools increasingly participate in interpretive processes, the ontological and epistemological boundaries of these traditions require rethinking.

Sustained Relevance Amid AI Advances

Despite the exponential rise of computational tools, constructivist and interpretivist traditions remain indispensable to research in the digital era. AI systems, while capable of identifying patterns across vast datasets, cannot replicate human reflexivity, empathy, or contextual judgment (Bender et al., 2021). For instance, sentiment analysis may classify affective tones in text but cannot grasp the cultural meanings or power dynamics embedded within language use. Constructivist and interpretivist insights provide critical perspectives for interrogating AI's limitations. They foreground questions of meaning, situatedness, and values—dimensions often obscured by the technical abstraction of algorithms (Lupton, 2020). Moreover, they encourage researchers to interrogate not only what AI reveals but also what it conceals. Human interpretation remains essential in addressing issues of bias, representation, and interpretive ethics within algorithmic outputs (Raji & Buolamwini, 2019). In this sense, the continuing relevance of classical qualitative traditions lies both in preserving interpretive depth and in guiding responsible adoption of AI in research. These traditions offer an ethical compass for navigating the power asymmetries and epistemic uncertainties that computational systems introduce to inquiry.

Human-Centred Computational Methods

The growing integration of AI within qualitative research signals not a rupture but a transformation—a shifting methodological terrain where computational tools coexist with interpretive epistemologies. Natural language processing (NLP), automated coding, and machine learning techniques now enable researchers to work with datasets of unprecedented scale and complexity (Floridi & Chiriatti, 2020). For example, LLMs can assist in organising and summarising qualitative data, supporting the identification of emergent themes across thousands of texts. These methods extend analytic capacity but also challenge established notions of validity and interpretation. When algorithmic models identify patterns or generate codes, questions arise about whose meaning is being constructed—the researcher's, the participants', or the machine's. Such tensions underscore the need for "critical digital reflexivity," a commitment to understanding how computational systems influence epistemic practices (Selwyn, 2019).

The convergence of human-centred inquiry and computational methods thus requires a dual orientation: embracing innovation while safeguarding the interpretive values that define qualitative work. AI can function as both collaborator and object of inquiry—simultaneously enabling new analytic possibilities and demanding ethical scrutiny. By positioning AI not as a replacement for human interpretation but as a partner in meaning-making, researchers can explore hybrid methodologies that enrich qualitative analysis without undermining its philosophical foundations (Williamson et al., 2023).

Toward an Ethical and Reflexive Future

As qualitative researchers navigate this terrain, several issues come to the fore. The flexibility and rigour of classical qualitative approaches should guide how AI tools are integrated into analytic processes. Frameworks such as Thematic Analysis (Braun & Clarke, 2021) and Constructivist Grounded Theory (Charmaz, 2014) provide adaptable structures within which AI-assisted coding and data visualisation can operate. Ethical transparency must accompany technological innovation. As such, qualitative researchers must critically assess the provenance, composition, and representational implications of computational data. This involves maintaining openness about how AI contributes to interpretation and ensuring that participants' voices are not obscured within algorithmic mediation.

Finally, AI literacy should be recognised as an essential competency for contemporary qualitative scholars. Understanding how computational systems process, classify, and generate language is key to maintaining interpretive credibility (Kasneci et al., 2023). Such literacy allows researchers to engage ethically and reflexively with AI, treating it as both a methodological tool and a cultural phenomenon that reshapes social inquiry. By balancing interpretive depth with computational innovation, qualitative research can remain both relevant and transformative in the age of AI. Ultimately, the goal is not to mechanise interpretation but to reimagine it—to craft a form of inquiry where human empathy and algorithmic capability coalesce into a richer, more nuanced understanding of the social world.

Chapter Aims and Structure

This chapter examines the evolving relationship between qualitative research methodologies and AI, positioning this relationship as both a methodological innovation and a philosophical turning point. It explores how classical qualitative traditions rooted in constructivism and interpretivism continue to underpin meaning-making in the digital age, even as AI introduces new modes of data generation, interpretation, and reflexivity. The central focus lies in understanding how human-centred inquiry can coexist, collaborate, and critically engage with computational systems in ways that uphold the interpretive depth, ethical rigour, and reflexive integrity foundational to qualitative research. Throughout the chapter, empirical illustrations of AI-supported qualitative research are explicitly linked back to the Human–AI Hybrid Epistemology Model and the proposed governance and reflexivity framework, clarifying how these conceptual elements are instantiated in practice.

The chapter analyses core sites of integration between AI and qualitative methods, including NLP, LLMs, automated transcription, AI-assisted coding, thematic analysis, and computational grounded theory, illustrating how these techniques enable work with datasets that exceed human-scale while raising questions about depth, authenticity, and researcher agency. It then examines emerging hybrid workflows, advanced analytic strategies such as sentiment analysis and AI-supported conversation analysis, and new research environments including virtual moderation, computational ethnography, and post-coding inquiry. The chapter also systematically addresses key technical limitations (e.g., multimodal data handling, model drift, output uncertainty, cross-lingual performance) alongside evaluation metrics such as researcher-AI code alignment, theme stability, and audit trail completeness to ensure methodological rigour.

A sustained emphasis is placed on ethical integrity and trustworthiness in AI-mediated qualitative research, addressing informed consent, data privacy in algorithmic ecosystems, algorithmic bias and opacity, and the reworking of credibility, dependability, and reasoning about utility under conditions of probabilistic, opaque systems. Human-centred AI and ethical design are presented as scaffolding for governance frameworks that link fairness, transparency, accountability, and human oversight to qualitative standards of legitimacy.

The chapter also develops the notion of AI positionality and researcher–AI co-production, arguing that AI systems function as situated epistemic agents whose contributions must be documented, scrutinised, and reflexively negotiated through transparent workflows, validation logs, and critical examination of bias and explainability. Reflexivity is reframed as a shared practice of accountability distributed across human and algorithmic actors, with particular attention to how documentation, audit trails, and explainable AI techniques sustain interpretive responsibility. In addition to its conceptual focus, the chapter draws selectively on empirical illustrations of AI-supported qualitative research in domains such as education, health, and digital sociology, highlighting how hybrid workflows are enacted in practice and with what methodological consequences.

Finally, the chapter advances AI literacy and critical digital reflexivity as core competencies for contemporary qualitative scholars, outlining pedagogical strategies for embedding technical, ethical, and epistemological understanding of AI into methods training. It sketches future trajectories for hybrid qualitative–computational paradigms, evolving governance regimes, and the changing role of the researcher as a methodological conductor, reaffirming human interpretation as the orienting value in co-produced, AI-supported qualitative inquiry.

AI IN QUALITATIVE METHODOLOGIES

AI tools are increasingly woven into qualitative methodologies in ways that extend and unsettle established practices such as Grounded Theory and Thematic Analysis (Braun & Clarke, 2021; Charmaz, 2014). Rather than replacing these traditions, AI tends to inhabit specific stages of the research cycle—transcription, coding, pattern detection, and summarisation—creating hybrid workflows where computational assistance coexists with human interpretation (Bender et al., 2021; Kasneci et al., 2023). The key methodological question is not whether AI can "do" qualitative analysis, but how its capabilities are aligned with, and constrained by, the epistemological commitments of interpretive inquiry (Lupton, 2020).

In Grounded Theory, AI tools are most integrated into early and intermediate analytic cycles, where large bodies of textual data must be coded, compared, and revisited iteratively (Charmaz, 2014). Automated transcription and NLP can expedite the move from audio to analysable text and generate preliminary code suggestions or clusters of related concepts, which researchers then interrogate against the constant-comparative logic of Grounded Theory (Guetterman et al., 2018). For example, studies in health research have used topic modelling and automated clustering to screen and compare hundreds of interview transcripts on patient experiences, with researchers subsequently re-entering the data to refine categories, pursue negative cases, and ground emergent concepts in participants' language. This kind of design illustrates how AI-enabled pattern detection can support, but not replace, memo writing, theoretical sampling, and human-led theory construction.

Within Thematic Analysis, AI is often used to support, rather than supplant, Braun and Clarke's phases of familiarisation, coding, theme development, and refinement (Braun & Clarke, 2021). LLMs and other NLP-based systems can propose initial codes, group similar segments, and generate candidate thematic summaries across extensive datasets, offering a form of high-volume pattern recognition that can sit alongside manual, inductive coding (Kasneci et al., 2023). Empirical work in education and learning analytics, for instance, has used LLM-based tools to generate provisional codes and cross-case thematic summaries from thousands of student reflections, with human analysts iteratively modifying or rejecting machine-suggested themes to ensure alignment with theoretical frameworks and contextual knowledge.

Taken together, these examples demonstrate how interpretive authority remains with the researcher, even as AI contributes to managing corpus scale and surfacing potential patterns. Researchers may employ AI outputs as heuristic devices—starting points that are then evaluated, adapted, or rejected in light of theoretical frameworks, research questions, and contextual knowledge (Bender et al., 2021; Williamson et al., 2023). Crucially, interpretive authority resides in the human analyst's judgment about what constitutes a meaningful theme, how it is defined, and how it is evidenced, ensuring that AI contributes to, but does not dictate, the analytic narrative (Braun & Clarke, 2021; Lupton, 2020).

Across both Grounded Theory and Thematic Analysis, therefore, effective integration of AI depends on clear methodological governance: explicit decisions about when and why AI is used, critical scrutiny of its outputs, and transparent reporting of human–machine collaboration in the production of qualitative knowledge (Kasneci et al., 2023; Williamson et al., 2023). The following sections outline: core computational techniques, AI-enhanced qualitative workflows, advanced data assessment methods, and emerging research environments, with a focus on their implications for qualitative methodology. In terms of the Human–AI Hybrid Epistemology Model, these designs enact a human interpretive core surrounded by computational augmentation within an explicit ethical frame.

Core Computational Techniques

NLP, LLMs, and related AI techniques are reshaping qualitative research by enabling analysis and new forms of interpretation at different stages of the research workflow (Bender et al., 2021; Floridi & Chiriatti, 2020). These developments influence not only how data are processed and organised but also how meaning, depth, and reflexivity are conceptualised within interpretive traditions (Kasneci et al., 2023; Williamson et al., 2023). This section examines three computational techniques influential in qualitative inquiry: NLP methods for textual analysis, LLMs as emergent interpretive collaborators, and automated transcription systems that transform data preparation practices. Together, these techniques illustrate how computational tools are increasingly embedded within qualitative workflows while raising important methodological and epistemological considerations. Taken together, these core computational techniques operationalise the Human–AI Hybrid

Epistemology Model by situating AI as an augmentative layer governed by ethical oversight, rather than as an autonomous locus of interpretive authority.

NLP for Text Analysis

NLP offers methods for preprocessing and analysing text, such as tokenisation, lemmatisation, topic modelling, and vector-based representations, which allow scholars to work with corpora that would be infeasible to handle using manual coding alone (Guetterman et al., 2018; Snelson, 2016). These techniques support pattern detection across documents, enabling researchers to explore discursive structures, narrative trajectories, and conceptual associations in datasets that include interview transcripts, survey responses, or social media posts (Guetterman et al., 2018; Snelson, 2016). In qualitative work, NLP is frequently used to support initial coding, identify recurring concepts, and conduct cross-case comparisons, while interpretive reading remains central for contextualising and validating computational outputs (Guetterman et al., 2018). Studies consistently underline that although NLP can surface dominant themes efficiently, it tends to miss local meanings, and contextual cues, so human-led interpretation is required to maintain trustworthiness (Guetterman et al., 2018).

In one line of work on social media-based qualitative inquiry, researchers have combined topic modelling with manual interpretive coding to examine evolving public narratives around health crises; the models surface recurring topics and discursive clusters, while qualitative reading is used to interpret local meanings, contextual nuance, and power dynamics within and across topics. Such studies exemplify how NLP extends corpus reach while preserving interpretive judgement. In terms of the Human–AI Hybrid Epistemology Model, this work demonstrates computational augmentation at the analytic periphery, with human interpretation retaining epistemic centrality, though infrastructural constraints such as limited API access in low-resource regions and cross-cultural challenges like idiom detection remain significant hurdles alongside successes in pattern detection. Technical limitations include poor handling of multimodal data (e.g., images in social media posts) and language-specific performance drops; evaluation typically involves inter-coder agreement between human and AI outputs alongside qualitative validation of thematic coherence.

LLMs and Interpretive Processes

LLMs, trained on very large text corpora, can perform tasks such as summarisation, classification, and conversational interaction, offering new possibilities for interpretive work in qualitative research (Bender et al., 2021; Floridi & Chiriatti, 2020). Recent empirical studies show that LLMs can assist with generating thematic summaries from interview data, drafting preliminary analytic memos, and supporting the organisation of large textual datasets, thereby accelerating early analytic phases (Kasneci et al., 2023; Williamson et al., 2023). At the same time, LLM behaviour is shaped by training data, model architecture, and prompt design, which means that outputs can reproduce societal biases, omit minority perspectives, or generate plausible but incorrect interpretations (Bender et al., 2021; Raji & Buolamwini, 2019). From an interpretivist standpoint, LLMs act less as neutral analytic devices and more as additional meaning-making agents whose contributions must be critically scrutinised and situated within reflexive accounts of the research process (Kasneci et al., 2023; Williamson et al., 2023).

Recent classroom-based studies have used LLM-generated summaries of open-ended student feedback as a starting point for thematic analysis, showing that while machine summaries approximate broad thematic contours, human analysts are needed to refine theme boundaries, attend to minority voices, and interrogate how institutional power relations shape the data. These designs illustrate both successes in scalability and persistent challenges such as data sovereignty in cross-national studies and unequal access to high-performance computing. Taken together, they demonstrate the promise and limits of treating LLMs as interpretive collaborators, corresponding to the model's depiction of AI as a peripheral, rather than central, epistemic agent. LLM limitations extend to multimodal integration (e.g., limited vision-language models for video transcripts) and hallucination risks; rigorous evaluation uses metrics like semantic similarity to gold-standard human codes and iterative prompt refinement tracked via audit logs. Methodological risks include model drift—where LLM performance degrades on out-of-distribution data—and output uncertainty, requiring researchers to track confidence scores and validate against gold-standard human interpretations through iterative audit trails.

Automated Transcription and Data Prep

Automated speech recognition systems have dramatically reduced the time and cost of generating transcripts from interviews, focus groups, and observational recordings, thereby expanding the scale and diversity of data that qualitative researchers can feasibly analyse (Lupton, 2020). Contemporary tools incorporate features such as speaker diarisation, punctuation restoration, and domain-adapted language models, which increase transcription accuracy and reduce manual cleaning and formatting (Lupton, 2020). Nonetheless, accuracy remains uneven across accents, languages, and recording conditions, and automated transcription can misrepresent or omit marginalised voices, requiring systematic checking and correction to maintain ethical and epistemic integrity (Raji & Buolamwini, 2019).

Ethnographic projects in clinical and organisational settings have reported substantial gains in efficiency from AI-powered transcription yet also document the need for careful human review where domain-specific jargon, overlapping talk, or accent variation in non-Western contexts produce transcription errors that could alter analytic interpretations. Here, the empirical record underscores that automated transcription reconfigures, but does not eliminate, the labour of close listening and checking. These cases highlight both efficiency gains and persistent challenges around linguistic diversity and computational infrastructure, reflecting the governance-oriented strand of the chapter's framework where infrastructural efficiencies are balanced against ethical oversight and accountability for representational accuracy. Beyond linguistic challenges, transcription tools struggle with multimodal inputs like non-verbal cues in video; evaluation metrics include Word Error Rate (WER) benchmarked against manual transcripts, plus qualitative checks for contextual fidelity.

AI-Enhanced Research Workflows

AI-enhanced research workflows extend the role of computational techniques beyond data preparation and analysis to active participation in qualitative sense-making (Guetterman et al., 2018; Kasneci et al., 2023). By integrating AI tools into core analytic practices—such as coding, thematic analysis, and theory development—researchers can manage larger datasets, accelerate iterative processes, and surface patterns that might otherwise remain obscured (Bender et al., 2021; Williamson et al., 2023). These workflows do not replace interpretive labour but reconfigure it, shifting researcher effort toward critical evaluation, reflexive judgment, and theoretical integration (Charmaz, 2014; Lupton, 2020).

The Researcher–AI Co-Production Matrix in Qualitative Analysis Workflow (Figure 2), visually maps how AI tools intersect with traditional stages of qualitative research, highlighting points at which human oversight, ethical scrutiny, and reflexive decision-making are required to maintain interpretive rigour (Braun & Clarke, 2021; Kasneci et al., 2023). The Matrix conceptualises qualitative workflows according to the degree of automation and the distribution of interpretive responsibility, distinguishing between manual analysis, AI-assisted coding, AI-guided exploration, and reflexive hybrid co-production (Kasneci et al., 2023; Williamson et al., 2023). By foregrounding interpretive responsibility rather than technical capability, the matrix identifies highly automated arrangements with limited human oversight as a methodological risk zone, and positions reflexive hybrid co-production as the most robust configuration for interpretive inquiry (Kasneci et al., 2023; Williamson et al., 2023).

Figure 2. Researcher-AI Coproduction Matrix

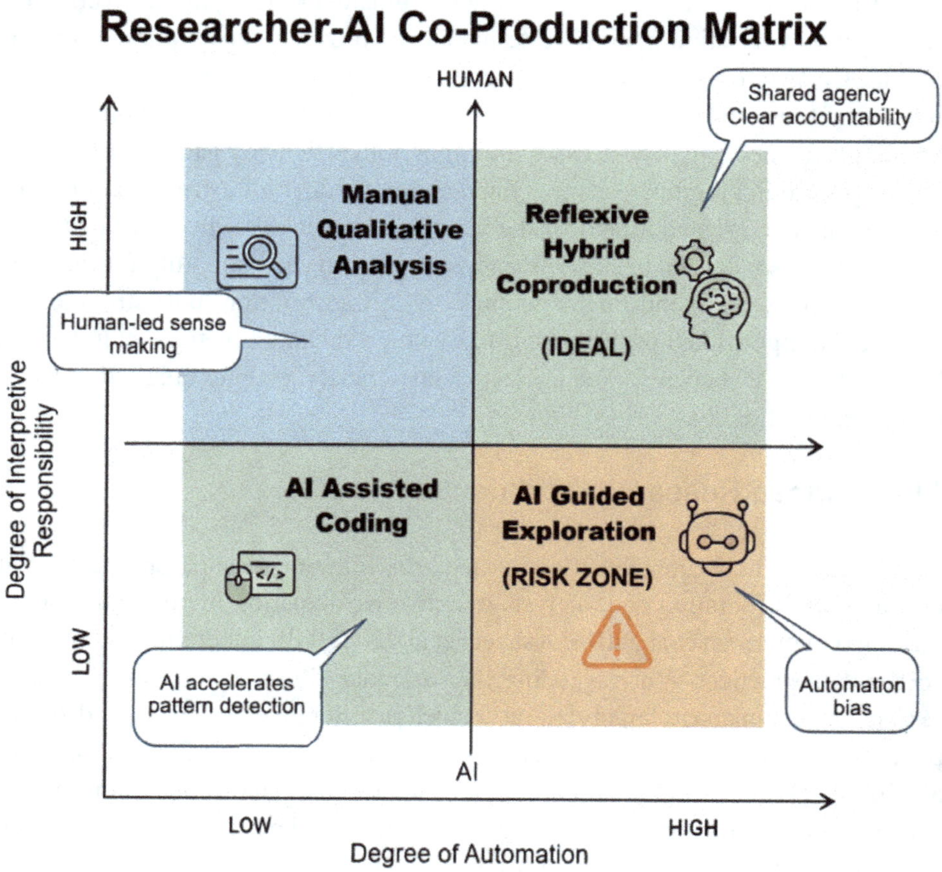

The following sections examine how AI supports coding, thematic analysis, and Grounded Theory approaches, while highlighting the methodological and epistemic considerations that arise when algorithmic assistance is embedded within qualitative research practice (Braun & Clarke, 2021; Charmaz, 2014).

AI-Assisted Coding

AI-assisted coding tools combine NLP and machine learning to generate preliminary code labels for large volumes of text, supporting rapid familiarisation with extensive datasets and enabling more efficient handling of repetitive coding work (Guetterman et al., 2018; Snelson, 2016). Researchers can train models on a subset of manually coded data and apply them to additional material, treating AI-generated codes as provisional suggestions that require human review, refinement, and theoretical interpretation (Guetterman et al., 2018). This approach can increase consistency in code application and highlight previously overlooked clusters of data, but it also risks reinforcing surface-level lexical regularities at the expense of deeper contextual meanings unless critically overseen by experienced qualitative researchers (Snelson, 2016; Williamson et al., 2023).

AI-Assisted Thematic Analysis

In thematic analysis, AI tools can facilitate clustering of semantically related segments, ranking of topics by prevalence, and automated generation of candidate theme labels across large corpora (Braun & Clarke, 2021; Kasneci et al., 2023). Emerging studies demonstrate that LLM-based systems can produce thematic summaries of qualitative data—such as expert interviews or student reflections—that approximate human-identified patterns and offer useful starting points for interpretive elaboration (Kasneci et al., 2023; Williamson et al., 2023). When combined with transparent documentation of prompts, model parameters, and analytic decisions, AI-supported thematic analysis can increase procedural transparency and reproducibility (Braun & Clarke, 2021). However, the interpretive authority remains with the researcher, who must triangulate AI-suggested themes with manual coding, theoretical commitments, and contextual knowledge to avoid uncritical reliance on algorithmic outputs (Braun & Clarke, 2021; Lupton, 2020).

Computational Grounded Theory

Grounded Theory's cycles of coding, constant comparison, and emerging theoretical development can be augmented by computational tools that accelerate pattern detection and category refinement (Charmaz, 2014; Snelson, 2016). Machine learning models can support the identification of recurring concepts, co-occurring categories, and negative cases across datasets, providing analytic prompts that researchers can interrogate within iterative theory-building (Charmaz, 2014; Guetterman et al., 2018). LLMs and other NLP techniques can also contribute to memo generation by suggesting possible relationships among codes or highlighting underexamined segments of the corpus (Kasneci et al., 2023). Nevertheless, fundamental Grounded Theory principles—such as openness to emergent categories, close engagement with participants' meanings, and sustained reflexivity—must be maintained by researchers who critically assess how computational routines shape conceptualisation and theoretical saturation (Charmaz, 2014; Lupton, 2020).

Advanced Data Assessment

Advanced data assessment methods extend AI-supported qualitative analysis beyond organisation and coding toward deeper analytic engagement with affect, interaction, and meaning-making. These approaches use computational techniques to surface latent patterns in emotion, discourse, and social interaction that are difficult to detect through manual analysis alone, particularly in large or complex datasets. By integrating algorithmic assessments with interpretive scrutiny, researchers can explore how feelings, conversational dynamics, and relational structures operate across contexts while maintaining qualitative sensitivity. The following subsections examine sentiment analysis and AI-enhanced conversation analysis as complementary tools that enrich qualitative inquiry when applied reflexively and in dialogue with established methodological traditions.

Across sentiment analysis and AI-supported conversation analysis, empirical studies both confirm the framework's claim that AI extends, rather than replaces, interpretive labour and underscore the need to refine its ethical dimension around representation, affect, and power.

Sentiment Analysis of Affective Dimensions

Sentiment analysis algorithms classify texts according to affective valence (e.g., positive, negative, neutral) and can be extended to more detailed emotional categories, offering an efficient way to explore affective patterns in qualitative datasets (Guetterman et al., 2018; Snelson, 2016). When applied to interviews, open-ended survey responses, or online discussions, these tools can reveal shifts in collective mood, identify emotionally charged narratives, and contribute to mixed method designs that link aggregated affective scores to in-depth qualitative interpretation (Guetterman et al., 2018). Yet the models often struggle with sarcasm, irony, cultural idioms, and context-specific emotional expressions, making human oversight necessary to interpret sentiment scores and associate them with participants situated experiences (Lupton, 2020; Snelson, 2016).

For example, mixed-method studies of online support groups have used sentiment trajectories to identify periods of heightened distress, which are then explored through close qualitative reading of posts to understand how participants narrate coping, solidarity, and conflict. Such work illustrates how computational affective patterning and interpretive depth can be combined.

Conversational Analysis with AI (CAAI)

The application of AI to conversation analysis extends micro-analytic approaches by using computational tools to examine turn-taking, sequence organisation, and interactional patterns at scale (Peräkylä, 2004; Snelson, 2016). Speech-processing and NLP methods can segment talk into turns, annotate speaker roles, and detect recurring patterns, enabling researchers to explore how institutional norms, power relations, or cultural practices shape conversational structures in large audio or video corpora (Peräkylä, 2004). LLM-based tools can assist by generating conversational summaries or classifying interaction types, but fine-grained judgments about politeness, alignment, resistance, or identity work still depend on interpretive expertise and detailed contextual reading (Peräkylä, 2004; Lupton, 2020). CAAI thus illustrates how computational techniques can amplify conversation analytic work while preserving the central role of human interpretation.

In studies of institutional encounters—such as helpline calls or telehealth consultations—AI-supported conversation analysis has been used to automatically segment large corpora into turns and identify recurrent interactional formats, which conversation analysts then examine qualitatively to understand how power, alignment, and resistance are negotiated in situ. These empirical designs highlight how computational scale and micro-analytic sensitivity can be brought together under a conversation-analytic lens.

Emerging Research Environments

Emerging research environments reflect the growing entanglement of qualitative inquiry with digital infrastructures and AI-driven systems that reshape where, how, and with whom research is conducted. Beyond analytic support, AI increasingly participates in data generation, fieldwork practices, and post-analytic reflection, challenging established distinctions between researcher, participant, and tool. These environments extend qualitative research into virtual, platform-mediated, and hybrid spaces, raising new ethical, epistemological, and methodological questions. The following sections examine how virtual moderators, computational ethnography, and hybrid post-coding approaches reconfigure qualitative workflows, highlighting both their methodological potential and the reflexive responsibilities they impose on interpretive research traditions.

Virtual Moderators and Conversational AI

Virtual moderators and conversational agents are increasingly used to conduct qualitative interviews, digital focus groups, and diary studies, sometimes operating at scales not achievable with human facilitators alone (Kasneci et al., 2023; Williamson et al., 2023). These systems can follow discussion guides, adapt follow-up questions to participants' responses, and generate preliminary analytic summaries, thereby reducing researcher workload and enabling data collection across diverse locations and time zones (Kasneci et al., 2023). Evidence from education and market research suggests that some participants feel comfortable sharing sensitive information with AI moderators, particularly when they perceive these agents as non-judgmental and anonymous (Kasneci et al., 2023; Selwyn, 2019). However, issues of rapport, informed consent, data security, and the subtle influence of system design on participants' narratives demand robust ethical oversight and explicit methodological reflexivity (Raji & Buolamwini, 2019; Williamson et al., 2023).

Computational Ethnography

Computational ethnography extends ethnographic sensibilities to digital spaces by using digital trace data—such as clickstreams, social media posts, and platform logs—as fieldwork materials that can be analysed through NLP, network analysis, and other AI-based methods (Lupton, 2020; Pink et al., 2015). These approaches enable the mapping of interaction patterns, temporal dynamics, and cultural practices in online communities at scales that surpass traditional participant observation, while still drawing on ethnographic principles like thick description and long-term engagement (Pink et al., 2015). Working with digital traces raises complex ethical questions about visibility, consent, anonymity, and the public–private boundary, as well as epistemological questions about how computational "views from above" shape what researchers recognise as the field (Lupton, 2020; Pink et al., 2015). As a result, computational ethnography requires integrated ethical frameworks and explicit reflection on how digital infrastructures and analytic tools condition what can be seen and said.

Hybrid Methodologies and Post-Coding Inquiry

Hybrid methodologies combine human-led qualitative techniques with computational tools across multiple stages of research, from data collection and coding to synthesis and dissemination (Braun & Clarke, 2021; Kasneci et al., 2023). In post-coding inquiry, AI can be used to re-analyse coded datasets by visualising relationships among codes, suggesting alternative thematic configurations, or simulating how different theoretical lenses might reorganise the material, thereby supporting meta-analytic reflection and secondary interpretation (Kasneci et al., 2023; Williamson et al., 2023). These designs position AI as both collaborator and object of critique, encouraging researchers to examine how different configurations of human and machine agency produce distinct readings of the same corpus (Lupton, 2020; Williamson et al., 2023). For interpretive traditions, hybrid and post-coding approaches reinforce the importance of transparency, reflexivity, and epistemic humility, acknowledging that computational and human perspectives each illuminate facets of complex social realities.

Ethical Integrity and Trustworthiness

In AI-mediated qualitative research, informed consent must go beyond traditional disclosures to explain how intelligent systems collect, process, and repurpose participant data (Ahmed, 2024; Marshall & Naff, 2024). Participants should receive clear, jargon-free information about the role AI will play in recording, transcribing, analysing, or even generating interaction during the study. This includes identifying which tools are used, what data they access, and where those data are stored or transferred (Williams, 2025; University of Surrey, 2024). Such expanded transparency is essential for respecting participant autonomy. AI systems often involve complex, opaque infrastructures—cloud platforms, proprietary models, and third-party Application Programming Interfaces (APIs)—that extend far beyond the immediate researcher–participant relationship. Acknowledging these additional actors and the ways they may interact with participant data strengthens accountability and ethical integrity (Marshall & Naff, 2024; Xiao et al., 2023). At the same time, researchers must also confront operational constraints such as software reliability, transcription accuracy, version changes in commercial models, and the possibility that tool updates may alter analytic outputs over time.

The following sections examine key dimensions of this ethical landscape: informed consent in AI-mediated data collection, data privacy and storage in algorithmic ecosystems, algorithmic bias and the black-box problem, re-evaluating credibility and dependability, the utilitarian challenge, and human-centred AI and the importance of ethical design. These concerns are not uniform across contexts: what counts as responsible AI use in a well-resourced university lab may differ from the realities of community-based research, low-bandwidth settings, or multilingual fieldwork in the Global South. Including contextualised applications is therefore essential because the risks and benefits of AI are shaped by local infrastructure, language diversity, institutional capacity, and regulatory environments.

Informed Consent in AI Data Collection

AI-mediated consent interfaces themselves, such as chatbots that present information sheets or answer questions, create additional ethical tensions. Experimental work suggests that conversational agents can increase comprehension of consent materials and reduce participant burden in online settings, but they also risk narrowing consent to information transfer while sidelining its relational and trust-building dimensions (Rothstein, 2023; Xiao et al., 2023). When AI tools moderate interviews or collect trace data in digital environments, researchers must guard against de facto "silent capture" of data—situations where individuals are unaware that an AI system is involved or do not fully understand downstream uses such as model training, synthetic data generation, or cross-platform linkage (Williams, 2025; Tuten, 2025).

Ethical practice therefore requires layered consent that is ongoing, revocable, and specific about secondary uses, with options to opt out of AI-based processing while still participating where feasible (Marshall & Naff, 2024; University of Surrey, 2024). In practice, this may mean using different consent protocols in a low-literacy rural interview setting than in a digitally native urban sample, and tailoring explanations to the technological familiarity of the participant group.

Informed consent in AI-mediated contexts is inseparable from broader questions of trustworthiness in qualitative research. Credibility and dependability depend on disclosing not only how data were gathered, but how AI systems shaped what was recorded, transcribed, filtered, and highlighted for interpretation (Ahmed, 2024; University of Surrey, 2024). Documenting AI involvement in consent materials, protocols, and reporting—alongside clear routes for questions, complaints, and data withdrawal—positions participants as partners rather than mere data sources, sustaining ethical integrity as computational tools become embedded in qualitative inquiry (Williams, 2025; Friese, 2024). This is especially important when AI is used in multilingual interviews, where transcription errors or translation drift can alter meaning in subtle but consequential ways.

Data Privacy in Algorithmic Ecosystems

Algorithmic ecosystems—complex constellations of data brokers, platforms, models, and APIs—intensify long-standing concerns about data privacy and storage by multiplying the actors, flows, and purposes attached to personal information. These environments routinely aggregate data from diverse sources, including research participation, platform activity, and third-party datasets, making it difficult for individuals to understand where their data travel, how long they are stored, and for what downstream uses they may be repurposed (Lu, 2022; Nguyen, 2024). Technical opacity compounds this problem: even where organisations disclose that they use "privacy-preserving" techniques, external stakeholders often lack meaningful ways to verify whether protections such as anonymisation or minimisation are effective in practice (Lu, 2022). A practical challenge is that researchers may depend on commercial transcription or coding platforms whose data-processing terms are difficult to audit, especially when those services operate across jurisdictions with different privacy laws.

Within AI-intensive research, privacy risks are closely tied to storage architectures. Cloud-based AI services typically require uploading audio, text, or metadata to remote servers, where copies may persist beyond the immediate research project and be integrated into broader training pipelines or monitoring systems (Samuel & Wassenaar, 2024). This introduces layered exposure, since security depends not only on the primary research institution but also on commercial providers and their subcontractors. Data breaches, model inversion attacks, and the re-identification of supposedly anonymised data illustrate how stored datasets can be reconstructed or linked to other repositories, undermining assurances of confidentiality (Nguyen, 2024; Sun et al., 2025). These operational risks are particularly salient in cross-border collaborations, where data may move between servers hosted in Europe, North America, or Asia, creating compliance challenges for local ethics boards and institutional review processes.

Algorithmic Bias and Opacity

Algorithmic bias and the black-box problem sit at the centre of contemporary debates about explainability because they jointly undermine fairness, accountability, and trust in AI systems. Algorithmic bias refers to systematic patterns of error that produce unfair or discriminatory outcomes for particular groups, typically because models are trained on data that encode historical inequities or reflect skewed sampling and labelling practices (Barocas & Selbst, 2016; Mehrabi et al., 2021), particularly since a foundational challenge is ensuring data quality, as the maxim "garbage in, garbage out" remains an enduring truth (Nakhla, 2026). In high-stakes domains such as hiring, credit scoring, or risk assessment, these biases can amplify existing social hierarchies, misclassify minoritised communities, and create feedback loops in which biased outputs are re-ingested as training data, further entrenching disparity (Mehrabi et al., 2021). From an ethical standpoint, addressing such bias is not just a technical optimisation task but a question of whose experiences are represented, whose errors are tolerated, and whose interests are prioritised in system design (Benjamin, 2019). In qualitative research, this means that AI-assisted coding tools may under-recognise dialectal speech, culturally specific idioms, or regionally distinct terminology, particularly in multilingual or postcolonial research contexts.

The black-box problem intensifies these concerns by making it difficult—even for system developers—to trace how complex models, especially deep neural networks and large language models, arrive at particular outputs (Castelvecchi, 2016; Burrell, 2016). When the internal decision logic is opaque, stakeholders cannot easily contest harmful decisions, audit models for disparate impact, or understand which features drive classifications and predictions (Burrell, 2016). Addressing these challenges demands comprehensive governance structures promoting transparency, accountability, and equity (Nakhla, 2026). Central to this tension is balancing model interpretability against predictive performance (Nakhla, 2026). Operationally, this means researchers may need to compare AI outputs against manual coding in diverse cases—for instance, comparing English-language datasets from North American participants with transcripts from bilingual participants in Latin America or South Asia—to assess whether the model's confidence masks systematic omission.

For qualitative and mixed-methods researchers, the combination of algorithmic bias and black-box opacity raises acute methodological and ethical challenges. Analysts must not only assess whether AI-assisted coding or classification replicates social biases but also document how model opacity may have shaped what patterns were surfaced, which voices were foregrounded, and which were systematically misrepresented or silenced (Benjamin, 2019; Selbst & Powles, 2017). Embedding critical scrutiny of bias and explainability into research design, analysis, and reporting is therefore essential to sustaining trustworthiness and accountability in AI-supported inquiry. This scrutiny should also be contextual: an AI workflow that performs adequately in a standardised English-language interview corpus may fail in Indigenous, multilingual, or community-based studies where meaning is carried through idiom, code-switching, and culturally embedded reference.

Re-evaluating Credibility and Dependability

Trustworthiness in the age of AI requires rethinking classical qualitative criteria such as credibility, dependability, and confirmability considering systems that are probabilistic, adaptive, and partially opaque. Credibility no longer pertains solely to the adequacy of data and the persuasiveness of interpretations; it now also depends on whether AI components meet broader requirements for trustworthy systems, including human oversight, fairness, transparency, robustness, privacy, and accountability (European Commission High-Level Expert Group on AI, 2019). Dependability, traditionally concerned with the stability and traceability of research procedures, must likewise extend to versioned models, changing training data, and continuous updates that can subtly alter analytic outputs over time (Li et al., 2024). A technical challenge here is reproducibility: a model that produces one output today may yield a different one after a vendor update tomorrow, even when the researcher uses the same prompt and the same transcript.

From this perspective, credibility in AI-supported qualitative work rests on making visible how algorithmic tools shape each stage of the inquiry. Researchers are expected to document model choice, configuration, prompts, and data flows so that others can evaluate the plausibility of interpretations and the risks of hidden bias or drift (Li et al., 2024). This aligns with emerging frameworks that emphasise transparency and explainability as preconditions for trust—not in the sense of fully revealing complex internal mechanics, but in offering intelligible accounts of what the system does, its limitations, and how its outputs were integrated into human-led analysis (European Commission High-Level Expert Group on AI, 2019; Schlicker et al., 2025). In practice, this may involve methodological appendices on AI usage, audit trails that log interactions with tools, and triangulation strategies that compare AI-assisted results with manual coding or alternative models. For researchers working in international or multi-site projects, these documentation practices are especially valuable because they make it possible to compare how AI behaved across different datasets, languages, and institutional settings.

Dependability, meanwhile, requires lifecycle-aware governance. Guidance from standards bodies stresses that trustworthiness must be designed, evaluated, and monitored across development, deployment, and use, rather than inferred from performance at a single point in time (National Institute of Standards and Technology, 2023). For qualitative researchers, this means attending to how updates to transcription services, language models, or analytic platforms could affect longitudinal studies, re-analyses, or secondary use of archived data. Maintaining detailed technical documentation, locking critical analyses to specific model versions, and periodically reassessing AI components against project aims become central to preserving a dependable evidentiary base. In this way, traditional qualitative notions of trustworthiness are not abandoned but expanded, coupling interpretive rigour with explicit stewardship of the socio-technical infrastructures through which data are processed and interpreted. In resource-constrained settings, however, researchers may lack access to version-locking or enterprise-grade logging, which makes low-cost governance practices—such as local backups, manual audit notes, and human verification of summaries—especially important.

The Utilitarian Challenge

The utilitarian challenge in AI centres on designing and governing systems so that aggregate benefits are maximised while harms and unfair burdens are minimised across different groups. Utilitarianism evaluates actions by their consequences for overall welfare, which translates in AI contexts into questions about how predictive accuracy, efficiency, and scalability are weighed against risks such as surveillance, displacement, or the amplification of social inequalities (McGraw, 2022; Mitov, 2021). A purely outcome-focused stance can make high-performing but socially harmful systems appear ethically acceptable if they generate net gains, prompting critics to argue for stronger constraints that recognise rights, dignity, and non-domination alongside aggregate utility (Floridi et al., 2018; Whittlestone et al., 2019). This is particularly relevant in qualitative research, where a tool that saves time may still distort context, flatten meaning, or disadvantage participants whose voices are less legible to the model.

Attempts to operationalise utilitarian reasoning in AI ethics often foreground beneficence—improving health, safety, or access to services—while also emphasising non-maleficence, justice, and accountability as necessary correctives to simple "greatest good for the greatest number" calculations (Floridi et al., 2018). Frameworks that integrate utilitarian considerations propose multi-level assessments: at the data level (privacy and representativeness), the algorithmic level (bias, interpretability, robustness), and the systems level—risk, reliability, and distribution of benefits and burdens— (Mitov, 2021; Emdad et al., 2023). In practice, this demands explicit attention to who gains from AI deployment, who absorbs the residual risks, and whether those exposed to potential harm can consent, contest, or benefit from compensatory measures (McGraw, 2022; Whittlestone et al., 2019). For example, an AI transcription tool may benefit the research team by reducing labour, but if it consistently misreads accents or code-switching, then the efficiency gain comes at the expense of epistemic fairness.

Fairness becomes a central pressure test for utilitarian claims in AI. An intervention that improves average outcomes may still be ethically flawed if it systematically disadvantages already marginalised communities or treats some individuals as expendable "costs" of optimisation (Benjamin, 2019). Contemporary debates therefore push toward "constrained utilitarianism," where outcome maximisation is bounded by fairness, non-discrimination, and respect for persons (Floridi et al., 2018; Whittlestone et al., 2019). For researchers and practitioners, this entails embedding distributive analyses into impact assessments, publishing evidence of subgroup performance, and engaging affected communities in defining what counts as a benefit worth pursuing and a harm that cannot be justified by aggregate gains (Benjamin, 2019; McGraw, 2022). This matters across global contexts because the same AI system may be welcomed as a time-saving tool in one setting and experienced as an extractive or exclusionary technology in another.

Human-centred AI (HCAI) and Ethical Design

HCAI frames artificial intelligence as a socio-technical system designed to protect and promote human values, rather than as an autonomous optimisation engine. HCAI emphasises that AI should augment human capabilities, preserve autonomy, and respect fundamental rights, including privacy, non-discrimination, and dignity (European Commission High-Level Expert Group on AI, 2019). Within this orientation, ethical design is not an optional "add-on" but a structuring principle that shapes how problems are formulated, how data are collected, and how models are deployed and governed (UNESCO, 2021). A more balanced reading of AI in qualitative research therefore requires attention not only to ethics, but also to the operational realities of implementation: cost, infrastructure, training, multilingual support, maintenance, and the need for human review at every stage.

Ethical design in HCAI usually rests on a cluster of intertwined principles: fairness, transparency, accountability, and human oversight. Fairness demands deliberate attention to how datasets, model architectures, and evaluation metrics affect different social groups, particularly those historically marginalised or under-represented (UNESCO, 2021). Transparency involves offering intelligible explanations of system purpose, limitations, and decision logic at levels appropriate to different stakeholders, from end users to regulators (European Commission High-Level Expert Group on AI, 2019). Accountability requires clearly assigned responsibility for AI-enabled actions and outcomes, including mechanisms for contestation, remedy, and redress when harms occur (UNESCO, 2021). Human oversight, finally, ensures that consequential decisions remain subject to human judgment, especially in high-stakes domains such as healthcare, education, or criminal justice (European Commission High-Level Expert Group on AI, 2019). In qualitative research, these principles can be enacted through contextualized applications such as pilot-testing AI tools on locally produced transcripts, co-designing workflows with community researchers, and explicitly comparing AI-assisted results with manual interpretations before drawing conclusions.

From a methodological perspective, HCAI calls for participatory and iterative design processes that involve users and affected communities throughout the AI lifecycle. This includes co-specifying requirements, co-interpreting model outputs, and revising systems in response to lived experience, rather than treating users as passive recipients of technological solutions (UNESCO, 2021). Such practices align closely with qualitative and interpretive traditions, which foreground situated knowledge, reflexivity, and dialogue. By embedding these commitments into design and governance, HCAI seeks to ensure that AI systems do not merely function effectively, but also contribute to just, inclusive, and trustworthy socio-technical futures (European Commission High-Level Expert Group on AI, 2019; UNESCO, 2021). For diverse global audiences, the most relevant ethical question is therefore not simply whether AI can be used in qualitative research, but under what infrastructural, linguistic, institutional, and cultural conditions it should be used—and how researchers can document those conditions transparently.

REFLEXIVITY AND INTERPRETIVE RESPONSIBILITY

Reflexivity and interpretive responsibility take on renewed significance when qualitative analysis is mediated by AI. Researchers are now accountable not only for their own positionalities but also for the epistemic stances embedded in computational systems, from training data biases to model design choices (Crawford, 2021; Gebru et al., 2021).

AI Positionality and Bias

AI systems are never neutral. Their algorithms reflect the standpoints embedded in their training data, model designs, and the contexts in which they are developed and used (Crawford, 2021). In this sense, AI functions as an epistemic agent—one that encodes the cultural, historical, and political assumptions of its makers and data sources. Large language models, for instance, often reproduce dominant Western, English-centric norms (Bender et al., 2021). This can marginalise diverse dialects, local meanings, and perspectives, thereby reinforcing mainstream narratives while muting the dissenting or ambiguous voices so vital for qualitative inquiry.

Researchers should therefore approach AI not as an objective tool but as a *situated knower*—a participant in the production of meaning. This perspective calls for careful examination of training datasets, prompting practices, and model outputs, alongside critical reflection on how these align or conflict with the researcher's own positionality. Consider, for example, sentiment analysis models trained on consumer reviews. Such models may fail to detect irony, critique, or ambiguity common in academic writing. These misreadings should not be dismissed as minor technical glitches but recognised as epistemic signals—opportunities to recalibrate analytic strategies and openly reflect on the interpretive consequences of AI collaboration.

Algorithmic Transparency and Accountability

Ensuring transparency in algorithmic decision-making means making visible each stage of computation—from data preprocessing to the final outputs—that shapes qualitative findings (Raji et al., 2020). In practice, this involves keeping detailed records of prompts, model versions, parameter settings, and intermediate outputs so that the analytic process can be audited or, when relevant, replicated (Gebru et al., 2021). When researchers take responsibility for how AI systems are used and openly report limitations such as biased training data or mismatched domains, they strengthen accountability and enable others to evaluate the reliability and boundaries of the analysis (Raji et al., 2020).

Including methodological appendices that document code repositories, decision logs, and explainable-AI visualisations—such as SHAP plots, which show which features most strongly affect model outputs—can further clarify how algorithms contribute to coding or theme development (Lundberg & Lee, 2017). These transparency practices help close the gap between opaque computational processes and qualitative interpretation. They also position human–AI collaboration as a rigorous and credible research approach rather than as a novelty or superficial use of technology (Gebru et al., 2021).

Reflexivity and Human Validation

Reflexivity helps researchers critically evaluate how AI-generated interpretations align with, distort, or overlook participants' meanings and the study's theoretical goals (Finlay, 2021). It anchors human involvement by encouraging researchers to continually question the relationship between computational outputs and qualitative intent. Validation, in this context, becomes an ongoing process of comparison—checking AI results against raw data, seeking participant feedback where possible, and conducting additional rounds of human coding. These steps help identify where models might be overly confident, too simplistic, or shaped by culturally narrow assumptions (Braun & Clarke, 2022). The researcher's interpretive judgment is especially vital when meaning is ambiguous or context-dependent—as in instances of irony, metaphor, or silence—where lived experience and contextual understanding carry more weight than statistical patterns (Finlay, 2021).

Keeping detailed validation logs can further support transparency. These should note when and why AI-generated codes or themes were kept, modified, or rejected, and how the researcher's own positionality influenced these choices (Finlay, 2021). Such reflexive documentation strengthens auditability and invites readers to engage critically with the analytic process. Ultimately, it ensures that AI serves to extend, rather than dilute, the emancipatory and critical goals that underpin qualitative research (Braun & Clarke, 2022).

AI LITERACY AND COMPETENCE

AI literacy equips qualitative researchers with the knowledge to critically engage computational tools without ceding interpretive authority (Selwyn, 2019). This competence spans technical understanding of algorithms, ethical awareness of biases, and reflexive skills to document hybrid workflows. Far from mere "tool proficiency," it fosters discernment about when AI augments depth versus when it risks superficiality, ensuring methodological rigour amid technological enthusiasm (Kasneci et al., 2023). Competent researchers treat AI as a collaborator requiring oversight, not an oracle demanding deference.

AI Literacy for Qualitative Researchers

AI literacy for qualitative researchers encompasses three intertwined dimensions: conceptual grasp of machine learning principles, practical facility with tools like NLP and LLMs, and critical evaluation of their epistemological implications (Selwyn, 2019). Conceptually, researchers must understand probabilistic reasoning, training data dependencies, and output stochasticity to contextualise AI-generated insights within interpretive paradigms. Practically, this means selecting appropriate models, crafting effective prompts, and interpreting metrics like confidence scores or perplexity.

Critically, literacy demands interrogation of AI's philosophical assumptions—its positivist leanings versus qualitative constructivism—and recognition of limitations such as hallucination or cultural myopia (Bender et al., 2021). For instance, thematic extraction via LLMs requires validating outputs against thick description and member perspectives, not accepting surface patterns as authoritative. This definition rejects reductive "button-pushing" training in favour of holistic competence that integrates technical savvy with methodological scepticism, positioning researchers as stewards of hybrid knowledge production (Kasneci et al., 2023; Selwyn, 2019).

Digital Reflexivity and Transparency

Critical digital reflexivity involves habitual self-examination of how AI tools mediate perception, selection, and interpretation of data, mirroring traditional reflexivity but extended to socio-technical entanglements (Lupton, 2023). Researchers cultivate this by maintaining analytic journals that log tool choices, prompt iterations, and moments of discord between human intuition and algorithmic suggestion, surfacing hidden influences on findings.

Technical transparency complements reflexivity through systematic disclosure: versioned model citations, reproducibility scripts, and visualisation of decision pathways (e.g., attention weights in transformers). Development occurs via iterative practice—piloting AI workflows on subset data, peer debriefing hybrid analyses, and publishing methodological appendices that demystify black-box contributions (Raji et al., 2020). Workshops emphasising "prompt engineering as theory-building" bridge technical and interpretive skills, while frameworks like datasheets for datasets ensure upstream transparency (Gebru et al., 2021). Together, these practices transform opacity into auditability, empowering researchers to claim ownership of co-produced knowledge while inviting scrutiny (Lupton, 2023).

Embedding AI Literacy

Pedagogical strategies must embed AI literacy within core qualitative methods courses, using experiential, scaffolded approaches that mirror research cycles (Selwyn, 2019). Start with unplugged activities dissecting algorithmic logic through role-play (e.g., simulating bias propagation), progressing to hands-on labs applying LLMs to canonical datasets like interview transcripts, followed by reflexive debriefs critiquing outputs. Case studies of AI successes and failures—such as biased sentiment analysis in cross-cultural studies—foster critical discernment, paired with assignments requiring hybrid analysis reports that juxtapose manual and AI coding (Kasneci et al., 2023). Collaborative projects simulate real-world co-production, with peer review emphasising transparency documentation. Guest sessions from ethicists and developers contextualise tools within power structures.

Longitudinally, capstone modules integrate AI literacy into thesis supervision via checklists for tool auditing and ethical impact statements. Online repositories of open-source prompts and validation protocols democratise access, while certification pathways validate competence. These strategies ensure AI literacy permeates training as reflexive practice, not ancillary skill, preparing researchers for ethical hybrid scholarship (Selwyn, 2019).

CASE STUDIES OF AI IN QUALITATIVE RESEARCH WORKFLOWS

The three papers show that AI is beginning to play a practical role across several stages of qualitative research, but they also show that the value of AI depends on bounded use, transparency, and sustained human judgement. Taken together, the studies move beyond abstract discussion by providing concrete examples of where AI can help in workflow support, where it struggles, and how researchers can preserve trustworthiness.

Cook et al. (2025) provide one of the clearest empirical demonstrations of AI support in qualitative analysis. Their study included an exploratory case in which ChatGPT-4 was applied to three existing narrative datasets. In this case, the model was able to generate accurate brief summaries, showing that AI can be useful for rapid familiarisation with text. However, when the authors attempted more ambitious tasks, the initial prompt was insufficient, and the model did not reliably produce the desired qualitative output. This finding is important because it shows that AI may support preliminary engagement with data, but it does not automatically perform higher-level interpretive work such as thematic analysis. Cook et al. also identify a broad range of AI-supported qualitative activities reported in the literature, including inductive coding, deductive coding, thematic analysis, computational grounded theory, discourse analysis, transcription, translation, and analysis of large datasets. Their review suggests that AI is most useful when it helps with repetitive, structured, or large-scale tasks, while the researcher remains responsible for interpretation, meaning-making, and methodological decisions.

A second concrete example comes from Kirsten et al. (2025), who interviewed 15 researchers with qualitative data analysis experience to understand where AI could fit into real workflows. Participants saw promise in AI support for pre-processing data, onboarding new researchers to unfamiliar datasets, and helping coders build consensus. These are practical workflow moments rather than abstract future possibilities. For example, AI could help a research team organise raw interview material before formal coding begins, or support comparisons between coders when different interpretations emerge. Importantly, the interviewees did not frame AI as a replacement for analysis. Instead, they imagined a continuum of assistance ranging from small workflow efficiencies to much greater autonomy, with the strongest support given to AI functions that reduce friction without removing researcher control. The paper therefore provides a realistic example of how AI may be inserted into collaborative research practice: not as an autonomous analyst, but as a supportive layer that helps researchers move through stages of the process more efficiently.

A third case study is offered by Bakharia et al. (2025), who describe a trustworthy workflow for qualitative analysis using large language models. Their paper is explicitly structured around moving from transcripts to themes, which makes it especially useful as a concrete example. Rather than asking the model to replace the analyst, the authors design a workflow in which the LLM helps derive an initial set of themes from text data while the researcher maintains control over validation and interpretation. This is significant because it shows a partial division of labour: the AI assists with early theme generation, but the human researcher checks whether those themes are credible, relevant, and grounded in the data. The workflow is intended to preserve trustworthiness by ensuring that AI suggestions are traceable and that analytic decisions remain reviewable. In practical terms, this is a model for how researchers can use LLMs to handle the first pass through transcripts while still keeping the interpretive centre of gravity with the human analyst.

Across the three papers, a common pattern emerges. AI appears most helpful in qualitative workflows when it supports summarisation, organisation, retrieval, pre-processing, or early-stage theme generation. It is less convincing when asked to perform the full interpretive labour of qualitative inquiry on its own. The studies also converge on a core methodological principle: AI should be used with human oversight, not as an autonomous substitute for sense-making. This means that concrete AI use in qualitative research is best understood as augmentation rather than automation. The workflow examples in these papers show that AI can reduce time and support analytical movement, but the final responsibility for depth, nuance, and trustworthiness still belongs to the researcher.

FUTURE DIRECTIONS

Hybrid Qualitative–Computational Paradigms

Hybrid paradigms merge qualitative depth with computational scale, creating methodologies where human interpretation guides algorithmic exploration (Paulus et al., 2017). Future developments will likely centre on multimodal AI that integrates text, audio, video, and sensor data, enabling richer ethnographic analysis of digital lifeworld (Johnson & Parry, 2015). Agentic workflows—autonomous AI systems that iteratively query, refine, and synthesise across datasets—promise to accelerate theoretical saturation while preserving researcher oversight (Chatzichristos, 2025).

Anticipated paradigms include "computational thick description," where LLMs generate contextual narratives from pattern clusters, and "reflexive autoethnography with AI mirrors," using models to surface researcher biases in real-time. These approaches demand new validity criteria: hybrid triangulation across human-AI outputs, longitudinal model drift monitoring, and participatory validation with data subjects (Paulus et al., 2017). Ethical hybrids will prioritise open-source pipelines and community-governed datasets to democratise access beyond elite institutions. As quantum computing matures, probabilistic simulations could model interpretive multiplicity, challenging singular truths central to constructivism (Johnson & Parry, 2015). Ultimately, these paradigms reposition qualitative inquiry as orchestrator of computational ensembles, amplifying nuance at planetary scales while safeguarding epistemic humility.

AI Research Ethics and Governance

Ethical frontiers demand governance frameworks that operationalise principles across AI research lifecycles (Jobin et al., 2019). Policy implications include mandatory algorithmic impact assessments, akin to environmental reviews, requiring pre-deployment audits of bias, privacy erosion, and epistemic harms (Ebers, 2023). International standards bodies are developing "research AI passports"—digital certificates documenting model provenance, training ethics, and performance equity—to facilitate cross-jurisdictional trust.

Future governance will emphasise dynamic consent models, allowing participants revocable control over AI re-uses, and "right to interpretation," entitling data subjects to contest algorithmic framings (Zuboff, 2019). Policy levers include funding mandates tying grants to open reproducibility and public scholarship on hybrid methods (Jobin et al., 2019). At supra-national levels, UNESCO's AI ethics instruments foreshadow binding treaties addressing data colonialism in global qualitative corpora. These frameworks position researchers as ethical entrepreneurs, co-designing governance that balances innovation with justice amid accelerating automation (Ebers, 2023).

Researchers Role in Automated Analysis

Researchers will transition from solo interpreters to "methodological conductors," orchestrating AI ensembles while curating human insight (Selwyn, 2022). This evolution demands meta-skills: prompt orchestration, model ensemble design, and "interpretation hygiene" to mitigate automation bias. Core competencies shift toward strategic intervention—knowing when to intervene in black-box processes—and narrative synthesis, weaving computational exhaustiveness with phenomenological depth (Kvale & Brinkmann, 2015).

Automation liberates researchers for higher-order pursuits: theory provocation, cross-paradigm dialogue, and public sensemaking. Yet role erosion looms if interpretive labour commoditises into gig-prompting; counterstrategies include "human-in-the-loop" certification and guild-like professional bodies enforcing standards (Selwyn, 2022). Future researchers embody "cyborg scholarship," their cognition augmented yet distinctly human through embodied reflexivity and ethical discernment. Educational pipelines must prioritise philosophical grounding alongside technical fluency, ensuring automation serves emancipation rather than deskilling (Kvale & Brinkmann, 2015).

CONCLUSION

Human interpretation remains the irreducible core of qualitative inquiry, even as AI scales analytic capacity to unprecedented levels. Computational tools excel at pattern detection and exhaustiveness, yet they cannot replicate the contextual empathy, ethical discernment, and theoretical creativity that define interpretive scholarship (Denzin & Lincoln, 2011). AI-augmented research thrives precisely because human researchers serve as critical interlocutors—interrogating algorithmic outputs, infusing cultural nuance, and safeguarding participant meanings against probabilistic reductionism. This symbiosis reaffirms qualitative traditions' commitment to multiplicity, reflexivity, and social justice, ensuring technology amplifies rather than supplants the humanistic project of understanding lived experience.

Rigour in hybrid methodologies demands methodological transparency: documenting model choices, prompt engineering, validation protocols, and human interventions to enable auditability and replication (Braun & Clarke, 2022). Reflexivity extends beyond personal positionality to encompass socio-technical entanglements—reflexively accounting for how AI's embedded biases, training epistemologies, and black-box opacities shape interpretive pathways. Responsibility, finally, positions researchers as ethical stewards, balancing innovation with accountability through informed consent, bias audits, and participatory governance that honours data subjects' agency (Lincoln & Guba, 1985).

Future scholarship must cultivate AI literacy as foundational competence, embedding critical digital reflexivity within training curricula and professional standards. As hybrid paradigms mature, qualitative researchers will evolve into methodological orchestrators—designing human-AI workflows that preserve interpretive depth while harnessing computational breadth. This trajectory promises enriched inquiry: planetary-scale ethnographies, real-time reflexive autoethnographies, and democratised access to sophisticated analysis beyond elite institutions. Yet enduring vigilance is required. The temptation to defer to algorithmic authority risks epistemic deskilling and the erosion of qualitative rigour. By reaffirming human interpretation as sovereign—guided by rigour, reflexivity, and responsibility—scholars can navigate the computational turn toward more equitable, trustworthy, and transformative social science. In this hybrid future, qualitative research does not merely adapt to AI; it reasserts its emancipatory potential, ensuring technology serves understanding rather than supplanting it (Selwyn, 2019).

REFERENCES

Ahmed, S. K. (2024). The pillars of trustworthiness in qualitative research. *Journal of Medicine, Surgery, and Public Health, 57*, 101753. DOI: 10.1016/j.glmedi.2024.100051

Bakharia, A., Shibani, A., Lim, L.-A., McCluskey, T., & Buckingham Shum, S. (2025). *From transcripts to themes: A trustworthy workflow for qualitative analysis using large language models*. CEUR Workshop Proceedings. https://ceur-ws.org/Vol-3995/LLMQUAL_paper1.pdf

Barocas, S., & Selbst, A. D. (2016). Big data's disparate impact. *California Law Review, 104*(3), 671–732.

Bender, E. M., Gebru, T., McMillan-Major, A., & Shmitchell, S. (2021). On the dangers of stochastic parrots: Can language models be too big? *Proceedings of the 2021 ACM Conference on Fairness, Accountability, and Transparency*, 610–623. DOI: 10.1145/3442188.3445922

Benjamin, R. (2019). Race after technology: Abolitionist tools for the new Jim code. *Polity*.

Braun, V., & Clarke, V. (2021). *Thematic analysis: A practical guide*. SAGE Publications. https://www.torrossa.com/it/resources/an/5282292

Burrell, J. (2016). How the machine 'thinks': Understanding opacity in machine learning algorithms. *Big Data & Society, 3*(1), 1–12. DOI: 10.1177/2053951715622512

Castelvecchi, D. (2016). Can we open the black box of AI? *Nature, 538*(7623), 20–23. DOI: 10.1038/538020a PMID: 27708329

Charmaz, K. (2014). *Constructing grounded theory* (2nd ed.). SAGE Publications., https://www.torrossa.com/en/resources/an/5019293

Chatzichristos, G. (2025). Qualitative research in the era of AI: A return to positivism? *International Journal of Qualitative Methods, 24*, 16094069251337583. Advance online publication. DOI: 10.1177/16094069251337583

Cook, D. A., Ginsburg, S., Sawatsky, A. P., Kuper, A., & D'Angelo, J. D. (2025). Artificial intelligence to support qualitative data analysis: Promises, approaches, pitfalls. *Academic Medicine, 100*(10), 1134–1149. https://pubmed.ncbi.nlm.nih.gov/40560241/. DOI: 10.1097/ACM.0000000000006134 PMID: 40560241

Crawford, K. (2021). *Atlas of AI: Power, politics, and the planetary costs of artificial intelligence*. Yale University Press.

Denzin, N. K., & Lincoln, Y. S. (Eds.). (2011). *The SAGE handbook of qualitative research* (4th ed.). SAGE Publications.

Ebers, M. (2023). The European Commission's Proposal for an Artificial Intelligence Act. In *Research Handbook on EU Internet Law* (pp. 271–292). Edward Elgar Publishing., DOI: 10.4337/9781803920887.00022

Emdad, F. B., Ho, S. M., Ravuri, B., & Hussain, S. (2023). *Towards a unified utilitarian ethics framework for healthcare artificial intelligence*. arXiv. https://doi.org// arXiv.2309.14617DOI: 10.48550

European Commission High-Level Expert Group on AI. (2019). *Ethics guidelines for trustworthy AI*. European Commission.

Finlay, L. (2021). Thematic analysis: the 'good', the 'bad'and the 'ugly'. *European Journal for Qualitative Research in Psychotherapy*, *11*, 103–116. https://openjournals .ljmu.ac.uk/ejqrp/article/view/3062

Floridi, L., & Chiriatti, M. (2020). GPT-3: Its nature, scope, limits, and consequences. *Minds and Machines*, *30*(4), 681–694. DOI: 10.1007/s11023-020-09548-1

Floridi, L., Cowls, J., Beltrametti, M., Chatila, R., Chazerand, P., Dignum, V., Luetge, C., Madelin, R., Pagallo, U., Rossi, F., Schafer, B., Valcke, P., & Vayena, E. (2018). AI4People—An ethical framework for a good AI society: Opportunities, risks, principles, and recommendations. *Minds and Machines*, *28*(4), 689–707. DOI: 10.1007/s11023-018-9482-5 PMID: 30930541

Friese, S. (2024, May 2). Navigating ethics in AI-driven qualitative research: Privacy, confidentiality and consent. *Qeludra.* https://qeludra.com/blog/research-ethics -ai-privacy-confidentiality

Gebru, T., Morgenstern, J., Vecchi, B., Vaughan, J. W., Wallach, H., Daumé, H.III, & Crawford, K. (2021). Datasheets for datasets. *Communications of the ACM*, *64*(12), 86–94. DOI: 10.1145/3458723

Geertz, C. (1973). Chapter 1/Thick Description: Toward an interpretive theory of culture. *The interpretation of cultures: Selected essays*, 3-30.

Guba, E. G., & Lincoln, Y. S. (1994). Competing paradigms in qualitative research. In Denzin, N. K., & Lincoln, Y. S. (Eds.), *Handbook of qualitative research* (pp. 105–117). SAGE Publications.

Guetterman, T. C., Chang, T., DeJonckheere, M., Basu, T., Scruggs, E., & Vydiswaran, V. V. (2018). Augmenting qualitative text analysis with natural language processing: Methodological study. *Journal of Medical Internet Research, 20*(6), e231. https://www.jmir.org/2018/6/e231/. DOI: 10.2196/jmir.9702 PMID: 29959110

Jobin, A., Ienca, M., & Vayena, E. (2019). The global landscape of AI ethics guidelines. *Nature Machine Intelligence, 1*(9), 389–399. DOI: 10.1038/s42256-019-0088-2

Johnson, C. W., & Parry, D. C. (2015). *Fostering social justice through qualitative inquiry*. Angeles: Routledge.

Kasneci, E., Sessler, K., Küchemann, S., Bannert, M., Dementieva, D., Fischer, F., Gasser, U., Groh, G., Günnemann, S., Hüllermeier, E., Krusche, S., Kutyniok, G., Michaeli, T., Nerdel, C., Pfeffer, J., Poquet, O., Sailer, M., Schmidt, A., Seidel, T., & Kasneci, G. (2023). ChatGPT for good? On opportunities and challenges of large language models for education. *Learning and Individual Differences, 103*, 102274. DOI: 10.1016/j.lindif.2023.102274

Kirsten, E., Buckmann, A., Lassak, L., Borgert, N., Mhaidli, A., & Becker, S. (2025). *From assistance to autonomy: A researcher study on the potential of AI support for qualitative data analysis*. arXiv. https://arxiv.org/abs/2501.19275

Kvale, S., & Brinkmann, S. (2015). *InterViews: Learning the craft of qualitative research interviewing* (3rd ed.). SAGE.

Li, Y., Wang, S., & Zhang, H. (2024). Developing trustworthy artificial intelligence: A three-dimension framework of human–AI trust. *Patterns (New York, N.Y.), 5*(4), 100987. DOI: 10.3389/fpsyg.2024.1382693 PMID: 38694439

Lincoln, Y. S., & Guba, E. G. (1985). *Naturalistic inquiry*. SAGE Publications. DOI: 10.1016/0147-1767(85)90062-8

Lu, S. (2022). Data privacy, human rights, and algorithmic opacity. *California Law Review, 110*, 2087. https://repository.law.umich.edu/cgi/viewcontent.cgi?article=1003&context=research

Lundberg, S. M., & Lee, S.-I. (2017). A unified approach to interpreting model predictions. *Advances in Neural Information Processing Systems, 30*, 4765–4774. https://proceedings.neurips.cc/paper/2017/hash/8a20a8621978632d76c43dfd28b67767-Abstract.html

Lupton, D. (2020). *Data selves: More-than-human perspectives*. Polity Press.

Marshall, D. T., & Naff, D. B. (2024). The ethics of using artificial intelligence in qualitative research. *Journal of Empirical Research on Human Research Ethics; JERHRE, 19*(3), 92–102. DOI: 10.1177/15562646241262659 PMID: 38881315

McGraw, D. K. (2022). Ethical responsibility in the design of artificial intelligence (AI) systems. *International Journal on Responsibility, 6*(1), 45–70. DOI: 10.62365/2576-0955.1114

Mehrabi, N., Morstatter, F., Saxena, N., Lerman, K., & Galstyan, A. (2021). A survey on bias and fairness in machine learning. *ACM Computing Surveys, 54*(6), 1–35. DOI: 10.1145/3457607

Mitov, A. (2021). *Ethical use of artificial intelligence through the utilitarianism theory* [Bachelor's thesis, University of Twente].

Nakhla, G. (2026). From Statistical Analysis to Deep Learning: Modern Approaches in Quantitative Decision-Making. In Derici, S. (Ed.), *Advancements in Numerical Analysis and Quantitative Decision-Making*. IGI Global., DOI: 10.4018/979-8-3373-6746-0.ch003

National Institute of Standards and Technology. (2023). *Artificial intelligence risk management framework (AI RMF 1.0)*. U.S. Department of Commerce.

Nguyen, K. (2024). *Enhancing data privacy in artificial intelligence: A study on corporate practices and regulatory compliance* (Master's thesis, University of Turku, Finland). UTUPub Institutional Repository. https://www.utupub.fi/bitstream/handle/10024/178070/Nguyen_Khanh_Thesis.pdf?sequence=1

Paulus, T. M., Lester, J. N., & Dempster, P. G. (2017). *Digital tools for qualitative research*. SAGE Publications.

Peräkylä, A. (2004). Conversation analysis. *Qualitative research practice*, (18), 165-179. https://www.torrossa.com/en/resources/an/4912495#page=172

Pink, S., Horst, H., Lewis, T., Hjorth, L., & Postill, J. (2015). Digital ethnography: Principles and practice. SAGE Publications. https://www.torrossa.com/en/resources/an/5017934

Raji, I. D., & Buolamwini, J. (2019). Actionable auditing: Investigating the impact of publicly naming biased performance results of commercial AI products. *Proceedings of the 2019 AAAI/ACM Conference on AI, Ethics, and Society*, (pp. 429–435). DOI: 10.1145/3306618.3314244

Raji, I. D., Smart, A., White, R. N., Mitchell, M., Gebru, T., Hutchinson, B., Smith-Loud, J., Theron, D., & Barnes, P. (2020). Closing the AI accountability gap: Defining an end-to-end framework for internal algorithmic auditing. *Proceedings of the 2020 Conference on Fairness, Accountability, and Transparency*, 33–44. DOI: 10.1145/3351095.3372873

Rothstein, M. A. (2023). Should chatbots be used to obtain informed consent for research? *The Journal of Law, Medicine & Ethics, 51*(1), 52–60. https://onlinelibrary.wiley.com/doi/full/10.1002/eahr.500190 PMID: 37988278

Samuel, G., & Wassenaar, D. (2024). Informed consent and AI transcription of qualitative data. *Journal of Empirical Research on Human Research Ethics; JERHRE, 20*(1–2), 3–5. DOI: 10.1177/15562646241296712 PMID: 39600037

Schlicker, N., Müller, S., & Krämer, N. (2025). How do we assess the trustworthiness of AI? Introducing the Trustworthiness Assessment Model (TrAM). *Computers in Human Behavior, 152*, 108256. DOI: 10.1016/j.chb.2025.108671

Schwandt, T. A. (1994). Constructivist, interpretivist approaches to human inquiry. In Denzin, N. K., & Lincoln, Y. S. (Eds.), *Handbook of qualitative research* (pp. 118–137). Sage Publications.

Selbst, A. D., & Powles, J. (2017). Meaningful information and the right to explanation. *International Data Privacy Law, 7*(4), 233–242. https://proceedings.mlr.press/v81/selbst18a.html

Selwyn, N. (2019). *Should robots replace teachers?: AI and the future of education*. John Wiley & Sons.

Selwyn, N. (2022). *Education and technology: Key issues and debates* (3rd ed.). Bloomsbury Academic. DOI: 10.5040/9781350145573

Snelson, C. (2016). Qualitative and mixed methods social media research. *International Journal of Qualitative Methods, 15*(1), 1–15. DOI: 10.1177/1609406915624574

Sun, W., Li, Y., & Zhao, J. (2025). Trustworthy AI on safety, bias, and privacy: A survey. *AI Review, 44*(2), 1–38. https://arxiv.org/pdf/2502.10450

Tuten, T. (2025, November 12). Expanding informed consent in the age of synthetic data and digital twins. *Quirks Media.* https://www.quirks.com/articles/expanding-informed-consent-in-the-age-of-synthetic-data-and-digital-twins

UNESCO. (2021). *Recommendation on the ethics of artificial intelligence*. United Nations Educational, Scientific and Cultural Organization.

University of Surrey. (2024, October 17). Ethics of AI in qualitative data analysis (QDA). University of Surrey CAQDAS Networking Project. https://www.surrey.ac.uk/computer-assisted-qualitative-data-analysis/qual-ai/ethics-ai-qda

Whittlestone, J., Nyrup, R., Alexandrova, A., & Cave, S. (2019). The role and limits of principles in AI ethics: Towards a focus on tensions. *Proceedings of the 2019 AAAI/ACM Conference on AI, Ethics, and Society*, 195–200. DOI: 10.1145/3306618.3314289

Williams, B. (2025, February 25). Informed consent in AI-driven qualitative research. *Insight7*. https://insight7.io/informed-consent-in-ai-driven-qualitative-research/

Williamson, B., Eynon, R., Knox, J., & Davies, H. (2023). Critical perspectives on AI in education: Political economy, discrimination, commercialization, governance and ethics. In *Handbook of artificial intelligence in education* (pp. 553–570). Edward Elgar Publishing., DOI: 10.4337/9781800375413.00037

Xiao, Z., Li, T. W., Karahalios, K., & Sundaram, H. (2023, April). Inform the uninformed: improving online informed consent reading with an AI-powered Chatbot. In *Proceedings of the 2023 CHI Conference on Human Factors in Computing Systems* (pp. 1-17). DOI: 10.1145/3544548.3581252

Zuboff, S. (2019). *The age of surveillance capitalism: The fight for a human future at the new frontier of power*. PublicAffairs. https://www.taylorfrancis.com/chapters/edit/10.4324/9781003320609-27/age-surveillance-capitalism-shoshana-zuboff

Section 2
Approaches to Artificial Intelligence

Chapter 7
Qualitative Inquiry in the Age of Algorithms:
A Factorial Systematic Review of AI's Impact on Management Control Systems

Mouhcine Rhouiri
http://orcid.org/0009-0007-7115-871X
Sidi Mohamed Ben Abdellah University, Morocco

Hicham Saidi
http://orcid.org/0009-0008-0764-8201
Sidi Mohamed Ben Abdellah University, Morocco

Mehdi Bensouda
http://orcid.org/0000-0002-7320-3795
Sidi Mohamed Ben Abdellah University, Morocco

ABSTRACT

This chapter explores how the integration of AI into Management Control Systems is impacting qualitative research. An integrative systematic review of 483 Scopus articles using IRaMuTeQ reveals four clusters: (1) decision support for optimizing industrial processes and logistics, (2) methodological frameworks for multi-criteria Decision-Making addressing uncertainty and subjectivity, (3) bibliometric analyses and systematic literature reviews mapping the field's evolution, and (4) AI-driven strategic decision-making and organizational transformation, including the implications of algorithmic authority. The results indicate a tension between algorithmic efficiency and qualitative depth. To preserve human values in algorithm development and application, we recommend Human-in-the-Loop frameworks that retain

DOI: 10.4018/979-8-3373-7640-0.ch007

t © 2026, IGI Global Scientific Publishing. Copying or distributing in print or electronic forms without written permission of IGI Global Scientific Publishing is prohibited. Use of this chapter to enerative artificial intelligence (AI) technologies is expressly prohibited. The publisher reserves all rights to license its use for generative AI training and machine learning model development.

human value systems. Finally, this research presents a methodological taxonomy for understanding how to work with AI-enhanced Management Control Systems in a manner that aligns technology with human values.

INTRODUCTION

Today's decision-support and organizational management ecosystems are rapidly evolving through the intersection between advanced computing power and growing complexity of socio-technical systems. Over the last decade, the dialogue surrounding Decision Support Systems (DSS), and Operations Research, has focused on the Industrial Revolution 4.0 model of achieving operational efficiency through automation, big data analytic, and algorithm optimization. Today, as these technologies reach the strategic levels of organizations, a need for critical reflection has emerged regarding the direction of technology. The goal is to move from technology that merely provides an organization with computational capacity to technology that offers a broader vision, centered on humans, explainability, and ethical validity (Krause et al., 2021). This shift marks the beginning of a new methodological era, in which the human decision-maker becomes a central partner in a "hybrid intelligence" loop where technology complements judgment rather than replacing it; an area being explored by human-machine synergy (Zhang et al., 2025), and collaborative AI clinical tools (Singer et al., 2022).

Despite this promise of this new paradigm, the integration of AI into execution support presents not only technical implementation issues but also other challenges related to value creation for an organization. Increasing the use of "opaque" or "black box" models for strategic management is changing how expertise, accountability, and ethical frameworks are defined. When using "canonical" Multi-Criteria Decision Making (MCDM) models in uncertain environments, many have raised concerns about the validity of algorithms, "algorithmic determinism" reducing human autonomy, a gap between theory and how theories can be operationalized and governance structures related to who makes the final decision (Zuboff, 2019; Arrieta et al., 2020). These concerns are sustained by empirical evidence. AI systems designed to assist employees may unintentionally shift power towards management, therefore providing increased control over employees (Monod et al., 2023; Monod et al., 2024), AI systems used in public employment services continue to show significant disconnects between developing trustworthy AI and the realities of implementing AI with respect to transparency (Berman et al., 2024). While the literature provides great insights into either algorithms or management on their own, there exists a need for a comprehensive understanding of how qualitative, quantitative, and human agency intersect when AI is applied to help organizations identify, analyze, and ultimately execute on decisions.

An integrated systematic review on the growing body of literature exploring artificial intelligence (AI), management control, and qualitative methodologies is provided by this research. A body of literature with 483 publications was examined using IRaMuTeQ via Descending Hierarchical Classification (DHC), to better understand how qualitative inquiry is being used and transformed in relation to the integration of AI into management control and decision-making processes.

Specifically, this chapter answers three questions:

> **RQ1:** *How are key qualitative research approaches (case studies, grounded theory, thematic analysis, etc.) being modified for AI impacts on management control systems?*
> **RQ2:** *What major challenges (epistemological and ethical) to qualitative approaches to studying AI (black box problem, algorithm bias, positionality) are reported by researchers?*
> **RQ3:** *What emerging frameworks (Human-Centered AI (HCAI), computational ethnography, hybrid methodologies) are being proposed to maintain reflexivity, trustworthiness, and connection to the interpretative tradition within AI-driven management control research?*

The remainder of the chapter is organized as follows: we initiated our analysis by exploring the theoretical framework: bounded rationality, socio-technical systems theory and the human-in-loops paradigm. The second part is devoted to the history of the evolution of quantitative decision support systems (DSS) toward intelligent systems that take the qualitative dimension into account, highlighting the tension between mathematics and human experience. Third, a description of the systematic review process and the textual analysis using IRaMuTeQ to implement the analytical methods of review and tessellation (i.e., partitioning the corpus into structured clusters). Fourth, a presentation of the four groups identified by DHC. Finally, conclusions on the implications for theory and practice, the identification of key methodological concerns and solutions, and a discussion of potential future research directions.

BACKGROUND

AI and Big Data analytics are proliferating in organizational decision-making, creating unprecedented opportunities for computational efficiency and predictive capability alongside equally persistent challenges such as the opacity of algorithmic reasoning, devaluation of tacit human knowledge, and erosion of employee autonomy in automated environments (Kellogg, et al. 2020). In response to these challenges, a wave of "human-centric" and "explainable" methodological approaches (e.g. Hybrid Intelligence, Fuzzy Logic, and Qualitative Case Studies) is emerging as a powerful counter to these challenges, offering promise for improving decision quality, ensuring ethical alignment, and reshaping the future of work (Shneiderman 2020; Power 2007). This intervention represents a fundamental shift in corporate responsibility and establishes a new "nexus" between algorithmic rationality and social realities, explored in the context of public sector DSS (Caruso, et al. 2023) and trustworthy AI in labor markets (Berman, et al. 2024).

In both industrial and administrative work environments, there exists intrinsic asymmetries between the ability of organizations to have visibility into the "soft constraints," the tacit knowledge that experts possess, and the various heuristic nuances that experts utilize to inspect, assess, and navigate the complexities within any circumstance or context (Silva, 2009). The use of traditional "hard" operations research (OR) methodologies to create mathematically-based models of problems creates a tension, as although these mathematical models can model massive volumes of data, they have difficulty incorporating or dealing with "vagueness," as well as the context and fluidity within which many of the organizations' actual decision-making processes take place (Belton & Stewart, 2002). The result is that even though the OR mathematical models may identify the optimal models based on quantitative measurements, the implementation of those optimal models in practice is unsuccessful because the qualitative aspects of the workplace have not been considered and result in user resistance to implementing the ideal model, therefore making the resultant system fragile. This gap exist at assembly line feeding (Baller et al. 2020) and in the field of port logistics where mathematical prediction models must be validated against operational data in order to be used successfully (Maldonado et al. 2019).

Despite the recognition of the value of incorporating qualitative insights into quantitative models, there are substantial gaps. Traditional approaches to Decision Support are based on static, single loop learning and rigid mathematical frameworks. In a dynamic environment, traditional Decision Support approaches fail at times to adapt to unpredictable disruptions leading to "rigidity traps" where systems cannot adapt to unanticipated disruptions (Maldonado et al., 2019). Some of the barriers to the adoption of "human-aware" intelligent solutions are: the "black box" nature of current AI undermines trust in AI and support for the use of intelligent support tools; there is a lake of standards and universal framework for evaluating "soft" outcomes such as user empowerment, fragmentation exists between the technical and social sciences (Faraj et al., 2018), and there is a rapid increase in AI innovations, which creates additional vulnerabilities as rapid innovations disrupt professional definitions of work and establish a precarious relationship between people and algorithmic authority. Research on the dislocation between traditional and digitized roles of management accountants (Rautiainen et al., 2024; Andreassen, 2020) and bank controllers (Rautiainen et al., 2024) exemplifies this disruption.

On the one hand, algorithmic-based support tools improve prediction accuracy and increase efficiency in operational processes. On the other hand, algorithmic-based tools raise new methodological and ethical concerns. For illusrration, MCDM tools, such as Preference Ranking Organization METHod for Enrichment of Evaluations (PROMETHEE) or Analytic Hierarchy Process (AHP), used for standardizing complex evaluations, frequently do not account for the cognitive bias of the decision-maker who is reluctant to assign definite weights and therefore creates a risk of "false precision" (Kahraman, 2008). Refinements of MCDM methods that have addressed this limitation include David Envelopment Analysis (DEA) with imprecise weights as inputs (Karasakal et al., 2022) and evidential reasoning for quantifying qualitative uncertainty (Yang et al., 2009). Moreover, algorithmic support solutions for management control create additional opacity, making it difficult to contest the rationality of a black box evaluation of employee performance (Burrell, 2016). These limitations of existing methodologies require methodological adaptation to ensure that the introduction of technology will invariably augment what humans can will. Integrated frameworks that have emerged in response to these limitations include fuzzy logic and DEMATEL (Decision-Making Trial and Evaluation Laboratory) for sustainable planning (Jeong & Ramírez-Gómez, 2018), and cognitive maps for CSR implementation (Poplawska, et al., 2015), represent attempts to address these challenges.

In the context of these complexities, the study bridges the gap between theory and practice by using critical analysis to analyze the relationship between digital technology, decision-making methodologies, and organizational behavior for the purpose of identifying implications for researchers, designers, and executives to create more equitable and human-centered intelligent DSS. It combines empirical and critical analysis to produce recommendations for conducting responsible research with qualitative rigor, in conjunction with deploying technology.

THEORETICAL AND EPISTEMOLOGICAL LANDSCAPE

To understand how to interpret this work, it is necessary to locate the results in a larger theoretical and epistemological context surrounding Decision Support Systems (DSS), Management Control and Operations Research. Artificial Intelligence (AI) is currently being rapidly incorporated into organizations, which is a technological update. However, it also represents a new epistemology regarding how to make decisions: it is creating a new definition of what constitutes valid knowledge, proof of truth, and control. Hence, the need for a contemporary critical analysis requires the combination of three disparate but related traditions: the mathematical basis for creating optimized results, cognitive components involved in human decision making and socio-scientific considerations raised by algorithmic governance. The below discussion of these theoretical foundations provides insight into the movement from positivistic models of rationality to the current socio-technical model. By doing so, we will be able to understand more fully the ongoing tension that exists between the accuracy of algorithms and the qualitative characteristics associated with human agents in an organizational setting.

The Evolution of Decision Support: From Mathematical Optimization to Hybrid Intelligence

The origin of DSS can be found in the post-war science management paradigm and Operations Research (OR), using a positivist philosophy to model many complex organizational problems through deterministic or stochastic mathematics (Power, 2002). The initial focus of DSS was to increase human cognitive limitations through the application of computational capability for data processing and algorithmic optimization, which was an early representation of Simon's (1997) 'bounded rationality' notion. The main unit of analysis was the algorithm and the procedure used to assess the success of the methodology was based upon the efficiency of the model to either minimize costs or maximize utility based on a defined set of constraints (Arnott & Pervan, 2014).

However, with the emergence of Big Data analytics and the use of advanced Artificial Intelligence (AI), this has brought about a major epistemological transformation towards Hybrid Intelligence. We are now moving away from "model-driven" DSS methodologies, which use pre-programmed logic for decision making, to "data-driven" DSS that use patterns in data to learn from (Power, 2007). In this new paradigm, instead of just automating decision making, it is important to form symbiotic partnerships between human beings and AI, with human beings making sure that all the contextual and ethical considerations are identified and the complexity involved in computationally making different options available, like logistics routing (Maldonado et al., 2019). This development of Hybrid Intelligence is challenging the traditional approach to OR through providing evidence that optimality as a solution is not only a mathematical number, but rather it is a dynamic entity that is obtained through a synergistic interaction between the human being and the machine (Krause et al., 2021).

The Methodological Tension: Precision vs. Human Ambiguity

A major theme in the literature is the methodological tension associated with the operationalisation of subjective human judgement into rigid quantitative frameworks. MCDM remains a fundamental component of DSS research; however, it continually struggles with epistemological validity issues. Classical MCDM methods require that decision-makers assign precise numerical weighting to vague or ambiguous criteria. As Belton and Stewart (2002) note, imposing "false precision" on complex strategic issues can produce skewed results, decrease confidence in their outcomes, and create a disconnect between model results and decision-maker mental models.

In addition, the rapid growth of "black-box" AI models (e.g., Deep Learning and ensemble-based models) now found in many areas of modern management have increased these concerns. Although these models have higher predictive accuracy than equivalent models, they generally lack the level of transparency required by high-stakes decision-making environments (Arrieta et al., 2020). With this lack of transparency, there is a significant validity threat to DSS: if the decision-maker does not understand or cannot trust the rationale for an algorithm's recommendation, then the DSS does not function as a support tool. Consequently, there is a growing body of methodological research examining Fuzzy Set Theory (Zadeh, 1965) and Evidential Reasoning (Yang, 2001). The focus of these methodological approaches has been to mathematically model uncertainty and imprecision such that the logic of the tools used in decision-making can be more closely aligned with the uncertainty and vagueness of human thought (Kahraman, 2008). This represents a shift in focus from finding a single "optimal" solution to identifying multiple "satisfactory" solutions that allow for and assume vagueness on the part of the decision-maker.

The Sociotechnical Turn: Power, Ethics, and Organizational Control

DSS research has expanded from a traditional mathematical and algorithmic focus to include a sociotechnical perspective that examines the impact of AI on organisational workflows as a non-neutral technology. The implementation of AI technology will have a fundamental impact on the dynamics of power, labour processes, and ethical frameworks. Researchers using sociological frameworks to analyse the DSS have described AI-based DSS systems as a means or tools of "digital Taylorism" that enhance management control and surveillance through increased efficiency (Zuboff, 2019). For instance, algorithm-driven management systems include logistics or HR management systems can devalue employees' social capital ans treating them as nodes of an optimisation algorithm (Möhlmann et al., 2021). This turn highlights the risks of dehumanization and the erosion of employee autonomy when algorithms dictate workflows without human input (Kellogg et al., 2020).

Therefore, the "Human-in-the-Loop" (HITL) design concept has emerged not only as a technical requirement but also as an ethical necessity (Shneiderman, 2020). The sociotechnical perspective posits that a DSS can be effective only when it is designed to maintain human agency. This leads to a shift from purely technical metrics (i.e., speed and accuracy) to qualitative metrics evaluating the user's empowerment, trust in the user, and "algorithmic legibility" (i.e., the ability of the user to understand and contest the logic of the system) (Burrell, 2016).

The Fragmentation of the Field and the Need for Qualitative Synthesis

The field of Intelligent Decision Support currently exhibits extensive fragmentation and division along traditional disciplinary lines. Computer Science continues to advance in the area of algorithmic based innovations, Operations Research places an emphasis on mathematical methods of optimization, while both Management and Information Systems emphasize the strategic adoption and subsequent usage of Intelligent Decision Support Systems (Faraj et al., 2018). As a result, there are many distinct streams of research within the Intelligent Decision Support field that continue to develop independently of one another without the use of an overarching theoretical framework. As noted by Gregor (2006), the Information Systems community has struggled with the development of cumulative theory and has resulted in the creation of disparate and disconnected case studies that do not contribute to the construction of a unified body of knowledge.

This fragmentation among the various methodologies employed by researchers provides a significant gap in methodological rigor and integrative validity within this body of research. There is a wealth of technical proofs and disparate case studies available. However, there is a dearth of research that explores the convergence or conflict between the various methodologies when implemented in practice. There exists a critical need for a synthesis that interrogates the research paradigms behind each methodology (Tranfield et al., 2003). This need for a synthesis supports the necessity for an Integrative Systematic Review using qualitative textual analysis. An Integrative Systematic Review will assist in identifying the intellectual structure of the Intelligent Decision Support field, assisting in determining where quantitative rigor intersects with qualitative reality, and identifying unresolved methodological debates that are occurring within the field.

RESEARCH DESIGN & DATA COLLECTION

This chapter applies an Integrative Systematic Review (ISR) of the qualitative methodology. This approach combines the breadth of an integrative literature review—which includes empirical case studies, theoretical models, and expert documents—with the systematic review's rigor (Lubbe et al., 2020 The research has been designed to produce a nuanced understanding of the intersection of management control systems, AI, and qualitative methodologies particularly relating to the interpretative nature of how AI transforms organisational governance and human agency. This aligns with longitudinal qualitative case studies exploring power transitions when implementing AI (Monod et al., 2023; Monod et al., 2024) and interpretative studies examining changing roles within management accountants due to digitalisation (Rautiainen et al., 2024; Andreassen, 2020). Additionally, the need for integrative approaches is supported by research using multiple qualitative methodologies (including grounded theory) to develop unified success factors for big data (Zhao et al., 2025) and thematic analysis to assess AI use within different sectors (Wider et al., 2025).

Data Collection

The research was conducted using the SCOPUS database, which is chosen because it includes many disciplines with a comprehensive list of academic journals that contain high quality articles in business, management and accounting (SUBJAREA "BUSI"). The literature review involves a Boolean search that is intended to find literature that relates to the three aspects of qualitative research methodology, artificial intelligence technology, and management control systems. To accomplish this Boolean search, all of the synonyms for each aspect of the research were included::

- For aspect of **Qualitative Methodology**: Qualitative Methodology, Case Study, Grounded Theory, Thematic Analysis, Interpretivism?
- For aspect of **Artificial Intelligence**: Artificial Intelligence, Machine Learning, Large Language Model (LLM), Natural Language Processing (NLP)
- For aspect of **Management Control**: Management Control, Management Accounting, Performance Management, Budgetary Control, Organizational Control Systems, Decision Support.

The final search string used was: *TITLE-ABS-KEY ("qualitative methodology" OR "qualitative research" OR "qualitative inquiry" OR "case study" OR "grounded theory" OR "thematic analysis" OR "interpretivism" OR "reflexivity")*

AND TITLE-ABS-KEY ("artificial intelligence" OR "AI" OR "machine learning" OR "large language models" OR "LLM" OR "natural language processing" OR "NLP" OR "AI-enhanced research") AND TITLE-ABS-KEY ("management control" OR "management accounting" OR "performance management" OR "budgetary control" OR "strategic control" OR "organizational control systems" OR "decision support") AND PUBYEAR > 2007 AND PUBYEAR < 2026 AND (LIMIT-TO (SUBJAREA, "BUSI"))

The targeted approach limited the review to a small group of articles which provided a common point between qualitative research, artificial intelligence and managerial control. By limiting the search to articles written in the BUSI domain provided this review with articles pertaining to organizational issues and eliminated strictly technical articles. This review period was from 2008 to 2025, allowing observation of long-term evolutions. Finally, this search yielded **483 articles**, ideally suited for qualitative textual analysis using IRaMuTeQ.

Procedure for Selection of Studies

The process to select the studies was conducted in accordance with the PRISMA 2020 protocol (Page et al., 2021), thus ensuring both transparency and validity, particularly in the realm of AI adoption (Wider et al., 2025) and systematic reviews in the area of digital employees (Pandya & Boukareva, 2025).

Figure 1 outlines how a multi-stage process was developed to select the studies through a series of steps that ultimately provided a final set of studies:

- The identification of 8,935 documents was the first step.
- Secondly, the Screening & Refinement was Applicated on BUSI subject area to limit excluded purely technical papers, reducing this corpus to 483 articles.
- Finally, the eligibility means that All abstracts and full text were screened to confirm peer-reviewed publications, English language and an explicit focus on AI integration into the area of managerial control, thus including both qualitative and/or methodological contributions.

Figure 1. Adapted PRISMA flow diagram.

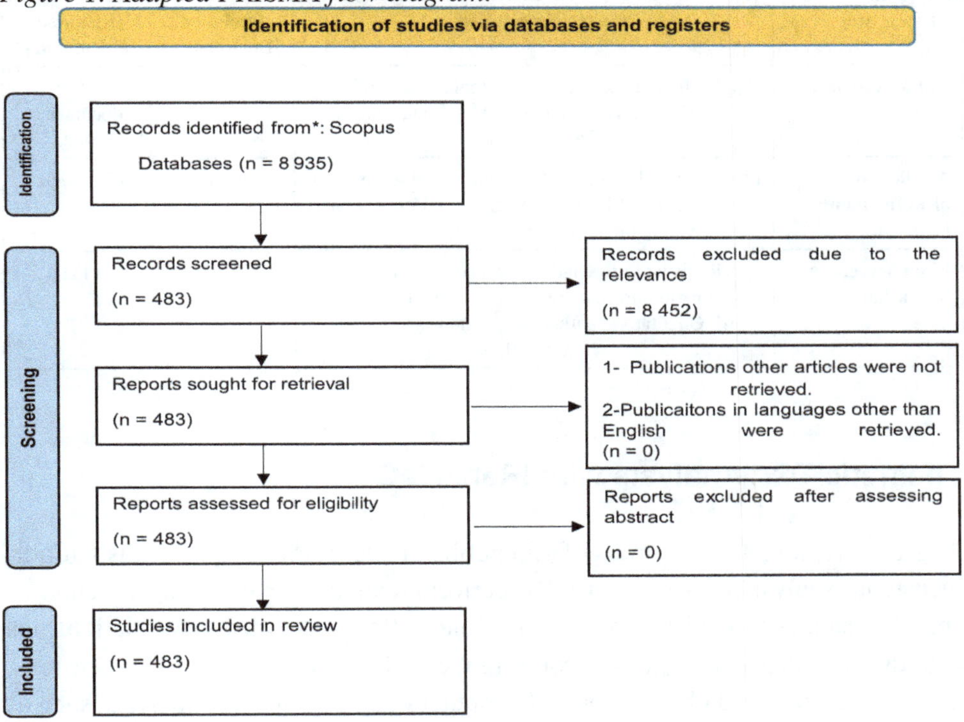

Corpus Overview

The final corpus of 483 articles was compiled from relevant business, management, and accounting disciplines over the period (2008-2025) in order to help address some of the cross-cutting challenges associated with "Intelligent Management Control" that occur at the intersection of algorithmic optimization, human intuition, and ethical governance. The interdisciplinary nature of these studies allows for a more complete understanding of the relationship between qualitative research methodologies and institutional theory (Rautiainen et al, 2024) or via the application of grounded theory to explore the socio-technical dynamics associated with DSS implementation (Nagle et al., 2024; Zhao et al., 2025).

Table 2. Filtering processes to identify the final corpus of articles

Process Step	Criteria Applied	Excluded Documents	Included Documents
Initial Research	Application of comprehensive Boolean search string on SCOPUS (Qualitative + AI + Management Control terms).		8 935 documents found
Application of Refinement Criteria	Limitation to Business, Management and Accounting subject area (SUBJAREA "BUSI"). Publication timeframe (2008-2025).	8 452	483 articles
Final Screening by Authors	Review of titles and abstracts (with full-text verification) to ensure focus on the intersection of AI, Qualitative Methodology, and Management Control.	0	483

Thematic Text Analysis with IRaMuTeQ

After creating a final corpus of 483 publications, IRaMuTeQ which is multidimensional analysis's texts was used to perform textual thematic analysis. This enabled a systematic identification of themes and patterns, as well as new relationships regarding AI-related management control's methodology and qualitative dimensions, in a reproducible and objective way. Computational tools for text analysis, such as IRaMuTeQ are increasingly being adopted for qualitative synthesis rigor in management research. The use of these tools in this way is being seen across multiple fields, e.g., as in research using Topic Modelling to analyze larger corpora (Xue et al., 2025; Sahin et al., 2025). The analysis consisted of the following stages:

- **Cleaning the corpus**: The corpus was cleaned by removing stop words, punctuation, and formatting irregularities to accomplish isolating the relevant vocabulary for the domain of research.
- **Lemmatization:** A lemmatization procedure was used to aggregate the respective forms of a word (e.g., "decide", "decides" and "decision"; "manage", "manager" and "management") to a single root form, thereby providing greater coherence in the results of the analysis.
- **Descending Hierarchical Classification (DHC):** is a specific method that allows for segmentation of the corpus into homogeneous thematic classes according to co-occurrence of words (Reinert, 1993), thereby allowing for the distinct thematic areas of the academic debate to emerge. DHC is similar to the clustering techniques used in other areas of decision support research, e.g., K-means is used for market segmentation (Lei & Moon, 2015) and Density Based Clustering is used for detection of anomalies (Coelho e

Silva & Murça, 2023). The findings of the classification are illustrated and discussed in the results section.

- **Supporting analyses**: Similarity analysis and keyword clouds were also developed to illustrate the frequency of co-occurrence and provide a snapshot of the dominant vocabulary for the field. This illustrative method was also relied upon to visualize the findings of decision support systems research related to urban planning (Coutinho-Rodrigues et al., 2011) and to study risk perception on social media (Shook & Turner, 2016).

RESULTS OF THE TEXT ANALYSIS

According to the results of the analysis, we can approach a complete and comprehensive overview of multi-diversified domains interconnected under a multidimensional framework governing the converging areas of methodological modelling, operational logistical delivery and artificial intelligence that governs decision making within organisations. The Descending Hierarchical Classification (DHC) uncovered four major areas of focus contained within the hierarchical structure that represent themes associated with moving beyond traditional research toward intelligent decision-making ecosystems supported by the integration of AI technologies into business processes.

Figure 2. Distribution of descending hierarchical classes

Source: Release of IRaMuteQ 0.8a7

This analysis shows that:

- The first area of focus contains the most percentage (34.7%) dedicated to examining methods related to the development of various types of methodological frameworks, including Multi-Criteria Decision-Making (MCDM) methods such as PROMETHEE, AHP and various mathematical models employed by decision support systems.
- The second area of focus contains the second highest percentage of all of the four areas analysed (28.1%) associated with the application of DSS to operational/ industrial activities, such as transportation, supply chain management, port management and planning infrastructure.
- The third area of focus contains the third highest percentage among the four analysed (21.9%) that is concerned with the strategic integration of artificial intelligence (AI) and Big Data (BD) as part of management, as well as how AI is redefining and transforming existing (traditional) business processes and control systems used by management/ organisations.
- The fourth area of focus includes analysis of the evolution of the field through the application of bibliometrics and literature review methods in order to develop understanding of the identified research trends within the field as well as research gaps.

Collectively, these clusters represent an evolution toward the emergence of an integrated decision-making model whereby the mathematical optimisation processes used by both operationally driven and AI/technologically driven strategic transformation will now be incorporated into the new integrated decision-making paradigm. The following sections will describe each cluster in greater detail through references to selected peer-reviewed journal articles that comprise the body of scientific literature discussed previously.

Class 1: Decision Support for Industrial Operations and Logistics Optimization (28.1%)

This class illustrates the practical use of decision support systems through examples of the operation of both physical and industrial environments, including those related to logistics, supply chain management, or optimizing operational workflows that include items such as trucks, ports, stacking, and containers. The research has demonstrated the deployment of mathematical models as tools to help solve complex problems of scheduling and resource allocation. Maldonado et al. (2019) propose a decision support system designed to predict container dwell time at ports that reduce the number of times an imported container must be moved when stacking in a port for efficiency. Baller et al. (2020) develop a mixed-integer linear programming model to optimise material supply policy for automotive "assembly line feeding problem" that decreases total costs in comparison to the traditional method of planning via manual means. Legato and Mazza (2018) have developed a decision support modelling approach to support integrated handling of containers within transshipment hubs, which utilises both queuing network models and discrete-event simulation tools. Silva (2009) described a scheduling system for producing synthetic fibre via the application of formal rules combined with ad hoc means of decision-making, and is an example of the importance of maintaining flexibility in decision-making by human planners. Other research has sought to optimise the operations of humanitarian supply chains (Sahebjamnia et al., 2017), the routing of emergency evacuation of people from dangerous locations (Shahparvari et al., 2016), the operation of container terminal facilities (Ursavas, 2014; Ding et al., 2023) and the transportation of hazardous materials (Romero et al., 2016).

Class 2: Methodological Frameworks and Multi-Criteria Decision Making (34.7%)

This class evaluates both algorithmic and mathematical foundations of MCDM decision support systems. Numerous keywords like decisions, criterion, model, and propose dominate the core literature related to the methodological basis of these types of research. This type is used to develop better mathematical tools for dealing with uncertainty (DeSmet & Doan, 2018) and conflicting preferences. Karasakal et al. (2022) present new ranking methods that apply PROMETHEE to create a ranking system for addressing imprecise threshold weights/criteria using DEA methods. Yang et al. (2009) conducted a study on how evidential reasoning can help prioritize the voices of customers by combining both qualitative/quantitative market survey data into one framework. The challenge of handling uncertainty is addressed through inverse optimization techniques as a method for conducting weight sensitive analysis of PROMETHEE-based systems (Doan & De Smet, 2018), gray-based multi-criteria decision analysis systems (Maghrabie et al., 2019), or fuzzy set-based theories used in multi-expert MCDM systems (Noor-E-Alam et al., 2011).

Lolli et al. (2017) developed a combination AHP-based group decision support system that included clustering model and knapsack model for requalifying public buildings. Integrated frameworks include GIS-based green supply chain models for carbon price uncertainty (Ren et al., 2020) and fuzzy DEMATEL-MARCOS models of humanitarian supply chain issues (Wang et al., 2024). This class has a strong emphasis on how successful decision support systems heavily depend on good methodological development, with respect to mathematical models and aggregation techniques, supported by research on belief-rule based systems (Yang et al., 2018; Zhou et al., 2013).

Class 3: Bibliometric Analysis and Systematic Literature Reviews (15.3%)

This class takes a great look at the current status of meta-analysis as it relates to knowledge reviews in a number of areas including mapping research, evaluating how much knowledge is produced, and collating existing knowledge. This class of literature contains words like "research", "article", "review" and "bibliographic" indicating that this cluster is serving more of a reflective than an active role. Xue et al. (2025) have produced a comprehensive bibliometric analysis covering evolutionary trends relative to the development of AI-driven firm-wide management such as their leading countries, institutions and evolving thematic shifts towards the human resource and financial management applications. Zhao et al. (2025) have taken a grounded theory-based literature review and extended the ICO model into Big Data using their Unified Theory of Intelligent Decision Support. Kambur and Yildirim (2023) have provided a systematic review detailing the evolution from traditional to intelligent HR management whereby over 5,500 articles were analyzed allowing for mapping of the impacts of AI, chatbots and Industry 4.0 on HR-related functions. The review of literature illustrated in this cluster demonstrates the value of systematic reviews in determining the current state of knowledge, identifying gaps in the literature and providing directions for future research, as evidenced in the literature reviews of selecting a MCDA method (Cinelli et al., 2020) and analyzing AI applications in law (Singh & Arora, 2025). Additionally, requests for research on how to integrate ESG risks (Crawford & Nilsson, 2023), emphasize the future look of this meta-analytical type work.

Class 4: AI-Driven Strategic Management and Organizational Transformation (21.9%)

It is evident that this course embodies the convergence of advanced artificial intelligence and strategic business management. The keywords 'artificial intelligence', 'AI', 'management' and 'big data' illustrate an emphasis on the use of cognitive technologies on how organizations are affected by, process, control and manage resources in ways that go beyond operational efficiencies. Caruso et al. (2023) provide an example of an AI application that addresses the issue of KPI overload in public organizations, demonstrating how machine learning can assist in fraud detection and aiding in internal control. Mobo (2025) discusses how AI is being used in green HR practices, including AI based recruitment and performance management for developing sustainable workforces. Similar evidence can be found in research on HR technology trends, where the use of AI, machine learning and blockchain are changing the way HR is performed (Lakshmi, 2025; Gupta et al., 2025). Monod et al. (2023) give a critical account of the power dynamics associated with CRM by examining how AI systems that are intended to assist employees may inadvertently lead to an increase in management control and alter the symbolic and cultural capital of the organization. Monod et al. (2024) build upon this work by tracking the change in AI tools from being used to empower workers to being used to control managers. Studies on the use of trustworthy AI in public sector organizations (Berman et al., 2024), AI driven DSS in the automotive sector (Mantouzi & Youssef, 2025), and the use of AI in facilitating cloud computing and blockchain within healthcare (Kulkov et al., 2024) illustrate that the future of DSS will involve not only the calculation of data but also the strategic integration of AI into organizations and therefore necessitating a complete understanding of socio-technical dynamics (Arnaboldi et al., 2022).

DISCUSSION OF RESULTS

Basing on the result of this integrative systematic study using a Descending Hierarchical Classification, four clusters illustrate the intricate progression of research methodologies moving through the aegis of Artificial Intelligence. The clusters also illustrate the plurality of methodologies that exist, from a pure traditional positivist/quantifiable modelling perspective to a pluralistic type of methodological approach, including the blending of qualitative data, design science and computational text analyses. These areas provide the structure in which four different methodological axes can be defined: methodological hybridisation of optimisation and case study research (class 1); mathematical operationalisation of subjective judgment (Class 2); AI-enabled tools for qualitative synthesis (Class 3); and the re-emergence of interpretive case studies for the assessment of the effect of AI (Class 4).

The Methodological Hybridization of Operations Research

The first axis focuses on the changing nature of research design within industrial logistics as the field has shifted from traditional mathematical modelling methodologies to more experimental designs predicated on the principles of Design Science Research (DSR). Unlike traditional operations research methods that derive their "rigor" from theorem proof, DSRs derive their "rigor" from iterative method application to real-world case studies.

Moving from deductive to abductive reasoning

As illustrated by Caruso et al. (2023), a shift exists from deductive modelling methods to abductive modelling methods which validate the use of artificial intelligence (AI) techniques against public organisation data to mitigate "overload of key performance indicators (KPI)". This methodology is accomplished by combining "hard" quantitative predictive methods with "soft" empirical field data and is also demonstrated in logistics-related field studies of port operations (Legato & Mazza, 2018) and sustainable transportation management (Shi et al., 2019).

Accommodating for soft constraints

Kallestrup et al. (2014) and Baller et al. (2020) challenge the existing operations research framework that attempts to determine a single optimal solution and argue instead for the development of heuristic approaches that can accommodate "soft constraints," or informal knowledge, that cannot be quantified. Kallestrup et al. (2014) developed a framework for hierarchical planning in oil refining that takes into account the realities of organisational process and acknowledges that human planners possess tacit knowledge that cannot be represented by algorithms. This has been documented in studies that used simulation and optimisation techniques for developing disaster-reduction plans that ensure business continuity (Schätter et al., 2019) and broiler production supply chains (Brevik et al., 2020). Also, the need to integrate empirical knowledge into predictive analytics to develop truck waiting times (Hill & Böse, 2017) and reverse logistics decision support systems has been supported by case study research (Lechner & Reimann, 2020).

The Operationalization of Subjectivity in MCDM

The second axis of MCDA research consists of research related to methodological problems with the MCDM process and how it creates tensions between quantitative accuracy and the subjective nature of human preferences. Therefore, the research agenda for this axis is based on the valid application of MCGM methods given their inherent lack of precision.

Bridging the Qualitative-Quantitative Gap:

Cinelli et al. (2022) and Yang et al. (2009) address the "garbage in, garbage out" phenomenon that occurs when exact numerical inputs are forced into the process. According to Cinelli et al. (2022), the use of inappropriate methods for MCDA in energy systems can produce suboptimal decision-making and therefore should select methods commensurate with the problem's characteristics. Their use of evidential reasoning is considered an innovative approach to address the inherent fuzziness of data used in MCDA processes. This is a central concern of Doan and De Smet (2018), who proposed the use of inverse optimization for weight sensitivity analysis of PROMETHEE II, since they recognize that precise weight information is difficult to obtain and are therefore able to provide robust decision support in environments such as wildfire management by implementing a ranking stability test under different weight change scenarios (Suarez et al., 2024).

Behavioral Operationalization:

In pursuit of a more robust ordinal regression model based on nonlinear preferences, Luli et al. (2024) used Bézier curves to improve the selection process for insulating materials. This refinement reflects a methodological shift toward behavioral operations research, which involves adapting mathematical models for use with the psychological profiles of decision-makers. This trend can also be observed in approaches that deconstruct and reconstruct the collective decision-making process (Bregar, 2011; Yannacopoulos et al., 2014). The methodological approach to modeling uncertainty in human judgments through the integration of fuzzy set theories illustrates this trend further and provides examples of its application, such as supplier evaluation (Noor-E-Alam et al., 2011) and during the pandemic for public transportation management (Deveci et al., 2022).

AI as a Qualitative Research Tool (Meta-Analysis)

The third component of this study centers on the research methodology focus on the use of AI and bibliometric tools to review and summarize extant literature about mis classifications of research articles. This cluster provided evidence that researchers are moving away from descriptive narrative reviews towards computational forms of qualitative analysis.

Automated Qualitative Synthesis:

Sidahmed and Pascale (2024) used a literature review to explore how strategic decision support systems were used in managing crises such as COVID19. They Identified critical interdependencies among various sectors. This is reflected in Al-Alawi and Albinali (2024), who undertook a systematic literature review to analyze employee engagement framework-models through performance analytics and examine how HRM integrates AI and human judgment.

Grounded Theory in the Era of Big Data:

Nagle et al. (2024) utilized a Grounded Theory methodology on articles in the news to capture social-technical aspects of VAR Implementation in football. This indicates the development of a new frontier where qualitative coding and theory development can be used to classify and analyze unstructured text data combining the qualitative depth of the data types along with breadth of the data types with a large number of records. This class demonstrates that AI is a valid methodology to assist in conducting knowledge synthesis. Likewise, Siering et al. (2017) utilized cluster analysis of SEC litigation reports to formulate a taxonomy of financial market manipulations. Methods such as topic modeling and text mining to analyze large corpora of data, as illustrated by Ting et al. (2014) in logistical data mining on food supply chains and by Niu et al. (2016) in event-based predictions of financial markets provide further evidence of the use of computational analytical methods to conduct qualitative and meta-analysis research.

The Return to Interpretive and Sociological Inquiry

Research has increased significantly over time for the "black box" nature of AI in Management, with researchers looking to qualitative case studies to discover how AI affects organizations using Critical Social Theory. As algorithms get more complex, most of the current literature is lacking in the qualitative realm of case studies to aid in understanding the organizational impact of what cannot be captured with quantitative measurement.

Interpretive Case Studies:

Mackrell et al. (2009) used qualitative interpretive case study design informed by Rogers' innovation-decision model and Orlikowski's technology-in-practice model. Using this method allows researchers to observe how users reflexively adapt to changes in technology based on their needs. They found that successful implementation is often considerably different than originally anticipated, and many of these alternative uses of technology would not fit within traditional metrics for measuring success. Hartmann (2011) provides further evidence of the importance of focusing on the dynamics of the implementation process. He did a longitudinal case study where he showed that the goals and processes for implementing a 4D decision support system were constantly negotiated by members of the project team.

Phenomenological Approaches to AI Impact:

Crandall et al. (2025) use a design science research methodology to create a chatbot for school counselors. They focus on phenomenology by examining user-driven design curiosities, user satisfaction, and user intent to continue. Their qualitative research methodology examines the user experiences of DSS in non-technical fields. Mixed method research is necessary to determine the success of AI by comparing the technical effectiveness of AI technology to the autonomy of employees using it and the ethical implication of the AI solution. Literature on hotel training workforce for AI technology demonstrates this by sharing case studies of how AI improves productivity and addresses the issue of high employee turnover rates (Kumar et al., 2025). Additionally, the need for qualitative research to understand how the digital future is changing the professional roles of workers such as management accountants supports the incorporation of interpretive methodologies (Secinaro et al., 2024; Korhonen et al., 2021). Lastly, the case study research on e-government procurement DSS by Soliman & El-Barkouky (2020) provide evidence that qualitative case study methods provide knowledge of system design through historical performance data and expertise from stakeholders to support the design.

Finally, these four classes of methods indicate that decision support systems research methodology is experiencing a significant methodological pluralism. The rigid distinctions between quantitative modeling and qualitative inquiry are beginning to disappear and researchers increasingly use hybrid methods (Class 1), develop mathematical models for qualitative subjectivity (Class 2), incorporate AI in qualitative syntheses (Class 3), and rely on critical social theory for interpreting AI results (Class 4). All this indicates that rigorous research in the AI era requires a combination of methods: computational power for managing complexity, and qualitative depth for ensuring humanity.

ISSUES, CONTROVERSIES, AND PROBLEMS

Despite the revolutionary effects and possible future uses of Decision Support Systems (DSS) and Artificial Intelligence (AI), there are many underlying issues related to legislation systems. Qualitative issues include: validity of measurement, operationalization of human behaviour, and fragmentation of theories. There are many difficulties in being able to measure “soft” power dynamics, the reductionist approach to using quantitative models, the validity gap between the values produced from algorithms compared to their actual values in the field, and having no universally accepted frameworks to tie the whole process together.

The "Invisible Metrics" Problem: Methodological Blindness to Power Dynamics

The adequacy of existing metrics for assessing the socio-ecological impact of AI/ automated systems is one central issue of dispute. Although quantitative methods have adequately assessed efficiency mainly (Caruso et al., 2023) these metrics have a significant degree of "methodological blindness" to the impacts on human agency. Qualitative research has identified that AI systems designed to empower workers are acting as tools of "digital Taylorism" and increasing management's control over workers (Monod et al., 2024). Berman et al. (2024) showed that there are significant disparities between theoretical and practical definitions of trustworthy AI regarding transparency. The issue is focused on threats to validity in positivist designs. If researchers continue to limit their focus to the standard measurable output, they will validate efficiency whilst ignoring the "symbolic violence" enacted on workers. Arnaboldi et al. (2022) argue that an emergent conceptual framework is necessary to connect the technical, organisational, social, and individual dimensions of research.

The Reductionism of Quantification: Validity in MCDM

The operationalization of subjective judgement is a source of considerable concern. Classical MCDM studies commonly assume that human preferences can be accurately quantified (Cinelli et al., 2020; Yang et al., 2009), but this suffers from the problems of reductionism, which involves cramming rich, qualitative, and context-dependent materials into fixed boxes. The danger of being "garbage in, garbage out" is profound and the authors of Maghrabie et al. (2019) address it through a new gray-based MCDA approach designed to incorporate, both subjectively and objectively, uncertainty. Similarly, integrated frameworks for the selection of green suppliers using QFD and MCDM have been proposed (Yazdani et al., 2017), and a comprehensive taxonomy of MCDA processes is found in Cinelli et al. (2020), both aimed at addressing this area of weakness. The methods community is divided; should they continue to refine current mathematical models to include uncertainty as fuzziness, or should they change direction toward qualitative/ descriptive decision-making?

Contextual Blindness: The Gap Between Algorithmic Models and Field Reality

Literature describes a huge difference between the controlled environment (laboratory) and the uncontrolled environment (field). Trucco and Lapko (2023) show how deterministically optimized pipe-laying projects will exceed their budgets due to unanticipated weather. This is a problem of external validity in the literature.

In many studies, human behaviour in dynamic situations, like humanitarian relief chains, is chaotic (Sahebjamnia et al., 2017). Thus, there is a controversy regarding the research design itself; is it appropriate to pursue mathematically optimal solutions when they fall apart with the unpredictable nature of the real world? The successful execution of a DSS for vendor-managed inventory (Borade & Sweeney, 2015) was only possible through a case study that compared the model against actual supply chain usage data.

Theoretical Fragmentation and the Absence of a Unifying Paradigm

Rapid advances in AI research have led to methodological siloing and confusion within epistemology. As identified by Zhao et al. (2025) and Xue et al. (2025), there exists fragmentation in the AI knowledge domain as computer scientists pursue the idea of algorithmic efficiency; managers pursue the idea of strategy and operations researchers pursue the idea of logistics; however, they lack a cohesive, common framework—the "Success Model"—to link them together. Such discrepancies in the theoretical underpinning create barriers to building cumulative knowledge. If this continues, AI will fall into a state of having only isolated research studies as opposed to a progressive discipline, as shown by the varied use of an ontology-based framework (El-Gayar & Deokar, 2013) compared to one based on argumentation theory (Janjua & Hussain, 2012). Practitioners have been forced to adopt AI based on "hype-driven" rather than on evidence-based strategies. The same is true of generative AI, according to Abras et al. (2025).

SOLUTIONS AND RECOMMENDATIONS

The literature offers solutions to methodological problems, including threats to validity and reductionist approaches, as advances in research design, theory integration, and evaluation methods, that will bridge the divide between the quantitative algorithmic precision of ML and the qualitative contextual depth of action research.

Adopting Participatory Design and Action Research Methodologies

In order to address the "power shift paradox," researchers need to adopt PD and AR methodologies. Silva (2009) observes systems that build in a "decision room" for human planners, thereby documenting human tacit (knowledge) that cannot be captured within algorithms. Singer et al. (2022) construct a practical model for developing an ML-based clinical decision support tool, showing the necessity of engaging in an iterative and collaborative development for all stakeholders involved in using the tools. By using qualitative methods, researchers will validate that AI systems have been validated to enhance rather than diminish human agencies in decision-making. There is also research on the successful adoption of agriculture DSSs by users who successfully implemented the technology by reflexively adapting it to meet their evolving needs using longitudinal case studies (Mackrell et al., 2009).

Integrating Mixed-Methods for the Operationalization of Vagueness

To overcome the "false precision" concern, the literature recommends a transition towards Mixed-Methods Research (MMR) and Fuzzy Logic methods. Karasakal et al. (2022) suggest the combination of DEA with PROMETHEE, in order to mathematically model qualitative uncertainty while maintaining the nuances associated with them. Lolli et al. (2017) present an innovative method of using Bézier curves to create preference functions based on reference points, allowing decision-makers to articulate their preferences without defining parameters a priori. An example of using fuzzy set theory to model imprecision is demonstrated in the development of fuzzy-AHP models for contractor prequalification (Hosny et al., 2013) and fuzzy decision support systems for supply-chain disturbance management (Nunes et al., 2011). Overall, the recommendation is to abandon the requirement of 'exact inputs' as a methodological assumption, and instead focus on developing methodologies that can effectively handle imprecise information by treating vagueness as legitimate data rather than as noise.

Theoretical Synthesis and Systematic Mapping as Research Tools

Utilizing a theoretical framework that unifies all theories will help researchers deal with the fragmentation found in theoretical research. In expanding the Information System (IS) Success Model into the realm of Big Data, Zhao et al. (2025) have provided a holistic view across disparate studies. The researcher community should routinely utilize Systematic Literature Reviews (SLR) and Bibliometric Analysis for research evaluation purposes. Xue et al. (2025) show how bibliometric tools allow researchers to understand the semantic structure of the field by identifying the main thematic areas within AI-assisted enterprise management. Another great example comes from Cinelli et al. (2020), who created a comprehensive taxonomy of Multi-Criteria Decision Analysis (MCDA) processes by using a review of case studies. Through the use of meta-analysis type methods, researchers can find areas where research is needed, negatively influence their research through "hype," and align their research with one unified scientific agenda.

Developing Ethical and Longitudinal Evaluation Frameworks

To help counteract overload of important performance indicators (KPIs), Caruso et al. (2023) recommend implementing management by exception, where the artificial intelligence (AI) system flags anomalies and alerts the analyst, rather than drowning the analyst in data. The implementation of this design will require research designs that measure cognitive load on employees and decision quality over time. The ethical implementation of AI in finance can be achieved through conducting longitudinal qualitative case studies to examine the long-term sociological impacts of AI. Rautiainen et al. (2024) demonstrate the value of this by showing how management accountants' identity as a member of the profession becomes "fluid" based on the effect of digitalization on his/her work. In another study performed by Berman et al. (2024), the researchers used an empirical analysis of a labor market decision support system (DSS) in Sweden to illustrate the difference between the theoretical conception of trustworthy artificial intelligence and its actual outcomes in practice. By integrating governance protocols into the research design, researchers can investigate the impact of AI on the strategic role of human managers and define "success" as the extent to which human judgement has been enhanced. In this way, the validity of both operational efficiency and employee empowerment can be established.

CONCLUSION

The integration of three scientific fields "Decision Support Systems", 'Artificial Intelligence' and "operational research", and systematic analysis of the literature have produced a comprehensive overview of the evolution of DSS research methodology. The descending hierarchical classification analysis produced four clusters that exemplify how DSS has transitioned from a monolithic positivist to a pluralistic socio-technical model. Class 1 demonstrated a trend toward hybrid logistics design that incorporates both quantitative optimization and qualitative field-based constraints. Class 2 identified the methodological challenge posed by handling imprecision and subjectivity in multi-criteria decision making. Class 3 identified the rise of AI-based meta-analysis as a way of navigating scientific literature. Class

4 established the need for qualitative and critical inquiry for evaluating the social and political implications of using AI in management.

The overall conclusion of this methodology review is that the quality of point-in-time DSS has moved from being primarily dependent on mathematical sophistication to now being dependent on the quality of the design of the research used to create and build them. The literature shows a clear change from algorithm-based 'black box' determinism to a pluralistic methodological view in which quantitative models are being combined with qualitative insight about human behavior and organizational context. Despite these changes, this research also points out that the transition has substantial epistemological implications including 'method blindness' when looking at power dynamics, challenges associated with operationalizing vague human preferences, and the continuing gap between external validities of theoretical models and the operational reality of those same models. Overall, the research support the conclusion that future research will need to combine robust methodologies and provide mixed methodologies for the creation of robust decision support systems linking technology advancement to organizational agility, and human centric values.

FUTURE RESEARCH DIRECTIONS

While this study has synthesized the existing literature and given the ever-evolving nature of this field, there are several opportunities for future research based on specific methodologies.

First, longitudinal and organizational ethnographic studies should be conducted to examine the long-term impact of "Human-in-the-Loop" (HITL) systems, using diary studies and participant observation to examine cognitive load and identify "tipping points" at which human intervention becomes critical.

Second, research employing critical methodologies and interpretive case studies should examine the "algorithmic contract" between employers and employees, and explore how autonomy can be operationalized as an objective metric in the evaluation of decision support systems (DSS). Practical research approaches may provide opportunities for the co-design of AI systems to counter digital "Taylorism."

Third, this field requires a link between hard operations research (OR) and soft management information systems (MIS), and future research should provide a "unified theory of intelligent decision support" by synthesizing these two paradigms and using multilevel modeling techniques to account for both algorithmic variability and organizational strategy simultaneously.

Fourth, future research should shift from designing for stability to designing for adaptability using simulation-based methods, including agent-based modeling (ABM) and system dynamics, to generate "resilience analytics" by testing the resilience of systems under worst-case scenarios.

Finally, there are opportunities for methodologically based research that bridge normative decision theory with behavioral science, through the development of methodologies related to Behavioral OR that consider the impact of cognitive biases on the use of mathematical decision-making models while the developing field of Neuro-Information Systems (Neuro-IS) is exploring the use of physiological data to adjust DSS interfaces based on the user's cognitive state.

REFERENCES

Abras, J., Burnay, C., & Faulkner, S. (2025). Can Generative AI Mitigate Strategic Decision-Making Complexity? An Empirical Exploration of RAG-Augmented Decision-Support Systems. In *Lecture Notes in Business Information Processing* (Vol. 548 LNBIP, pp. 3-19). Springer.

Al-Alawi, A. I., & Albinali, F. A. (2024). Unveiling the Retention Puzzle for Optimizing Employee Engagement and Loyalty Through Analytics-Driven Performance Management: A Systematic Literature Review. In *2024 ASU International Conference in Emerging Technologies for Sustainability and Intelligent Systems (ICETSIS)* (pp. 292-296). IEEE. DOI: 10.1109/ICETSIS61505.2024.10459383

Andreassen, R.-I. (2020). Digital technology and changing roles: A management accountant's dream or nightmare? *Journal of Management Control*, *31*(3), 209–238. DOI: 10.1007/s00187-020-00303-2

Arnaboldi, M., de Bruijn, H., Steccolini, I., & Van der Voort, H. (2022). On humans, algorithms and data. *Qualitative Research in Accounting & Management*, *19*(3), 241–254. DOI: 10.1108/QRAM-01-2022-0005

Arrieta, I. B., Díaz-Rodríguez, N., Del Ser, J., Bennetot, A., Tabik, S., Barbado, A., & Herrera, F. (2020). Explainable Artificial Intelligence (XAI): Concepts, taxonomies, opportunities and challenges toward responsible AI. *Information Fusion*, *58*, 82–115. DOI: 10.1016/j.inffus.2019.12.012

Baller, R., Hage, S., Fontaine, P., & Spinler, S. (2020). The assembly line feeding problem: An extended formulation with multiple line feeding policies and a case study. *International Journal of Production Economics*, *222*, 107489. Advance online publication. DOI: 10.1016/j.ijpe.2019.09.010

Belton, V., & Stewart, T. J. (2002). *Multiple criteria decision analysis: An integrated approach*. Kluwer Academic Publishers. DOI: 10.1007/978-1-4615-1495-4

Berman, A., de Fine Licht, K., & Carlsson, V. (2024). Trustworthy AI in the public sector: An empirical analysis of a Swedish labor market decision-support system. *Technology in Society*, *76*, 102471. Advance online publication. DOI: 10.1016/j.techsoc.2024.102471

Borade, A. B., & Sweeney, E. (2015). Decision support system for vendor managed inventory supply chain: A case study. *International Journal of Production Research*, *53*(16), 4789–4818. DOI: 10.1080/00207543.2014.993047

Bregar, A. (2011). Investigation of the aggregation-disaggregation approach to multi-criteria negotiations: Consolidation of simulation and case studies. In *Proceedings of the 11th International Symposium on Operational Research in Slovenia (SOR 2011)* (pp. 117-124).

Brevik, E., Lauen, A. Ø., Rolke, M. C. B., Fagerholt, K., & Hansen, J. R. (2020). Optimisation of the broiler production supply chain. *International Journal of Production Research*, *58*(17), 5218–5237. DOI: 10.1080/00207543.2020.1713415

Burrell, J. (2016). How the machine 'thinks': Understanding opacity in machine learning algorithms. *Big Data & Society*, *3*(1), 1–12. DOI: 10.1177/2053951715622512

Caruso, S., Bruccoleri, M., Pietrosi, A., & Scaccianoce, A. (2023). Artificial intelligence to counteract "KPI overload" in business process monitoring: The case of anti-corruption in public organizations. *Business Process Management Journal*, *29*(4), 1227–1248. DOI: 10.1108/BPMJ-11-2022-0578

Cinelli, M., Burgherr, P., Kadziński, M., & Słowiński, R. (2022). Proper and improper uses of MCDA methods in energy systems analysis. *Decision Support Systems*, *163*, 113848. DOI: 10.1016/j.dss.2022.113848

Cinelli, M., Kadziński, M., Gonzalez, M., & Słowiński, R. (2020). How to support the application of multiple criteria decision analysis? Let us start with a comprehensive taxonomy. *Omega*, *96*, 102261. Advance online publication. DOI: 10.1016/j.omega.2020.102261 PMID: 33746337

Coelho e Silva, L., & Murça, M. C. R. (2023). A data analytics framework for anomaly detection in flight operations. *Journal of Air Transport Management*, *110*, 102409. Advance online publication. DOI: 10.1016/j.jairtraman.2023.102409

Coutinho-Rodrigues, J., Simão, A., & Antunes, C. H. (2011). A GIS-based multicriteria spatial decision support system for planning urban infrastructures. *Decision Support Systems*, *51*(3), 720–726. DOI: 10.1016/j.dss.2011.02.010

Crandall, K., Noteboom, C., & Smith, K. (2025). Designing curiosity-driven AI chatbots for non-technical fields: A design science approach to supporting school counselors. *Issues in Information Systems*, *26*(3), 1–14.

Crawford, J., & Nilsson, F. (2023). Integrating ESG Risks into Control and Reporting: Evidence from Practice in Sweden. In King, T., & Lehner, M. (Eds.), *Handbook of Big Data and Analytics in Accounting and Auditing* (pp. 255–277). Springer., DOI: 10.1007/978-981-19-4460-4_12

Deveci, M., Pamucar, D., Gokasar, I., Delen, D., & Martínez, L. (2022). A fuzzy Einstein-based decision support system for public transportation management at times of pandemic. *Knowledge-Based Systems, 252*, 109414.

Ding, Y., Zhang, Z., Chen, K., Ding, H., Voss, S., Heilig, L., Chen, Y., & Chen, X. (2023). Real-Time Monitoring and Optimal Resource Allocation for Automated Container Terminals: A Digital Twin Application at the Yangshan Port. *Journal of Advanced Transportation, 2023*, 6909801. Advance online publication. DOI: 10.1155/2023/6909801

Doan, N. A. V., & De Smet, Y. (2018). An alternative weight sensitivity analysis for PROMETHEE II rankings. *Omega, 80*, 166–174. DOI: 10.1016/j.omega.2017.08.017

El-Gayar, O., & Deokar, A. (2013). A semantic service-oriented architecture for distributed model management systems. *Decision Support Systems, 55*(1), 374–384. DOI: 10.1016/j.dss.2012.05.046

Faraj, S., Pachidi, S., & Sayegh, K. (2018). Working and organizing in the age of the learning algorithm. *Organization Studies, 39*(5-6), 621–642.

Gregor, S. (2006). The nature of theory in information systems. *Management Information Systems Quarterly, 30*(3), 611–642. DOI: 10.2307/25148742

Gupta, D., Gupta, M., Balusamy, B., Dhanaraj, R. K., & Gupta, P. M. (2025). *Digital HR: Technologies for HR Transformation and Performance Improvement.* CRC Press., DOI: 10.1201/9781032619651

Hartmann, T. (2011). Goal and process alignment during the implementation of decision support systems by project teams. *Journal of Construction Engineering and Management, 137*(12), 1134–1141. DOI: 10.1061/(ASCE)CO.1943-7862.0000389

Hill, A., & Böse, J. W. (2017). A decision support system for improved resource planning and truck routing at logistic nodes. *Information Technology and Management, 18*(3), 241–251. DOI: 10.1007/s10799-016-0267-3

Hosny, O., Nassar, K., & Esmail, Y. (2013). Prequalification of Egyptian construction contractors using fuzzy-AHP models. *Engineering, Construction, and Architectural Management, 20*(4), 381–405. DOI: 10.1108/ECAM-09-2011-0088

Janjua, N. K., & Hussain, F. K. (2012). Web@IDSS - Argumentation-enabled Web-based IDSS for reasoning over incomplete and conflicting information. *Knowledge-Based Systems, 32*, 9–27. https://doi.org/DOI: 10.1016/j.knosys.2011.09.009

Jeong, J. S., & Ramírez-Gómez, Á. (2018). Development of a web graphic model with fuzzy-decision-making Trial and Evaluation Laboratory/Multi-criteria-Spatial Decision Support System (F-DEMATEL/MC-SDSS) for sustainable planning and construction of rural housings. *Journal of Cleaner Production, 199*, 584–592. DOI: 10.1016/j.jclepro.2018.07.227

Kahraman, C. (Ed.). (2008). *Fuzzy multi-criteria decision making: Theory and applications with recent developments*. Springer Science & Business Media. DOI: 10.1007/978-0-387-76813-7

Kallestrup, K. B., Lynge, L. H., Akkerman, R., & Oddsdottir, T. A. (2014). Decision support in hierarchical planning systems: The case of procurement planning in oil refining industries. *Decision Support Systems, 68*, 49–63. DOI: 10.1016/j.dss.2014.09.003

Kambur, E., & Yildirim, T. (2023). From traditional to smart human resources management. *International Journal of Manpower, 44*(3), 422–452. DOI: 10.1108/IJM-10-2021-0622

Karasakal, E., Eryılmaz, U., & Karasakal, O. (2022). Ranking using PROMETHEE when weights and thresholds are imprecise: A data envelopment analysis approach. *The Journal of the Operational Research Society, 73*(9), 1978–1995. DOI: 10.1080/01605682.2021.1963195

Kellogg, K. C., Valentine, M. A., & Christin, A. (2020). Algorithms at work: The new controllability. *The Academy of Management Annals, 14*(1), 202–231. DOI: 10.5465/annals.2018.0174

Korhonen, T., Selos, E., Laine, T., & Suomala, P. (2021). Exploring the programmability of management accounting work for increasing automation: An interventionist case study. *Accounting, Auditing & Accountability Journal, 34*(2), 253–280. DOI: 10.1108/AAAJ-12-2016-2809

Krause, J., Perer, A., & Ng, K. (2021). Interacting with predictions: Visual inspection of black-box machine learning models. *Proceedings of the 2021 CHI Conference on Human Factors in Computing Systems*, 1-13.

Kulkov, I., Kulkova, J., Leone, D., Rohrbeck, R., & Menvielle, L. (2024). Standalone or run together: Artificial intelligence as an enabler for other technologies. *International Journal of Entrepreneurial Behaviour & Research, 30*(8), 2082–2105. DOI: 10.1108/IJEBR-02-2023-0169

Kumar, S., Talukder, M. B., & Manoharan, G. (2025). Integrating Technology with Workforce Training and Management in the Hotel Industry. In *Managing the Smart Revolution in Travel, Tourism, and Hospitality* (pp. 55-68). Emerald Publishing. DOI: 10.1108/978-1-83662-330-420251005

Lakshmi, H. (2025). Trends in HR Technology: AI, Machine Learning, and Blockchain. In *Transforming Organizational Culture Through Meta-Driven Human Resources* (pp. 427–464). IGI Global Scientific Publishing. https://doi.org/DOI: 10.4018/979-8-3373-0720-6.ch015

Lechner, G., & Reimann, M. (2020). Integrated decision-making in reverse logistics: An optimisation of interacting acquisition, grading and disposition processes. *International Journal of Production Research*, *58*(19), 5786–5805. DOI: 10.1080/00207543.2019.1659518

Legato, P., & Mazza, R. M. (2018). A decision support system for integrated container handling in a transshipment hub. *Decision Support Systems*, *108*, 45–56. DOI: 10.1016/j.dss.2018.02.004

Lei, N., & Moon, S. K. (2015). A Decision Support System for market-driven product positioning and design. *Decision Support Systems*, *69*, 82–91. DOI: 10.1016/j.dss.2014.11.010

Lolli, F., Ishizaka, A., Gamberini, R., Rimini, B., Balugani, E., & Prandini, L. (2017). Requalifying public buildings and utilities using a group decision support system. *Journal of Cleaner Production*, *164*, 1081–1092. DOI: 10.1016/j.jclepro.2017.07.031

Lubbe, B. A., et al. (2020). *Reference regarding Integrative Systematic Review (ISR) methodology*. (Citation conceptuelle basée sur le texte fourni).

Mackrell, D., Kerr, D., & von Hellens, L. (2009). A qualitative case study of the adoption and use of an agricultural decision support system in the Australian cotton industry: The socio-technical view. *Decision Support Systems*, *47*(2), 143–153. DOI: 10.1016/j.dss.2009.02.004

Maghrabie, H. F., Beauregard, Y., & Schiffauerova, A. (2019). Grey-based Multi-Criteria Decision Analysis approach: Addressing uncertainty at complex decision problems. *Technological Forecasting and Social Change*, *146*, 366–379. DOI: 10.1016/j.techfore.2019.05.031

Maldonado, S., González-Ramírez, R. G., Quijada, F., & Ramírez-Nafarrate, A. (2019). Analytics meets port logistics: A decision support system for container stacking operations. *Decision Support Systems*, *121*, 84–93. DOI: 10.1016/j.dss.2019.04.006

Mantouzi, S., & Youssef, S. (2025). The role of AI-driven Decision Support Systems (DSS) in enhancing organizational performance: A qualitative study of Morocco's automotive industry. In *Transparency in AI-Assisted Management Decisions* (pp. 397–422). IGI Global Scientific Publishing. https://doi.org/DOI: 10.4018/979-8-3373-1737-3.ch012

Mobo, F. D. (2025). AI in green HR practices: Sustainable workforce management. In *Convergence of AI, Education, and Business for Sustainability* (pp. 195–214). IGI Global Scientific Publishing. https://doi.org/DOI: 10.4018/979-8-3373-1917-9.ch009

Möhlmann, M., Zalmanson, L., Henfridsson, O., & Lyytinen, K. (2021). Strategic responses to digital surveillance in the gig economy: The politics of platform work. *Long Range Planning*, *54*(3), 102054.

Monod, E., Lissillour, R., Köster, A., & Jiayin, Q. (2023). Does AI control or support? Power shifts after AI system implementation in customer relationship management. *Journal of Decision Systems*, *32*(3), 542–565. DOI: 10.1080/12460125.2022.2066051

Monod, E., Mayer, A.-S., Straub, D., Joyce, E., & Qi, J. (2024). From worker empowerment to managerial control: The devolution of AI tools' intended positive implementation to their negative consequences. *Information and Organization*, *34*(1), 100498. Advance online publication. DOI: 10.1016/j.infoandorg.2023.100498

Nagle, T., Sammon, D., & Pope, A. (2024). Exploring the socio-technical dynamics of VAR implementation and use. *Journal of Decision Systems, 33*(sup1), 47–62. https://doi.org/DOI: 10.1080/12460125.2024.2354598

Niu, D., Yuan, H., Wang, M., & Xu, W. (2016). Event-driven data mining methods for large-scale market prediction: A case study of an agricultural products company. In **Proceedings of the 2nd ACM SIGSPATIAL International Workshop on the Use of GIS in Emergency Management (EM-GIS 2016)** (pp. 1-7). ACM. DOI: 10.1145/3017611.3017618

Noor-E-Alam, M., Lipi, T. F., Ahsan Akhtar Hasin, M., & Ullah, A. M. M. S. (2011). Algorithms for fuzzy multi expert multi criteria decision making (ME-MCDM). *Knowledge-Based Systems, 24*(3), 367–377. https://doi.org/DOI: 10.1016/j.knosys.2010.10.006

Nunes, I. L., Figueira, S., & Cruz Machado, V. (2011). Evaluation of a supply chain performance using a fuzzy decision support system. In *IEEE International Conference on Industrial Engineering and Engineering Management* (pp. 1735–1739). IEEE. https://doi.org/DOI: 10.1109/IEEM.2011.6118213

Page, M. J., McKenzie, J. E., Bossuyt, P. M., Boutron, I., Hoffmann, T. C., Mulrow, C. D., & Moher, D. (2021). The PRISMA 2020 statement: An updated guideline for reporting systematic reviews. *BMJ (Clinical Research Ed.)*, *372*(71). PMID: 33782057

Pandya, B., & Boukareva, B. (2025). Empowerment of digital start-ups in the gig economy through digital employees – A systematic review. *International Journal of Entrepreneurship and Innovation*, 14657503251400965. Advance online publication. DOI: 10.1177/14657503251400965

Poplawska, J., Labib, A., & Reed, D. M. (2015). A hybrid multiple-criteria decision analysis framework for corporate social responsibility implementation applied to an extractive industry case study. *The Journal of the Operational Research Society*, *66*(9), 1491–1505. DOI: 10.1057/jors.2014.116

Power, D. J. (2002). *Decision support systems: Concepts and resources for managers*. Quorum Books.

Power, D. J. (2007). Specifying an expanding conceptualization of the field of decision support systems. *Decision Support Systems*, *43*(3), 1224–1230.

Rautiainen, A., Scapens, R. W., Järvenpää, M., Auvinen, T., & Sajasalo, P. (2024). Towards fluid role identity of management accountants: A case study of a Finnish bank. *The British Accounting Review*, *56*(4), 101341. Advance online publication. DOI: 10.1016/j.bar.2024.101341

Ren, H., Zhou, W., Guo, Y., Huang, L., Liu, Y., Yu, Y., Hong, L., & Ma, T. (2020). A GIS-based green supply chain model for assessing the effects of carbon price uncertainty on plastic recycling. *International Journal of Production Research*, *58*(6), 1705–1723. DOI: 10.1080/00207543.2019.1693656

Romero, N., Nozick, L. K., & Xu, N. (2016). Hazmat facility location and routing analysis with explicit consideration of equity using the Gini coefficient. *Transportation Research Part E, Logistics and Transportation Review*, *89*, 165–181. DOI: 10.1016/j.tre.2016.02.008

Sahebjamnia, N., Torabi, S. A., & Mansouri, S. A. (2017). A hybrid decision support system for managing humanitarian relief chains. *Decision Support Systems*, *95*, 12–26. DOI: 10.1016/j.dss.2016.11.006

Sahin, C., Ar, I. M., & Baki, B. (2025). An integrated approach on new product development process with data mining: A case study on smartphone design. *International Journal of System Assurance Engineering and Management*, *16*(4), 1485–1500. DOI: 10.1007/s13198-025-02763-y

Schätter, F., Hansen, O., Wiens, M., & Schultmann, F. (2019). A decision support methodology for a disaster-caused business continuity management. *Decision Support Systems, 118*, 10–20. DOI: 10.1016/j.dss.2018.12.006

Secinaro, S., Calandra, D., Lanzalonga, F., & Biancone, P. (2024). The role of artificial intelligence in management accounting: An exploratory case study. In *Digital Transformation in Accounting and Auditing: Navigating Technological Advances for the Future* (pp. 207-236). Springer.

Shahparvari, S., Abbasi, B., Chhetri, P., & Abareshi, A. (2016). Vehicle routing and scheduling for bushfire emergency evacuation. In *IEEE International Conference on Industrial Engineering and Engineering Management* (Vol. 2016-January, pp. 696–700). IEEE. https://doi.org/DOI: 10.1109/IEEM.2015.7385737

Shi, Y., Arthanari, T., Liu, X., & Yang, B. (2019). Sustainable transportation management: Integrated modeling and support. *Journal of Cleaner Production, 212*, 1381–1395. DOI: 10.1016/j.jclepro.2018.11.209

Shneiderman, B. (2020). *Human-centered AI*. Oxford University Press.

Shook, E., & Turner, V. K. (2016). The socio-environmental data explorer (SEDE): A social media–enhanced decision support system to explore risk perception to hazard events. *Cartography and Geographic Information Science, 43*(5), 427–441. DOI: 10.1080/15230406.2015.1131627

Sidahmed, E., & Pascale, Z. (2024). Strategic Support Systems for Crisis Management: A Literature Review. *Foundations and Trends in Information Systems, 8*(1), 1–65. DOI: 10.1561/2900000025

Siering, M., Clapham, B., Engel, O., & Gomber, P. (2017). A taxonomy of financial market manipulations: Establishing trust and market integrity in the financialized economy through automated fraud detection. *Journal of Information Technology, 32*(3), 251–269. DOI: 10.1057/s41265-016-0029-z

Silva, C. (2009). Combining ad hoc decision-making behaviour with formal planning and scheduling rules: A case study in the synthetic fibre production industry. *Production Planning and Control, 20*(7), 636–648. DOI: 10.1080/09537280903034321

Simon, H. A. (1997). *Administrative behavior: A study of decision-making processes in administrative organization* (4th ed.). Free Press.

Singer, S. J., Kellogg, K. C., Galper, A. B., & Viola, D. (2022). Enhancing the value to users of machine learning-based clinical decision support tools: A framework for iterative, collaborative development and implementation. *Health Care Management Review, 47*(2), E21–E31. DOI: 10.1097/HMR.0000000000000324 PMID: 34516438

Singh, R., & Arora, N. (2025). The Application of Artificial Intelligence in Law: A Bibliometric Analysis. *Prabandhan: Indian Journal of Management, 18*(7), 58–71. DOI: 10.17010/pijom/2025/v18i7/174565

Soliman, K., & El-Barkouky, N. (2020). An E-government Procurement Decision Support System Model for Public Private Partnership Projects in Egypt. In *Lecture Notes in Business Information Processing* (Vol. 384 LNBIP, pp. 112-124). Springer. DOI: 10.1007/978-3-030-46224-6_9

Suarez, D., Gomez, C., Medaglia, A. L., Akhavan-Tabatabaei, R., & Grajales, S. (2024). Integrated Decision Support for Disaster Risk Management: Aiding Preparedness and Response Decisions in Wildfire Management. *Information Systems Research, 35*(2), 609–628. DOI: 10.1287/isre.2022.0118

Ting, S. L., Tse, Y. K., Ho, G. T. S., Chung, S. H., & Pang, G. (2014). Mining logistics data to assure the quality in a sustainable food supply chain: A case in the red wine industry. *International Journal of Production Economics, 152*, 200–209. DOI: 10.1016/j.ijpe.2013.12.010

Tranfield, D., Denyer, D., & Smart, P. (2003). Towards a methodology for developing evidence-informed management knowledge by means of systematic review. *British Journal of Management, 14*(3), 207–222. DOI: 10.1111/1467-8551.00375

Trucco, P., & Lapko, Y. (2023). Planning Pipe-laying Projects Under Uncertainty: A Simulation Approach. In *2023 IEEE International Conference on Industrial Engineering and Engineering Management (IEEM)* (pp. 1667–1671). IEEE. https://doi.org/DOI: 10.1109/IEEM58616.2023.10406602

Ursavas, E. (2014). A decision support system for quayside operations in a container terminal. *Decision Support Systems, 59*(1), 312–324. DOI: 10.1016/j.dss.2014.01.003

Wang, W., Chen, Y., Wang, Y., Deveci, M., Cheng, S., & Brito-Parada, P. R. (2024). A decision support framework for humanitarian supply chain management – Analysing enablers of AI-HI integration using a complex spherical fuzzy DEMATEL-MARCOS method. *Technological Forecasting and Social Change, 206*, 123556. Advance online publication. DOI: 10.1016/j.techfore.2024.123556

Wider, W., Ho, C. P., Wu, C., Lee, H. P. T., & Tee, M. (2025). Adopting AI for HRM in the Malaysian hotel industry: A Delphi study. *Journal of Hospitality and Tourism Technology*, 1–21. Advance online publication. DOI: 10.1108/JHTT-03-2025-0202

Xue, L., Satpayeva, Z., Kangalakova, D., & Özen, E. (2025). Trends of artificial intelligence-driven enterprise management development: A bibliometric analysis. *Problems and Perspectives in Management, 23*(4), 1–12. DOI: 10.21511/ppm.23(4).2025.01

Yang, J. B. (2001). Rule and utility based evidential reasoning approach for multiattribute decision analysis under uncertainties. *European Journal of Operational Research, 131*(1), 31–61. DOI: 10.1016/S0377-2217(99)00441-5

Yang, J. B., Xu, D. L., Xie, X., & Maddulapalli, A. K. (2009). Multicriteria evidential reasoning decision modelling and analysis: Prioritising voices of the customer. In *51st Annual Conference of the Operational Research Society 2009* (pp. 61-89).

Yang, L.-H., Wang, Y.-M., Liu, J., & Martínez, L. (2018). A joint optimization method on parameter and structure for belief-rule-based systems. *Knowledge-Based Systems, 142*, 220–240. https://doi.org/DOI: 10.1016/j.knosys.2017.11.039

Yannacopoulos, D., Spyridakos, A., & Tsotsolas, N. (2014). Robustness Analysis in Multicriteria Disaggregation - Aggregation Approaches for Group Decision Making. *Lecture Notes in Business Information Processing, 184 LNBIP*, 167–180. https://doi.org/DOI: 10.1007/978-3-319-11364-7_15

Yazdani, M., Chatterjee, P., Zavadskas, E. K., & Hashemkhani Zolfani, S. (2017). Integrated QFD-MCDM framework for green supplier selection. *Journal of Cleaner Production, 142*, 3728–3740. DOI: 10.1016/j.jclepro.2016.10.095

Zadeh, L. A. (1965). Fuzzy sets. *Information and Control, 8*(3), 338–353. DOI: 10.1016/S0019-9958(65)90241-X

Zhang, Z., Li, Z., & Wang, Z. (2025). Human-machine synergy in spatial decision-making: A multi-scenario SDSS prototype for high-rise buildings. *Cities (London, England), 165*, 106090. Advance online publication. DOI: 10.1016/j.cities.2025.106090

Zhao, X., Dai, H., Hu, T. E., Cheng, H. K., & Zhang, P. (2025). Unlocking big data success in the AI-driven era: Toward a unified theory for intelligent decision support. *Decision Support Systems, 194*, 114468. DOI: 10.1016/j.dss.2025.114468

Zhou, Z.-G., Liu, F., Jiao, L.-C., Zhou, Z.-J., Yang, J.-B., Gong, M.-G., & Zhang, X.-P. (2013). A bi-level belief rule based decision support system for diagnosis of lymph node metastasis in gastric cancer. *Knowledge-Based Systems, 54*, 128–136. https://doi.org/DOI: 10.1016/j.knosys.2013.09.001

Zuboff, S. (2019). *The age of surveillance capitalism: The fight for a human future at the new frontier of power*. PublicAffairs.

KEY TERMS AND DEFINITIONS

Algorithmic Management: This refers to the supervision, evaluation, and discipline of workers using data-driven algorithms. This form of management raises fundamental methodological questions about the transfer of decision-making authority from humans to computers, the "black box," and workers' loss of autonomy (digital Taylorism).

Black Box AI: Algorithms that are opaque to their users. This indicates a problem with the validity of operational research (OR) methods, as stakeholders cannot understand or question the algorithm they are using, which necessitating Explainable AI (XAI).

Data Envelopment Analysis (DEA): is a non-parametric, linear programming method used to assess the relative efficiency of *Decision-Making Units (DMUs)*—such as banks, hospitals, or schools—that operate with multiple inputs and outputs. By constructing an efficiency frontier from best-performing units, DEA benchmarks others against this standard, highlighting inefficiencies and areas for improvement.

Digital Taylorism: Applying the classical Taylorist approach of dividing labor into highly precise mechanical tasks and applying information technologies. This is an important framework for studying the negative social impacts of AI adoption, and for examining the loss of worker dignity versus the potential gains in profitability.

Fuzzy Logic / Evidential Reasoning: Mathematical frameworks for approximate rather than exact inference. The authors present these frameworks as key methodological responses to the limitations they identify in the classical model of multi-criteria decision-making (MCDM). Researchers now recognize that they cannot create exact analogues of human preferences. Inductive reasoning uses cumulative cases (mathematically similar to Bayesian algorithms) rather than interactions with individual humans to generate approximate knowledge.

Human-in-the-Loop (HITL): a design model that posits that all decision-making processes are punctuated by human interactions at critical points in the procedure. The authors recommend the HITL model as both an ethical and practical standard, enabling a better restructuring of the decision-making process and validating the outputs of decision support systems.

Hybrid Intelligence: Where DSS systems are not merely a support for human decision-making nor solely human-centered, but rather a blend of the best of human capabilities (slowness and ethics) and computer capabilities (exponential computation and scalability). It represents a sign of evolution from algorithmic optimization.

Multi-Criteria Decision Making (MCDM): A branch of operations research that uses value-based criteria to generate an output (AHP, PROMETHEE, and others). The authors criticize classical MCDM, asserting that it relies on "precise" rather than "fuzzy" inputs, leading to the dominance of FITUS protocols and highlighting the need for fuzzy MCDM variables.

Soft Constraints: Constraints in a problem that are precise or interchangeable, as opposed to "hard constraints" that must be followed literally. In the methodological section of the chapter, "flexible constraints" represent all the unstated and informal rules and contextual nuances that human planners rely on but are excluded from the results generated by strict and superficial mathematics, creating a gap in implementation.

Chapter 8
A Qualitative Phenomenological Approach to Exploring Lecturers' Perceptions of AI-Powered Learning Analytics in Higher Degree by Research Programmes

Ivan Marcelo Arguello
http://orcid.org/0000-0003-4749-1225
Torrens University, Australia

ABSTRACT

This article outlines a methodological framework for a qualitative phenomenological study examining university lecturers' perceptions of AI-powered Learning Analytics (AI-LA) in Higher Degree by Research programmes in Adelaide, Australia. Grounded in an interpretative, constructivist paradigm, the research explores lecturers' lived experiences and meanings attributed to this emerging technology within Education 4.0. Using purposive and convenience sampling, 14 lecturers from varied disciplines were recruited, allowing data saturation. Data were gathered through semi-structured interviews and analysed thematically with NVivo following Braun and Clarke's guidelines. The phenomenological approach, informed by Moustakas and

DOI: 10.4018/979-8-3373-7640-0.ch008

nt © 2026, IGI Global Scientific Publishing. Copying or distributing in print or electronic forms without written permission of IGI Global Scientific Publishing is prohibited. Use of this chapter to enerative artificial intelligence (AI) technologies is expressly prohibited. The publisher reserves all rights to license its use for generative AI training and machine learning model development.

Zahavi, enabled deep insight into human-technology interactions. The study offers a replicable methodological model for educational technology research and informs the design of AI-LA tools, contributing to ongoing discourse in higher education.

INTRODUCTION

Artificial Intelligence-powered Learning Analytics (AI-LA) is rapidly reshaping higher education by enabling data-informed, personalised and adaptive learning experiences that align closely with the core principles of Education 4.0 (Siemens, 2013). In Higher Degree by Research (HDR) programmes, where supervision is long-term, intensive and highly individualised, AI-LA provides supervisors with real-time visibility into student progress, engagement patterns, research activity and milestone achievement. Lecturers can identify early signs of difficulty, deliver timely and context-specific interventions, and adjust guidance to match each candidate's unique intellectual and personal trajectory (Pi et al., 2023). These capabilities hold particular promise in HDR contexts, where candidates often face isolation, extended timelines, high-stakes outputs and complex skill development demands. This empirical investigation draws on in-depth qualitative interview data to examine lecturers' perceptions of AI-powered Learning Analytics within Higher Degree by Research programmes.

Yet the successful integration of AI-LA into HDR supervision depends far more on how lecturers perceive and experience the technology than on its technical features alone. Faculty members serve as the primary gatekeepers of adoption. Their attitudes, prior experiences with digital tools, ethical concerns and expectations directly determine whether AI-LA becomes a valuable ally in mentoring or an unwelcome burden that disrupts established relational dynamics (Crompton et al., 2024). In HDR settings—marked by deep mentor–candidate trust, intellectual autonomy and the creative uncertainty of doctoral research—the human dimension assumes even greater importance. Lecturers must continually balance the benefits of predictive analytics against the need to preserve authenticity, agency and pedagogical judgement.

This study investigates lecturers' perceptions of AI-powered Learning Analytics in HDR programmes at universities in Adelaide, Australia—a city widely recognised as a centre of educational innovation with strong traditions of research-intensive supervision. Adelaide's academic institutions actively pilot advanced technologies while upholding rigorous standards of HDR mentoring across disciplines ranging from education and engineering to social sciences. This diversity allows the research to capture how institutional culture, disciplinary norms and individual experiences intersect to shape technology acceptance. By focusing on this vibrant and progressive academic environment, the study illuminates the interplay between local context and global trends in technology-enhanced doctoral education.

Although learning analytics and AI applications in higher education have attracted substantial scholarly attention, the majority of research focuses on undergraduate and coursework settings, where emphasis lies on technical development, performance prediction, retention strategies and large-scale implementation. Far less attention has been directed toward HDR supervision, where the unstructured, creative and deeply relational nature of doctoral research presents distinct opportunities and challenges. Faculty voices—particularly their lived experiences, ethical apprehensions and practical considerations—remain underrepresented in the literature. This gap leaves institutions without clear, evidence-based guidance on how to integrate AI-LA responsibly into research degrees (Săseanu et al., 2024; Campbell, 2025).

The present study directly addresses that gap. It asks: What are the perceptions of university lecturers regarding AI-powered Learning Analytics in Higher Degree by Research programmes, and how do their lived experiences shape these perceptions? To answer this question, the research pursues three interconnected objectives:

- first, to illuminate lecturers' subjective understandings, expectations and interpretations of AI-LA within their supervisory practice.
- second, to identify the contextual, institutional and disciplinary factors that influence their readiness to adopt the technology.
- third, to derive evidence-based implications for the ethical, effective and sustainable implementation of AI-LA in HDR environments.

The investigation adopts a qualitative phenomenological approach grounded in a constructivist paradigm. Phenomenology enables a deep exploration of the essence of lecturers' lived experiences, uncovering the personal meanings they attribute to AI-LA in the relational and intellectual space of HDR supervision (Moustakas, 1994; Zahavi, 2021). Constructivism complements this perspective by recognising that perceptions emerge through ongoing interactions with tools, contexts and colleagues, making it particularly suited to capturing the dynamic ways faculty construct their views of emerging technologies (Creswell & Poth, 2017). The study draws on complementary theoretical frameworks—the Technology Acceptance Model (Davis, 1989), Theory of Planned Behaviour (Ajzen, 1991) and Diffusion of Innovations to interpret the findings and situate them within broader processes of technology adoption.

By centring lecturers' perspectives, this work contributes original insights to the Education 4.0 discourse. It foregrounds human factors that technical studies often overlook and provides practical guidance for institutions seeking to implement AI-LA responsibly. The following sections outline the theoretical framework, describe the methodological design, present the key findings from thematic analysis, offer an in-depth discussion of the results, and conclude with recommendations for policy, practice and future research.

LITERATURE REVIEW

The Fourth Industrial Revolution has brought significant changes to education, giving rise to Education 4.0, which emphasises personalised learning through technology. AI-powered Learning Analytics (AI-LA) represents one of the most promising applications within this framework.

These trends align directly with the book's theme of responsible AI integration in higher education by demonstrating that successful implementation of AI-LA in HDR programmes depends not only on technical capability but on faculty perceptions, ethical governance and compatibility with the relational nature of doctoral supervision.

Historical Background

Education 4.0 is explicitly based on learner-centricity and personalised learning experiences. AI transforms these possibilities by customising learning to individual needs. The capacity of AI to create intelligent systems capable of learning, reasoning and problem-solving has evolved from early foundations in the 1960s–1990s (Hinton, 1990\) to the adaptive learning platforms of today.

Technological Foundation for AI-Powered LA

Big Data, Educational Data Mining (EDM) and Learning Analytics (LA) form the technological base. While EDM analyses historical data retrospectively, LA emphasises real-time analysis to enable immediate interventions. AI-powered LA builds on these foundations by analysing student data and behavioural patterns to identify at-risk students, personalise learning experiences and develop targeted interventions (Vashishth et al., 2024).

Perceptions and Technology Adoption Frameworks

Faculty perceptions of AI-powered LA in the pre-implementation stage are shaped by three established models: the Technology Acceptance Model (TAM) (Davis, 1989), the Theory of Planned Behaviour (TPB) (Ajzen, 1991) and the Diffusion of

Innovations (DOI). These frameworks help explain how lecturers evaluate usefulness, ease of use, social norms and compatibility with existing supervisory practices.

Faculty Perceptions of AI-Powered LA in HDR Programmes

Limited research has examined faculty perceptions specifically in Higher Degree by Research (HDR) contexts. Existing studies highlight concerns regarding data privacy, student autonomy and potential algorithmic bias. The present study addresses this gap by focusing on Adelaide universities.

Theoretical Framework

The literature review above provides the empirical and theoretical foundation for the integrated framework presented below.

This study integrates constructivism, phenomenology, and three established technology adoption models—the Technology Acceptance Model (TAM), the Theory of Planned Behaviour (TPB), and the Diffusion of Innovations (DOI)—while enriching them with contemporary perspectives on human–AI co-creation and ethical data governance. This combination provides a multi-layered conceptual scaffold for understanding how university lecturers interpret and experience AI-powered Learning Analytics (AI-LA) in Higher Degree by Research (HDR) supervision within the context of Adelaide universities and the broader Education 4.0 agenda.

Constructivism: Meaning-Making in HDR Supervision

We ground the study in constructivism to explain how lecturers actively construct meanings about AI-LA from their prior digital practices, disciplinary epistemologies, institutional routines, and personal supervisory craft. Supervisors do not passively accept analytics outputs; they interpret them through the lens of what they already know works in doctoral mentoring—diagnosing disengagement, calibrating feedback cycles, monitoring progress toward milestones, and nurturing intellectual independence. The same predictive dashboard can therefore appear to one lecturer as a natural extension of data-informed practice and to another as a potential misfit with the dialogic, thesis-centred work that defines HDR (Creswell & Poth, 2017; Siemens, 2013). This constructivist stance anticipates heterogeneous perceptions across disciplines and programmes in Adelaide: education faculty may value analytics for reflective practice, engineering supervisors for milestone tracking, and social sciences lecturers for relational nuance. The framework thus positions perception not as a fixed attitude but as an ongoing, context-sensitive process of meaning-making (Sampieri & Mendoza, 2018).

Phenomenology: Lived Textures of Judgement and Care

A phenomenological orientation allows us to examine the lived textures of supervision when AI-LA enters the relationship—how a risk alert reorients a supervision meeting, when an indicator aligns (or clashes) with professional intuition, and how trust and care evolve over long doctoral timelines. By applying the phenomenological reduction (epoché), we bracket preconceptions and foreground first-person accounts as legitimate evidence about what counts as good supervisory practice in HDR (Moustakas, 1994; Van Manen, 2016; Zahavi, 2021). Lecturers describe not only cognitive responses to dashboards but also emotional and relational shifts: the unease of receiving automated flags about a candidate's progress, the relief of early insight into wellbeing concerns, or the tension between data-driven prompts and the embodied judgement that has guided successful completions for years. Phenomenology is particularly apt for HDR, where supervision extends beyond technical milestones to encompass emotional labour, intellectual companionship, and the gradual formation of scholarly identity (Heidegger, 1962).

Technology Acceptance Model (TAM): Usefulness, Ease, and Trust as Acceptance Drivers

TAM predicts adoption through two primary constructs: perceived usefulness (the degree to which a person believes the technology will enhance performance) and perceived ease of use (the degree to which it is seen as effortless) (Davis, 1989). In HDR practice, lecturers judge analytics useful when they surface actionable prompts—for example, sustained engagement decline that warrants a supportive check-in—yet less useful when outputs fail to reflect the non-linear, creative nature of thesis work. Ease of use matters equally: visualisations must be interpretable without heavy cognitive overhead, and integration into existing workflows (e.g., thesis management platforms) must feel seamless. Education research shows that trust and perceived risk heavily condition acceptance in data-intensive tools: opacity, uncertain data provenance, and fairness concerns can undermine adoption even when usefulness appears high (Crompton et al., 2024). In doctoral contexts, usefulness is inherently practice-relative—LMS traces that prove predictive in coursework often lose diagnostic power in the unstructured thesis phase.

Theory of Planned Behaviour (TPB): Attitudes, Norms, and Perceived Behavioural Control

TPB clarifies how three elements shape behavioural intention: attitudes (evaluation of the behaviour), subjective norms (perceived social pressure), and perceived behavioural control (belief in one's ability to perform the behaviour) (Ajzen, 1991). Lecturers adopt AI-LA more readily when they see it augmenting—not replacing—scholarly judgement, when local norms (departmental culture, institutional policy, peer discourse) encourage "data-informed" improvement, and when they feel capable of using the tools responsibly. In pre-implementation settings typical of Adelaide HDR programmes, perceived behavioural control often emerges as the hinge variable that converts positive attitudes into sustained use. Supervisors frequently cite lack of training, time, or institutional support as barriers to control, while departmental encouragement or visible success stories among colleagues can strengthen norms (Venkatesh et al., 2016).

Diffusion of Innovations (DOI): Attributes That Support Diffusion in HDR Ecosystems

DOI identifies five attributes influencing adoption rates: relative advantage, compatibility, complexity, trialability, and observability. In HDR supervision, compatibility with trust-based, dialogic mentoring and the bespoke, non-linear nature of research proves decisive; complexity—both cognitive (interpreting outputs) and procedural (integrating tools)—can depress use; and supervisors consistently request trialability (low-stakes pilots, opt-in indicators) and observability (discipline-specific exemplars of successful use) before broader diffusion. These conditions suggest that small-scale pilots anchored in real supervisory practice, combined with reflective debriefs, are more likely to facilitate adoption than top-down mandates.

HUMAN–AI CO-CREATION: AUGMENTING JUDGEMENT, NOT DISPLACING CARE

We frame AI-LA as a collaborator that augments rather than displaces supervisory judgement and care. Co-creation means treating indicators as prompts for proportionate, humane action: a declining-engagement alert becomes the starting point for a supportive conversation rather than a definitive diagnosis. Effective co-creation design exposes model uncertainty, reveals data lineage, and allows supervisors to annotate context or override suggestions with reasoned explanations. Such features preserve intellectual autonomy, strengthen the supervisor–candidate relationship, and leverage pattern-sensitivity at scale (Fischer et al., 2022). In HDR, where supervision blends cognitive guidance with emotional and motivational support, co-creation aligns technology with the relational core of doctoral mentoring.

Ethical Data Governance: Legitimacy and Duty of Care

Ethical governance is not an add-on; it is the enabling condition for legitimate AI-LA in HDR. Supervisors consistently call for purpose limitation, role-based access, transparent model logic, contestability of outputs, and secure stewardship of student records. In Australia, national standards codify these expectations and require institutions to provide policy-aligned training for responsible use (National Health and Medical Research Council, 2018). Clear governance improves trust, mitigates perceived risk, and makes acceptance models (TAM/TPB) operational in practice (Crompton et al., 2024).

Synthesis: From Constructs to Supervisory Action

Constructivism explains how supervisors make meaning from local practice; phenomenology captures the lived negotiation of attention, trust, and care; TAM models tool–task alignment (helpfulness, usability, trust); TPB situates adoption within social capacity (values, expectations, confidence); and DOI specifies diffusion conditions (fit, complexity, pilots, exemplars). Grounded in Adelaide HDR supervision, this framework provides a robust scaffold for interpreting lecturers' perceptions and for designing analytics that genuinely augment supervisory judgement within Education 4.0 priorities (Davis, 1989; Ajzen, 1991).

Methods

This section outlines the methodological approach adopted to explore university lecturers' perceptions of AI-powered Learning Analytics (AI-LA) within Higher Degree by Research (HDR) programmes in Adelaide, Australia. The study employed a qualitative phenomenological design grounded in a constructivist paradigm to capture the depth and complexity of lecturers' lived experiences with this emerging technology. The subsections below detail the research design, participant selection, data collection procedures, and data analysis techniques, ensuring methodological rigour and alignment with the study's objectives.

RESEARCH DESIGN

The research was anchored in a qualitative phenomenological approach, which is particularly suited to investigating the subjective experiences and meanings that individuals attribute to a phenomenon (Moustakas, 1994). Phenomenology, as a methodological framework, enables an in-depth exploration of how lecturers perceive and interact with AI-LA in the context of HDR supervision, capturing the essence of their lived experiences rather than imposing external categories or preconceived theories (Van Manen, 2016). This approach was deliberately chosen over other qualitative methodologies, such as grounded theory—which aims to generate new theory from data—or ethnography—which focuses on describing cultural contexts—because phenomenology prioritises the first-person perspective and seeks to uncover the underlying structures of experience as they present themselves to consciousness (Creswell & Poth, 2017). By suspending assumptions about AI-LA and allowing lecturers' accounts to lead, the design remained open to the full range of meanings, tensions, and possibilities that emerge when a novel technology intersects with established supervisory practices.

Figure 1. Conceptual Map of the Research Design

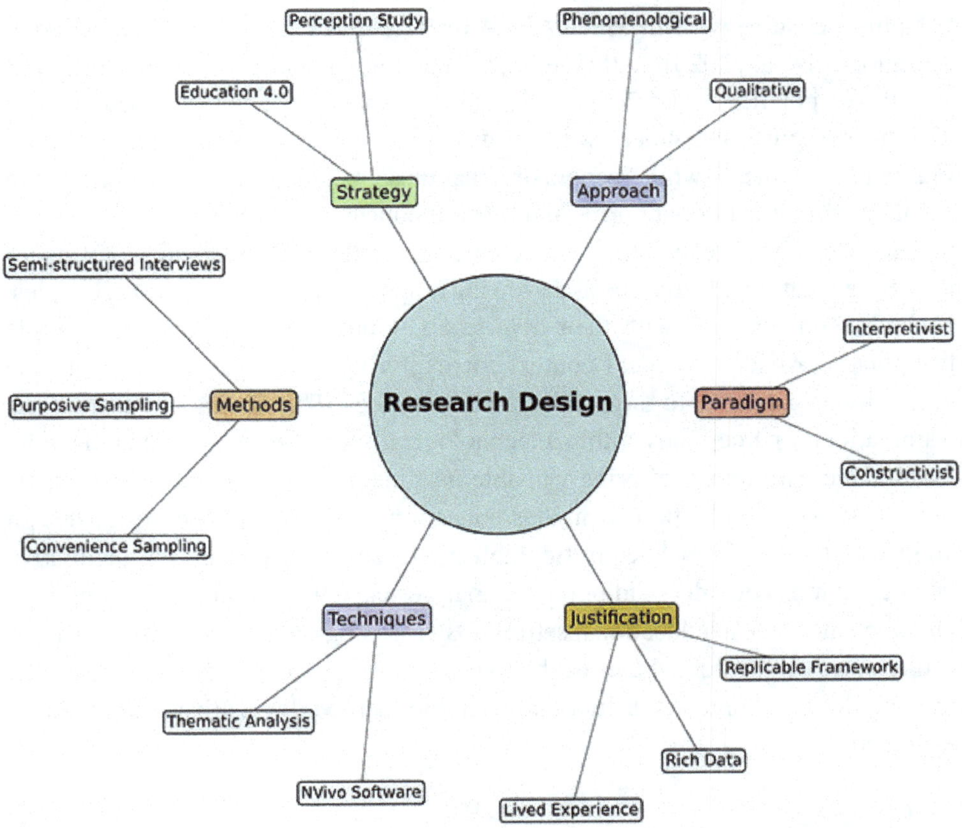

This radial mind map illustrates the research design framework, with "Research Design" at the centre. Key elements include a qualitative, phenomenological approach within an interpretivist, constructivist paradigm, purposive and convenience sampling, semi-structured interviews, and thematic analysis using NVivo software. The radial layout and colour-coded sections—green for strategy, orange for methods, blue for techniques and approach, yellow for justification, and pink for paradigm—highlight their interconnections.

The study's design is firmly grounded in a constructivist paradigm, which posits that knowledge is socially constructed and that individuals create meaning through their interactions with the world around them (Sampieri & Mendoza, 2018). This paradigm aligns seamlessly with the phenomenological focus on subjective interpretation, as it acknowledges that lecturers' perceptions of AI-LA are shaped by their personal histories with digital tools, their professional identities as supervisors, and the socio-cultural environment of HDR programmes in Adelaide. Constructivism supports the study's aim to generate rich, context-specific insights rather than seeking universal generalisations, which is essential for understanding the nuanced, discipline-sensitive ways in which AI-LA may (or may not) be adopted in doctoral education (Pi et al., 2023). In this sense, the paradigm respects the heterogeneity of supervisory practice across fields—education lecturers may emphasise pedagogical alignment, engineering supervisors technical precision, and social sciences faculty relational dynamics—while recognising that these meanings are co-constructed in dialogue with institutional norms and broader educational trends.

The research was conducted in Adelaide, Australia, a city recognised for its innovative educational landscape and growing emphasis on Education 4.0 principles. Adelaide's universities, with their diverse HDR programmes spanning multiple disciplines, provided an ideal context for exploring how lecturers navigate the integration of AI-LA into their supervisory practices. This setting allowed for the examination of perceptions within a technologically progressive yet pedagogically traditional environment, offering valuable insights into the challenges and opportunities of AI-LA adoption in higher education. The phenomenological design aligns with the IGI Global Scientific Publishing's emphasis on rigorous qualitative methods that provide deep, idiographic analyses capable of illuminating complex human phenomena in educational settings. By focusing on the lived experiences of lecturers, this study contributes to the journal's mission of advancing educational research through context-sensitive, transferable methodologies that bridge theory and practice.

Participants

Participants were selected using a combination of purposive and convenience sampling, strategies that are well-suited to qualitative phenomenological research aiming to capture information-rich cases (Patton, 2015). The initial target was to recruit up to 30 lecturers involved in HDR programmes across Adelaide universities, but data saturation was achieved with a smaller group, as no new themes or substantive insights emerged from additional interviews (Guest et al., 2006). Saturation was determined through ongoing, iterative analysis during data collection, ensuring that the sample size was sufficient to provide a comprehensive and trustworthy understanding of the phenomenon under investigation.

Inclusion criteria required participants to be actively involved in supervising HDR students and to have some experience with digital teaching or management tools, though prior exposure to AI-LA was not mandatory. This criterion ensured that participants could offer informed reflections on the potential integration of AI-LA into their supervisory practices, even if their direct experience with the technology was limited. Convenience sampling, facilitated through professional networks, institutional contacts and departmental email lists, enabled the recruitment of lecturers who were accessible and willing to participate—a common and pragmatic approach in qualitative studies focused on specific professional communities (Merriam & Tisdell, 2015).

The final sample comprised 14 lecturers from diverse disciplines, including education, engineering and social sciences. This disciplinary diversity enhanced the transferability of the findings by capturing a range of perspectives on how different academic fields perceive and engage with educational technologies (Popović et al., 2022). Participants brought varied levels of technological familiarity and supervisory experience, enriching the dataset with contrasting viewpoints on AI-LA's relevance, utility and potential challenges in HDR contexts.

Data Collection

Data were collected through in-depth semi-structured interviews, a method central to phenomenological research as it allows participants to express their experiences in their own words and on their own terms (Creswell & Poth, 2017). Each interview lasted between 45 and 60 minutes and was conducted either in person or via secure video conferencing, depending on participant availability and preference. All interviews were audio-recorded with participants' explicit consent and transcribed verbatim to ensure accuracy and fidelity in subsequent analysis.

The interview guide was developed iteratively based on the study's objectives and theoretical framework. It incorporated open-ended questions designed to explore lecturers' familiarity with AI-LA, perceived benefits and challenges, ethical considerations, and attitudes toward adoption. Sample questions included:

- Can you describe your experiences with digital tools in HDR supervision?

- What potential benefits do you see in using AI-powered Learning Analytics for HDR students?
- What concerns do you have about integrating AI-LA into your teaching practices?

The semi-structured format provided flexibility to probe deeper into emergent topics while maintaining focus on the phenomenon of interest. This approach also facilitated the co-construction of meaning between the researcher and participants, a key tenet of constructivist research that respects the active role of both parties in generating knowledge (Sampieri & Mendoza, 2018; Merriam & Tisdell, 2015). Interviews were conducted in a supportive, non-judgemental atmosphere to encourage open and reflective responses.

Data Analysis

The data were analysed using thematic analysis, following the six-phase reflexive process outlined by Braun and Clarke (2006): (1) familiarisation with the data, (2) generation of initial codes, (3) searching for themes, (4) reviewing themes, (5) defining and naming themes, and (6) producing the report. This method was selected for its flexibility and compatibility with phenomenological research, as it allows for the systematic identification of patterns of meaning across participants' narratives while remaining grounded in their lived experiences (Van Manen, 2016).

NVivo software was employed to facilitate systematic coding and theme organisation. The analysis began with multiple close readings of the transcripts to achieve deep immersion in the data. Initial codes were generated inductively, capturing key concepts and recurring ideas related to lecturers' perceptions of AI-LA. These codes were then grouped into broader, latent themes through iterative comparison and contrast across interviews. The process involved constant movement between the data, codes and emerging themes to ensure that interpretations remained faithful to participants' accounts. The final themes were defined and named to reflect the essence of lecturers' experiences, with carefully selected illustrative quotes used to substantiate and bring the findings to life. This rigorous, iterative approach ensured that the analysis stayed firmly grounded in the participants' voices, aligning with the phenomenological aim of capturing the structures of lived experience (Moustakas, 1994).

Results

This section presents an expanded and analytically enriched account of the findings derived from the qualitative phenomenological analysis of faculty interviews.

Building on the initial thematic structure, the revised Results section deepens the empirical interpretation by (a) examining relationships between themes, (b) foregrounding tensional and divergent perspectives, and (c) drawing more explicitly on NVivo-supported analytical outputs, including code co-occurrence, relative density, and cross-theme patterns. Throughout this section, participants' voices (P1–P14) remain central, and interpretation is confined to empirically grounded meaning-making rather than theoretical explanation.

Thematic Overview and Analytical Orientation

The NVivo-assisted thematic analysis generated five interrelated themes:

a) Centrality of Learning and Students,
b) Understanding and Using AI-Powered Analytics,
c) Role in Teaching, Support, and Supervision,
d) Contextual Factors within the University and HDR Environment, and
e) Challenges, Ethics, and Future Considerations.

While these themes are presented as discrete analytical categories, the coding structure and subsequent matrix queries revealed substantial interconnections among them. Rather than operating independently, themes frequently co-occurred within the same interview segments, indicating that faculty perceptions of AI-powered Learning Analytics (AI-LA) in HDR programmes are relational, layered, and context-dependent.

Automated sentiment coding indicated a greater number of positively valenced references (324) than negatively valenced ones (160). However, closer qualitative inspection showed that positive and negative orientations often appeared within the same narrative accounts, suggesting that faculty perspectives are better characterised as conditionally supportive rather than unambiguously favourable or resistant. This pattern of ambivalence is a defining empirical feature of the dataset and recurs across all major themes.

Word frequency analysis further contextualised these findings. The prominence of terms such as learning, students, analytics, using, and teaching underscores that participants consistently framed AI-LA through pedagogical and relational priorities, rather than through technological novelty or managerial efficiency. This lexical emphasis provides an important interpretive backdrop for the thematic findings that follow.

a) Centrality of Learning and Students as an Organising Empirical Lens

Across all interviews, the most dominant and consistently coded theme was the Centrality of Learning and Students. NVivo frequency data showed that references to learning and students not only outnumbered all other terms but also appeared across nearly all thematic clusters, indicating that this theme functions as an organising empirical lens through which other perceptions are filtered.

Participants repeatedly evaluated AI-LA in terms of its perceived capacity to support student development, progression, and wellbeing within HDR programmes. Rather than discussing analytics as neutral tools, faculty framed them in normative terms—asking whether, how, and under what conditions such systems genuinely enhance the doctoral learning journey. As P5 explained, AI-LA was only meaningful insofar as it enabled supervisors to "see the student more clearly" rather than merely track activity.

Importantly, this student-centred framing was not uniform across all stages of HDR candidature. Several participants distinguished between earlier, more structured phases and the later thesis-focused stages. For example, P4 expressed scepticism about the applicability of conventional learning analytics once doctoral work becomes less visible within institutional systems. This differentiation introduces an internal boundary within the theme itself, suggesting that the perceived relevance of AI-LA is contingent on the nature of doctoral work at a given moment.

NVivo co-occurrence analysis revealed that Centrality of Learning and Students most frequently overlapped with Understanding and Using AI Analytics and Challenges and Ethics. This pattern indicates that concerns about data interpretation, system design, and ethical implications are not peripheral but directly tied to faculty commitments to student-centred supervision. In empirical terms, learning and students do not merely represent one theme among others; they actively shape how all other themes are understood.

b) Understanding and Using AI-Powered Analytics: Between Awareness and Uncertainty

The second major theme captures faculty members' evolving understanding of AI-powered analytics and their practical implications for HDR supervision. While most participants demonstrated general awareness of AI and analytics discourse, their accounts revealed significant variation in conceptual clarity, confidence, and perceived readiness to engage with such systems.

NVivo coding showed high-density intersections between analytics, using, data, and artificial intelligence, indicating that participants often discussed these concepts together rather than in isolation. However, these discussions frequently involved clarification, distinction, or qualification. For instance, P1 explicitly differentiated between generative AI and analytics-driven machine learning, highlighting that faculty awareness does not necessarily translate into operational understanding.

Several participants articulated a sense of inevitability regarding the integration of AI-LA, yet simultaneously expressed uncertainty about how such systems would function in day-to-day supervisory practice. This tension was particularly evident in statements that acknowledged the potential of AI-LA while questioning its current maturity or relevance for HDR contexts. As P13 noted, analytics could be valuable if they make "the invisible visible", but only if the indicators meaningfully reflect doctoral activity rather than superficial engagement metrics.

From an empirical standpoint, NVivo matrix queries showed that references to understanding and effective systems frequently co-occurred with expressions of cautious optimism. This suggests that positive orientations toward AI-LA are often conditional upon system design, interpretability, and contextual fit, rather than based on abstract enthusiasm for AI as such.

c) Role in Teaching, Support, and Supervision: Augmentation with Limits

Faculty perceptions of the role of AI-LA in teaching, support, and supervision were characterised by a consistent emphasis on augmentation rather than substitution. Participants described multiple potential applications, including identifying students who may require additional support, informing reflective teaching practices, and managing supervisory workloads across multiple candidates.

NVivo density analysis showed that support and teaching were among the most frequently coded terms within this theme, while supervision appeared less often but carried substantial interpretive weight. This distribution reflects the hybrid nature of HDR roles, where teaching, mentoring, and pastoral care intersect.

Notably, participants did not conceptualise AI-LA as an autonomous decision-making system. Instead, analytics were framed as prompts or signals that require human interpretation. P9 highlighted that data-backed insights could make difficult conversations with candidates more grounded, while P6 emphasised that AI could assist with structure and monitoring but could not replace professional judgement.

At the same time, this theme contained clear boundary statements. Several participants explicitly rejected the idea that analytics could capture the full complexity of doctoral research. NVivo coding revealed that references to help frequently co-occurred with qualifiers such as depends, if, or as long as, indicating that perceived benefits are framed within explicit limits. These limits are empirically grounded in concerns about over-simplification and loss of relational nuance.

d) Contextual Factors: Institutional Structures and HDR Specificity

The theme of Contextual Factors within the University and HDR Environment highlights how faculty perceptions are shaped by institutional arrangements, disciplinary norms, and the distinctive characteristics of HDR programmes. NVivo coding showed frequent references to university, HDR, research, and management, underscoring that AI-LA is not perceived in a vacuum but within existing organisational ecosystems.

A recurring empirical pattern was the perceived contrast between institution-level adoption strategies and local supervisory realities. P1 described analytics as primarily driven by management for programme monitoring and improvement, suggesting a top-down orientation. In contrast, P5 emphasised the centrality of trust and unmediated relationships in HDR supervision, questioning whether analytics could meaningfully intervene in such contexts.

NVivo co-occurrence analysis indicated that contextual factors strongly intersected with challenges and ethics, suggesting that institutional conditions significantly influence how risks and concerns are perceived. Participants frequently linked questions of adoption to the availability of training, alignment with disciplinary practices, and recognition of HDR-specific work patterns. Empirically, this theme underscores that acceptance or resistance is not solely individual but structurally situated.

e) Challenges, Ethics, and Productive Tensions

Although positive orientations toward AI-LA were prevalent, the analysis also revealed a robust set of tensional and critical perspectives. These do not function as outright rejection but rather articulate conditions under which AI-LA may become problematic.

Key concerns clustered around trust, data provenance, ethical use, and training gaps. NVivo coding showed that references to ethics and risk frequently appeared alongside discussions of data and information sources. P3 articulated concern about the authenticity and traceability of data, while P11 stressed the necessity of clear ethical rules to sustain trust between students, supervisors, and institutions.

Importantly, lack of training emerged as a cross-cutting issue. P14's emphasis on insufficient preparation for both staff and students was echoed across multiple interviews, and NVivo matrix queries showed strong overlap between training, understanding, and ethical concerns. This pattern suggests that ethical apprehensions are not abstract but closely linked to perceived competence and institutional support.

Participants consistently highlighted ethical concerns around data provenance, algorithmic transparency and the preservation of trust in long-term supervisory relationships, concerns that echo the literature on faculty perceptions in pre-implementation stages.

Rather than constituting marginal dissent, these tensions represent integral components of faculty sense-making. Empirically, they delineate the conditions under which AI-LA might either enhance or undermine HDR supervision, reinforcing the complexity of faculty perceptions.

INTERIM SYNTHESIS OF EMPIRICAL FINDINGS

Taken together, the findings indicate that faculty perceptions of AI-powered Learning Analytics in HDR programmes are structured around a core commitment to student-centred learning, mediated by evolving understanding of AI, shaped by supervisory roles, and constrained by institutional and ethical contexts. The themes are empirically interconnected, and tensions within and between them form a central feature of the data rather than an anomaly.

This Results section has deliberately foregrounded these relationships and tensions to provide a robust empirical foundation for the interpretive and theoretical analysis that follows in the Discussion section.

The achievement of data saturation after 14 in-depth interviews served as a critical validation threshold for this phenomenological inquiry. According to the principles of Moustakas (1994) and the iterative process of Braun and Clarke (2006), the recurrence of themes—specifically the tension between institutional monitoring and pedagogical autonomy—indicated that the essential 'essence' of the faculty experience had been captured. This robustness allows for a high degree of qualitative reliability. The 14 participants, representing a cross-section of Adelaide's academic landscape, provided a dense narrative tapestry that reveals why AI-LA adoption is not a linear technical decision but a complex socio-pedagogical negotiation.

Ethical Considerations

This study adhered to stringent ethical standards to safeguard participant wellbeing, ensure research integrity and maintain public trust in the inquiry process. Ethical approval was granted under code 0291 by the Human Research Ethics Committee at Torrens University Australia, in full compliance with the Australian Code for the Responsible Conduct of Research (National Health and Medical Research Council, 2018). The approval process required detailed submission of the study protocol, interview guide, information sheet, consent form and risk assessment, confirming that the design minimised potential harm while maximising benefit to participants and the field.

Informed consent formed the cornerstone of ethical practice. Each participant received a comprehensive information sheet well in advance of the interview, clearly explaining the study's purpose (to explore perceptions of AI-powered Learning Analytics in HDR supervision), procedures (semi-structured interviews lasting 45–60 minutes), potential risks (minimal, primarily time commitment and discussion of sensitive topics such as data privacy concerns), benefits (contribution to improved understanding of educational technology adoption), and rights (including voluntary participation and the right to withdraw at any stage without consequence or explanation). Consent forms were signed electronically or in person before any data collection began, and participants were given the opportunity to ask questions and seek clarification. Copies of the signed forms were securely stored alongside pseudonymised data.

To protect participant identities and confidentiality, pseudonymisation was applied immediately after transcription. All names, institutional affiliations and any potentially identifying details (such as specific programme names or unique supervisory experiences) were replaced with neutral codes (e.g., P1, P2). Audio recordings were deleted after transcription verification, and transcripts were stored on password-protected university servers accessible only to the research team. Data will be retained for five years in line with Australian research regulations, after which they will be permanently and securely destroyed. No raw data will be shared publicly, and any quotations used in publications will be carefully reviewed to prevent deductive disclosure.

Voluntary participation was actively reinforced throughout. No financial or professional incentives were offered, as these could introduce coercion, particularly in an academic context where power dynamics might influence willingness to participate. Interviews were conducted in a supportive, non-judgemental environment to encourage open and honest responses, especially when discussing sensitive topics such as concerns over data privacy in AI-LA or perceived threats to supervisory autonomy. The researcher remained attentive to participant comfort, offering breaks or redirection if discussions became distressing.

Ethical considerations extended beyond procedural compliance to encompass respect for participants' professional expertise and vulnerability. Lecturers shared candid reflections on emerging technologies that may challenge established practices; the study therefore prioritised reciprocity by ensuring that findings would be disseminated in ways that could inform institutional policy and professional development. This commitment to beneficence, respect for persons and justice guided every stage of the research process (National Health and Medical Research Council, 2018).

Validity and Reliability

To ensure methodological rigour, this study employed multiple strategies to establish credibility, transferability, dependability and confirmability, the four criteria widely accepted for assessing trustworthiness in qualitative research (Lincoln & Guba, 1985). These measures were essential to validate the phenomenological approach and to provide readers with confidence that the findings authentically represent lecturers' lived perceptions of AI-powered Learning Analytics (AI-LA) in HDR programmes.

Credibility, the confidence that findings accurately reflect participants' experiences, was strengthened through triangulation and prolonged engagement. Data from the fourteen semi-structured interviews were cross-verified to identify consistent patterns across disciplines and levels of technological familiarity. Prolonged engagement with the transcripts—multiple close readings and iterative coding—allowed the researcher to develop deep familiarity with the data and to discern subtle nuances in lecturers' accounts (Creswell & Poth, 2017). Member checking further enhanced credibility: preliminary themes and selected quotations were shared with a subset of five participants, who confirmed the accuracy of interpretations and provided additional context where needed.

Transferability, the extent to which findings may apply to other contexts, was supported by thick description of the research setting and participants. Detailed accounts of Adelaide's innovative yet pedagogically traditional higher education landscape, the diversity of disciplines represented (education, engineering, social sciences) and the specific characteristics of HDR supervision enable readers to judge the relevance of the methodology and findings to similar settings, such as other Australian or international universities adopting AI-LA (Merriam & Tisdell, 2015). The study's focus on pre-implementation perceptions within a defined geographic and institutional context further facilitates assessment of transferability.

Dependability, ensuring consistency and transparency of the research process, was achieved through systematic documentation and external checks. A comprehensive reflective journal recorded all methodological decisions, challenges encountered during data collection and analysis, and adjustments made in response to emerging insights. Peer debriefing sessions with two colleagues experienced in qualitative research provided an independent audit of coding decisions and thematic development, helping to identify potential biases and refine interpretations (Sampieri & Mendoza, 2018). The audit trail—comprising raw transcripts (pseudonymised), coding structures, NVivo exports and reflective notes—allows another researcher to follow the analytical pathway and verify its logical coherence.

Confirmability, the assurance that findings stem from the data rather than researcher bias, was upheld by grounding all interpretations in participants' narratives.

NVivo facilitated systematic coding and theme organisation, minimising subjective influence through transparent code definitions and constant comparison across interviews. Direct quotations were extensively used to substantiate each theme, allowing readers to assess the fit between raw data and reported findings (Braun & Clarke, 2006). Reflexivity—ongoing self-examination of the researcher's assumptions and positionality—was documented in the reflective journal to further safeguard against undue influence on the analysis.

These strategies collectively ensured that the study's findings were robust, credible and faithful to the participants' lived experiences. They align with the rigorous qualitative standards expected by IGI Global Scientific Publishing's and provide a solid foundation for the empirical claims presented in subsequent sections.

The phenomenological richness and the thick description of the Adelaide HDR context enhance transferability, compensating for the modest sample size while providing actionable insights for similar research-intensive environments, as detailed in the study of Adelaide universities in the literature review.

Limitations

This study, while methodologically robust, is subject to several limitations inherent to its qualitative phenomenological design. Firstly, the relatively small sample size, while justified by data saturation ensuring comprehensive insights, limits the breadth of perspectives compared to larger, more diverse samples (Guest et al., 2006), may limit the breadth of perspectives compared to larger samples. The use of purposive and convenience sampling, while appropriate for capturing in-depth experiences, restricts statistical generalisability, as the findings are specific to the context of Adelaide universities and their HDR programmes.

Secondly, the reliance on self-reported data through semi-structured interviews introduces the potential for self-selection and recall biases, as participants may have provided socially desirable responses or incomplete reflections on their experiences with AI-LA (Creswell & Poth, 2017). The pre-implementation focus of the study, exploring perceptions prior to widespread AI-LA adoption, further limits insights into post-implementation experiences, which could reveal additional challenges or benefits.

Finally, the specific context of Adelaide's higher education landscape, while a strength for contextual depth, may constrain the direct applicability of the findings to other regions or educational settings with differing technological or cultural dynamics (Popović et al., 2022). Despite these limitations, the study's methodological transparency and focus on transferability ensure its value for informing similar qualitative inquiries into educational technology adoption.

These limitations do not diminish the contribution of the study; rather, they reflect the depth-oriented aims of phenomenological inquiry and reinforce the value of examining AI-LA adoption through lived experience in a clearly defined institutional context.

Discussion

Overview and Analytical Positioning

This Discussion interprets the empirical findings presented in the Results section in relation to the study's research objectives and the broader body of scholarship on AI-powered Learning Analytics (AI-LA) within higher education. Rather than reiterating thematic findings, the discussion focuses on how and why faculty perceptions take the forms observed, paying particular attention to the Higher Degree by Research (HDR) context, where supervision is relational, non-linear, and deeply embedded in academic identity and trust.

The discussion is structured around four interrelated analytical dimensions that emerged from the findings:

(1) the primacy of student-centred values as an organising principle,
(2) the gap between awareness and operational confidence in AI analytics,
(3) the framing of AI-LA as augmentation rather than substitution, and
(4) the centrality of context, ethics, and institutional mediation in shaping acceptance.

These dimensions are examined through the combined lenses of the Technology Acceptance Model (TAM), the Theory of Planned Behaviour (TPB), and Diffusion of Innovations (DOI), while remaining grounded in the phenomenological accounts of participating faculty.

Taken together, the findings demonstrate that faculty engagement with AI-powered Learning Analytics in HDR contexts is shaped less by technological capability than by alignment with pedagogical values, relational trust, and ethical legitimacy. This positions HDR supervision as a distinct adoption environment that cannot be adequately explained through generic learning analytics assumptions.

Student-Centred Values as the Dominant Interpretive Frame

A defining contribution of this study is the empirical demonstration that faculty perceptions of AI-LA in HDR programmes are not primarily technology-driven, but value-driven, with student learning and wellbeing functioning as the dominant interpretive frame. While student-centredness is widely cited in the learning analytics literature, the present findings show that, in HDR contexts, this orientation takes on a qualitatively different form.

Faculty did not assess AI-LA in terms of efficiency gains or predictive accuracy alone, but rather in relation to its capacity to support doctoral learning trajectories that are inherently idiosyncratic, creative, and temporally uneven. This helps explain why positive attitudes towards AI-LA were consistently conditional rather than unconditional. From a TPB perspective, favourable attitudes were present, but they were bounded by concerns about alignment with doctoral pedagogy. Similarly, TAM's perceived usefulness construct appears here as context-sensitive, varying according to the stage of candidature and the visibility of doctoral work within institutional systems.

This finding extends existing learning analytics research by illustrating that perceived usefulness cannot be assumed to operate uniformly across educational levels. In HDR supervision, usefulness is evaluated not against standardised performance indicators (Davis, 1989), but against the preservation of intellectual autonomy, scholarly identity formation, and relational trust. As such, AI-LA tools designed for coursework environments may struggle to gain traction in HDR contexts unless they are explicitly reconfigured to support these deeper pedagogical commitments.

These findings suggest that established adoption constructs, such as perceived usefulness, require contextual reinterpretation in HDR supervision, where value is assessed not through performance optimisation but through the preservation of intellectual autonomy, scholarly identity formation, and supervisory trust.

Rather than representing an extension of coursework-based teaching, HDR supervision emerges from the data as a pedagogical domain with its own evaluative logics, temporal rhythms, and relational demands, which fundamentally reshape how analytics are interpreted and judged.

AWARENESS WITHOUT MASTERY: UNDERSTANDING AI ANALYTICS IN HDR PRACTICE

The findings reveal a persistent gap between conceptual awareness of AI-powered analytics and operational confidence in their use. Faculty participants demonstrated familiarity with AI discourse and acknowledged its growing institutional presence, yet many expressed uncertainty regarding how analytics would meaningfully integrate into their supervisory practice. This distinction between awareness and proficiency is analytically significant.

From the perspective of TPB, this gap reflects limited perceived behavioural control. Even where attitudes toward AI-LA were broadly positive, uncertainty about system functionality, data relevance, and interpretability constrained faculty willingness to engage. TAM further clarifies this dynamic, as perceived ease of use emerged as a critical moderating factor: analytics were viewed favourably when imagined as intuitive and supportive, but problematically when associated with opaque dashboards or administratively driven metrics.

Importantly, this uncertainty was not framed as resistance to innovation, but as a request for epistemic clarity and professional legitimacy. Faculty sought to understand not only how AI-LA systems operate, but also how their outputs should be interpreted within the intellectual and ethical responsibilities of HDR supervision. This suggests that professional development initiatives focused solely on technical training may be insufficient unless they also address interpretive judgement and contextual application.

Augmentation, Not Replacement: Reconfiguring the Supervisory Role

A consistent theme across the findings was the rejection of AI-LA as a substitute for human judgement, alongside a cautious openness to its use as a supportive augmentation. Faculty articulated a clear boundary between data-informed insight and pedagogical decision-making, emphasising that analytics should prompt reflection rather than dictate action.

This positioning aligns closely with contemporary discussions of human–AI collaboration in education, but the present study adds nuance by situating this collaboration within the emotionally and intellectually demanding space of HDR supervision. Faculty valued analytics insofar as they could surface early signals of disengagement or difficulty, yet they resisted any implication that doctoral progress could be reduced to quantifiable indicators.

From a DOI perspective, this finding highlights compatibility as a decisive factor in adoption. AI-LA systems perceived as misaligned with the relational and dialogic nature of supervision were viewed sceptically, regardless of their technical sophistication. Conversely, tools framed as enhancing supervisory awareness without undermining autonomy were seen as potentially compatible with HDR practice. This underscores the importance of framing AI-LA not as an evaluative mechanism, but as a reflective aid embedded within existing supervisory relationships.

Context, Power, and Institutional Mediation

The discussion of contextual factors reveals a structural tension between institutional adoption strategies and local supervisory realities. Faculty frequently described AI-LA as a management-led initiative oriented towards programme oversight, quality assurance, and risk mitigation. While such objectives are institutionally legitimate, they can conflict with the values and practices that define HDR supervision.

This tension has important implications for TPB's subjective norms construct. In the pre-implementation phase observed in this study, institutional expectations appeared to exert greater influence than peer norms. However, this influence did not automatically translate into behavioural intention, particularly where faculty perceived a misalignment between institutional priorities and supervisory values.

DOI further illuminates this dynamic by highlighting limited trialability and observability. Faculty had few opportunities to experiment with AI-LA in low-stakes HDR contexts or to observe discipline-specific exemplars of effective use. As a result, adoption remained abstract and policy-driven rather than practice-led. These findings suggest that without meaningful opportunities for local experimentation and peer learning, institutional AI strategies risk being perceived as externally imposed rather than pedagogically grounded.

Ethics, Trust, and the Conditions of Legitimate Adoption

The findings indicate that ethical considerations operate not as secondary constraints on AI-LA adoption, but as structural conditions that determine whether such systems are perceived as legitimate within HDR supervision.

Ethical considerations emerged not as secondary concerns, but as foundational conditions for the legitimate use of AI-LA in HDR programmes. Faculty concerns about data provenance, algorithmic transparency, and surveillance were tightly coupled with questions of trust—both in the technology itself and in the institutions deploying it.

Crucially, ethical apprehensions were intertwined with issues of competence and support. Lack of training amplified concerns about misuse, misinterpretation, and unintended consequences, suggesting that ethical risk is perceived as higher when faculty feel underprepared. This finding challenges approaches that treat ethics as a compliance issue rather than as a lived, practice-based concern.

The discussion thus reframes ethics from a peripheral barrier to a structuring dimension of acceptance. In HDR contexts, where supervision is grounded in trust and mutual responsibility, any technology perceived as undermining these foundations is likely to encounter resistance. Ethical governance, transparency, and shared understanding therefore function not merely as safeguards, but as enablers of adoption.

Integrative Interpretation and Contribution to Knowledge

Taken together, the findings suggest that faculty perceptions of AI-powered Learning Analytics in HDR programmes are shaped by a complex interplay of values, understanding, role identity, institutional context, and ethical judgement. The combined application of TAM, TPB, and DOI demonstrates that no single model adequately captures this complexity in isolation.

This study contributes to the literature by empirically demonstrating that contextual compatibility and ethical legitimacy can outweigh perceived technological advantages in shaping adoption decisions. It also extends learning analytics scholarship by foregrounding HDR supervision as a distinct pedagogical domain that challenges assumptions derived from coursework-based settings.

By centring faculty lived experience, the study highlights the necessity of designing AI-LA systems that are not only technically robust, but also pedagogically aligned, ethically grounded, and responsive to the relational realities of doctoral education.

By foregrounding faculty meaning-making, this study demonstrates that successful integration of AI-powered Learning Analytics in HDR programmes depends not on predictive sophistication, but on the capacity of systems to support reflective judgement, ethical responsibility, and relational supervision.

The strategic utility of this research lies in its capacity to serve as a decision-making framework for Higher Education leadership. By identifying 'Compatibility' (DOI) and 'Subjective Norms' (TPB) as the primary influencers of adoption, this paper provides a roadmap for the ethical implementation of Education 4.0. It warns against 'Management-Led' implementations that ignore the 'Relational Analytics' required in doctoral supervision. For software developers, the utility is found in the demand for 'Explainable AI' (XAI), ensuring that predictive insights are transparent and facilitate the supervisor-candidate bond rather than automating it.

CONCLUSION

This study set out to explore university lecturers' perceptions of AI-powered Learning Analytics (AI-LA) within Higher Degree by Research (HDR) programmes in Adelaide universities, with particular attention to how these perceptions are shaped by supervisory practice, institutional context, and ethical considerations. By adopting a qualitative phenomenological approach, the research foregrounded faculty lived experiences and meaning-making processes, offering a nuanced account of how emerging analytics technologies are understood, negotiated, and conditionally accepted in doctoral education.

Summary of Key Findings

The findings demonstrate that faculty perceptions of AI-LA in HDR contexts are neither uniformly positive nor resistant, but fundamentally conditional and relational. At the centre of these perceptions lies a strong commitment to student-centred learning, which functions as an organising principle through which all other considerations are evaluated. Lecturers consistently assessed AI-LA not as a neutral technological innovation, but in terms of its perceived capacity to support doctoral candidates' learning trajectories, wellbeing, and scholarly development.

While participants acknowledged the growing institutional prominence of AI and analytics, a clear distinction emerged between general awareness and practical confidence. Faculty expressed interest in the potential of AI-LA to surface early indicators of disengagement or difficulty, yet also articulated uncertainty about how such systems would operate meaningfully within the non-linear, largely invisible nature of doctoral research. This gap between awareness and operational readiness emerged as a critical factor shaping cautious engagement.

The role of AI-LA was predominantly framed as augmentative rather than substitutive. Lecturers valued analytics as prompts that could inform reflection and support decision-making, but strongly resisted any implication that supervisory judgement, relational trust, or intellectual mentorship could be automated. This boundary was particularly pronounced in HDR supervision, where academic identity, autonomy, and personalised guidance are central to the doctoral experience.

Contextual and institutional factors further shaped these perceptions. AI-LA was frequently perceived as driven by management-level objectives related to programme oversight and risk management, creating tension with the lived realities of supervision at the local level. Ethical concerns—particularly around trust, data provenance, transparency, and training—were not peripheral but emerged as foundational conditions for any legitimate implementation of AI-LA in HDR programmes.

Contributions of the Study

This research makes several contributions to the field of learning analytics and higher education.

First, it contributes empirical insight into a context that remains underexplored in the literature. While much learning analytics research focuses on coursework-based environments, this study demonstrates that HDR supervision constitutes a distinct pedagogical domain with different evaluative criteria, temporal structures, and relational demands. The findings show that assumptions derived from undergraduate or taught postgraduate contexts cannot be straightforwardly transferred to doctoral education.

Second, the study offers a conceptual contribution by illustrating how established models of technology adoption operate under conditions of high relational intensity and professional autonomy. The findings suggest that in HDR contexts, compatibility with pedagogical values and ethical legitimacy may outweigh perceived technological advantage. This challenges more linear or instrumental interpretations of technology acceptance and diffusion, highlighting the need for context-sensitive application of adoption frameworks.

Third, the study contributes methodologically by demonstrating the value of phenomenological inquiry for examining faculty engagement with educational technologies. By centring lived experience rather than system performance, the research surfaces tensions, boundaries, and conditions of acceptance that are often obscured in survey-based or purely technical studies.

Implications for Practice and Policy

The findings carry several implications for universities considering the implementation of AI-powered Learning Analytics in HDR programmes.

For institutional practice, the study suggests that AI-LA initiatives should prioritise alignment with supervisory values rather than efficiency alone. Systems are more likely to be accepted when they are designed to support reflective supervision, respect doctoral autonomy, and operate transparently within existing relational practices. Generic analytics dashboards developed for coursework contexts may be insufficient or inappropriate for HDR supervision unless they are meaningfully adapted.

Training emerged as a critical enabling condition. Professional development should extend beyond technical instruction to include interpretive guidance, ethical reflection, and context-specific use cases relevant to HDR supervision. Without such support, even well-intentioned analytics initiatives risk being perceived as burdensome or intrusive.

At the policy level, the findings underscore the importance of clear ethical governance frameworks. Data use, access rights, and the purposes of analytics must be explicitly articulated to build trust among faculty and candidates alike. Ethical clarity should precede, rather than follow, technological deployment, particularly in high-stakes research environments.

Directions for Future Research

This study also points to several avenues for future research. Longitudinal studies examining how faculty perceptions evolve following hands-on experience with AI-LA would provide valuable insight into how initial ambivalence translates into sustained practice or resistance. Comparative research across institutions or national contexts could further illuminate how local governance, disciplinary cultures, and policy environments mediate adoption.

Future work could also explore doctoral candidates' perspectives on AI-LA, particularly how analytics-informed supervision is experienced from the student side.

Examining points of convergence and divergence between supervisor and candidate perceptions would contribute to more holistic and ethically grounded system design.

Finally, design-oriented research involving co-creation between faculty, students, and system developers could help translate the conditional acceptance identified in this study into analytics tools that genuinely support HDR learning without undermining its relational foundations.

Concluding Remarks

In conclusion, this study demonstrates that the integration of AI-powered Learning Analytics into HDR programmes is not primarily a technical challenge, but a pedagogical, ethical, and relational one. Faculty acceptance depends less on the promise of automation than on the capacity of analytics systems to align with deeply held values of supervision, trust, and scholarly development. By foregrounding faculty lived experience, this research contributes to a more nuanced understanding of how AI-LA can be responsibly and meaningfully integrated into doctoral education within the evolving landscape of Education 4.0.

These recommendations are designed specifically for institutions and faculty developers seeking responsible and context-sensitive integration of AI-LA in HDR programmes.

REFERENCES

Ajzen, I. (1991). The theory of planned behavior. *Organizational Behavior and Human Decision Processes*, *50*(2), 179–211. DOI: 10.1016/0749-5978(91)90020-T

Braun, V., & Clarke, V. (2006). Using thematic analysis in psychology. *Qualitative Research in Psychology*, *3*(2), 77–101. DOI: 10.1191/1478088706qp063oa

Creswell, J. W., & Poth, C. N. (2017). *Qualitative Inquiry and Research Design: Choosing Among Five Approaches* (4th ed.). SAGE.

Crompton, H., Burke, D., & Lin, Y.-C. (2024). The use of generative AI in education: A systematic review. *arXiv*, 2405.11800. https://doi.org//arXiv.2405.11800DOI: 10.48550

Davis, F. D. (1989). Perceived usefulness, perceived ease of use, and user acceptance of information technology. *Management Information Systems Quarterly*, *13*(3), 319–340. DOI: 10.2307/249008

Fischer, C., Pardos, Z. A., & Baker, R. S. (2022). Learning analytics and educational data mining: Towards communication and collaboration. *Journal of Learning Analytics*, *9*(2), 1–7. DOI: 10.18608/jla.2022.7921

Guest, G., Bunce, A., & Johnson, L. (2006). How many interviews are enough? An experiment with data saturation and variability. *Field Methods*, *18*(1), 59–82. DOI: 10.1177/1525822X05279903

Heidegger, M. (1962). *Being and Time*. Harper & Row.

Lincoln, Y. S., & Guba, E. G. (1985). Naturalistic Inquiry. *Sage (Atlanta, Ga.)*.

Merriam, S. B., & Tisdell, E. J. (2015). *Qualitative Research: A Guide to Design and Implementation* (4th ed.). Jossey-Bass.

Moustakas, C. (1994). Phenomenological Research Methods. *Sage (Atlanta, Ga.)*.

National Health and Medical Research Council. (2018). *Australian Code for the Responsible Conduct of Research*. Commonwealth of Australia. https://www.nhmrc.gov.au/about-us/publications/australian-code-responsible-conduct-research-2018

Patton, M. Q. (2015). *Qualitative Research & Evaluation Methods* (4th ed.). SAGE.

Pi, Z., Yang, B., Yang, J., & Liu, X. (2023). Students' interaction patterns with learning videos integrated into the learning management system and their impact on learning outcomes. *British Journal of Educational Technology*, *54*(4), 1049–1068. DOI: 10.1111/bjet.13232

Popović, N., Popović, T., & Dragović, I. (2022). Use of an interculturally enriched collaboration script in computer-supported collaborative learning in higher education. *Technology, Pedagogy and Education*, *31*(3), 349–365. DOI: 10.1080/1475939X.2022.2027267

Sampieri, R. H., & Mendoza, C. P. (2018). *Metodología de la investigación*. McGraw-Hill.

Siemens, G. (2013). Learning analytics: The emergence of a discipline. *The American Behavioral Scientist*, *57*(10), 1380–1400. DOI: 10.1177/0002764213498851

Van Manen, M. (2016). *Researching Lived Experience: Human Science for an Action Sensitive Pedagogy* (2nd ed.). Routledge.

Venkatesh, V., Thong, J. Y. L., & Xu, X. (2016). Unified theory of acceptance and use of technology: A synthesis and the road ahead. *Journal of the Association for Information Systems*, *17*(5), 328–376. DOI: 10.17705/1jais.00428

Zahavi, D. (2021). Applied phenomenology: Why it is safe to ignore the epoché. *Continental Philosophy Review*, *54*(2), 259–273. DOI: 10.1007/s11007-019-09463-y

APPENDIX

Interview Guide

The following interview guide was used to structure the semi-structured interviews with lecturers, designed to explore their perceptions of AI-powered Learning Analytics (AI-LA) in Higher Degree by Research (HDR) programmes. The questions were developed based on the study's objectives and theoretical framework, allowing flexibility to probe emergent topics.

1. Can you describe your experiences with digital tools in HDR supervision?
2. What potential benefits do you see in using AI-powered Learning Analytics for HDR students?
3. What concerns do you have about integrating AI-LA into your teaching practices?
4. How familiar are you with AI-powered Learning Analytics, and how has this influenced your supervisory approach?
5. What role do you believe institutional support plays in adopting AI-LA in HDR programmes?
6. How do you perceive the ethical implications of using AI-LA, particularly regarding data privacy?
7. What factors might influence your willingness to adopt AI-LA in your supervisory role?

Example of NVIVO Coding

The following table illustrates a sample of the coding process conducted in NVivo, based on Braun and Clarke's (2006) thematic analysis framework. It includes initial codes, emergent themes, and illustrative quotes from the interviews.

Table 1.

Initial Code	Emergent Theme	Illustrative Quote
Technological optimism	Perceived benefits	"AI-LA could help me track student progress more efficiently, saving time for feedback."
Ethical concerns	Ethical concerns	"I worry about who has access to student data and how it might be used."
Lack of training	Adoption barriers	"Without proper training, I wouldn't know how to use AI-LA effectively."

Chapter 9
Rethinking Qualitative Inquiry With AI:
Thematic Analysis of Digital Consumer Interactions

Wahyu Wibowo
http://orcid.org/0009-0000-5176-7103
Sunan Kalijaga State Islamic University, Yogyakarta, Indonesia

Alda Fildza Amima
http://orcid.org/0009-0007-0240-7874
Sunan Kalijaga State Islamic University, Yogyakarta, Indonesia

Nesha Rizky Ashari
Sunan Kalijaga State Islamic University, Yogyakarta, Indonesia

Dafelia Qodhari
http://orcid.org/0009-0002-9691-3519
Sunan Kalijaga State Islamic University, Yogyakarta, Indonesia

Riska Prasasti
http://orcid.org/0009-0008-9330-640X
Sunan Kalijaga State Islamic University, Yogyakarta, Indonesia

ABSTRACT

This chapter rethinks qualitative inquiry in the era of generative artificial intelligence by examining thematic analysis of digital consumer interactions. It argues that large language models can broaden the scope and speed of qualitative work through data acquisition support, pre-processing, segmentation, pre-coding, and pattern mapping, yet they cannot substitute human interpretation, sensitivity to context,

DOI: 10.4018/979-8-3373-7640-0.ch009

ıt © 2026, IGI Global Scientific Publishing. Copying or distributing in print or electronic forms without written permission of IGI Global Scientific Publishing is prohibited. Use of this chapter to enerative artificial intelligence (AI) technologies is expressly prohibited. The publisher reserves all rights to license its use for generative AI training and machine learning model development.

and theoretical justification. Building on reflexive thematic analysis, the chapter clarifies epistemological stakes, distinguishes AI as a tool versus collaborator, and highlights risks such as decontextualization, automation bias, and opacity. It then proposes a hybrid workflow that keeps researchers accountable for coding decisions, theme development, and audit trails, while using AI to manage volume and surface candidate patterns. The chapter concludes with guidance for trustworthy and ethically responsible AI-augmented thematic analysis.

INTRODUCTION

The advancement of Artificial Intelligence (AI), particularly in the domains of Natural Language Processing (NLP) and Large Language Models (LLMs), has introduced transformative changes to qualitative research methodologies. Generative models such as ChatGPT, Gemini, and similar systems now enable researchers to process and analyze extensive text corpora far more efficiently than traditional manual methods, while still allowing precise identification of significant themes and patterns of meaning. This evolution is particularly meaningful because qualitative research has historically relied on human interpretation, including close reading, reflexive thought, and sensitivity to social and cultural contexts embedded in the data.

AI's capabilities in summarizing text, segmenting datasets, suggesting coding candidates, and detecting repetitive patterns elevate it beyond a simple technical tool. Instead, AI serves as a powerful assistant in interpretive processes, potentially reshaping the workflow of qualitative research. Methodological literature increasingly references AI-enhanced qualitative research, which frames AI as a research assistant deployed at specific stages of analysis such as initial coding, thematic exploration, and early-stage pattern recognition, without relinquishing interpretive authority to automated systems.

The integration of AI into qualitative research mirrors its broader expansion as an analytical infrastructure across multiple sectors. Studies demonstrate that AI is progressively embedded within intelligent manufacturing systems, cybersecurity frameworks, healthcare, educational platforms, and sustainable information management. These applications underscore that AI is no longer perceived solely as a tool for technical automation but as a mechanism to support analysis and informed decision-making in a wide range of knowledge domains (Ayyaswamy, 2026b, 2026a; Kannan et al., 2025; N. Kathirvel, Ayyaswamy, & Gunasekaran, 2025; Rithik & Kathirvel, 2025).

Empirical research further supports AI's utility in thematic analysis, showing that it enhances the speed and scope of data processing while preserving the reflective quality that characterizes qualitative inquiry. For instance, Goyanes et al. (2025)

demonstrate that generative models can facilitate the development of more reliable thematic analysis protocols through structured stages—from data preparation to validation of results. These findings are particularly relevant given that traditional thematic analysis has often been criticized for procedural variability and dependence on the researcher's consistency.

AI proves especially useful in labor-intensive early stages of qualitative analysis, such as data segmentation, initial code generation, and preliminary mapping of relationships among units of meaning. Nonetheless, such efficiency gains must be contextualized within the epistemological framework of qualitative research, where meaning is constructed through the interaction of empirical data, social context, and the researcher's interpretive reasoning. In this sense, AI should enhance analytical rigor and transparency rather than replace the conceptual foundations that justify qualitative knowledge claims.

The urgency to adapt qualitative research practices intensifies in the context of big qualitative data—large-scale, complex data generated from human interactions in digital spaces. In the study of digital consumer behavior, such data include social media comments, online forum discussions, product reviews, and transcripts of customer service interactions. These data are naturalistic, temporally situated, and rich in evaluative and emotional content. However, their volume and heterogeneity challenge the core assumptions of traditional qualitative methods, which typically involve intensive reading of a limited number of documents.

When the textual unit reaches tens or hundreds of thousands of documents, human constraints on time, labor, and cognitive attention become significant methodological obstacles. AI provides a solution by enabling effective coding and theme identification at scale, without diminishing the need for interpretive assessment. Moreover, the continual generation of digital interactions in sectors like hospitality and online advertising ensures a steady stream of textual data suitable for qualitative analysis (Ayyaswamy, Kathirvel, & Vianny, 2025; Kathirvel et al., 2025).

A notable conceptual framework in this domain is the Guided AI Thematic Analysis (GAITA) approach proposed by Nguyen-Trung (2025). GAITA emphasizes that AI should function as a system guided by the researcher, not as an autonomous decision-maker. Researchers maintain control over key stages such as data familiarization, code formation, and theme development, while AI accelerates these processes through temporary generative outputs. This framework mitigates the risk of transferring interpretive authority from researchers to automated systems, which could otherwise compromise conceptual justification and auditability in academic research.

Similar concerns about human oversight have emerged in applied AI systems, including industrial cybersecurity, intelligent library automation, and robotics-based manufacturing. Across these domains, human supervision remains essential to ensure reliable analytical decisions (Ayyaswamy et al., 2025b; Ayyaswamy et al., 2025c; Ayyaswamy et al., 2025d).

Despite its potential, LLMs have limitations in text coding that require realistic acknowledgment. Han et al. (2025) report that while generative models can automate some aspects of analysis, they struggle to understand complex contexts, particularly in longer or nuanced documents. Coding accuracy and completeness tend to decline as document length and complexity increase.

Methodologically, this underscores the necessity of active researcher engagement in evaluating, revising, and interpreting AI-generated outputs. Evidence from other AI applications, such as disease detection and digital health systems, also demonstrates that AI outputs require human verification to prevent misclassification and contextual misinterpretation (Ayyaswamy et al, 2025e; Ayyaswamy et al., 2025f). Consequently, integrating AI into qualitative research demands a hybrid approach that combines computational power with human interpretive expertise.

Reimagining qualitative research in the AI era does not imply abandoning the discipline's core principles. Instead, it necessitates reconstructing methodological workflows to accommodate large-scale digital data and advanced computational capabilities. Thematic analysis remains particularly relevant, offering a flexible framework for identifying patterns of meaning in textual data, both descriptively and interpretively.

When combined with AI, thematic analysis can be expanded to support broader data exploration, mapping inter-theme relationships, and reinforcing initial coding stages. Yet, the researcher must remain the final authority in interpretation to ensure findings are valid, contextually grounded, and theoretically meaningful (Bowden et al., 2025). Qualitative research validity depends not only on procedural consistency but also on the clarity of interpretive argumentation, transparency of analytical procedures, and accurate placement of findings within empirical contexts. In contemporary literature on educational leadership, AI-driven entrepreneurship, and AI-based governance systems, the central question is no longer whether AI can be used in research, but how to integrate it responsibly and transparently into knowledge production (Ayyaswamy et al., 2025g; Ayyaswamy et al., 2025h).

Building on this understanding, integrating AI in thematic analysis emerges as a strategic step to optimize qualitative research, especially in analyzing complex, high-volume digital consumer interactions. AI is not merely a tool for efficiency; when used methodologically and reflectively, it enriches analysis through the synergy of computational processing and human interpretation.

This chapter, therefore, seeks to make conceptual, epistemological, and empirical contributions to the discourse on AI integration in qualitative research. By employing thematic analysis on digital consumer interaction data, it aims to provide a systematic, reflective methodological foundation for researchers. Such a framework allows AI to be leveraged productively while preserving the interpretive integrity of qualitative research in the era of large-scale digital data (Kalanda & Cheboi, 2025).

RETHINKING QUALITATIVE INQUIRY: THEORETICAL AND METHODOLOGICAL FOUNDATIONS

Thematic analysis is now understood not merely as a technical procedure for grouping codes but as an interpretive practice that positions the researcher as the primary agent in meaning-making. From the outset, Braun and Clarke rejected the notion that thematic analysis could be applied mechanically as a neutral method; every analytical decision carries epistemological and theoretical implications, from determining relevance, constructing codes, to formulating themes (Braun & Clarke, 2006). This perspective is reinforced by the concept of reflexive thematic analysis, which asserts that themes are not simply discovered in data but are analytical constructs generated through the researcher's active, reflective, and contextual engagement (Braun & Clarke, 2021; Cena et al., 2024). Therefore, the quality of thematic analysis is evaluated less by adherence to technical procedures and more by the coherence of interpretive argumentation, the clarity of the relationship between data and themes, and the researcher's transparency regarding theoretical positions and analytical choices.

Within the interpretive tradition, thematic analysis assumes that social reality is not merely a collection of measurable facts but a meaningful experience shaped through language, interaction, and context. Analysis thus examines not only what participants say but also how meaning is formed, negotiated, and associated with specific social situations. This tradition emphasizes the importance of the lifeworld, context, and researcher involvement in understanding social practices (Schwartz-Shea & Yanow, 2013). In its operational logic, reflexive thematic analysis aligns with hermeneutical movements: researchers move iteratively between parts and the whole, between quotations and narratives, and between initial codes and themes until a consistent interpretive understanding emerges (Belhassen & Caton, 2009). Consequently, a robust theme is not the product of mechanical aggregation but of rigorous scientific argumentation. Researchers must justify why a pattern is significant to the research question, how it differs from other themes, and how it relates to the theoretical framework employed.

Thematic analysis is often praised for its flexibility, as it can be applied across various philosophical paradigms. However, this flexibility requires clarity of paradigm. Many studies claim to use thematic analysis but treat it as a paradigm-free technical recipe, producing themes that appear neat yet are epistemically weak. Recent literature confirms that thematic analysis should demonstrate coherence among research questions, ontological and epistemological assumptions, sampling strategies, and the process of theme construction (Cena et al., 2024). Braun & Clarke (2022) further emphasize that reflexivity, reasoning, and transparency of analytical decisions underpin rigor, not mere procedural compliance. In this sense, thematic analysis constitutes a methodological practice informed by explicit paradigms, which must be consistently articulated. This perspective aligns with cross-disciplinary discussions in AI governance, leadership, and ethical entrepreneurship, which stress human accountability in decision-making (Ayyaswamy et al., 2025g; Ayyaswamy et al., 2025h).

This stance has direct methodological implications for understanding the six phases of thematic analysis. Braun and Clarke's model comprises data familiarization, initial coding, theme search, theme review, theme definition and naming, and report writing. While often interpreted linearly, Braun and Clarke highlight that these phases are conceptual guides for navigating iterative, reflective, and nonlinear processes (Braun & Clarke, 2006; Braun & Clarke, 2021). In reflexive practice, familiarization is ongoing, evolving throughout analysis. Codes are temporary analytical tools, revised, combined, or discarded as conceptual understanding deepens. Themes are not merely large categories of similar codes but structures of meaning that analytically address research questions. Theme review thus involves both empirical fit and conceptual coherence (Ahmed et al., 2025).

The most pressing challenge in contemporary thematic analysis arises with digitalization. Researchers accustomed to in-depth interviews, observations, and field notes increasingly confront massive, layered, fast-moving, and semantically rich data, including social media interactions, online forums, digital archives, learning platforms, and audiovisual recordings. This shift raises a crucial epistemological question: how can interpretive depth be maintained when data volume and semiotic complexity increase? In digital consumer behavior studies, for example, meaning is conveyed through text, emojis, images, memes, tone, platform contexts, and interaction patterns. Digitalization across sectors, including hospitality and digital advertising, continues to produce rich semiotic footprints that challenge traditional manual methods (Ayyaswamy et al., 2025; Kathirvel et al., 2025).

To address these challenges, researchers increasingly use Computer-Assisted Qualitative Data Analysis Software (CAQDAS) such as NVivo, ATLAS.ti, or MAXQDA. These tools support coding, archiving, decision tracking, and audit trails. However, methodological literature cautions that software is not epistemologically neutral; the way software displays and groups data influences how researchers conceptualize themes (Braun & Clarke, 2021; Fryer, 2022). This concern is amplified as CAQDAS incorporates AI features, potentially creating the illusion that analysis can be accelerated without interpretive involvement. Paulus & Marone (2025) suggest that such design can affect the researcher's perception of qualitative analysis.

The integration of NLP and LLMs offers both opportunities and risks. NLP supports transcription automation, data cleaning, and preliminary summaries, which reduces administrative burden and allows researchers to focus on conceptual work Hayes (2025) However, transcription is not neutral; it already involves interpretive decisions regarding speech, pauses, emotion, and context. Therefore, improved speech-to-text accuracy does not remove the need for researcher validation, especially in complex or multimodal interactions (Yan & Pourdavood, 2024).

In the later stages, LLMs are often used for pre-coding, i.e. generating initial code candidates through specific prompts. Cook et al. (2025) view AI-assisted pre-coding as an analytic scaffold, which is an initial buffer that helps researchers see possible patterns without having to start completely from scratch. However, these benefits must be balanced with methodological vigilance. Wheeler (2025) reminds of the danger of reification, which is the tendency to treat AI-generated code as an objective category. In fact, such outputs should be positioned as analytical hypotheses that need to be tested, compared, revised, or rejected. This principle is in line with reflexive thematic analysis which rejects the idea that themes can be discovered automatically from data (Braun & Clarke, 2021). This need for human oversight is also evident in a variety of applied AI systems—for example, industrial cybersecurity, manufacturing robotics, and intelligent library management—where AI is useful for accelerating processes, but still requires human interpretive and evaluative control (Ayyaswamy et al, 2025b; Ayyaswamy et al., 2025c; Ayyaswamy et al., 2025d).

At this point, the distinction between topic modeling and thematic interpretation becomes important. Topic modeling works through probabilistic logic and statistical associations between words to extract the structure of topics in a large corpus, making them useful as an initial exploration. Odacioglu et al. (2024) suggests that this technique can help identify early focus areas in large text data, especially when combined with a theoretical framework. However, topic modeling is not synonymous with thematic analysis. Thematic analysis is oriented towards meaning, context, social relations, and theoretical significance, while topic modeling tends to produce a collection of words that often appear together, which are not necessarily conceptually meaningful to the research question. If researchers rely too much on computational outputs, there is a risk that meaning is reduced to just the most frequently popped up topics, while nuance, ambiguity, and minority voices are lost.

Conceptual frameworks become increasingly important as AI begins to be treated not just as a tool, but as an analytical partner. The difference between AI as a tool and AI as collaborator must be methodologically affirmed. As a tool, AI helps with transcription, summaries, early pattern tracking, or code candidate drafting. As a collaborator, AI began to influence the proposal of the theme, the preparation of interpretive narratives, and even the direction of claims. This shift requires clear boundaries, as qualitative research remains an interpretive craft that includes methodological, theoretical, and ethical decisions that demand human responsibility (Smith et al., 2025). Dalam konteks ini, prinsip human in the loop bukan hanya prosedur teknis, melainkan prinsip epistemik: peneliti harus tetap memegang kendali atas apa yang dianggap bermakna, bagaimana tema dibentuk, dan bagaimana klaim pengetahuan dipertanggungjawabkan.

Sejumlah studi telah mulai mengoperasionalisasikan prinsip ini dalam bentuk evaluasi dan praktik baik. Goyanes et al. (2025), for example, designing a comparison between manual analysis, CAQDAS, and AI coding services to assess the consistency of AI-assisted thematic analysis results. Chen et al. (2025) also tested how AI summaries with and without human supervision affected data quality in virtual focus groups. In addition, Siiman et al. (2023) It shows that context-rich and theory-oriented prompts tend to produce outputs that are more aligned with the researcher's interpretation while increasing transparency because they can be reported and reused. Findings from the application of AI to health systems, cybersecurity, and AI-based feature selection also show a similar pattern: AI is most valuable when it functions as a supervised support system, not as a substitute for human decisions (Kathirvel et al, 2025a; Kathirvel et al., 2025b).

Thus, the rearrangement of qualitative research at the level of theoretical and methodological foundations cannot be reduced to a matter of efficiency alone. The main challenge is maintaining interpretive integrity when researchers are dealing with large scales of digital data, increasing semiotic complexity, and the presence of AI that can speed up and blur the analytical process. Reflexive thematic analysis is becoming increasingly important not as a technology-rejecting method, but as a practice that ensures meaning remains contextually understood, critically, and accountably in the midst of a wave of automation. Within this framework, AI can be a very powerful exploratory tool, even to some extent it can serve as a co-interpreter, but only if the role of the researcher as an epistemic controller is firmly stated, consistently executed, and supported by adequate methodological transparency.

DIGITAL CONSUMER INTERACTIONS AS A RICH QUALITATIVE DATA SOURCE

Digital transformation is broadly understood as the integration of digital technology into various business functions and processes that then transform the way organizations create value, manage customer relationships, and make data-driven decisions (Singh et al., 2022). In this landscape, digital platforms no longer function simply as transaction channels, but also as social spaces where consumers build meaning, evaluate experiences, and negotiate their perceptions of products and services. Therefore, digital platforms are increasingly understood as a strategic qualitative data source to understand dynamic, spontaneous, and layered consumer interactions (La Ragione & Risitano, 2026). For consumer behavior and marketing research, this development opens up methodological opportunities because consumer-generated digital texts can be treated as social artifacts that reflect cognitive, emotional, and relational processes in the consumption experience.

Empirically, digital consumer interaction is a very valuable form of data because it often contains a direct evaluation of product attributes, service quality, price, seller reputation, and brand credibility (Krishen et al., 2021). Unlike structured surveys or interviews, these data are usually produced naturally without the direct intervention of the researcher, thus having a higher ecological validity because the expressions that appear are close to the actual experience and are not limited by predetermined response categories (Pinto et al., 2025). In addition, the digital context allows researchers to capture the dynamics of consumer perceptions and emotions more quickly and close to the moment of experience.

The richness of digital interaction data is also seen in the variety of channels and their forms. E-commerce reviews often take the form of personal narratives that contain product evaluations, user experiences, satisfaction or disappointment, as well as explicit and implicit recommendations (Cruz-Cárdenas et al., 2025). Komentar media sosial, di sisi lain, berfungsi sebagai ruang diskursif tempat opini kolektif terbentuk melalui interaksi sosial dan electronic word of mouth (Kübler & Hennig-Thurau, 2025). In this space, consumers not only judge products, but also influence each other through support, criticism, humor, irony, or debate. Meanwhile, chat-based customer service conversations, online forums, and app conversation logs provide a more process-oriented interaction trail, allowing researchers to trace how the consumer experience evolves through questions, clarifications, negotiations, and responses.

This kind of interaction is increasingly important as the development of communication technology has changed the way consumers relate to digital services, including expectations patterns, response speeds, and satisfied experience standards (Arinto et al., 2025). In the digital ecosystem, consumers are also dealing with sociotechnical features that mediate their experience, such as recommendation systems, application interfaces, chatbots, and automated services. At the same time, the ability of digital systems to generate text, sound, and images in response to commands is shaping the way consumers judge an experience (Kaplan-Rakowski et al., 2023). Thus, the consumption experience is no longer just a relationship between humans and products, but also involves relationships between humans, platforms, algorithms, and automated systems. In this context, AI is not only present as a technical infrastructure, but also helps shape the user experience and their interpretation of digital services (Kannan et al., 2025; Rithik & Kathirvel, 2025).

Conceptually, the use of digital interaction as a source of qualitative data is becoming increasingly central to the rethinking qualitative inquiry agenda because its scale, speed, and complexity demand adjustment of analytical approaches. In e-commerce, for example, customer satisfaction, loyalty, and trust are important factors for the sustainability of digital businesses (Jahan et al., 2025). Digital interactions provide textual evidence of how satisfaction and dissatisfaction are formed, how trust arises through narratives of reliability, transparency, and risk perception, and how loyalty appears in patterns of repurchases, brand advocacy, or recommendations to other users. A similar pattern can also be seen in the hospitality and digital advertising sectors, where the footprint of online interaction is an important indicator to read the formation of business culture, service perception, and the dynamics of user trust (Ayyaswamy et al., 2025; Kathirvel et al., 2025).

The value of digital interaction data is getting stronger because of its enormous volume. Digital platforms allow millions of interactions to occur every day, resulting in the formation of big qualitative data that provides the opportunity to observe a wide variety of consumer experiences, identify relatively consistent thematic patterns, and compare dynamics across time, across platforms, and even across cultural contexts. At the methodological level, this large scale allows researchers to see dominant patterns while also finding theoretically meaningful minority variations, provided that sampling strategies and analytical procedures are consciously designed. However, this large volume also demands auxiliary tools capable of efficiently managing data, including AI for preprocessing, selection, and pattern exploration. In this context, AI is understood as a transformational force that provides tools to read complex data and generate new value for stakeholders (Wang et al., 2025). Its relevance to qualitative research lies in its ability to help researchers handle data scales that are no longer realistic if analyzed entirely manually. This argument is in line with studies showing how AI is used in cybersecurity, feature selection, and intelligent decision systems as tools that expand human analytical capacity, rather than replace it (Kathirvel et al., 2025a; Kathirvel, et al., 2025b).

Another dimension that makes digital interaction qualitatively rich is multimodality. Unlike conventional text data, digital interactions often combine text with emojis, images, short videos, stickers, links, or screenshots. In comments and reviews, emojis can serve as markers of emotions, while images or visual evidence can reinforce the narrative of the experience. That is, meaning is not only present in the verbal content, but also in the visual symbols and context of the platform where the interaction takes place. In this sense, digital consumer interaction is a complex semiotic experience, involving layered attention, emotion, identity, and the production of meaning (Kozinets & Gretzel, 2025). From a marketing perspective, this kind of digital communication also shapes brand presence, trust, and brand equity, as consumer perceptions are built not only by the company's official message, but also by conversations between consumers (Pathak & Kaushik, 2024).

However, the use of digital interaction as qualitative data also presents methodological challenges. First, data noise: not all digital interactions are relevant, substantive, or meaningful for research purposes. Brief comments, duplicate content, sarcasm, off-topic discussions, or trend-driven responses can blur the analysis. In addition, the ease with which digital systems produce text, sound, and images also increases the likelihood of inauthentic or low-quality content (Kaplan-Rakowski et al., 2023). Second, anonymity and limitations of socio-demographic context limit the ability of researchers to contextualize findings based on age, social class, gender, or cultural background, especially with the presence of fake reviews, automated accounts, and manipulative content; Therefore, checking the credibility of the data, triangulation, and reflective recognition of data limitations are important (Boyes & Watson, 2022). Third, representation: digital participation is uneven because it is influenced by access, digital literacy, platform preferences, and community norms, so that available data can be more representative of certain groups than others (Smith et al., 2025). However, in the qualitative tradition, this inequality of participation can also be read as an object of analysis in itself to look at the dominant discourse and the social mechanisms that determine who is heard and who is excluded.

Considering these strengths and challenges, digital consumer interactions remain highly relevant for examining satisfaction, trust, loyalty, emotions, product preferences, and the dynamics of digital experiences. Satisfaction and dissatisfaction arise through evaluative narratives; trust through the language of reliability, transparency, and risk; while emotions are seen in word choice, intensity of expression, and visual symbols. Furthermore, digital interactions capture the consumer experience as a moving process, allowing researchers to examine how perceptions develop through information exposure, feature updates, service changes, and community discourse (Zhang et al., 2024). Therefore, digital consumer interaction is a rich, naturalistic, large-scale, multimodal, qualitative source of data that is highly relevant for thematic analysis that is adaptive to complexity without losing interpretive depth.

AI-ASSISTED THEMATIC ANALYSIS FRAMEWORK

Qualitative research has long been seen as an important approach to understanding complex social phenomena. Through interviews, focus groups, observations, and document analysis, researchers can explore experiences, emotions, values, and meanings revealed through participants' language and actions. However, this process often requires considerable time and resources, especially as data becomes increasingly large and complex as in the digital context. One of the main challenges of qualitative research is how to process large amounts of data without losing the depth and reflexivity of the analysis.

Advances in artificial intelligence, particularly large language models (LLMs), offer great potential to transform the way thematic analysis is conducted. This technology can speed up the abridgment, initial grouping, and formation of code candidates that previously demanded a lot of time and human effort. However, AI integration cannot be understood as a substitute for human interpretation. LLMs generate outputs through statistical associations based on large training data. This output may appear coherent, but it still risks ignoring the context, cultural nuances, and power relations that are at the center of qualitative interpretation. Therefore, the use of AI must still place human control at every step of interpretation. This limitation is increasingly evident in non-English-language data that have different linguistic and cultural characteristics, which often makes it difficult for AI to capture more complex meanings (Aspers & Corte, 2025).

To address these challenges, the AI-assisted thematic analysis framework is proposed as a step-by-step model that maintains a balance between computational efficiency and interpretive integrity. This framework positions AI as an auxiliary tool under the control of researchers, with the affirmation that key analytical decisions must still be determined through reflexivity and theoretical justification. In general, the framework consists of five interrelated, iterative, and character-adjustable stages, including digital consumer interaction data: data acquisition, AI-based pre-processing, LLM-assisted initial coding, human-led thematic refinement, and validation and trustworthiness layer.

The first stage is data acquisition, which greatly determines the quality of the entire analysis. This stage includes collecting data from a variety of relevant sources, such as transcripts of customer service conversations, social media comments, product reviews, and online forums. At this stage, the researcher must confirm the context and conditions of data production, for example whether the data arises spontaneously and how the platform's norms shape consumer expression. Data from social media or e-commerce needs to be treated with caution because it is not essentially created for research purposes. Therefore, this stage is a critical point to ensure that the data used is indeed feasible for thematic analysis and can be interpreted according to the research context.

The second stage is AI-based pre-processing, which is preparing data so that it is ready for analysis without losing its naturalistic character. In the context of digital consumer interaction, this stage includes data cleansing, duplication removal, and initial mapping of the corpus structure to reduce noise. Eliminating duplication and filtering out irrelevant data is important to prevent repetitive patterns that can obscure the analysis. At this stage, researchers can also utilize clustering techniques to map the variation of conversations based on product categories, communication channels, or time periods. Additionally, AI can help add sentiment cues as an initial clue about areas that are rich in emotions or conflicts. However, the decision regarding what is considered noise, the data retained, and the direction of theme selection remains in the hands of the researcher. Thus, AI helps efficiency, but conceptual decisions remain human.

The third stage is LLM-assisted initial coding, which is the use of AI to help with initial coding. In thematic analysis, the initial encoding aims to break down the raw data into smaller, relevant units of meaning. LLMs can be used to summarize data segments, propose code candidates, and provide probabilistic coding based on possible patterns in the data. Although LLMs are capable of generating code quickly and systematically, AI-generated code should be treated as temporary analytical hypotheses that still need to be tested and sharpened by researchers. AI can recognize linguistic patterns, but it is often not sensitive enough to cultural contexts, nuances of meaning, or broader theoretical frameworks. Therefore, AI outcomes should always be evaluated and reconsidered in the context of the data and research objectives.

The fourth stage, human-led thematic refinement, is at the heart of the entire process. At this stage, the researcher takes full control to refine and affirm the AI-generated theme. The initial encoding results are retested against the raw data through iterative interpretive cycles, i.e. the back and forth between data, code, and themes to ensure consistency and relevance. This process can involve theme redefinition, when overlapping themes are combined, or overly broad themes are broken down into more specific subthemes. Researchers can also use comparative coding to examine thematic patterns across different data groups, such as between platforms or between periods. At this stage, triangulation can also be done through comparison of data sources, theoretical perspectives, or discussions between researchers to strengthen the reliability and validity of findings (Smith et al., 2025). This stage confirms that AI can help unlock the possibilities of patterns, but the formation of meaning remains human-led.

The final stage is the validation and trustworthiness layer, which aims to ensure that the results of the analysis remain credible and accountable. At this stage, the researcher compiles a trail audit by documenting each step of the analysis, including the prompts used, examples of AI outputs, decisions to accept or reject AI results, and the rationale behind the theme revision. Coder consensus can be achieved through team discussions or cross-examination between researchers to maintain consistency of interpretation. Researchers also need to note how their reflexivity affects the analysis process, especially when using AI. Verification of AI results is important to prevent automation bias, which is the tendency to accept AI output without adequate criticism. Thus, AI continues to function as a tool that supports the verification of findings through matching with raw data, so that the results of the analysis remain valid, accurate, and in accordance with the research context.

Overall, the AI-assisted thematic analysis framework offers a systematic and flexible method for managing and analyzing large and complex qualitative data. This framework not only speeds up data processing, but also maintains the depth of interpretation that is at the core of qualitative research. By placing AI as a tool that works under the control of researchers, this framework ensures that the quality of analysis is not eroded by automation. The integration of AI in thematic analysis, thus, opens up opportunities to expand the reach

HYBRID CODING WORKFLOW: INTEGRATING AI PUTPUT WITH HUMAN INTERPRETATION

The increasing use of large language models (LLMs) in qualitative research drives the need for workflows that explicitly divide roles between computational automation and the interpretive responsibilities of researchers. In thematic analysis of large-scale digital data, the most methodologically accountable approach is not to leave the coding to machines, but rather to design a hybrid coding workflow that puts AI outputs as initial analytical support, while the researcher retains conceptual authority over the meaning, context, and claims of knowledge. This principle is in line with the argument that generative AI essentially generates probability-based texts that appear coherent, but lack situational understanding and cannot replace the theoretical justifications that are at the core of qualitative research (Roberts et al., 2024; Nguyen & Welch, 2026).

The Hybrid Qualitative Workflow (HQW) model can be understood as a work design that organizes the relationship between humans and AI into a clear, reflective, and auditable division of functions. In the early layer, AI is used to speed up work that is mechanical and repetitive, such as corpus segmentation, data unit summarizing, initial grouping, and code candidate proposal. In the next layer, the researcher leads the interpretive process by re-examining the raw data, testing the feasibility of the code, reading the social and platform context, and affirming the final theme through analytical argumentation. Thus, AI is not placed as the final interpreter, but rather as an exploration tool that helps researchers manage the scale of the data without sacrificing depth. This approach is in line with the Guided AI Thematic Analysis framework which emphasizes that data familiarization, code template formation, and theme development must still be led by the researcher's reflexivity (Nguyen-Trung, 2025). Similar principles are also seen in various domains of applied AI, where AI works most effectively when used as a support system under human supervision, for example in cybersecurity, industrial automation, and AI-based systems analysis (Ayyaswamy et al., 2025; Kathirvel, et al., 2025; Rithik & Kathirvel, 2025).

In HQW, the role of AI is most powerful in at least two main functions. First, find the initial pattern. In heterogeneous digital consumer interaction data, AI can help identify discourse repetitions, emotionally charged segments, or language variations that mark satisfaction, dissatisfaction, trust, and preferences. However, these patterns should be treated as preliminary clues, not as final findings, as their status is still tentative and vulnerable regardless of context. Second, accelerate data segmentation. AI can break down large corpus into more manageable units based on topic, product, communication channel, or time period, allowing researchers to perform more targeted deep readings. This practice is in line with the finding that transparent automation should delegate mechanical tasks to the system while maintaining a human-centered interpretation space (Gao et al., 2025). In the context of big digital data, these initial functions are critical because they help researchers save analytical energy for stages that really demand conceptual judgment.

On the contrary, the role of humans in HQW is irreplaceable and determines the epistemic quality of research. First, the researcher is responsible for the interpretation of meaning. The theme in thematic analysis is not just a word aggregation or statistical grouping, but an analytical construct that must answer the research question and connect with the theoretical framework. Therefore, researchers must assess whether the AI's proposed code candidates do indeed represent relevant patterns of meaning or just linguistic regularity that seems prominent. Second, the researcher must ensure the relevance of the context. Digital data is often full of irony, sarcasm, community norms, and the context of the platform that shapes the meaning of speech. LLMs may seem fluent, but they are still susceptible to decontextualization; Therefore, researchers need to relink the code to the data segments, reread the context of the conversation, and test the suitability of the interpretation with the social situation that gave birth to the data. Third, the researcher sets the final theme. The decision to combine, break down, reject, or maintain a theme is an argumentative decision that must be scientifically accountable and should not be transferred to machines.

To clarify when AI is used and when humans are dominant, HQW can be run in three levels of engagement. At the AI-assisted level, AI functions as an assistant for summaries, segmentation, and code proposals, while researchers still perform substantive coding, theme development, and narrative writing of findings. This is the most secure configuration because the risk of automation bias is lower and the auditability is higher. At the AI-augmented level, AI begins to be involved in theme candidate development or thematic consistency testing across data segments. This level is useful for big qualitative data, but it demands more rigorous documentation, especially towards prompts, outputs, and revision decisions. Studies show that context-rich and theory-oriented prompts tend to produce better outputs as well as be more transparent because they can be reported and reused (Siiman et al., 2023). As for the AI-dependent level, which is when AI becomes the main encoder and the researcher only checks the final results, it should not be recommended. This model risks moving the center of interpretation from the researcher to the machine, increasing decontextualization, and weakening epistemic justification (Nguyen & Welch, 2026).

In order for the integration of AI and humans in HQW to remain ethical and methodologically strong, several principles need to be used as working standards. First, transparency by design: researchers should note prompts, model versions, usage parameters, sample outputs, as well as reasons for accepting or rejecting AI suggestions as part of the audit trail. Second, human accountability: all thematic claims must be able to be linked back to the raw data through relevant citations and clear reasoning. Third, context preservation: important metadata such as time, platform, type of interaction, and conversation context need to be stored so that interpretation is not detached from the data production situation. Fourth, awareness bias: researchers must reflexively assess possible biases related to the dominance of certain language patterns, representation of majority groups, and the invisibility of minority voices. These principles are in line with the need for AI systems that are auditable, transparent, and remain under human control, as seen in studies on AI governance, ethical strategies, and decision support systems (Ayyaswamy, 2026b; Kathirvel et al., 2025a).

Ultimately, hybrid coding workflows are not just a technical compromise between speed and precision, but rather an epistemic design to handle the scale of digital data without losing the interpretive character of qualitative research. By placing AI on mechanical and exploratory tasks, as well as placing humans on conceptual, contextual, and argumentative tasks, HQW maintains rigor, transparency, and scientific responsibility while remaining responsive to the reality of big qualitative data. Within this framework, AI expands the work capacity of researchers, but the quality of knowledge remains determined by how the researcher directs, assesses, and accounts for the interpretation process.

METHODOLOGICAL & ETHICAL CHALLENGES IN AI-AUGMENTED QUALITATIVE RESEARCH

The use of artificial intelligence, particularly large language models (LLMs), in qualitative research marks a significant methodological shift. AI allows researchers to process huge volumes of text data, such as consumer reviews, social media comments, online forums, and other digital interactions, that were previously difficult to manually analyze in a reasonable amount of time. In this context, AI offers computational capacity to accelerate data cleansing, corpus segmentation, initial grouping, and code candidate proposals. However, such increased technical capacity does not automatically guarantee the quality of interpretation. In qualitative research, methods are not just technical procedures, but epistemic frameworks that determine how meaning is produced, tested, and accounted for. Therefore, the integration of AI in thematic analysis must be read critically so as not to shift the process of reflective interpretation into a computational operation that relies solely on statistical patterns.

One of the biggest methodological challenges is algorithmic bias. AI does not work in a neutral space, but is trained on data that reflects power relations, the dominance of certain languages, and the inequality of cultural representation. In the study of digital consumer behavior, this bias can influence what themes are considered important, which data segments appear to be dominant, and whose experiences are marginalized. AI may be more likely to recognize patterns that often appear in dominant discourse, such as urban consumer experiences, global languages, or majority communication styles, while minority group experiences, local contexts, and non-standard expressions become less visible. This risks creating the illusion of objectivity, as if the theme that emerges is a direct reflection of empirical reality, when in fact it has gone through algorithmic filtering that is full of hidden assumptions (Wei et al., 2025). Therefore, researchers need to actively evaluate how AI frames data, rather than simply accepting thematic outputs as natural outcomes. This need for human oversight is also affirmed in applied AI studies in the fields of cybersecurity, healthcare, and analytics systems, which show that AI is most valuable when it is treated as a supervised support system, rather than as an autonomous decision-maker (Ayyaswamy, et al., 2025; Kathirvel, et al, 2025).

The second methodological challenge is the black box nature of most LLMs. In qualitative research, the quality of analysis is measured not only by the final result, but also by the traceability of the process: how the code is built, why certain categories are combined, and how the main themes are formulated. When AI generates summaries, codes, or themes without adequate explanation of grouping logic or meaning priority, researchers will have a hard time building a transparent trail audit. Sharma et al. (2024) asserts that the lack of interpretability of AI systems can hinder critical evaluation and reduce confidence in research results, especially in academic contexts. Therefore, in AI-assisted thematic analysis, AI should not be positioned as an independent analyst, but rather as a tool that must be continuously examined, tested, and directed by researchers. This is where documentation of prompts, model versions, usage parameters, sample outputs, and decisions to accept or reject AI proposals becomes critical.

The next issue is related to dependability and confirmability, two quality indicators that are central to qualitative research. Dependability demands process consistency, while confirmability demands that findings be traceable back to empirical data. The use of AI poses new challenges because model output can change due to small variations in prompts, instruction contexts, or system updates. Adel & Alani (2025) suggests that LLMs can produce different interpretations of the same dataset simply because of small changes in the formulation of the question. In thematic analysis, these conditions can lead to unstable themes if AI procedures are not strictly documented. Therefore, AI-assisted research demands an expansion of methodological practice: researchers must record prompts, model versions, dates of use, sample outputs, and interpretive revision rationales so that the process can still be examined and assessed by other researchers.

Another important limitation is the sensitivity of the cultural context. LLMs are trained on a large corpus, but are not always able to capture the cultural nuances, implicit meaning, or historical background that give meaning to language. In the study of digital consumer behavior, this becomes a serious problem when AI fails to understand local humor, irony, community language styles, or non-literal expressions that users commonly use. In fact, in thematic analysis, meaning is often hidden in such a context. Therefore, even though AI is able to help large-scale analysis, the cultural sensitivity of researchers is still needed to keep the findings socially relevant and inseparable from the real-life context. Within this framework, AI should be positioned as an initial exploration tool, while contextual interpretation remains an area of human responsibility.

Another methodological challenge that is increasingly discussed is hallucination, which is when AI generates information, code, or themes that seem convincing but are not fully supported by available data. In qualitative research, this risk increases when AI is asked to synthesize meaning, infer patterns, or make high-level inferences from ambiguous data. Adel dan Alani (2025) suggests that hallucinations often arise when models are geared toward generalizing or explaining something that is actually not adequately supported by empirical evidence. In thematic analysis, this can result in a neat-sounding but data-weak theme. Therefore, all AI outputs should be treated as temporary analytical hypotheses that must be retested through raw data reading, theoretical reflection, and interpretive discussion by researchers.

In addition to methodological issues, the use of AI in qualitative research also presents ethical issues that are no less important. One of the most basic is data privacy. AI-assisted thematic analysis often involves collecting and processing large amounts of data, including online comments, product reviews, social media posts, and other digital interaction trails. Even if much of that data is publicly available, its use can still pose a risk of privacy breaches, especially if the data can be linked back to a specific individual. In this context, the principles of privacy-by-design and data minimization are essential to ensure that research does not compromise users' privacy rights (Ratti et al., 2025). Researchers need to consider whether identities should be anonymized, the extent to which live quotes are safe to use, and how to store data securely when processed through AI systems.

Another ethical issue is data misrepresentation. Because LLMs work with linguistic patterns and probabilities, rather than with real-life experiences, AI can generate interpretations that appear coherent but do not fully represent the meaning that participants intend. In thematic analysis, this means that AI has the potential to simplify the complexity of the consumer experience, reduce ambiguity, or ignore minority voices that are theoretically important. As a result, the knowledge generated can be inaccurate or even misleading (Fundira & Mbohwa, 2025). This risk is especially relevant when research uses public data from social media or online forums, which are often considered free to use simply because they are accessible. In fact, open access does not automatically remove ethical considerations. Public data remains in a specific social context, and its use on a scale with AI can have impacts that data owners could not have imagined (Ratti et al., 2025).

Finally, the issue of algorithmic justice becomes increasingly central when AI is used to analyze social data. AI models can reinforce existing inequalities in training data, resulting in interpretations that are more representative of dominant groups and blurring the experiences of marginalized groups. In thematic analysis, this means that the themes formed can be biased against the majority vote, while the minority perspective is reduced or lost altogether. Therefore, human oversight of algorithmic bias, documentation of analytical decisions, and transparent ethical audits are essential requirements for AI to be used fairly and inclusively. The literature on AI ethical strategies and systems governance also leads to the same conclusion: the quality of AI use is not measured by how far humans are replaced, but by how clearly human responsibility is maintained (Ayyaswamy et al., 2025b; Ayyaswamy et al., 2025f).

Thus, the methodological and ethical challenges in AI-assisted qualitative research cannot be seen as mere additional issues. At the heart of the question is how to harness AI's capacity to handle large scales of digital data without sacrificing interpretive integrity, methodological accountability, cultural sensitivity, and ethical responsibility. Within this framework, AI can be a very useful tool, but only if its use is placed under the reflective control of the researcher, supported by adequate documentation, and driven by a commitment to fairness and qualitative knowledge quality.

IMPLICATIONS, FUTURE DIRECTIONS, AND CONCLUSION

The integration of artificial intelligence (AI) in qualitative research has important methodological implications, especially for the study of digital consumer behavior which is increasingly characterized by large data volumes, speed of interaction, and complexity of contexts. AI allows researchers to manage a vast and diverse corpus of data more efficiently, so that thematic analysis can be applied at a scale that was previously difficult to reach. However, the main implication of this development lies not in technical acceleration alone, but in the need to redesign qualitative workflows while maintaining human reflexivity, context sensitivity, and interpretive responsibility. In this case, AI is most valuable when positioned as a support system that expands the analytical capacity of researchers, rather than as a substitute for qualitative reasoning.

Going forward, AI-assisted qualitative research needs to move towards a more human-centered, transparent, and auditable approach. One important direction is the development of post-coding qualitative inquiry, which is a practice in which the initial output of AI is treated as exploration material that is then tested, sharpened, and reinterpreted by researchers. In addition, the integration of multimodal analysis is also an important agenda, considering that digital interaction is no longer just in the form of text, but also emojis, images, sounds, and other visual elements that form meaning simultaneously. In this context, AI can expand the scope of qualitative research, but the quality of knowledge remains determined by the researcher's ability to maintain the connection between data, context, theory, and interpretation.

Overall, the study confirms that the future of qualitative research does not lie in full automation, but rather in the careful synergy between the computational power of AI and human interpretive intelligence. Thus, AI can expand the possibilities of analysis without sacrificing the depth of meaning that is at the core of qualitative methodology.

REFERENCES

Adel, A., & Alani, N. (2025). Can Generative AI Reliably Synthesise Literature? Exploring Hallucination Issues in ChatGPT. *AI & SOCIETY 2025 40:8, 40*(8), 6799–6812. DOI: 10.1007/s00146-025-02406-7

Ahmed, S. K., Mohammed, R. A., Nashwan, A. J., Ibrahim, R. H., Abdalla, A. Q. M., Ameen, B. M., & Khdhir, R. M. (2025). Using Thematic Analysis in Qualitative Research. *Journal of Medicine, Surgery, and Public Health*, *6*(2), 100198. DOI: 10.1016/j.glmedi.2025.100198

Arinto, A. D., Sitanggang, D. H., Syauqi, M. F., & Fahlevi, M. (2025). The Role of Time Pressure, Behavioral Control and Human Interaction in Shaping E-Commerce Continuance Intention Across Product Categories in Indonesia. *Journal of Electronic Business & Digital Economics*, 1–23. DOI: 10.1108/JEBDE-11-2024-0049

Aspers, P., & Corte, U. (2025). Evaluating ChatGPT in Qualitative Thematic Analysis with Human Researchers in the Japanese Clinical Context and Its Cultural Interpretation Challenges: Comparative Qualitative Study. *Journal of Medical Internet Research*, *27*(2), 139–160. DOI: 10.1007/s11133-019-9413-7 PMID: 40273439

Ayyaswamy, K. (2026a). AI Technologies for Improving Education Systems. Https://Services.Igi-Global.Com/Resolvedoi/Resolve.Aspx?Doi=10.4018/979-8-3373-3236-9.Ch003, 65–92. DOI: 10.4018/979-8-3373-3236-9.ch003

Ayyaswamy, K. (2026b). Artificial Intelligence Strategies for Cleaner and Greener Manufacturing. Https://Services.Igi-Global.Com/Resolvedoi/Resolve.Aspx?Doi=10.4018/979-8-3373-2352-7.Ch002, 29–59. DOI: 10.4018/979-8-3373-2352-7.ch002

Ayyaswamy, K., Gobalakrishnan, N., & Kathirvel, N. (2025). Cyber Security in Industrial Automation Using AI. Https://Services.Igi-Global.Com/Resolvedoi/Resolve.Aspx?Doi=10.4018/979-8-3373-3241-3.Ch024, 509–540. DOI: 10.4018/979-8-3373-3241-3.ch024

Ayyaswamy, K., Gobinath, V. M., & Kathirvel, N. (2025). AI-Based Healthcare Systems for Pets and Birds. Https://Services.Igi-Global.Com/Resolvedoi/Resolve.Aspx?Doi=10.4018/979-8-3693-7483-2.Ch015, 419–447. DOI: 10.4018/979-8-3693-7483-2.ch015

Ayyaswamy, K., Gobinath, V. M., Kathirvel, N., & Maheswaran, C. P. (2025). AI-Based Automated Disease Detection Based on Symptoms Healthcare for Pets and Birds. Https://Services.Igi-Global.Com/Resolvedoi/Resolve.Aspx?Doi=10.4018/979-8-3693-8472-5.Ch014, 311–345. DOI: 10.4018/979-8-3693-8472-5.ch014

Ayyaswamy, K., Gobinath, V. M., Sathya, V., Kathirvel, N., & Anthony, R. A. (2025). The Role of Robotics in Smart Manufacturing: Increasing Efficiency and Reducing Costs. Https://Services.Igi-Global.Com/Resolvedoi/Resolve.Aspx?Doi=10.4018/979-8-3693-7673-7.Ch014, 315–356. DOI: 10.4018/979-8-3693-7673-7.ch014

Ayyaswamy, K., Gobinath, V. M., Senthil, M. S., & Kathirvel, N. (2025). Automating library management for efficiency: IoT-enabled smart libraries. *Project Management Information Systems: Empowering Decision Making and Execution*, 325–349. DOI: 10.4018/979-8-3373-0700-8.ch010

Ayyaswamy, K., Kathirvel, N., Gobinath, V. M., & Maheswaran, C. P. (2025). Effective Leadership Strategies for Enhancing Student Success in Higher Education. Https://Services.Igi-Global.Com/Resolvedoi/Resolve.Aspx?Doi=10.4018/979-8-3693-6437-6.Ch003, 61–91. DOI: 10.4018/979-8-3693-6437-6.ch003

Ayyaswamy, K., Kathirvel, N., & Vianny, M. M. (2025). Strategies for Combating Criminal Use and Abuse of Artificial Intelligence. Https://Services.Igi-Global.Com/Resolvedoi/Resolve.Aspx?Doi=10.4018/979-8-3693-7041-4.Ch009, 257–282. DOI: 10.4018/979-8-3693-7041-4.ch009

Ayyaswamy, K., Kathirvel, N., & C. P., M. (2025). Ethical AI Entrepreneurial Strategies for Business Intelligence and Blockchain. Https://Services.Igi-Global.Com/Resolvedoi/Resolve.Aspx?Doi=10.4018/979-8-3693-3478-2.Ch001, 1–26. DOI: 10.4018/979-8-3693-3478-2.ch001

Belhassen, Y., & Caton, K. (2009). Advancing Understandings: A Linguistic Approach to Tourism Epistemology. *Annals of Tourism Research*, *36*(2), 335–352. DOI: 10.1016/j.annals.2009.01.006

Bowden, J., Mohanakrishnan, M., & Thompson, A. R. (2025). Incorporating Artificial Intelligence in Qualitative Research: Exploring the Role of ChatGPT in Thematic Analysis. *Medical Science Educator*, *36*(1), 1–7. DOI: 10.1007/s40670-025-02587-2 PMID: 41939031

Boyes, H., & Watson, T. (2022). Digital Twins: An Analysis Framework and Open Issues. *Computers in Industry*, *143*, 103763. DOI: 10.1016/j.compind.2022.103763

Braun, V., & Clarke, V. (2006). Using Thematic Analysis in Psychology. *Qualitative Research in Psychology*, *3*(2), 77–101. DOI: 10.1191/1478088706qp063oa

Braun, V., & Clarke, V. (2022). Toward Good Practice in Thematic Analysis: Avoiding Common Problems and Be(Com)Ing a Knowing Researcher. *International Journal of Transgender Health*, *24*(1), 1–6. DOI: 10.1080/26895269.2022.2129597 PMID: 36713144

Braun, Virginia., & Clarke, Victoria. (2021). *Thematic Analysis : A Practical Guide*. https://books.google.com/books/about/Thematic_Analysis.html?hl=id&id=eMArEAAAQBAJ

Cena, E., Brooks, J., Day, W., Goodman, S., Rousaki, A., Ruby-Granger, V., & Seymour-Smith, S. (2024). Quality Criteria: General and Specific Guidelines for Qualitative Approaches in Psychology Research. A Concise Guide for Novice Researchers and Reviewers. *International Journal of Qualitative Methods*, *23*, 16094069241282843. Advance online publication. DOI: 10.1177/16094069241282843

Chen, H., Wang, Y., & Chang, C. (2025). Enhancing Qualitative Inquiry: AI-Assisted Focus Group Data Collection. *Qualitative Research Journal*, 1–17. DOI: 10.1108/QRJ-04-2025-0145

Cook, D. A., Ginsburg, S., Sawatsky, A. P., Kuper, A., & D'Angelo, J. D. (2025). Artificial Intelligence to Support Qualitative Data Analysis: Promises, Approaches, Pitfalls. *Academic Medicine*, *100*(10), 1134–1149. DOI: 10.1097/ACM.0000000000006134 PMID: 40560241

Cruz-Cárdenas, J., Ramos-Galarza, C., Giménez-Baldazo, M., & Palacio-Fierro, A. (2025). A Review of Consumer-to-Consumer Digital Information and Knowledge Sharing. *Management Decision*, *63*(13), 96–122. DOI: 10.1108/MD-04-2024-0869

Fryer, T. (2022). A Critical Realist Approach to Thematic Analysis: Producing Causal Explanations. *Journal of Critical Realism*, *21*(4), 365–384. DOI: 10.1080/14767430.2022.2076776

Fundira, M., & Mbohwa, C. (2025). AI Ethics in Banking Services: A Systematic and Bibliometric Review of Regulatory and Consumer Perspectives. *Discover Artificial Intelligence 2025 5:1*, *5*(1), 319-. DOI: 10.1007/s44163-025-00432-4

Gao, J., Shu, Z., Yeo, S. Y., Prakash, A., Huang, C.-M., Dredze, M., & Xiao, Z. (2025). Efficiency with Rigor! A Trustworthy LLM-powered Workflow for Qualitative Data Analysis. *Proceedings of Make Sure to Enter the Correct Conference Title from Your Rights Confirmation Emai (Conference Acronym 'XX)*, *1*. https://arxiv.org/abs/2501.00775

Goyanes, M., Lopezosa, C., & Jordá, B. (2025). Thematic Analysis of Interview Data with ChatPGT: Designing and Testing a Reliable Research Protocol for Qualitative Research. *Quality & Quantity 2025 59:6*, *59*(6), 5491–5510. DOI: 10.1007/s11135-025-02199-3

Han, Z., Tavasi, A., Lee, J. Y., Luzuriaga, J., Suresh, K., Oppenheim, M., Battaglia, F., & Terlecky, S. R. (2025). Can Large Language Models Be Used to Code Text for Thematic Analysis? An Explorative Study. *Discover Artificial Intelligence, 5*(1), 171-. DOI: 10.1007/s44163-025-00441-3

Hayes, A. S. (2025). Conversing with Qualitative Data: Enhancing Qualitative Research Through Large Language Models (LLMs). *International Journal of Qualitative Methods*, *24*, 16094069251322346. Advance online publication. DOI: 10.1177/16094069251322346

Jahan, E., Moniruzzaman, M., Yasmin, S., & Saiful Islam, A. H. M. (2025). E-Commerce and Consumers Satisfaction: Empirical Evidence from Digital Marketing System of Mango in Bangladesh. *Journal of Agriculture and Food Research, 24*(3), 102456. DOI: 10.1016/j.jafr.2025.102456

Kalanda, B. F., & Cheboi, A. J. (2025). Artificial Intelligence in the Analysis of Unstructured Qualitative Data: A Literature Review. *Advances in Social Sciences Research Journal*, *12*(08), 199–205. DOI: 10.14738/assrj.1208.19286

Kannan, H., Rodriguez, R. V., Rajaraman, S., & Pise, A. A. (2025). AI-Driven Breakthroughs in Antimicrobial Resistance. *AI-Driven Breakthroughs in Antimicrobial Resistance*, 1–406. DOI: 10.4018/979-8-3693-7550-1

Kaplan-Rakowski, R., Kaplan-Rakowski, R., Grotewold, K., Hartwick, P., & Papin, K. (2023). Generative AI and Teachers' Perspectives on Its Implementation in Education. *Journal of Interactive Learning Research*, *34*(2), 313–338. https://www.learntechlib.org/primary/primary/p/222363/. DOI: 10.70725/815246mfssgp

Kathirvel, A., Naren, A. K., & Saravanan, N. (2025). Entrepreneurship and Business Culture in Rural Area. *Digital Disruption in Hospitality, Sustainable Hospitality: A Global Perspective on People, Planet, and Profit*, 187–212. https://doi.org/DOI: 10.1108/978-1-83662-360-120251009

Kathirvel, N., Ayyaswamy, K., & Gunasekaran, K. (2025). Future Cyber Physical System Security. Https://Services.Igi-Global.Com/Resolvedoi/Resolve.Aspx?Doi=10.4018/979-8-3693-3579-6.Ch010, 239–276. https://doi.org/DOI: 10.4018/979-8-3693-3579-6.ch010

Kathirvel, N., Ayyaswamy, K., Saravanan, N., & Kaliappan, P. (2025). Mathematical Approach: AI-Based Feature Selection in Machine Learning. Https://Services.Igi-Global.Com/Resolvedoi/Resolve.Aspx?Doi=10.4018/979-8-3693-7205-0.Ch012, 253–280. DOI: 10.4018/979-8-3693-7205-0.ch012

Kozinets, R., & Gretzel, U. (2025). Netnography and Qualitative Digital Methods to Advance Service Theory and Practice. *Journal of Service Theory and Practice, 35*(4), 487–500. DOI: 10.1108/JSTP-07-2025-471

Krishen, A. S., Dwivedi, Y. K., Bindu, N., & Kumar, K. S. (2021). A Broad Overview of Interactive Digital Marketing: A Bibliometric Network Analysis. *Journal of Business Research, 131*, 183–195. DOI: 10.1016/j.jbusres.2021.03.061

Kübler, R. V., & Hennig-Thurau, T. (2025). Spatial Word of Mouth: How User Reviews in the Metaverse Shape the Success of Virtual Reality Apps. *Journal of Retailing, 53*(3), 297–318. DOI: 10.1016/j.jretai.2025.03.002

La Ragione, G., & Risitano, M. (2026). Rethinking Consumer Behaviour in a Green Digital Marketing Landscape. *Journal of Business Research, 202*, 115755. DOI: 10.1016/j.jbusres.2025.115755

Nguyen, D. C., & Welch, C. (2026). Generative Artificial Intelligence in Qualitative Data Analysis: Analyzing—Or Just Chatting? *Organizational Research Methods, 29*(1), 3–39. DOI: 10.1177/10944281251377154

Nguyen-Trung, K. (2025). ChatGPT in Thematic Analysis: Can AI Become a Research Assistant in Qualitative Research? *Quality & Quantity 2025 59:6, 59*(6), 4945–4978. DOI: 10.1007/s11135-025-02165-z

Odacioglu, E. C., Zhang, L., Allmendinger, R., & Shahgholian, A. (2024). Big Textual Data Research for Operations Management: Topic Modelling with Grounded Theory. *International Journal of Operations & Production Management, 44*(8), 1420–1445. DOI: 10.1108/IJOPM-03-2023-0239

Pathak, A. A., & Kaushik, K. (2024). Driving Consumer Engagement for Digital Payment Brands: Content Strategy on Instagram. *Journal of Retailing and Consumer Services, 79*, 103821. DOI: 10.1016/j.jretconser.2024.103821

Paulus, T. M., & Marone, V. (2025). "In Minutes Instead of Weeks": Discursive Constructions of Generative AI and Qualitative Data Analysis. *Qualitative Inquiry, 31*(5), 395–402. DOI: 10.1177/10778004241250065

Pinto, F., da Silva, C. F., Moro, S., & Aquino, P. (2025). Consumer-Controlled Digital Twin Architecture: How blockchain technology gives consumers control over their smart devices' digital twins and data. *Blockchain: Research and Applications, 100342*, 100342. Advance online publication. DOI: 10.1016/j.bcra.2025.100342

Ratti, E., Morrison, M., & Jakab, I. (2025). Ethical and Social Considerations of Applying Artificial Intelligence in Healthcare—A Two-Pronged Scoping Review. *BMC Medical Ethics 2025 26:1, 26*(1), 68-. DOI: 10.1186/s12910-025-01198-1

Rithik, G., & Kathirvel, A. (2025). The IoT of Robotics. *Artificial Intelligence-Enabled Digital Twin for Smart Manufacturing*, 279–303. DOI: 10.1002/9781394303601.ch13

Roberts, J., Baker, M., & Andrew, J. (2024). Artificial Intelligence and Qualitative Research: The Promise and Perils of Large Language Model (LLM) 'Assistance.'. *Critical Perspectives on Accounting*, *99*(6), 102722. DOI: 10.1016/j.cpa.2024.102722

Schwartz-Shea, P., & Yanow, D. (2013). Interpretive Research Design: Concepts and Processes. *Interpretive Research Design: Concepts and Processes*, 1–186. DOI: 10.4324/9780203854907

Sharma, Y., Saha, A., & Goldsack, J. C. (2024). Defining the Dimensions of Diversity to Promote Inclusion in the Digital Era of Health Care: A Lexicon. *JMIR Public Health and Surveillance*, *10*(1), e51980. DOI: 10.2196/51980 PMID: 38335013

Siiman, L. A., Rannastu-Avalos, M., Pöysä-Tarhonen, J., Häkkinen, P., & Pedaste, M. (2023). Opportunities and Challenges for AI-Assisted Qualitative Data Analysis: An Example from Collaborative Problem-Solving Discourse Data. *Lecture Notes in Computer Science (Including Subseries Lecture Notes in Artificial Intelligence and Lecture Notes in Bioinformatics), 14099 LNCS*, 87–96. DOI: 10.1007/978-3-031-40113-8_9

Singh, U. S., Singh, N., Gulati, K., Kumar Bhasin, N., Kumar, H., & Sreejith, P. M. (2022). A Study on the Revolution of Consumer Relationships as a Combination of Human Interactions and Digital Transformations. *Materials Today: Proceedings*, *51*, 460–464. DOI: 10.1016/j.matpr.2021.05.578

Smith, A., Swart, J., & Broersma, M. (2025). Bypassing Digital Literacy: Marginalized Citizens' Tactics for Participation and Inclusion in Digital Societies. *New Media & Society*, *27*(6), 3127–3145. DOI: 10.1177/14614448231220383

Smith, B., Roderick, M., Hockin-Boyers, H., Monforte, J., Williams, T. L., Jachyra, P., Dodd-Reynolds, C., & Phoenix, C. (2025). 25 Years of Qualitative Research in Sport and Exercise Psychology: How Did It Go? And What Now? *Psychology of Sport and Exercise*, *81*, 102958. DOI: 10.1016/j.psychsport.2025.102958 PMID: 40712908

Wang, J. J., Mao, L. L., & Mastromartino, B. (2025). Guest Editorial: Sport Business Research in the Era of Artificial Intelligence. *International Journal of Sports Marketing & Sponsorship*, *26*(4), 661–669. DOI: 10.1108/IJSMS-10-2025-376

Wei, X., Kumar, N., & Zhang, H. (2025). Addressing Bias in Generative AI: Challenges and Research Opportunities in Information Management. *Information & Management*, *62*(2), 104103. DOI: 10.1016/j.im.2025.104103

Wheeler, K. (2025). *How to Use Generative AI to Assist the Analysis of Qualitative Data*. How to Use Generative AI to Assist the Analysis of Qualitative Data., DOI: 10.4135/9781036217471

Yan, M., & Pourdavood, R. G. (2024). Faculty and Student Perspectives on Online Learning in Higher Education. *Education Sciences 2024, Vol. 14, 14*(8). DOI: 10.3390/educsci14080801

Zhang, L., Anjum, M. A., & Wang, Y. (2024). The Impact of Trust-Building Mechanisms on Purchase Intention towards Metaverse Shopping: The Moderating Role of Age. *International Journal of Human-Computer Interaction, 40*(12), 3185–3203. DOI: 10.1080/10447318.2023.2184594

Chapter 10

From Manual Coding to Machine Insight: Comparing Traditional Thematic Analysis With ChatGPT and NVivo in Qualitative Research

Kritika Sharma
http://orcid.org/0009-0001-2884-557X
Lovely Professional University, India

Harry Kaur
http://orcid.org/0000-0001-8276-2313
Lovely Professional University, India

ABSTRACT

This chapter examines the methodological shift in qualitative research from traditional manual thematic analysis to technology-assisted approaches using NVivo and AI-supported tools such as ChatGPT. Manual coding enables interpretive depth and reflexivity but is time-consuming and difficult to scale. NVivo strengthens analytical rigor through systematic coding and transparency while remaining dependent on researcher interpretation. ChatGPT supports exploratory analysis through summarization, preliminary coding, and reflexive questioning, yet raises methodological, epistemological, and ethical concerns related to validity, contextual sensitivity, bias, and data privacy. Rather than treating these approaches as competing methods, the chapter proposes a hybrid, reflexive framework that integrates human interpretive authority with digital and AI-based support tools, offering practical guidance for responsible and rigorous qualitative analysis.

DOI: 10.4018/979-8-3373-7640-0.ch010

ıt © 2026, IGI Global Scientific Publishing. Copying or distributing in print or electronic forms without written permission of IGI Global Scientific Publishing is prohibited. Use of this chapter to enerative artificial intelligence (AI) technologies is expressly prohibited. The publisher reserves all rights to license its use for generative AI training and machine learning model development.

INTRODUCTION

In the social sciences, qualitative research plays a significant role, as it allows scholars to study meanings, experiences, and social realities that cannot be adequately described using quantitative measures. Qualitative methods offer detailed insights into human behavior, culture, and social organization through interviews, focus group discussions, narratives, and field notes (Denzin, 2018). Thematic analysis has been one of the most commonly applied qualitative methods because it is flexible and can be used across different fields. The thematic analysis method has traditionally been based on manual coding, in which the researcher accesses text content by reading, re-reading, classifying, and analyzing patterns and themes (Creswell & Poth, 2016). Although this method enables the researcher to be contextually sensitive, reflexive, and culturally aware, particularly when studying marginalized or indigenous groups, it can be time-intensive and cognitively demanding. Manual coding is prone to inconsistency and researcher bias, especially when dealing with large data sets (Guest et al., 2020).

As digital technologies evolve, qualitative research has experienced an impressive methodological stream. Computer-Assisted Qualitative Data Analysis Software (CAQDAS) like NVivo has revolutionized the way researchers handle, code, and analyze qualitative data. These tools provide a systematic approach, increased transparency, and effective processing of vast amounts of textual content (Cabiddu et al., 2018). Nevertheless, even though software-assisted analysis enhances efficiency, it is highly reliant on the researcher's interpretive functions. In recent years, artificial intelligence (AI) tools, such as extensive language models like ChatGPT, have begun to shape the practices of qualitative research. Such tools can aid data summarisation, generate initial code, identify potential themes, and support reflective thought through natural language processing (Potančok & Radváková, 2023). The quality of qualitative analysis with AI raises significant methodological, epistemological, and ethical concerns, particularly regarding validity, sensitivity to context, data privacy, and the importance of human interpretation.

The chapter is a critical analysis of the shift from manual coding to machine-assisted insight, comparing three methods of qualitative thematic analysis: traditional manual methods, NVivo-assisted analysis, and AI-supported analysis using ChatGPT. Instead of treating these methods as mutually exclusive, the chapter considers how each approach can be used in qualitative inquiry. The relevance of the latter, specifically concerning social science research, such as research on indigenous and tribal populations, where ethical responsibility and contextual understanding are paramount, is given particular attention. In its attempt to provide a comparative and reflexive analysis, this chapter will guide the researchers towards a hybrid methodological framework that balances the depth of human interpretation with the efficiency of technology. Through this, it will contribute to ongoing discussions of what responsible incorporation of digital tools and artificial intelligence would entail in qualitative research.

CONCEPTUAL FOUNDATIONS OF THEMATIC ANALYSIS

Qualitative thematic analyzes is a well-known research technique focused on identifying, analyzing, and interpreting patterns in qualitative data, treating them as meaningful. The quality that is particularly appreciated is its flexibility, as it is not confined to a single theoretical or epistemological tradition, allowing it to be used across fields such as sociology, anthropology, education, psychology, and cultural studies (Braun & Clarke, 2016). As a result of this flexibility, the thematic analysis is commonly used in social science research aimed at investigating lived experiences, cultural practices, and social meanings.

In its simplest form, thematic analysis is a methodical, cyclical process of becoming acquainted with the data, coding, developing themes, and interpreting the data. Themes do not exist; they are constructed as the researcher works with the data (Braun & Clarke, 2019). This interpretive aspect sets thematic analysis apart from descriptive or automated methods, with the primary focus on the role the researcher plays in making meaning.

The ability to support both inductive and deductive methods of analysis is a key characteristic of thematic analysis. In inductive thematic analysis, themes are constructed from the data itself and are not influenced by preexisting theoretical assumptions (Patton, 2014). Such a strategy is especially useful in exploratory research and other studies that engage marginalized and indigenous people, where strict theoretical frameworks can conceal local meanings and lived experiences. On the other hand, deductive thematic analysis is informed by existing theories, concepts, or research questions, in which a researcher analyzes data using analytical frameworks (Fereday & Muir-Cochrane, 2006).

Although the deductive analysis is effective at increasing theoretical alignment, it can limit the discovery of new insights when used rigidly. Thematic analysis also works at various levels of interpretation, which are usually divided into semantic analysis and latent analysis. Semantic analysis is concerned with what is explicitly covered by the data, including the meanings explicitly expressed by the participants. Latent analysis, however, transcends the superficial meanings to trace assumptions, ideologies, and power arrangements that were concealed in the narratives (Braun & Clarke, 2006). This difference is particularly applicable to cultural and social studies, where oral narratives tend to represent larger structural and historical frames that might not be obvious.

Researcher reflexivity is a key aspect of rigorous thematic analysis. Because the researcher influences coding and theme development, their views, experiences, and social positioning, reflexivity demands constant self-reflection throughout the research process (Finlay, 2002). Reflexive practice is more transparent and ethical because it recognizes the co-construction of knowledge between the researcher and the data. It is especially so when conducting studies concerning tribal and indigenous populations, where unequal power balance and cultural misunderstanding present a serious threat to morality (J. D. Smith, 2012). Although thematic analysis is usually considered an easy and open form of analysis, one should not underestimate its analytical legitimacy. Thematic quality analyzes requires prolonged interaction with the data, cyclical coding, constant comparison, and meticulous development of themes to provide coherence and credibility.

If these steps are not adequately employed, thematic analysis will be reduced to a mere superficial or anecdotal approach rather than an analytically sound method. Theoretical principles of thematic analysis in the framework of modern qualitative studies are also rooted deeply in the interpretation and meaning-making of human beings. Although it has become clear that digital tools and artificial intelligence-driven systems are increasingly helpful in qualitative analysis, they cannot substitute for the epistemological contribution of the researcher (Knaflic, 2020). The practical and theoretical implications, as well as the differences between traditional manual dedication, software-assisted analysis with NVivo, and AI-assisted solutions like ChatGPT, rely on a clear understanding of their theoretical and methodological bases, making it vital to assess their differences in practice and implications.

Traditional Manual Coding in Qualitative Research

One of the oldest and most original methods of analyzing qualitative data is the traditional manual coding method. Under this approach, researchers interact with raw qualitative information, including interview transcripts, field notes, life histories, and observational records, and do not use digital software or artificial intelligence to filter these data. Manual coding aligns closely with the traditions of interpretivist and constructivist research, in which meaning is perceived as a social creation and situation-specific (Peel, 2020).

Manual thematic analysis usually starts with familiarization with the data, in which the researcher reviews and re-reads the data to gain a general sense of what it entails and what it relates to. The resulting involvement enables researchers to be sensitive to meanings, emotions, and nuances that could be hidden on the surface and therefore cultural (Braun & Clarke, 2006). The first layer of analytical reflection is usually in the form of notes and first impressions, which are sometimes recorded on the margins or in the research journal.

After familiarization, the researcher proceeds to initial coding, in which meaningful segments of text are marked with descriptive or interpretative codes. Such codes can be words, brief phrases, or even conceptual labels that represent essential features of the data relevant to the research questions (Saldaña, 2018). This process in manual coding is usually iterative, necessitating an unceasing movement of data parts and emerging concepts. Refinement, combination, or elimination of codes occurs during analysis. Once a significant number of codes are constructed, the researcher begins the theme-building process. It is the process of grouping related codes into larger patterns of meaning that represent repeated ideas in the dataset. Themes are not summative accounts of information, but are analytical representations of how participants interpret their experiences. (Braun & Clarke, 2019)

At this point, the researcher critically considers whether the themes are consistent, discrete, and sufficiently justified by data excerpts. Manual coding is one of its greatest strengths, thanks to its interpretive depth and sensitivity to context. Since the researcher is involved in working with the data all the time, manual analysis enables one to have a subtle insight into language, symbolism, emotion, and cultural practice (Patton, 2014). It is especially relevant when tribal, indigenous, or marginalized groups are used in qualitative research, as the meaning is usually hidden in oral tradition, local languages, and relationships that are not easily categorized or decoded into abstract forms.

Another major strength of manual coding is its focus on the researcher's reflexivity. Manual analysis prompts researchers to be critically aware of their assumptions, positionality, and role in the research process (Finlay, 2015). Memo writing and analytic journaling are also reflexive practices that help the researcher record their decision-making process, thereby increasing the transparency and credibility of the research. Traditional manual code has several limitations, despite its strengths. The time-consuming character of the process is one of the most important obstacles. Manual analysis of extensive qualitative data is a process that demands focus, significant time, and interaction, which may result in researcher burnout and inconsistency (Guest et al., 2011).

The larger the dataset, the less systematic the coding and classification of themes become. Subjectivity and researcher bias can also be found in the manual coding. The fact that the researcher's interpretive lens influences coding choices implies that personal assumptions or theoretical biases could shape the development of the theme (Peel, 2020). Although reflexivity can reduce this risk, it cannot eliminate it. Also, manual procedures usually do not provide the auditability and transparency of software-aided methods, and it is more challenging to demonstrate the analytical rigour to external reviewers. Additionally, there are problems with manual coding, such as the inability to code large quantities and the necessity to work as a team.

Standardized digital systems are not required when several researchers are involved, as inter-coder reliability and consistency are complex (J. D. Smith, 2012). Such shortcomings have led to the growing use of qualitative data analysis software, and most recently, software based on artificial intelligence. However, handwritten codes still have their place in qualitative research. Instead of being eliminated by technological changes, manual coding provides the epistemological basis for software-aided and AI-based analyzes. Knowledge of the principles, strengths, and limitations of manual thematic analysis is therefore necessary to be critical of more modern tools like NVivo and ChatGPT, which are discussed in the following sections of this chapter.

NVivo as a Tool for Qualitative Data Analysis

Computer-Assisted Qualitative Data Analysis Software (CAQDAS) has become a common practice due to the growing volume and complexity of qualitative data in social science research. NVivo has become one of the most popular software programs for analyzing qualitative data. NVivo is also designed to assist researchers in sorting, coding, accessing, and visualizing qualitative data, including transcripts of interviews and field notes, audiovisual data, and documents, without excluding the interpretive authority of the researcher (Kekez, 2019).

NVivo does not substitute for the analytical thinking needed in qualitative research; instead, it is facilitative, helping improve systematic data management and analytical transparency. NVivo helps researchers manage their qualitative data, which is a significant limitation of traditional manual coding: it is challenging to manage large qualitative datasets effectively (Elliott, 2018). It is most useful when conducting a qualitative study (large scale) or when carrying out multi-site research studies. One of the main features of NVivo is its coding system, which allows a researcher to label data with nodes that can be concepts, categories, or themes. The NVivo coding process is similar to manual thematic analysis, but it is enabled by digital accuracy and versatility.

Researchers can create, edit, combine, and rearrange nodes as their analysis develops, which is why the analytical process may be iterative and reflexive (Saldaña, 2016). This coding framework is iterative and supports both inductive and deductive methods of thematic analysis. NVivo also includes various analytical tools and queries, in addition to the code, that help organize patterns in the data. These are text search queries, word frequency queries, a query based on coded comparison, and a query based on matrix coding. These tools allow researchers to investigate connections among themes, participants' characteristics, and data sources, facilitating more in-depth analytical findings.

Visual means, such as models, charts, and coding maps, also contribute to the representation of intricate thematic relations in ways that are understandable and explicit. The key strength of NVivo is that it helps improve the rigor, transparency, and auditability of qualitative research. NVivo maintains a digital record of code decisions, revisions, and analysis actions, forming an audit trail that enhances the validity and reliability of qualitative results (Dolan et al., 2023). This aspect is especially valuable when conducting a collaborative research project, in which case consistency among multiple coders should be established and documented. NVivo also allows qualitative data analyzes in teams, with two or more researchers using the same data to examine coding consistency. Inter-coder reliability is another aspect that is frequently difficult to measure in a completely manual analysis but easy to assess with coding comparison tools (Castleberry & Nolen, 2018).

Consequently, NVivo is commonly preferred in institutional research where methodological accountability is a concern. Although these are the benefits, there are limitations to using NVivo. Among the most frequently mentioned ones is the high cost of learning how to use the software, which can put off a beginning researcher or someone with a low level of technical education (Silver & Lewins, 2014). Also, NVivo is commercial software that requires a financial investment, thus it is less accessible to the researcher working in a resource-bound context. The other issue of paramount importance is the danger of excessive mechanization in qualitative analysis. NVivo organizes the environment, which might encourage the researcher to focus on coding efficiency rather than interpretive depth and to break rich stories into data fragments.

NVivo may lead to a procedural approach to analysis that isolates the researcher from the contextual and emotional aspects of qualitative data when applied mindlessly. Also, NVivo does not necessarily address cultural sensitivity and ethical interpretation challenges, although it does enhance data organization. When it comes to native or tribal populations, the use of software without a profound contextual understanding can lead to de-contextualization or the misinterpretation of culturally hidden meanings (Sotiriadou et al., 2014). Consequently, NVivo should be applied reflexively and ethically, and the researcher must have constant interaction with the data. On the whole, NVivo is a powerful development in the field of qualitative data analysis, as it combines manual thematic analysis with the power of digital tools. It promotes systematic organization, analytical rigor, transparency, and maintains the interpretive power of the researcher. NVivo, however, is a tool, not an agent of analysis. It is necessary to understand its strengths and weaknesses before delving further into newer advances in AI-assisted qualitative analysis, such as ChatGPT, discussed in the next section.

ChatGPT and AI-Assisted Qualitative Analysis

In recent years, developments in the field of artificial intelligence (AI), specifically natural language processing (NLP), have already begun to have a pronounced impact on qualitative research practices. Large language models like ChatGPT are among the developments that form a new category of AI tools capable of producing human-like text, summarizing information, identifying patterns, and responding reflexively to prompts. However, unlike standard qualitative data analysis tools, ChatGPT is not a data management or coding tool but an AI-based analytical support tool that communicates with researchers in natural language (Siiman et al., 2023).

ChatGPT is increasingly used in qualitative research, especially at the initial stages of thematic analysis, such as familiarization with the data, summarizing the data, and exploratory coding. Scholars can feed interview quotes or anonymized textual information to receive condensed summaries, keyword clusters, or proposed thematic patterns. This feature is invaluable when working with massive volumes of data, and the initial immersion and familiarization can be time-consuming (Rannastu-Avalos & Pöysä-Tarhonen, 2023). ChatGPT provides researchers with a path to greater efficiency by speeding up initial analytical steps without necessarily binding oneself to rigid analytical typologies. The use of ChatGPT to produce provisional codes and themes is one of the most interesting applications in qualitative analysis. ChatGPT can propose possible code snippets for researchers to use in the first coding step based on linguistic patterns, repetition, and semantic similarity. These outputs, however, are not analytical conclusions but prompts that human beings think about and use in making decisions. Researchers note that AI-generated code should be critically analyzed and optimized by scholars to achieve conceptual consistency and contextual accuracy (Braun & Clarke, 2019).

ChatGPT can also be used as a reflexive support of qualitative research. Reflexive awareness and analytical depth can be increased by involving the model in dialogic questioning, by posing it to challenge emerging interpretations, or by finding other readings and researchers to do so. This dialogic application aligns with the reflexive traditions of qualitative inquiry, which reinforce interpretation by critically questioning the self (Zhang et al., 2023). In this respect, ChatGPT can act as a cognitive companion rather than an analytical tool. Despite these benefits, there are serious methodological and epistemological issues with applying ChatGPT in qualitative research. The lack of lived experience, cultural embeddedness, and contextual understanding is one of the model's main constraints. ChatGPT responds to language according to probable patterns rather than interacting directly with social reality. Consequently, it can fail to recognize the culturally specific meanings, symbolic expressions, or emotional nuances that inform qualitative interpretation.

An additional significant issue is the so-called AI hallucination phenomenon, in which the model produces outputs that seem plausible but lack empirical support or analytical validity. Such outputs can be misleading when analysts assume they are valid interpretations. This danger underscores the need to maintain human interpretive control throughout the analysis. The use of ChatGPT in qualitative analyzes is also ethically problematic. The issue of privacy and confidentiality of information is a sensitive matter, particularly when dealing with marginalized, tribal, or indigenous groups of people in a research study. When proper safeguards are not established, uploading interview transcripts or field data to AI sites can constitute a breach of ethical commitments regarding informed consent, data ownership, and community trust (Nguyen-Trung, 2025).

The scientists should thus be careful and strictly follow ethical principles when considering AI-assisted analysis. Epistemologically, ChatGPT questions the conventional knowledge production in qualitative research. Thematic analysis is essentially based on human meaning-making, reflexivity, and contextual interpretation. ChatGPT can support linguistic processing, but it will not be able to substitute for interpretive work in the context of which themes acquire analytical meaning (Braun & Clarke, 2006). This approach to using AI as an analytical tool would undermine the epistemic principles of qualitative research. However, in cases of reflection and transparent use, ChatGPT can supplement traditional and software-based methods of thematic analysis. Its advantages are that it is fast, available, and can be applied in the initial stages of exploration and reflection on the questions. ChatGPT could serve as a helpful aid rather than a limitation in the intensive quality analysis of early-career researchers and those with time constraints.

Overall, ChatGPT is an essential innovation in AI-based qualitative research, offering new opportunities for efficiency and reflexive interaction. Nevertheless, it should be supportive rather than determinative. A critical and ethically conscious attitude is necessary so that qualitative research can be deepened rather than reduced by AI-assisted analysis, making it less shallow, less credible, and less culturally sensitive. This negative stance serves as the basis for the comparative analysis of manual coding, NVivo, and ChatGPT in the following section.

COMPARATIVE ANALYSIS: MANUAL CODING, NVIVO, AND CHATGPT IN QUALITATIVE RESEARCH

The increased use of digital tools and artificial intelligence in qualitative studies requires them to be critically compared with traditional methods of thematic analysis. Manual coding, NVivo-assisted analysis, and AI-based tools like ChatGPT have distinct methodological orientations, each with its advantages and drawbacks.

Instead of treating the two approaches as competing options, this section of the paper compares them along major analytical dimensions, including efficiency, interpretive depth, reflexivity, transparency, and ethical implications (Sakaguchi et al., 2025).

Comparative Overview

Table 1. Comparative Overview of Manual Thematic Analysis, NVivo-Assisted Analysis, and ChatGPT-Assisted Analysis in Qualitative Research

Criteria	Manual Thematic Analysis	NVivo-Assisted Analysis	ChatGPT-Assisted Analysis
Role of Researcher	Central and interpretive	Central but system-supported	Supervisory and evaluative
Time Efficiency	Low	Moderate to high	Very high
Context Sensitivity	Very high	High	Limited
Reflexivity	Strong and continuous	Moderate	Weak unless guided
Transparency & Audit Trail	Limited	Strong	Limited
Scalability	Low	High	Very high
Ethical Risk	Minimal	Data storage concerns	High (Privacy, consent)
Suitability for Indigenous Studies	Very High	High (with reflexivity)	Low to Moderate

Source: Compiled by Author

Manual Coding as an Interpretive Benchmark

Qualitative inquiry is based on manual thematic analysis as an epistemological tool. The main advantage is that the researcher is more involved in data analysis, allowing for rich, culturally, and thoroughly reflexive interpretation. This method can be beneficial for research on marginalized and indigenous populations, where meanings are ingrained in local contexts, oral traditions, and social relations (May et al., 2022). Manual coding, however, is time-consuming, not scalable, and prone to fatigue and subjectivity, especially with large datasets.

NVivo as a Systematic Enhancement

NVivo lies between technological efficiency and manual interpretation in its methodology. It does not involve automation, but the systematic coding, data organization, and transparency of analyses are strengthened through digital means, i.e., nodes, queries, and visualizations (Houghton et al., 2017). NVivo helps minimize logistical pressures and facilitate collaborative research by providing audit trails and intercoder reliability. However, NVivo can also pose a risk of breaking down the narrative and stimulating procedural over interpretive participation when it is not employed critically.

ChatGPT as an AI-Assisted Support Tool

ChatGPT is based on an entirely new analytical logic, as the model operates on the idea of probabilistic language modelling rather than interpretive reasoning. Its advantages include speed, accessibility, and support for prior tasks, such as summarization, exploratory coding, and reflexive questioning (Daungsupawong & Wiwanitkit, 2024). Nevertheless, ChatGPT is not contextually aware, culturally embedded, or lived, which are the primary aspects of a qualitative analyzes. Data privacy, consent, and hallucination in AI are other ethical issues that limit the independent use of AI analysis.

Toward Complementarity Rather Than Replacement

The comparative analysis also reveals that a single approach is not sufficient. Manual coding is more in-depth and ethical, NVivo is more structured and transparent, and ChatGPT is more likely to be fast and supportive of exploration. Regarding AI tools as alternatives to human interpretation poses a threat to the epistemological basis of qualitative research. Instead, a complementary and reflexive combination of these methods can provide greater analytical rigor alongside interpretive integrity (Brynjolfsson & Milgrom, 2013). Such a comparative understanding preconditions a further discussion of ethical, methodological, and epistemological issues related to AI-assisted qualitative research, which is discussed in the next section.

Worked Illustrative Example of Manual, NVivo, and ChatGPT-Assisted Analysis

To further describe the practical differences between manual thematic analysis, NVivo-assisted analysis, and ChatGPT-supported analysis, this section provides a short hypothetical case with reference to interview data. It is illustrative and does not rely on empirical field data, but shows how qualitative material can be treated differently across different approaches to analysis. Suppose a researcher is conducting a study of semi-structured interview transcripts about the academic pressure participants in higher education experience.

In the manual thematic analysis, the researcher begins by repeatedly reading the transcripts to become familiar with the data. Through close interaction with participants' stories, original sets of codes are developed, including fear of failure, parental expectations, time pressure, and emotional exhaustion. These codes are tightened through writing memos and reflexive journals, and at some point, they are shown to be part of larger themes, such as performance anxiety and the emotional burden of achievement. The manual method enables the researcher to listen closely to the tone, pauses, and emotional displays in the narratives, facilitating subtle interpretation.

In the NVivo analysis, the researcher is expected to input the transcripts into the software and then create nodes based on the initial coding during manual familiarization. NVivo allows coded segments to be organized systematically and enables the researcher to visualize increasingly dense coding within and between interviews. The researcher can understand the relationship between changes in the expression of academic stress and certain demographic variables, such as year of study or field, by querying the matrix codes. NVivo thus enhances analytical transparency and efficiency while leaving the defining, refining, and interpretation of themes to the researcher's interpretive judgment.

In comparison, as an assistive technology, ChatGPT can be used by the researcher when anonymized interview text snippets are typed in and a summary or initial thematic recommendations are requested. ChatGPT can generate potential provisional themes, such as achievement pressure, because it can find linguistic similarities in terms of stress, pressure, and expectations. Although these outputs may help explore the area at the beginning stage, they are superficial and lack context. ChatGPT lacks emotional sensitivity, situational awareness, and the theoretical grounding of the researcher, and thus is not capable of generating themes that are analytically grounded in its own.

These three approaches differ in critical methodological ways. Manual analysis is time-consuming and difficult to scale; it prioritizes depth, reflexivity, and contextual sensitivity. NVivo promotes rigor, consistency, and transparency, especially with larger datasets, but it still requires the researcher's interpretive competence. ChatGPT is fast and provides an exploratory tool. However, it should be used with caution, as its results are generated from probabilistic language patterns rather than lived experience or contextual understanding.

Most importantly, this example shows that human interpretation will always be decisive in all approaches. Although NVivo and ChatGPT can facilitate organization and exploration, it is up to the researcher to apply analytical coherence and ethical judgment and to add meaning to the work. This practical example supports the main idea in the chapter that technology-assisted tools are to be used as supplementary means in the reflexive qualitative model, not as a replacement for interpretive expertise.

Expanded Operational Workflow of the Hybrid Qualitative Analysis Model

The hybrid qualitative analysis model presented in this chapter is expected to serve as a researcher-based, technology-aided workflow, integrating manual interpretation, qualitative software, and AI-aided tools in a consistent, ethically accountable way (Swain, 2018).

Instead of providing analysis as a linear or automated process, the workflow focuses on the flow between stages. It makes iterative movement a core characteristic of the entire process, leaving interpretive authority in the researcher's hands until the end. The initial phase of the working process is called the contextual immersion and analytical orientation. The researcher will interact with the raw qualitative data at this stage by repeatedly reading it and taking notes. It is not to write immediately without coding, but to get used to the storyline, emotional nuances, and the references to the context. Early impressions, questions, and theoretical sensitivities are recorded using analytic memos. This step establishes a conceptual and explanatory foundation for later methodological choices and helps avoid premature reliance on technology outputs.

The second phase focuses on the organized coding and data handling with NVivo. Once the researcher has reached a level of familiarity with the context, the data is imported into NVivo, and the first coding framework is built. Codes can be descriptive or interpretive, and they are refined through ongoing interaction with the data. The hierarchical node system in NVivo enables the researcher to group codes into categories and emergent themes without being restricted when eliminating or rearranging analytical structures. During this phase, it builds a clear audit trail by using memo links and coding logs, which contribute to the analytical rigor and reliability.

The third phase implies the analytical interrogation and pattern exploration. The query functions provided by NVivo, such as matrix coding and text search, are used to explore relationships among themes, cases, or data attributes. The tools will help detect convergences and divergences in the dataset and foster a deeper analytical understanding. Notably, query outcomes are viewed as cues to be interpreted rather than as grounds for analysis, which supports the researcher's judgment as the center stage.

The fourth phase is bounded AI-assisted exploration, wherein ChatGPT is deliberately used to reinforce reflexive cognition rather than primary analysis. At this stage, the researcher can apply AI tools to summarize thematic clusters, formulate additional analytical questions, or challenge emergent interpretations. To maintain ethical and methodological integrity, only anonymized, well-chosen excerpts are used, and AI results are reported as additional analytical commentary rather than as indicative results. This stage improves flexibility of thinking and prevents excessive dependence on automated suggestions.

The fifth step concerns reflexive synthesis and theoretical integration. In this case, knowledge obtained through manual work, analysis with NVivo, and AI-generated thoughts are combined into meaningful themes and analytical stories. Reflexivity is highlighted by regular memo writing and by interrogating oneself regarding analytical decisions. The researcher is vigilant about how the use of tools could have influenced interpretation, ensuring that the results are not based on technology but on data and theoretical consistency.

The last step of the workflow is the validation, documentation, and ethical disclosure procedures. Themes are discussed in accordance with their internal consistency and analytically precise data, with exemplary data samples selected to support interpretive assertions. NVivo and AI tools have a methodological role, which is clearly documented in the research report, thereby improving transparency and accountability. These ethical provisions concerning data management, application of AI, and analytical responsibility are discussed as part and parcel of methodological reporting.

This broadened scope of work shows that a hybrid qualitative model of analysis can be put into practice while preserving interpretive richness, methodological rigor, and ethical responsibility. The model offers a practical guide for qualitative researchers navigating the shifting methodological climate by treating technology as an enabling infrastructure rather than a guide.

ETHICAL, METHODOLOGICAL, AND EPISTEMOLOGICAL CONCERNS IN TECHNOLOGY-ASSISTED QUALITATIVE RESEARCH

The adoption of digital tools and artificial intelligence in qualitative research has also opened new opportunities for greater efficiency and analytical support. It has, however, also raised some complex ethical, methodological, and epistemological issues that need to be scrutinized. These issues are especially pertinent in social science studies involving marginalized, tribal, and indigenous groups, as ethical responsibility and awareness of context become key factors.

Ethical Concerns

Data privacy and confidentiality are among the major ethical concerns related to technology-assisted qualitative analysis. Qualitative studies usually consist of intimate biographical accounts, cultural information, and lived experience, which are exchanged in a relationship of trust. The increased use of software platforms and AI, particularly cloud-based systems like ChatGPT, raises questions about data storage, ownership, and unauthorized access (Brown et al., 2025). The researchers need to ensure that the use of these tools is ethically sound with respect to informed consent, anonymization, and data preservation.

The ethical issues regarding tribal and indigenous people in a study are not limited to individual consent but also to group ownership of knowledge. Indigenous researchers stress that cultural texts and traditional knowledge should not be treated as neutral information but rather as intellectual and cultural property of the community (V. Smith et al., 2012). Submission of such information to AI systems without direct community consent would constitute an infringement of ethical obligations, and extractive research methods would be recreated.

Methodological Concerns

In terms of methodology, the growing reliance on digital tools can lead to the proceduralization of qualitative analysis. The use of software-aided and AI-informed methods can lead more researchers to focus on speed, efficiency, and technical results rather than on in-depth data analyzes. This abatement jeopardizes the weakening of reflexivity, which is the primary focus of qualitative inquiry. The other methodological issue concerns excessive reliance on AI outputs. Even though tools like ChatGPT can help with summarization and preliminary coding, they rely on statistical trends rather than interpretation. The passivity of AI-generated themes in human analysis can lead to the superficial or false interpretation of the work (Braun & Clarke, 2019). Researchers should consequently retain interpretive power and treat AI outputs as temporary rather than final.

Epistemological Concerns

The application of AI in qualitative research papers brings up powerful epistemological issues concerning the production of knowledge. The qualitative inquiry is based on the premise that meaning is constructed through the interaction among the researcher, participants, and context (Lee-Johnson et al., 2023). On the contrary, AI systems produce text with no lived experience, no social positioning, and no cultural embeddedness. This conflict questions the conventional notions of who is a producer of knowledge during qualitative research. As renderings and meanings become increasingly influenced by algorithmic systems, there is a danger that qualitative analysis becomes less a reflexive meaningful-making process and more an automated pattern-recognition process. It could be a blow to the interpretivist blocs on the qualitative methodology.

Bias, Power, and De-Contextualization

The training of AI systems is performed on large datasets that reflect current social biases and power structures. Consequently, AI-assisted qualitative analysis can reproduce dominant narratives and exclude alternative or subaltern narratives. Such danger is especially sharp in the studies that refer to indigenous or marginalized people, where the acts of historically false representation and epistemic injustice are already present. Another essential issue is de-contextualization. Dissecting narratives into codes or summaries, particularly with automated tools, can remove the cultural, emotional, and relational context of qualitative data. Technology-aided analysis can narrow the depth of analysis of complex social facts, undermining ethical responsibility without attentive, reflexive interaction, and reducing them to abstract categories (Isik, 2025).

Toward Ethical and Reflexive Use

To address such issues, a reflexive, ethically informed approach to technology-assisted qualitative research is necessary. Researchers should be open about the use of digital tools and AI systems, and this should be documented, with the role of these systems in the analytical process identified. The application of AI tools, including their use in relation to data sensitivity and community consent, should be explicitly addressed in ethical review processes. This chapter does not oppose the use of technological tools. Instead, it proposes their use in a prudent and supplementary manner, with human interpretation, ethical judgment, and an understanding of the context as the key factors. It would be a way to ensure that technological innovation does not negate the epistemological integrity of qualitative research.

RELIABILITY, VALIDITY, AND TRUSTWORTHINESS IN TECHNOLOGY-ASSISTED QUALITATIVE ANALYSIS

Trustworthiness and the importance of rigor, transparency, and credibility are often used in place of reliability and validity in qualitative research, as a proxy for the positivist sense of replication. Since the use of digital tools has become increasingly prevalent in qualitative analysis (NVivo and AI-assisted systems such as ChatGPT), the issue of trustworthiness has become more complex and warrants methodological consideration once again. To achieve rigor in a technology-aided qualitative research study, a close correspondence among analytical instruments, researcher reflexivity, and epistemological assumptions must be maintained.

The framework of trustworthiness developed by Lincoln and Guba (1985) includes credibility, dependability, confirmability, and transferability, and is one of the most important references for measuring qualitative rigor. Such a structure comes in especially handy when evaluating technology-assisted analysis, as it prefigures human judgment, reflexive accountability, and procedural transparency.

Credibility

Credibility is the degree to which the qualitative findings reflect the meanings and experiences of the participants. In the manual thematic analysis, credibility is supported by extensive data analysis, repeated coding, and reflexive interpretation. The proximity of the researcher to textual content enables sensitivity to nuance, emotion, and context.

NVivo would increase credibility, as it facilitates the systematization of coding, the retrieval of parts of data, and the recording of analytical choices. Adjuncts such as node hierarchies, coding comparison queries, and visualizations assist the researcher in returning to and developing themes further, creating analytical coherence. Nevertheless, credibility is maintained, though it is based on the researcher's interpretive ability rather than on the software. In their turn, AI-assisted technologies like ChatGPT present different risks to credibility.

Although ChatGPT may provide summaries or propose temporary codes, they are generated on linguistic grounds rather than based on interpretative knowledge. Uncritically, AI-generated interpretations can distort participants' meanings or simplify complex stories. Thus, the credibility of AI-aided qualitative research requires that humans explicitly validate AI outputs.

Dependability

Dependability is the aspect of the research process that ensures consistency and logical interdependence. However, in manual analysis, it may be difficult to be reliable, especially in long-term research or when working with big data, since coding judgments can change with researcher fatigue or due to changing meanings in interpretation. NVivo makes the code substantially more reliable by providing a consistent digital space where decisions on coding, revisions, and the direction of analysis are appropriately documented. External reviewers can assess the consistency and rationality of the analytical process through an audit trail comprising coding histories, memos, and project logs. It is beneficial when conducting collaborative research, where consistency across many different analysts is required. But AI-assisted tools, on the other hand, make dependability difficult. Systems like ChatGPT produce outputs that can be more or less influenced by the development of prompts, model refinement, or contextual inputs. This inconsistency undermines the reproducibility of AI-assisted analysis and underscores the need for clear documentation of the application of AI tools in the research process.

Confirmability

Confirmability is a measure of how the research results are influenced by the data rather than by researcher bias or other external influences. Confirmability in qualitative research is achieved through reflexive activities such as memo writing, journaling, and transparent reporting of decisions made in the course of analysis. Confirmability can be supported and threatened by tools that involve technological assistance. NVivo is expected to increase confirmability because it enables researchers to identify themes from direct data snippets, thereby allowing interpretive claims to be tracked and validated. Nevertheless, AI-driven analysis comes with the problem of algorithmic bias, as AI-generated recommendations can incrementally shape the direction of the analysis without critical examination. To maintain confirmability in qualitative research using AI, researchers need to ensure clear boundaries between human decoding and machine outputs. AI tools are intended to be used as analytical tools rather than for authoritative interpretation, and their use must be clearly noted in methodological reporting.

Transferability

Transferability is a term that describes how qualified the findings of a qualitative study can be used in other settings. Transferability is based on detailed description and contextual clarity, rather than statistical generalization. Manual thematic analysis is better at generating thick descriptions, which enables readers to determine whether they can be applied to other settings. NVivo facilitates transferability by enabling systematic comparisons of cases, contexts, or participant groups, making it easier to see that patterns and variations exist in the data. But overuse of fragmented coding will result in a loss of narrative coherence unless it is compensated by interpretive synthesis. AI-assisted tools have little influence on transferability because they lack contextual understanding and cannot produce a thick description on their own. They are mainly helpful for facilitating exploratory analysis rather than for generating transferable qualitative information.

Maintaining Trustworthiness in Hybrid Approaches

To achieve credibility in technology-supported qualitative research, a reflexive, hybrid methodology in which human interpretation is central should be adopted. Scholars need to record the use of tools, explain the methodology, and provide a critical reflection on how digital and AI systems can influence the analysis. Qualitative rigor in an ever-more digital research setting requires transparency, reflexivity, and ethical accountability.

TOWARD A HYBRID MODEL OF QUALITATIVE ANALYSIS

The analysis of the three types of thematic analysis, the one performed manually and those performed with the participation of NVivo and ChatGPT, shows that none of the methods is adequate to meet the analytical, ethical, and epistemological requirements of modern qualitative research. Both approaches have their own advantages and serious drawbacks. To address these issues, this chapter presents a hybrid approach to qualitative analysis that combines human interpretive authority with technological assistance in a reflexive and ethically responsible way. The hybrid model is based on the fact that the human interpretation is impossible to replace in qualitative research. The processes of meaning-making, understanding something within its context, being culturally sensitive, and moral decision-making are intrinsically human processes grounded in lived experience and reflexivity (Denzin & Giardina, 2022). Digital tools and AI systems must be used as supportive tools rather than as self-reliant analytical tools. This model does not uphold technological determinism or methodological conservatism; instead, it follows a complementary approach.

Sequential Hybrid Workflow

The offered hybrid model is based on a sequential and reflexive workflow. To begin with, researchers engage in hand familiarization with qualitative data by reading transcripts, field notes, and narratives. It is necessary to engage in this kind of initial immersion to obtain the tone, emotion, cultural references, and contextual meaning that might not be extracted by automated processes.

Second, NVivo can facilitate systematic coding and data sorting. It is the point at which the researchers encode their interpretative intuitions into computer nodes and categories to enable them to be refined repeatedly, to support pattern comparison, and to provide a clear record of analytical choices. NVivo can be used as a methodological stabilizer that enhances rigor, auditability, and collaboration while maintaining the researcher's interpretive position.

Third, ChatGPT can be integrated into the discovery and reflection process. For example, the tool may help summarize large datasets, propose temporary codes, or suggest a different set of interpretive questions. Nonetheless, the AI-generated responses should be considered heuristic clues rather than analytical findings. These suggestions are subject to human judgment in validating, rejecting, or reworking.

Reflexivity as the Integrative Principle

Reflexivity works as the integrative principle of the hybrid model. In the course of the analytical process, scholars should critically consider the effects of tools on interpretation, decision-making, and representation. The use of memos, analytical journals, and a transparent reporting process will be imperative for documenting the roles of both software and AI in shaping the findings. This reflexivity plays a significant role in works with tribal and indigenous populations, where the scope of ethical responsibility extends beyond the rigor of methodological endeavor to include respecting cultural knowledge systems and communal values. The hybrid model makes it clear that technological convenience should not dominate the contextual sensitivity or moral responsibility.

Ethical Guardrails in the Hybrid Model

The hybrid framework has ethical protection. Qualitative data that are sensitive must not be uploaded mindlessly to AI platforms, and the researcher must adhere to the standards of informed consent and data security. When AI tools are implemented, methodological reporting must clearly specify their use to ensure transparency and maintain trust. The hybrid model also avoids the threats of de-contextualization, algorithmic bias, and over-automation by ensuring a clear distinction between human interpretation and machine aid. Technology will therefore enhance analytical potential without compromising the epistemological basis of the qualitative investigation.

Contribution of the Hybrid Model

The hybrid method proposed is part of the qualitative methodology, as it provides a practical, moderate approach to incorporating new technologies into established analytical practices. Instead of positioning AI as a substitute for qualitative expertise, the model introduces it as a companion in a reflexive, human-oriented research process. It allows the researcher to achieve technological efficiency while maintaining interpretive richness, ethical integrity, and cultural sensitivity.

IMPLICATIONS FOR SOCIAL SCIENCE RESEARCH

Such a combination of classic manual thematic analysis, any qualitative data analysis program like NVivo, and AI-based tools like ChatGPT has profound implications for the work of social science researchers. These consequences are not confined to the efficiency of the analysis but also to the rigor of the methodology, the researcher's reflexivity, and the ethical responsibility and future of qualitative research.

Methodological Implications

For social science researchers, the hybrid analytical framework suggested in the chapter offers a middle ground between deep analysis and technological support. The process of manual thematic analysis remains essential in close interaction with data, allowing the researcher to achieve the nuance, meaning, and complexity of qualitative stories. NVivo facilitates systematic coding, data organization, and precise documentation of analytical choices. Such selective application of AI-assisted tools, such as ChatGPT, also makes the methodology efficient, especially in the initial phases of the analysis, such as familiarization with the data, summarization, and exploratory coding. Reflexively, these instruments will assist researchers in working with large datasets and prompt human reflection rather than substituting for interpretation.

Implications for Research Practice

The increasing supply of digital and AI-based tools will prompt social science researchers to overcome the inflexible methodological dichotomies, such as manual and software-assisted analysis, and to employ flexible, context-sensitive research practices. The hybrid model proposed underscores that methodological decisions must be informed by the research questions, the nature of the data, and ethical issues, rather than by technological convenience in particular. This methodology upholds the relevance of reflexivity in qualitative studies. Researchers should also be critically aware of how tools influence the methods of analysis, interpretation, and data representation. Qualitative findings emphasize the importance of credibility and trustworthiness, which is best achieved by providing information on the role of software and AI tools in reporting.

Implications for Research Training and Capacity Building

There are also significant implications of the hybrid model for research training and capacity building, especially for early-career researchers. NVivo and AI systems have the potential to relieve the technical burden of qualitative analysis; however, they cannot replace a strong conceptual understanding of qualitative methodology. Unless proper training is provided, researchers may overuse automated results, leading to analysis in a superficial manner. Thus, qualitative research training programs should focus on methodological literacy, reflexive practice, and ethical awareness, in addition to technical skills. To maintain analytical rigor in a somewhat more digital research landscape, it is essential to develop researchers' capacity to assess the outputs generated by tools critically.

Implications for Academic and Policy Contexts

At the institutional level, the introduction of tools aided by AI to qualitative research provokes significant questions for academic publishing, the ethics committees of research, and policy frameworks. Transparency is becoming a necessity for journals and publishers that address the application of AI tools in research, especially in authorship, data processing, and analysis accountability. Hopefully, there should be clear rules and ethical principles that facilitate responsible AI implementation in qualitative research. The hybrid framework suggested in this chapter, through reflexivity, transparency, and human interpretive authority foregrounded, has added to the debate about responsible innovation and methodological integrity in social science research.

CONCLUSION

This chapter reviewed the shift in qualitative research methodologies in changing the manual approach to the use of technology-assisted techniques based on NVivo and AI-based (Chabot's like ChatGPT). The discussion has revealed that manual coding remains essential to provide interpretative depth and reflexive meaning-making, while digital and AI-assisted coding help address complexity, making analysis more efficient and transparent. The comparative analysis showed that NVivo enhances a rigorous approach to the qualitative process by organizing and making it auditable. In contrast, ChatGPT primarily supports exploratory and reflexive processes by aiding summarization, preliminary coding, and questioning. Nevertheless, software and AI will not substitute for the key role of the researcher in meaning construction, context interpretation, and ethical decision-making. An excessive dependency on automated results poses the threat of bankrupt analysis and epistemological watering down of qualitative research. The chapter does not present these methods as competing alternatives but instead proposes a hybrid model that incorporates human interpretive authority and technological assistance. NVivo and ChatGPT can also be applied in the reflexive and ethical use of traditional thematic analysis, allowing the researcher to balance both depth and efficiency. Finally, the responsible adoption of digital and AI tools is a development and not the overthrow of qualitative methodology, which once again confirms that the qualitative analysis that gains meaning is inherently human.

REFERENCES

Braun, V., & Clarke, V. (2006). Using thematic analysis in psychology. *Qualitative Research in Psychology, 3*(2), 77–101. DOI: 10.1191/1478088706qp063oa

Braun, V., & Clarke, V. (2016). (Mis) conceptualising themes, thematic analysis, and other problems with Fugard and Potts'(2015) sample-size tool for thematic analysis. *International Journal of Social Research Methodology, 19*(6), 739–743. DOI: 10.1080/13645579.2016.1195588

Braun, V., & Clarke, V. (2019). Reflecting on reflexive thematic analysis. *Qualitative Research in Sport, Exercise and Health, 11*(4), 589–597. DOI: 10.1080/2159676X.2019.1628806

Brown, K. S., Bender, S., Kasitz, D. L. H., & Martinez-Suarez, A. A. (2025). Why Academic Women Stay: A Thematic Analysis of Mentorship Experiences in Higher Education via the Job Embeddedness Theory. *Journal of Women and Gender in Higher Education, 18*(4), 393–417. DOI: 10.1080/26379112.2025.2571092

Brynjolfsson, E., & Milgrom, P. (2013). Complementarity in organizations. *The Handbook of Organizational Economics*, 11–55.

Cabiddu, F., Frau, M., & Moi, L. (2018). Exploring the role of NVivo software in marketing research. *Mercati e Competitività, 4*, 65–86.

Castleberry, A., & Nolen, A. (2018). Thematic analysis of qualitative research data: Is it as easy as it sounds? *Currents in Pharmacy Teaching & Learning, 10*(6), 807–815. DOI: 10.1016/j.cptl.2018.03.019 PMID: 30025784

Creswell, J. W., & Poth, C. N. (2016). *Qualitative inquiry and research design: Choosing among five approaches*. Sage publications.

Daungsupawong, H., & Wiwanitkit, V. (2024). Performance of ChatGPT as an AI-assisted decision support tool in medicine [comment]. *Acta Cardiologica, 79*(3), 413. DOI: 10.1080/00015385.2024.2322185 PMID: 38441117

Denzin, N. K. (2018). *The qualitative manifesto: A call to arms*. Routledge. DOI: 10.4324/9780429449987

Denzin, N. K., & Giardina, M. D. (2022). *Transformative visions for qualitative inquiry*. Routledge New York and London. DOI: 10.4324/9781003254010

Dolan, S., Nowell, L., & Moules, N. J. (2023). Interpretive description in applied mixed methods research: Exploring issues of fit, purpose, process, context, and design. *Nursing Inquiry, 30*(3), e12542. DOI: 10.1111/nin.12542 PMID: 36464974

Elliott, V. (2018). Thinking about the coding process in qualitative data analysis. *The Qualitative Report, 23*(11). Advance online publication. DOI: 10.46743/2160-3715/2018.3560

Fereday, J., & Muir-Cochrane, E. (2006). Demonstrating rigor using thematic analysis: A hybrid approach of inductive and deductive coding and theme development. *International Journal of Qualitative Methods, 5*(1), 80–92. DOI: 10.1177/160940690600500107

Finlay, L. (2002). Negotiating the swamp: The opportunity and challenge of reflexivity in research practice. *Qualitative Research, 2*(2), 209–230. DOI: 10.1177/146879410200200205

Finlay, L. (2015). Qualitative methods. *The Counselling and Psychotherapy Research Handbook, 55.*

Guest, G., MacQueen, K. M., & Namey, E. E. (2011). *Applied thematic analysis.* sage publications.

Guest, G., Namey, E., & Chen, M. (2020). A simple method to assess and report thematic saturation in qualitative research. *PLoS One, 15*(5), e0232076. DOI: 10.1371/journal.pone.0232076 PMID: 32369511

Houghton, C., Murphy, K., Meehan, B., Thomas, J., Brooker, D., & Casey, D. (2017). From screening to synthesis: Using nvivo to enhance transparency in qualitative evidence synthesis. *Journal of Clinical Nursing, 26*(5–6), 873–881. DOI: 10.1111/jocn.13443 PMID: 27324875

Isik, O. (2025). Qualitative Research Approaches and Data Collection Methods: Understanding Meaning and Experience. *Journal of Humanities and Education Development.* Https://Www.Academia.Edu/Download/125380321/Article_11_qualitative_research_approaches. *Pdf. Re-Trieved from* Https://Www. Academia. Edu/Download/125380321/Article_11_qualitative_research_approaches. *Pdf.*

Kekez, A. (2019). Qualitative data analysis in implementation and street-level bureaucracy research. In *Research handbook on street-level bureaucracy* (pp. 317–335). Edward Elgar Publishing. DOI: 10.4337/9781786437631.00032

Knaflic, C. N. (2020). Summarizing and Visualizing Data. *Research Methods for Public Health,* 229.

Lee-Johnson, Y. L., Fair, K., O'Connor, K., Rodney, T., Ono, J., & Dixon, T. (2023). Reflexivity for Restorying the Ontological and Epistemological Truths in Qualitative Research. *Impacting Education: Journal on Transforming Professional Practice, 8*(4), 43–49. DOI: 10.5195/ie.2023.339

May, C. R., Albers, B., Bracher, M., Finch, T. L., Gilbert, A., Girling, M., Greenwood, K., MacFarlane, A., Mair, F. S., May, C. M., Murray, E., Potthoff, S., & Rapley, T. (2022). Translational framework for implementation evaluation and research: A normalisation process theory coding manual for qualitative research and instrument development. *Implementation Science : IS*, *17*(1), 19. DOI: 10.1186/s13012-022-01191-x PMID: 35193611

Nguyen-Trung, K. (2025). ChatGPT in Thematic Analysis: Can AI become a research assistant in qualitative research? *Quality & Quantity*, *59*(6), 1–34. DOI: 10.1007/s11135-025-02165-z

Patton, M. Q. (2014). *Qualitative research & evaluation methods: Integrating theory and practice*. Sage publications.

Peel, K. L. (2020). A beginner's guide to applied educational research using thematic analysis. *Practical Assessment, Research & Evaluation*, *25*(1).

Potančok, M., & Radváková, V. (2023). The topic of Ai Chatbots in Higher Education. *IDIMT-2023: New Challenges for ICT and Management*.

Rannastu-Avalos[1], M., & Pöysä-Tarhonen, J. (2023). Opportunities and Challenges for AI-Assisted Qualitative Data Analysis: An Example. *Innovative Technologies and Learning: 6th International Conference, ICITL 2023, Porto, Portugal, August 28–30, 2023, Proceedings, 14099*, 87.

Sakaguchi, K., Sakama, R., & Watari, T. (2025). Evaluating ChatGPT in Qualitative Thematic Analysis With Human Researchers in the Japanese Clinical Context and Its Cultural Interpretation Challenges: Comparative Qualitative Study. *Journal of Medical Internet Research*, *27*, e71521. DOI: 10.2196/71521 PMID: 40273439

Saldaña, J. (2016). Goodall's verbal exchange coding: An overview and example. *Qualitative Inquiry*, *22*(1), 36–39. DOI: 10.1177/1077800415603395

Saldaña, J. (2018). Researcher, analyze thyself. *International Journal of Qualitative Methods*, *17*(1), 1609406918801717. DOI: 10.1177/1609406918801717

Siiman, L. A., Rannastu-Avalos, M., Pöysä-Tarhonen, J., Häkkinen, P., & Pedaste, M. (2023). Opportunities and challenges for AI-assisted qualitative data analysis: An example from collaborative problem-solving discourse data. *International Conference on Innovative Technologies and Learning*, 87–96. DOI: 10.1007/978-3-031-40113-8_9

Silver, C., & Lewins, A. (2014). *Using software in qualitative research: A step-by-step guide*.

Smith, J. D. (2012). Single-case experimental designs: A systematic review of published research and current standards. *Psychological Methods*, *17*(4), 510–550. DOI: 10.1037/a0029312 PMID: 22845874

Smith, V., Begley, C. M., Clarke, M., & Devane, D. (2012). Professionals' views of fetal monitoring during labour: A systematic review and thematic analysis. *BMC Pregnancy and Childbirth*, *12*(1), 166. DOI: 10.1186/1471-2393-12-166 PMID: 23270400

Sotiriadou, P., Brouwers, J., & Le, T.-A. (2014). Choosing a qualitative data analysis tool: A comparison of NVivo and Leximancer. *Annals of Leisure Research*, *17*(2), 218–234. DOI: 10.1080/11745398.2014.902292

Swain, J. (2018). *A hybrid approach to thematic analysis in qualitative research: Using a practical example*. Sage Research Methods. DOI: 10.4135/9781526435477

Zhang, H., Wu, C., Xie, J., Lyu, Y., Cai, J., & Carroll, J. M. (2023). Redefining qualitative analysis in the AI era: Utilizing ChatGPT for efficient thematic analysis. *ArXiv Preprint ArXiv:2309.10771*.

Chapter 11
Large Language Model

Angga Muhammad Galih
http://orcid.org/0009-0009-2998-9514
Universitas Esa Unggul, Indonesia

Binastya Anggara Sekti
http://orcid.org/0000-0001-5489-4888
Universitas Esa Unggul, Indonesia

Bartholomeus Dimanche Carl Beth
http://orcid.org/0009-0006-1088-9212
Universitas Esa Unggul, Indonesia

Gusti Muhamad Sardana
http://orcid.org/0009-0001-4281-2432
Universitas Esa Unggul, Indonesia

ABSTRACT

The integration of Large Language Models (LLMs) into qualitative research offers transformative capabilities in automated transcription, coding, and thematic analysis, yet challenges foundational principles including contextual understanding, researcher reflexivity, and interpretive depth. This examination critically analyzes LLM applications across research workflows, addressing the black-box problem, algorithmic bias, and ethical concerns regarding informed consent, data privacy, and participant rights. The study proposes hybrid frameworks positioning LLMs as augmentative tools while maintaining human interpretive authority, emphasizing transparent documentation, rigorous validation, and researcher AI literacy for responsible integration.

DOI: 10.4018/979-8-3373-7640-0.ch011

nt © 2026, IGI Global Scientific Publishing. Copying or distributing in print or electronic forms without written permission of IGI Global Scientific Publishing is prohibited. Use of this chapter to enerative artificial intelligence (AI) technologies is expressly prohibited. The publisher reserves all rights to license its use for generative AI training and machine learning model development.

1. INTRODUCTION

The landscape of the integration of Large Language Models (LLMs) into qualitative research marks a significant turning point in the evolution of interpretive methodologies, offering both unprecedented opportunities and complex challenges. As qualitative research traditionally emphasizes depth, context, and human-centered understanding, the introduction of AI-driven tools requires careful reflection on how these technologies align with established epistemological principles. LLMs, powered by advanced neural network architectures, have demonstrated remarkable capabilities in processing large volumes of textual data, identifying patterns, and generating coherent analytical outputs (Minaee et al., 2025). Their growing accessibility has made them attractive tools for researchers seeking to enhance efficiency and manage increasingly complex datasets. However, this technological advancement also raises important questions about the preservation of interpretive rigor and the role of human judgment in qualitative inquiry.

One of the most compelling advantages of LLMs lies in their ability to accelerate time-intensive research processes. Tasks such as transcription, coding, and thematic organization, which traditionally require extensive manual effort, can now be completed with significantly reduced time and labor. This efficiency enables researchers to engage with larger datasets and pursue more ambitious research designs, including mixed-methods approaches and longitudinal studies (Chakravorti et al., 2024). Moreover, LLMs can assist in identifying linguistic patterns and semantic relationships that may not be immediately apparent to human analysts, thereby supporting the discovery of new insights. Despite these benefits, the reliance on automated processes introduces the risk of superficial engagement with data, potentially compromising the depth of analysis that defines high-quality qualitative research.

At the same time, significant limitations inherent in LLMs must be critically acknowledged. These systems operate through statistical pattern recognition rather than genuine understanding, lacking the contextual awareness, cultural sensitivity, and experiential knowledge that human researchers bring to analysis (Asmawi & Alam, 2024). Qualitative research depends on the ability to interpret meaning within specific social, cultural, and historical contexts, a task that requires more than computational processing. The inability of LLMs to fully grasp these dimensions can lead to misinterpretations, especially in studies involving marginalized communities or culturally diverse populations. As highlighted by Erciyes (2020), interpretive depth in qualitative research emerges from immersion, reflexivity, and engagement with participants' lived experiences qualities that remain beyond the reach of current AI systems (Erciyes, 2020).

Another critical concern is the issue of transparency and the so-called "black-box" nature of LLMs. Unlike traditional analytical methods, where researchers can clearly articulate their reasoning processes, LLMs generate outputs through complex internal mechanisms that are not easily interpretable (Minaee et al., 2025). This lack of explicability poses challenges for auditability and trustworthiness, as other scholars may find it difficult to evaluate how conclusions were derived. The inability to fully trace or replicate AI-driven analytical decisions complicates peer review and raises questions about the reliability of findings. Consequently, researchers must develop new documentation practices and validation strategies to ensure that AI-assisted research maintains the standards of rigor expected in qualitative inquiry (Hitch, 2024).

Ethical considerations further complicate the integration of LLMs into qualitative research. Issues related to data privacy, informed consent, and algorithmic bias require careful attention, particularly when dealing with sensitive or personal data. Participants must be informed about how their data will be processed, including the involvement of AI systems, and researchers must ensure that data is handled securely and responsibly (Hasan, 2025). Additionally, biases embedded in training data can lead to skewed interpretations, reinforcing dominant narratives while marginalizing alternative perspectives. Addressing these challenges requires a commitment to ethical reflexivity and the development of guidelines that govern the responsible use of AI in research contexts.

In response to these opportunities and challenges, hybrid methodological frameworks have emerged as a promising approach to integrating LLMs into qualitative research. These frameworks position AI as an augmentative tool that supports, rather than replaces, human interpretation. By combining computational efficiency with human analytical expertise, researchers can leverage the strengths of both approaches while mitigating their respective limitations (Brailas, 2025). Such models emphasize the importance of maintaining human oversight, critical evaluation, and contextual sensitivity throughout the research process. As qualitative research continues to evolve in the age of AI, the key challenge lies in balancing innovation with integrity, ensuring that technological advancements enhance rather than diminish the depth, richness, and ethical foundations of qualitative inquiry.

2. LLM CAPABILITIES AND APPLICATIONS IN QUALITATIVE RESEARCH

Large Language Models (LLMs) have fundamentally transformed the landscape of qualitative research by offering advanced capabilities in language processing, interpretation, and automated data handling. Traditional qualitative methodologies such as in-depth interviews, focus groups, document analysis, and ethnographic observations depend heavily on the researcher's interpretive skills. These skills require substantial time, cognitive effort, and methodological precision. The emergence of LLMs expands the researcher's toolkit by enabling rapid processing of textual data, assisting in literature synthesis, and supporting structured analysis from early data preparation to advanced theoretical development. This section elaborates on how LLMs enhance each stage of qualitative inquiry, demonstrating their growing relevance in contemporary research ecosystems.

2.1 Natural Language Processing and Understanding in LLMs

LLMs excel in natural language processing and understanding, allowing them to interpret linguistic structures, contextual cues, and semantic relationships with remarkable depth. Unlike earlier computational approaches that relied on surface-level statistical patterns, modern LLMs can comprehend subtleties in sentence structure, emotional tone, speaker intent, and pragmatic context. These capabilities are particularly significant in qualitative research, where meaning is not derived merely from what is said but also from how it is articulated, in what context it is expressed, and what social, cultural, or ideological assumptions underlie the narration.

In qualitative interview analysis, for example, LLMs can parse lengthy transcripts, identify recurring themes, and interpret implicit meanings without losing the nuance embedded in the participant's voice. The models can detect shifts in attitude, contradictions in statements, and contextual references that might be overlooked in manual analysis. Their ability to recall earlier segments of the conversation also allows them to interpret data in a cohesive and longitudinal manner, aligning well with approaches such as narrative inquiry and discourse analysis (Hitch, 2024).

Furthermore, LLMs serve as intellectual companions that support critical reflexivity. Researchers can engage in dialogic exploration with the model, testing alternative interpretations of a transcript, reflecting on their positionality, or examining how different theoretical lenses shift the meaning of the same data. This interactive nature enhances the analytic depth and mitigates blind spots that often arise in individual interpretation. Nevertheless, the researcher remains responsible for validating and contextualizing the interpretations generated by the model, ensuring alignment with methodological and ethical standards.

2.2 Automated Transcription and Data Preparation

The early stage of qualitative research often involves demanding manual work such as transcribing interviews, cleaning raw data, anonymizing sensitive information, and segmenting textual content into analyzable units. LLMs significantly reduce the researcher's workload by offering automated and semi-automated tools that enhance the accuracy, consistency, and efficiency of data preparation (Minaee et al., 2025).

LLM-enabled transcription systems can convert audio recordings into coherent written text with improved recognition of speech patterns, accents, technical terms, and variations in voice quality. Their capacity to reconstruct punctuation and sentence structure ensures that the resulting transcripts are not only readable but also faithful to the original conversation (Wosny & Hastings, 2024). Once transcribed, LLMs can refine the text through normalization processes, such as removing filler words, correcting grammatical inconsistencies, and clarifying ambiguous phrasing while maintaining the integrity of the speaker's voice. This is particularly valuable in studies involving multilingual participants or natural conversational flow, where transcription fidelity is essential for reliable interpretation (Bennis & Mouwafaq, 2025).

Another critical contribution of LLMs is their ability to perform anonymization automatically. By detecting and replacing personal identifiers, institutional names, or location-specific information, the models support the ethical management of data while preserving contextual coherence (Kempny et al., 2025). In addition, LLMs facilitate the segmentation of lengthy transcripts into meaning units that align with analytic frameworks used in phenomenology, grounded theory, or thematic analysis. These preprocessing steps provide a structured foundation for subsequent coding and interpretation, allowing researchers to spend more time on analytic reasoning rather than mechanical data preparation (Wen et al., 2025).

2.3 AI-Assisted Literature Review and Synthesis

A comprehensive understanding of existing literature forms the backbone of high-quality qualitative research. Traditionally, this process demands extensive reading, conceptual mapping, and cross-referencing. LLMs accelerate and enrich these processes by offering advanced capabilities in summarizing complex arguments, synthesizing large bodies of knowledge, and assisting in theoretical integration.

LLMs can quickly distill the core arguments, methodologies, and findings of academic articles, enabling researchers to navigate large volumes of literature efficiently. Their ability to synthesize themes across multiple studies helps identify patterns, contradictions, and emerging debates within a field (Montes et al., 2025). For instance, when exploring digital transformation in education or community health practices, researchers can prompt the model to summarize theoretical positions, compare methodological approaches, or highlight trends that shape current discourse. The models also support the construction of conceptual frameworks by mapping relationships among theories, concepts, and empirical findings. This is particularly beneficial in interdisciplinary research, where conceptual boundaries may be fluid or contested. By generating theoretical explanations or linking concepts from different bodies of literature, LLMs provide intellectual scaffolding that helps researchers refine their research questions and analytical frameworks (Brailas, 2025).

2.4 Data Organization and Management

Qualitative research often generates large amounts of rich, unstructured data requiring systematic management to ensure analytic coherence. LLMs contribute substantially to organizing and structuring this data by offering automated coding, metadata generation, data categorization, and integrative knowledge management.

LLMs can identify thematic patterns within textual data and propose preliminary codes that can serve as starting points for manual refinement. These codes go beyond surface features, capturing nuanced sentiments, relational dynamics, and contextual meanings (Bennis & Mouwafaq, 2025). This capability supports grounded theory approaches where initial coding forms the basis for developing categories and theoretical constructs. Additionally, LLM-generated metadata enables researchers to navigate large datasets more effectively by attaching summaries, thematic descriptors, or contextual notes to individual documents or segments (Hitch, 2024).

Another advantage of LLMs is their capacity to build dynamic knowledge repositories. These systems integrate coded data, analytic memos, theoretical reflections, and evolving interpretations into a coherent structure. Such integration enhances the researcher's ability to track the development of analysis, link concepts across different data sources, and support triangulation. In team-based research, LLMs also help coordinate collaboration by summarizing discussions, tracking coding decisions, and maintaining analytic consistency among researchers.

2.5 Analytic Support: Coding, Interpretation, and Theory Building

Beyond preliminary organization, LLMs increasingly play a central role in deeper stages of qualitative analysis, including coding refinement, interpretive reasoning, and theory construction. Unlike simple automation tools, LLMs can actively participate in analytic sense-making by offering interpretive suggestions, questioning assumptions, and generating conceptual insights aligned with established methodological principles.

During the coding process, researchers can iteratively interact with LLMs to refine emerging categories. The model can highlight overlapping codes, propose hierarchical relationships, or identify segments of text that challenge the existing coding structure. This dialogic process mirrors collaborative peer debriefing, a technique commonly used to strengthen analytic rigor. In approaches like thematic analysis, researchers can ask the model to articulate the central organizing concept of a theme or evaluate whether evidence across data points coherently supports that theme. The model's capacity to examine large volumes of text enables it to reveal subtle patterns that might be overlooked in manual analysis.

Figure 1. Analytic Support: Coding, Interpretation, and Theory Building

3. LLMS IN QUALITATIVE DATA ANALYSIS: FROM CODING TO INTERPRETATION

The integration of Large Language Models into qualitative data analysis represents one of the most significant and contentious developments in contemporary research methodology. Data analysis constitutes the intellectual core of qualitative inquiry the process through which raw data transforms into meaningful insights, theoretical understanding, and scholarly contribution. Traditionally, this process has been intensely human-centered, requiring researchers to engage deeply with their data through iterative cycles of reading, reflection, coding, and interpretation. The introduction of LLMs into this sacred analytical space raises fundamental questions about the nature of interpretation, the role of the researcher, and the standards by which we evaluate analytical rigor (Roberts et al., 2024).

3.1 Traditional Coding Processes: Manual and Computer-Assisted Approaches

Before examining how LLMs are transforming qualitative data analysis, it is essential to understand the traditional approaches they are supplementing or potentially replacing. Coding the process of identifying, labeling, and categorizing patterns within qualitative data has long been recognized as both an art and a science. Manual coding, the foundational approach taught in qualitative methods courses, involves researchers reading through transcripts, field notes, or documents line by line, identifying meaningful segments, and assigning descriptive or interpretive labels. This process is inherently iterative; initial codes are refined, combined, or discarded as the researcher develops deeper understanding of the data.

The manual coding process serves multiple functions beyond simple data organization. It forces researchers into intimate familiarity with their data, enabling them to notice nuances, contradictions, and subtle patterns that might escape casual attention. The act of deciding what constitutes a codable moment and how to label it requires ongoing interpretive judgment shaped by theoretical sensitivity, research questions, and emerging insights. Through this immersive engagement, researchers develop what Charmaz describes as theoretical sensitivity the ability to perceive variables and relationships relevant to emerging theory.

The development of Computer-Assisted Qualitative Data Analysis Software (CAQDAS) such as NVivo, ATLAS.ti, and MAXQDA represented an earlier technological intervention in qualitative analysis. These tools did not replace human judgment but rather provided sophisticated organizational and retrieval capabilities. Researchers could code electronically, organize codes hierarchically, link related data segments, write analytical memos, and visualize relationships between codes. CAQDAS enhanced efficiency and facilitated handling of larger datasets while leaving interpretive decisions firmly in human hands (O'Kane et al., 2022).

3.2 LLM-Enhanced Coding: Automated Pattern Recognition

Large Language Models enter this analytical space with impressive capabilities for pattern recognition, text categorization, and semantic analysis. These systems can process vast quantities of textual data at speeds impossible for human researchers, identifying recurring phrases, concepts, and linguistic patterns with computational precision. When provided with qualitative datasets interview transcripts, open-ended survey responses, social media posts, or document collections LLMs can generate preliminary codes, suggest thematic categories, and identify relationships between different data segments based on semantic similarity and contextual patterns learned during their training.

The technical mechanisms underlying LLM coding capabilities warrant explanation. These models process language by converting words and phrases into high-dimensional numerical representations called embeddings, which capture semantic relationships. Words or concepts with similar meanings or that appear in similar contexts cluster together in this mathematical space. When analyzing qualitative data, LLMs leverage these semantic embeddings to identify content sharing similar meanings, even when expressed through different vocabulary. This enables them to recognize that “feeling overwhelmed,” “experiencing stress,” and “unable to cope” might represent instances of a broader theme related to psychological distress a connection that would require human interpretive work in traditional coding.

Moreover, modern LLMs can be fine-tuned or prompted to perform specific analytical tasks aligned with research objectives. Researchers can provide the model with coding instructions, examples of desired codes, or theoretical frameworks that should guide analysis. The LLM can then apply these guidelines across the entire dataset with perfect consistency a significant advantage over human coders whose attention and judgment may vary due to fatigue, mood, or evolving understanding. Some implementations allow for iterative refinement, where researchers review LLM-generated codes, provide feedback, and have the model re-analyze data with adjusted parameters, creating a form of human-AI dialogue.

The efficiency gains are substantial and undeniable. Tasks requiring weeks of human labor can be completed in hours or minutes. This acceleration potentially democratizes qualitative research by making larger-scale studies feasible for researchers lacking extensive team support or funding. It may also enable mixed-methods studies incorporating more substantial qualitative components than previously practical. For exploratory analysis or preliminary data examination, LLM-generated codes can provide useful starting points for deeper human investigation.

3.3 Thematic Analysis with AI Support

Thematic analysis the process of identifying, analyzing, and reporting patterns of meaning across a dataset represents a higher level of analytical abstraction than coding. While coding involves labeling specific data segments, thematic analysis requires researchers to identify broader patterns, construct meaningful categories, and develop conceptual frameworks that illuminate the phenomenon under investigation. This process typically involves organizing codes into potential themes, reviewing and refining these themes against the dataset, defining and naming final themes, and producing a coherent narrative that addresses research questions.

LLMs are increasingly being employed to support thematic analysis through various mechanisms. Topic modeling algorithms, now enhanced by LLM capabilities, can identify latent themes within large text collections by detecting co-occurrence patterns of semantically related terms. These approaches generate probabilistic distributions showing which topics appear most prominently across documents or data sources. More advanced implementations use LLMs' generative capabilities to produce human-readable theme descriptions, identify exemplary quotations representing each theme, and suggest relationships between themes.

Some researchers are experimenting with using LLMs as analytical collaborators in thematic development. They engage in iterative dialogue with the model, proposing potential themes and requesting the LLM to identify supporting or contradicting evidence within the dataset. The LLM can rapidly retrieve relevant segments, highlight patterns the researcher might have overlooked, and even suggest alternative thematic interpretations. This approach treats the LLM not as an autonomous analytical engine but as a sophisticated search and pattern-recognition tool that augments human analytical capabilities.

The potential benefits include more systematic coverage of large datasets, reduced risk of overlooking minority viewpoints or subtle patterns, and ability to identify connections across disparate data segments that human working memory limitations might miss. For multidisciplinary research teams, LLM-supported analysis might facilitate shared understanding by providing consistent analytical starting points that team members can collectively refine.

Yet thematic analysis with AI support raises profound methodological concerns. Themes are not simply “found” in data but are actively constructed by researchers through interpretive work informed by theoretical frameworks, research purposes, and epistemological commitments. This constructive process requires judgment about what constitutes a theme, which patterns are meaningful versus trivial, and how themes should be bounded and named. LLMs, lacking genuine understanding or theoretical positioning, cannot make these judgments authentically. Their theme suggestions reflect statistical prominence in training data rather than theoretical significance or contextual relevance to the specific research project.

3.4 Comparative Analysis: Human vs. LLM-Generated Themes

Emerging empirical research comparing human-generated and LLM-generated qualitative analyses provides valuable insights into the practical implications of AI integration. Several studies have provided identical datasets to both human researchers and LLMs, then compared the resulting codes, themes, and interpretations. These comparisons reveal important patterns about the strengths and limitations of each approach.

LLMs consistently demonstrate superior performance in identifying explicit, frequently-occurring patterns and maintaining coding consistency across large datasets. They excel at descriptive tasks requiring pattern recognition without deep interpretation for example, identifying that many participants mentioned financial stress, work-family balance challenges, or health concerns. The codes and themes they generate often show high coverage of the data, meaning few segments go unanalyzed. For certain research purposes, particularly exploratory descriptive studies or preliminary analysis of very large datasets, LLM performance may be adequate or even advantageous.

However, comparative studies also reveal systematic weaknesses in LLM analysis. Human researchers consistently outperform LLMs in several critical dimensions. They better recognize context-dependent meanings, where the significance of a statement depends on surrounding discourse, participant characteristics, or situational factors. They identify latent themes underlying patterns not explicitly stated but implied through what participants emphasize, avoid, or express indirectly. They detect contradictions and ambivalences within individual narratives, recognizing that participants may hold complex or conflicting views. They notice absences what is not said but might be expected given the research context. Perhaps most significantly, human-generated themes demonstrate greater theoretical sophistication and analytical depth. While LLMs produce themes that describe what the data contains, human researchers generate themes that interpret what the data means, connecting findings to broader theoretical conversations, identifying implications, and advancing conceptual understanding. Human analysis reveals the "so what" of research findings in ways that computational pattern recognition cannot achieve.

Interestingly, some comparative studies find that LLM analysis performs better with certain types of data than others. Data characterized by explicit, straightforward expression such as responses to structured interview questions about factual experiences may be more amenable to LLM analysis than data involving narrative complexity, emotional expression, or culturally-embedded meanings. This suggests that the appropriateness of LLM integration may vary by research design, population, and analytical goals rather than being universally applicable or inappropriate.

3.5 Hybrid Approaches: Human-AI Collaborative Analysis

Recognition of both the capabilities and limitations of LLMs has prompted methodological innovation in the form of hybrid approaches that strategically combine human and AI contributions. These collaborative frameworks aim to leverage computational efficiency for tasks where LLMs perform adequately while preserving human judgment for aspects of analysis requiring genuine interpretation, contextual understanding, and theoretical insight.

Several hybrid models are emerging in practice. One approach uses LLMs for preliminary coding of large datasets, generating an initial coding scheme that human researchers then review, refine, and deepen through close reading of key segments. This preserves human interpretive authority while reducing the sheer volume of initial coding labor. Another model employs LLMs for data organization and retrieval having the AI identify and cluster potentially related segments that researchers then examine to construct themes through traditional analytical processes. A third approach involves parallel analysis, where both human researchers and LLMs independently analyze the same data, then results are compared to identify points of convergence and divergence that warrant deeper investigation. Indeed, empirical work has begun to demonstrate the viability of such hybrid frameworks: a recent "LLM-in-the-loop" approach showed that LLM-assisted thematic analysis produced codebooks comparable in quality to human coders, while significantly reducing time and labor demands (Dai et al., 2023) .

More sophisticated hybrid approaches integrate LLM capabilities throughout an iterative analytical process. Researchers might begin with traditional immersive reading to develop sensitizing concepts, then use LLMs to identify all data segments potentially relevant to these concepts across a large dataset followed by human review that leads to conceptual refinement; another cycle of LLM-assisted retrieval; and ultimately human synthesis and interpretation. This iterative human–AI dialogue can be particularly powerful for large-scale studies where complete manual analysis would be prohibitively resource-intensive. This kind of approach has been demonstrated in recent studies comparing conventional thematic analysis with AI-driven coding: AI-augmented methods were able to match many human-identified themes and codes while boosting efficiency, although authors emphasize the need for careful human oversight to preserve interpretive depth and contextual nuance (Bennis & Mouwafaq, 2025).

The development of effective hybrid approaches requires clear protocols specifying when and how LLMs will be employed, what analytical decisions remain under human control, and how LLM contributions will be documented and evaluated. Transparency becomes paramount researchers must be explicit about which analytical steps involved AI assistance, what prompts or instructions were provided to the LLM, and how AI-generated outputs were validated or modified. This documentation enables readers to evaluate the trustworthiness of findings and other researchers to assess or replicate the analytical approach.

Hybrid approaches also demand new forms of researcher competency. Qualitative researchers must develop what might be termed "AI literacy" understanding how LLMs function, their capabilities and limitations, and how to craft effective prompts that yield useful analytical outputs. They must cultivate critical judgement about when to trust, modify, or reject LLM suggestions. They must learn to maintain their interpretive authority and theoretical grounding while engaging with a technological collaborator that may suggest patterns inconsistent with their developing understanding. As argued in recent literature, integrating AI into qualitative work does not simply "speed up" old processes: it requires researchers to gain new skills in prompt design and human-AI interaction, and to remain actively reflexive about the AI's outputs rather than treating them as neutral or final (Olla et al., 2024).

Ultimately, the most promising hybrid approaches maintain human researchers as the primary interpreters and analytical decision-makers while strategically deploying LLM capabilities for specific tasks aligned with AI strengths. They treat LLMs as sophisticated tools that can accelerate certain analytical processes (like preliminary coding, clustering, or data organization) but cannot replace the theoretical sensitivity, contextual understanding, and interpretive depth that constitute the essence of rigorous qualitative analysis. Recent empirical work comparing AI-assisted and traditional thematic analysis underscores this point: although AI-augmented methods can match many of the codes and themes identified by human analysts, authors consistently highlight the need for careful human oversight to preserve nuance, ensure context-appropriate interpretation, and avoid mis- or over-generalization (Bennis & Mouwafaq, 2025).

4. THE BLACK-BOX PROBLEM AND METHODOLOGICAL TRANSPARENCY

One of the most profound challenges posed by the integration of Large Language Models into qualitative research concerns the opacity of their internal operations commonly referred to as the "black-box problem." This challenge strikes at the heart of qualitative methodology's commitment to transparency, auditability, and reflexivity. In traditional qualitative research, methodological rigor depends substantially on researchers' ability to articulate and justify their analytical decisions, trace the pathways from raw data to conclusions, and enable others to evaluate the trustworthiness of their interpretive work. When critical analytical steps occur within computational systems whose decision-making processes remain fundamentally obscure, these foundational principles of qualitative inquiry face serious challenges that demand careful examination and methodological innovation.

4.1 Understanding LLM Decision-Making Processes

The black-box problem in Large Language Models (LLMs) stems from their complex neural network architecture, where decisions emerge from millions of learned parameters rather than explicit, interpretable rules. These models process language through layered mathematical transformations, generating outputs based on statistical patterns learned during training. As a result, even developers cannot fully explain why specific inputs produce particular outputs. In qualitative research, this opacity is problematic because coding and thematic analysis may be influenced by hidden biases or irrelevant correlations in training data. Researchers cannot trace or verify these internal processes, limiting their ability to critically evaluate, justify, or validate AI-generated interpretations.

4.2 Explicability and Auditability in AI-Enhanced Research

The black-box nature of LLMs poses significant challenges to explicability and auditability in qualitative research. Unlike traditional methods, where researchers can clearly justify analytical decisions and maintain transparent audit trails, AI-generated outputs lack accessible reasoning processes. Researchers may report themes identified by LLMs but cannot fully explain how or why these results were produced. This limits the ability of others to critically evaluate findings or replicate the analysis. Furthermore, factors such as model updates, proprietary systems, and variability in outputs reduce reproducibility. As a result, ensuring transparency and trustworthiness in AI-assisted qualitative research becomes increasingly difficult.

4.3 Maintaining Methodological Rigor and Transparency

Despite these significant and fundamental challenges, researchers employing LLMs in qualitative analysis are developing innovative strategies to maintain methodological rigor and transparency within the constraints imposed by AI integration. These approaches acknowledge the black-box problem as a real limitation while implementing compensatory practices that preserve as much explicability and auditability as possible under current technological circumstances.

One crucial strategy involves carefully limiting LLM involvement to specific, well-defined analytical tasks while maintaining human control over interpretive decisions that most require theoretical sensitivity, contextual understanding, and nuanced judgment. For example, researchers might use LLMs for initial data organization, automated transcription, or preliminary coding of straightforward content, but reserve theme development, interpretation of complex meanings, theoretical integration, and final analytical synthesis for human intellectual work. This bounded deployment restricts the scope of opaque decision-making to lower-level, more mechanical tasks where lack of full explicability poses less fundamental threat to overall study trustworthiness and credibility.

Another essential approach emphasizes rigorous validation of LLM outputs through multiple independent mechanisms. Researchers can employ systematic human review of substantial representative samples of LLM-generated codes to assess their appropriateness, accuracy, and consistency with research questions and theoretical frameworks. Recent empirical studies demonstrate that hybrid validation frameworks combining model confidence and diversity metrics with targeted human verification can significantly reduce manual workload while maintaining acceptable quality standards (Zhao & Liu, 2025).

Transparency demands extensive and detailed documentation that goes well beyond traditional qualitative methods reporting standards. As some methodological critics warn, the "black box" nature of generative-AI threatens interpretive accountability and risks obscuring whether findings represent participants' voices or artifacts of training data (Hasan, 2025). Therefore, hybrid models should treat LLMs as tools for data processing and organization, not as substitute interpreters. The human researcher must maintain interpretive authority, correct AI-generated errors, and embed contextual understanding and reflexivity into final analysis (Whitty, 2025).

4.4 Documentation Requirements for LLM-Assisted Analysis

Establishing standardized documentation for LLM-assisted qualitative research is essential to ensure transparency and methodological rigor. Researchers must clearly specify the technologies used, including model name, version, access method, and usage timeline, as these factors influence analytical outcomes. Equally important is defining the scope of AI involvement, detailing which tasks were AI-assisted and which remained under human control. Documentation should also include representative prompts, as prompt design directly shapes results. Finally, researchers must describe validation procedures, explaining how outputs were evaluated, revised, or rejected. Such comprehensive reporting enables readers to assess reliability, interpretive authority, and the overall credibility of AI-supported qualitative analysis.

Figure 2. Docuentations Requiements for LLM-Assisted Analysis

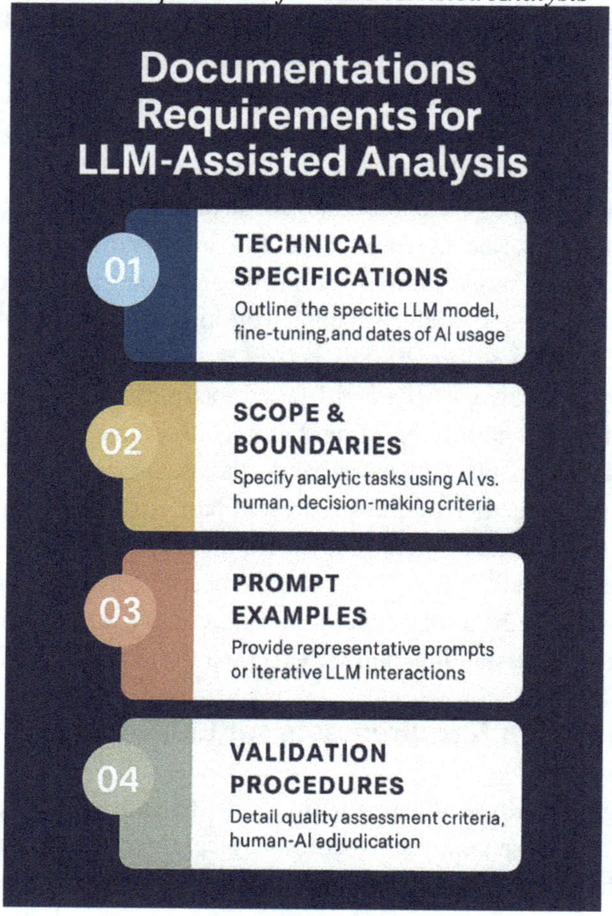

5. TRUSTWORTHINESS AND CREDIBILITY IN LLM-ENHANCED QUALITATIVE RESEARCH

Trustworthiness has long served as the foundation for assessing rigor in qualitative research. Lincoln and Guba's criteria credibility, transferability, dependability, and confirmability provide a framework for ensuring that interpretive findings are both meaningful and trustworthy without relying on positivist notions of validity or reliability. However, the incorporation of Large Language Models (LLMs) into qualitative research demands renewed examination of these criteria. As analysis becomes increasingly shaped by human-AI collaboration, researchers must reconsider how trustworthiness is conceptualized, operationalized, and evaluated.

This re-examination is not merely technical but epistemological, raising questions about interpretive authority, the nature of meaning-making, and the legitimacy of AI-assisted insights.

5.1 Reconsidering Lincoln and Guba's Trustworthiness Criteria

Lincoln and Guba's framework assumes that human researchers acting as the central interpretive instruments possess contextual understanding, theoretical sensitivity, and reflexive awareness. LLMs, by contrast, process language without lived experience, relational engagement, or genuine contextual understanding. As a result, the criteria remain conceptually valid but require significant methodological adaptation.

Credibility, traditionally established through prolonged engagement, triangulation, peer debriefing, negative case analysis, and member checking, becomes more complex when AI participates in interpretation. LLMs can simulate pattern recognition but cannot authentically grasp participants' perspectives. Transferability depends on thick description grounded in contextual immersion something LLMs cannot independently achieve because they lack cultural awareness and experiential understanding. Dependability, normally demonstrated through transparent methodological logs and audit trails, is challenged by the opaque, black-box nature of LLM decision-making. Confirmability becomes doubly complex, as researchers must now consider both their own biases and those embedded within the training data of AI systems.

5.2 Credibility: Member Checking and Triangulation in AI Contexts

Credibility is the most immediately affected quality criterion when LLMs enter qualitative analysis. Because LLM outputs may introduce subtle distortions or overlook contextual nuance, member checking becomes even more critical. Sharing interpretations with participants allows them to verify whether AI-assisted analyses resonate with their experiences. Ethical transparency also requires researchers to disclose when and how AI contributed to analysis, enabling participants to assess whether they are comfortable with its involvement. When appropriate, researchers may share both AI-generated and human-refined interpretations, inviting feedback on accuracy and resonance.

Triangulation must also be reconceptualized. Comparing AI-assisted coding with human coding can surface areas of agreement and divergence, allowing researchers to investigate discrepancies as opportunities for deeper insight. Similarly, investigator triangulation can now include an LLM as an "additional analyst," though its role must be critically assessed rather than accepted at face value. Theoretical triangulation, however, remains predominantly a human responsibility, as LLMs can imitate theoretical language but cannot authentically engage in theory-driven interpretation.

5.3 Dependability and Confirmability with LLM Integration

LLM integration complicates **dependability** because the internal workings of these models are not fully interpretable, making traditional auditability difficult. To preserve dependability, researchers must generate detailed process documentation, including logs of prompts, outputs, iterations, and decision rationales. Recording when and why AI suggestions were accepted or overridden provides transparency even when the model's internal reasoning remains unknown. Consistency checking rerunning similar prompts or analyzing the same dataset across different LLM sessions can further support dependability. Discrepant outputs should trigger closer human scrutiny or alternative methods.

Confirmability, the extent to which findings reflect participant realities rather than researcher bias, must now also encompass AI-related biases. Researchers should investigate what linguistic, cultural, or demographic patterns may be over- or underrepresented in an LLM's training data. This "algorithmic reflexivity" requires examining how model biases interact with human decisions and how they might distort findings. Confirmability audits may include explicit mapping of each interpretive claim to supporting data, as well as documentation of how AI outputs were incorporated or modified. This approach ensures that interpretations remain grounded in evidence rather than algorithmic tendencies.

5.4 Transferability: Contextual Understanding and Generalization

Transferability depends on rich contextual description enabling readers to judge whether findings apply to other settings. LLMs can help researchers extract contextual information from transcripts or field notes, ensuring comprehensive coverage of setting characteristics. However, identifying which contextual elements matter most for transferability requires human interpretive expertise.

The limitations of LLMs become particularly visible when working in culturally specific or marginalized contexts underrepresented in training corpora. AI may miss cultural nuance or misinterpret meaning, generating decontextualized analyses. Researchers must therefore validate LLM-assisted interpretations through careful human review.

Because LLMs are trained on broad datasets, they may inadvertently generalize beyond the boundaries of the study context. Humans must ensure that claims are appropriately bounded and avoid overstating generalizability simply because AI detected patterns. While LLMs may support cross-context comparative analysis, researchers must determine whether observed similarities reflect genuine cross-context patterns or algorithmic tendencies toward homogenization.

5.5 Reflexivity and Researcher Positionality in the Age of AI

Reflexivity remains central to qualitative rigor, but in LLM-enhanced research, it must extend beyond human positionality to include the positionality of the AI tools themselves. Researchers should reflect on how their backgrounds interact with AI outputs and how their assumptions shape the way they use LLMs. At the same time, they must examine how AI's training data, embedded biases, and cultural assumptions affect interpretation.

6. ETHICAL CHALLENGES AND CONSIDERATIONS

The integration of Large Language Models (LLMs) into qualitative research brings substantial benefits, but it also introduces a broad spectrum of ethical challenges that must be addressed with caution and critical awareness. As AI-driven tools become more deeply embedded in research workflows from data collection and transcription to analysis and interpretation researchers bear increasing responsibility to ensure that their methods remain ethically sound. Ethical considerations are not merely procedural requirements but foundational principles that safeguard participants, preserve academic integrity, and maintain public trust in scientific inquiry. This section outlines the key ethical areas that researchers must navigate when employing LLMs in qualitative studies.

6.1 Informed Consent in LLM-Enhanced Research

Informed consent becomes more complex in research environments that involve LLMs, particularly because participants may not fully understand how their data will be processed, transformed, or stored by AI systems. Traditional consent frameworks often assume a direct human–researcher relationship, yet the involvement of LLMs introduces additional layers of data handling that must be transparently communicated. Participants should be informed that their data may be processed by AI models, potentially stored on third-party servers, and used in ways that extend beyond conventional manual analysis. Researchers must articulate how AI tools function, the extent of automation, and the potential risks tied to algorithmic processing. Ensuring informed consent in this context requires clear, accessible explanations and opportunities for participants to ask questions or opt out of specific types of AI-driven analysis. Without such measures, consent runs the risk of being incomplete or ethically insufficient.

6.2 Data Privacy and Confidentiality Concerns

The use of LLMs in qualitative research raises significant concerns regarding data privacy and confidentiality, especially because many AI models operate through cloud-based infrastructures. When sensitive interview transcripts or personal narratives are processed by LLMs, there is a potential risk of unauthorized access, data breaches, or unintended retention of private information within model training logs (Kempny et al., 2025). Researchers must carefully evaluate the privacy policies of AI platforms, ensure compliance with data protection regulations, and implement robust anonymization procedures before uploading any qualitative data. Even anonymized data may inadvertently reveal identities through contextual clues, necessitating heightened vigilance. Furthermore, some LLMs may store prompts or user inputs to improve future model performance, which introduces additional ethical questions about data ownership and long-term exposure. Responsible practice requires minimizing identifiable information, using secure processing environments, and being transparent with participants about how their data is protected.

6.3 Algorithmic Bias and Its Impact on Interpretation

LLMs inherit biases from the datasets on which they are trained, and these biases can subtly influence the interpretation of qualitative data. When LLMs assist with coding, summarization, or thematic analysis, they may reinforce dominant cultural narratives, suppress minority perspectives, or misinterpret nuanced expressions rooted in specific social or linguistic contexts. Biases embedded in the model can lead to skewed interpretations that compromise the validity of research findings. For example, LLMs might inadvertently perpetuate stereotypes when analyzing sensitive topics such as race, gender, or socioeconomic status. Because qualitative research often involves marginalized communities, algorithmic bias can have particularly harmful effects. Researchers must approach AI-generated insights critically, triangulating findings with human judgment and reflecting on how the model's limitations may influence interpretive outcomes. Ethical practice requires ongoing bias monitoring, careful validation, and deliberate efforts to mitigate inequitable impacts in the analytic process.

6.4 Participant Rights and Data Sovereignty

Participant rights extend beyond consent and privacy to include broader concerns about data sovereignty, especially for communities that have historically experienced exploitation or misrepresentation. When LLMs process qualitative data, questions arise regarding who ultimately controls that data, how it can be used, and whether participants retain any agency over its future applications (Rashid et al., 2025). This issue is particularly salient for Indigenous communities, minority populations, and participants from regions with strong cultural expectations surrounding data ownership. AI-driven tools may inadvertently remove data from its cultural or political context, transferring it to corporate servers or global infrastructures that are not accountable to the communities represented in the research. Upholding participant rights requires adherence to culturally sensitive data governance practices, clear agreements about data storage and use, and respect for community-specific protocols related to data sovereignty. Ethical researchers must ensure that AI tools support, rather than undermine, the autonomy and dignity of participants (Kempny et al., 2025).

6.5 Academic Integrity: AI Assistance vs. AI Misuse

The growing accessibility of LLMs introduces new concerns regarding academic integrity, especially as boundaries between legitimate AI assistance and unethical AI misuse become increasingly blurred. While LLMs can support tasks such as summarizing literature, generating conceptual frameworks, or refining writing, overreliance on these tools risks compromising the originality and authenticity of scholarly work. One major ethical challenge is the potential fabrication of citations, concepts, or findings by AI systems, which could mislead researchers and contaminate academic discourse. Additionally, using LLMs to write large portions of analyses or interpretations may weaken the researcher's intellectual engagement with the data. Ethical scholarship requires transparency in disclosing the extent of AI use and ensuring that AI-generated outputs are rigorously verified and supplemented with critical human reasoning. AI should serve as a supportive tool rather than a replacement for scholarly judgment, reflective interpretation, or methodological rigor.

7. LIMITATIONS AND CRITICAL CONSIDERATIONS

While Large Language Models offer intriguing capabilities for qualitative research, a comprehensive and intellectually honest assessment must acknowledge their substantial limitations and the critical considerations that should guide decisions about their appropriate use. The enthusiasm surrounding AI technologies often overshadows serious concerns about what may be lost when computational approaches supplement or replace human interpretive work. This chapter section examines key limitations of LLM integration in qualitative research, addressing how these AI systems may compromise the contextual richness, cultural sensitivity, interpretive depth, and nuanced understanding that define excellent qualitative inquiry. By critically examining when and why LLMs may be inappropriate or inadequate for qualitative research purposes, we can develop more discerning approaches to their integration that preserve the distinctive values and epistemological commitments of interpretive methodology.

7.1 Loss of Contextual and Cultural Nuance

One of the most critical limitations of integrating Large Language Models (LLMs) into qualitative research lies in their inability to genuinely understand context and culture. Qualitative inquiry depends heavily on the principle of context-dependency, where meaning is shaped by historical, social, and cultural conditions (Asmawi & Alam, 2024). Human researchers interpret data by drawing on deep contextual awareness, including participants' backgrounds, power relations, and situational nuances. In contrast, LLMs rely on statistical pattern recognition derived from large datasets, identifying correlations rather than meaning (Minaee et al., 2025). This creates a fundamental gap between computational processing and human interpretation, as AI lacks the lived experience and cultural sensitivity required to fully grasp complex social phenomena (Erciyes, 2020).

This limitation becomes even more pronounced in culturally diverse or underrepresented contexts. Since LLM training data is largely dominated by Western and English-language sources, the models often fail to capture local meanings, cultural expressions, and implicit communication patterns. Subtle elements such as irony, indirect language, or culturally specific narratives may be misinterpreted or overlooked entirely. Furthermore, LLMs cannot recognize what is left unsaid silence, avoidance, or narrative structure which often carries significant meaning in qualitative data. As a result, reliance on LLMs risks producing interpretations that are superficial, culturally biased, and disconnected from the lived realities of participants.

7.2 Over-Reliance on AI: Risks to Interpretive Depth

The growing reliance on Large Language Models (LLMs) in qualitative research raises important concerns about diminishing researcher engagement with data and the potential loss of interpretive depth. Traditional qualitative inquiry depends on sustained immersion reading, reflecting, and revisiting data to uncover nuanced meanings. This process fosters deep familiarity and critical insight. However, the speed and efficiency of LLMs may encourage researchers to bypass this essential engagement, relying instead on rapid, AI-generated coding and summaries without fully understanding the data.

Such over-reliance risks producing research that is efficient but intellectually shallow, focusing on surface-level patterns rather than deeper meanings. It may also shift the analytical approach from discovery to mere validation of AI outputs, limiting creativity and critical thinking. This is especially concerning for novice researchers, who may miss opportunities to develop essential interpretive skills. Over time, excessive dependence on LLMs could weaken the collective methodological expertise of the research community, reducing the richness and rigor that define high-quality qualitative research.

7.3 Standardization vs. Contextual Sensitivity

The integration of LLM into qualitative research presents a tension between the tendency toward standardization and the need for contextual sensitivity. LLM operates with consistent and uniform patterns, which, on the one hand, enhances reproducibility, but on the other hand, potentially reduces the methodological flexibility that is a key characteristic of qualitative research. Qualitative approaches traditionally emphasize that methods should be tailored to the context, participants, and objectives of the study, rather than forced into a uniform analytical framework. As researchers begin to adapt research designs to accommodate AI capabilities, there is a risk of narrowing methodological variation, where only certain types of data and approaches are prioritized (Lec & Valenc, 2024).

Furthermore, the dominance of methods such as thematic analysis may become stronger due to their easier integration with LLM, while more interpretive approaches such as phenomenology or discourse analysis are potentially marginalized (Montes et al., 2025). LLM also tends to employ statistically based categorization patterns that can overlook the unique contextual nuances of each study (Minaee et al., 2025). As a result, the meaning that should emerge organically from the data can be displaced by an overly generic analysis structure. Therefore, researchers must maintain contextual sensitivity and ensure that the use of AI does not sacrifice methodological diversity and interpretive depth in qualitative research.

7.4 The Role of Tacit Knowledge and Human Intuition

Qualitative The excellence of qualitative research relies heavily on tacit knowledge and human intuition, a form of implicit knowledge that develops through experience and in-depth reflection. Experienced researchers are able to recognize hidden meanings, understand indirect cues, and determine when data requires further exploration. This ability cannot be reduced to explicit rules or algorithms because it involves the integration of experience, cultural sensitivity, and complex theoretical understanding. LLM, which relies on statistical patterns, lacks access to these dimensions of knowledge and therefore cannot fully replace the role of humans in in-depth interpretation (Minaee et al., 2025).

Furthermore, human intuition plays a crucial role in the creative aspects of qualitative research, such as generating new insights, connecting concepts innovatively, and developing theories that go beyond existing data. While LLM can assist in identifying patterns, the resulting output remains a combination of existing data, not the result of authentic creative thinking (Chakravorti et al., 2024). Human researchers possess the ability to capture subtle contextual meanings and generate rich interpretations based on empirical experience and critical reflection (Asmawi & Alam, 2024; Erciyes, 2020). irreplaceable in maintaining the quality and depth of qualitative research.

7.5 When Not to Use LLMs in Qualitative Research

Given these substantial limitations, researchers must exercise careful judgment in determining when LLM integration is inappropriate in qualitative research. Studies involving vulnerable populations or sensitive topics demand empathy, ethical awareness, and interpretive care that only human researchers can provide, as AI lacks the capacity to truly understand or respectfully represent deeply personal experiences. Similarly, research conducted within cultural contexts that are underrepresented in LLM training data carries a high risk of misinterpretation, bias, and loss of meaning, making AI use inadvisable. Methodologies such as phenomenology, narrative inquiry, and discourse analysis also require deep theoretical sensitivity and nuanced interpretation that current AI systems cannot replicate. In addition, small-scale studies gain minimal benefit from AI efficiency while risking reduced analytical depth. Finally, researchers without sufficient AI literacy should avoid LLM use, as uncritical reliance on generated outputs may compromise research validity and integrity.

8. BEST PRACTICES AND FUTURE DIRECTIONS FOR RESPONSIBLE LLM INTEGRATION IN QUALITATIVE RESEARCH

8.1 Ethical Foundations and Responsible Use of LLMs

The integration of Large Language Models (LLMs) into qualitative research must be rooted in a robust ethical framework that balances the potential for innovation with scientific responsibility. While LLMs can improve the efficiency of data analysis, accelerate the coding process, and facilitate thematic synthesis, their use also presents risks such as automation bias, misinterpretation, and reduced researcher reflexivity. Therefore, researchers must maintain epistemic responsibility by ensuring that AI output does not replace, but rather complements, human judgment. Transparency is a key principle, requiring researchers to explicitly explain the role of AI in every stage of the research, from transcription to analysis. Furthermore, data security must be strictly adhered to, particularly when using third-party platforms that potentially store sensitive data. Researchers must also be aware that LLMs can reinforce dominant narratives and disregard minority perspectives, necessitating a reflective approach in evaluating AI-generated results. Therefore, responsible use of LLMs focuses not only on regulatory compliance but also on a commitment to maintaining data integrity and participant dignity (Hasan, 2025).

8.2 Hybrid Methodological Frameworks for AI-Augmented Inquiry

A hybrid approach offers a strategic solution for integrating LLM into qualitative research without sacrificing interpretive depth. Within this framework, AI is positioned as a supporting tool, assisting with initial processes such as pattern identification, open coding, and conceptual summaries, while the researcher retains primary control over the interpretation and validation of findings. This approach allows for methodological flexibility by combining traditional techniques such as memoing, reflexive journaling, and manual analysis with the computational efficiency of AI. For example, LLM can be used to quickly group interview excerpts, but the researcher must still perform context-based reinterpretation. In grounded theory, AI can assist with axial coding, but the decision regarding theoretical saturation remains with the researcher. Thus, the hybrid framework not only increases efficiency but also maintains a balance between technological innovation and the epistemological principles of qualitative research (Hitch, 2024).

8.3 Ensuring Quality, Validity, and Analytical Rigor

The use of LLM in qualitative research requires the development of new strategies to maintain the validity and reliability of the analysis. One key approach is triangulation between AI results and human analysis, which allows for the identification of errors such as hallucinations or oversimplifications. Furthermore, the consistency of the output can be tested by running the same prompt on multiple iterations or different models. Documentation of the process is also crucial through an audit trail that records the use of the prompt, the AI results, and their influence on the research conclusions. Researcher reflexivity remains a key component, particularly in understanding how personal biases or assumptions influence interactions with AI. Furthermore, researchers must consider the cultural and linguistic appropriateness of the models used, especially in studies with diverse populations. All AI output should be critically verified before being used as the basis for analysis. With this approach, the integration of AI does not diminish the quality of the research but, when applied appropriately, strengthens methodological rigor.

8.4 Researcher Competency and Institutional Support Systems

The successful integration of LLMs depends heavily on researcher competency and adequate institutional support. Researchers need to develop digital literacy, including an understanding of how LLMs work, potential biases, and prompt engineering techniques. Furthermore, the ability to critically evaluate AI output is a crucial skill in the era of technology-driven research. Academic institutions also have a strategic role in providing guidelines, training, and policies related to the use of AI in research. Ethics committees need to update their protocols to include evaluations of AI use, particularly regarding data protection and transparency. Cross-disciplinary collaboration between researchers, AI experts, and ethicists is also key to building a resilient research ecosystem. With ongoing training and clear policies, researchers can optimally utilize LLMs without compromising ethical principles and academic quality (Chakravorti et al., 2024).

8.5 Future Directions: Innovation, Collaboration, and Methodological Evolution

Future developments in LLM demonstrate a deepening integration of qualitative research, marked by the emergence of domain-based models, multimodal systems, and increased algorithmic transparency. Models specifically trained in specific domains will produce more accurate and relevant interpretations, while multimodal capabilities enable richer data analysis, including audio, visual, and video. The relationship between humans and AI is also evolving into a dynamic collaboration through the concept of co-analysis, where researchers and AI work iteratively to develop insights. This collaboration enhances reflexivity and opens up new interpretive possibilities that were previously difficult to achieve. However, these developments also demand new methodological standards, including documentation of AI use, redefinition of validity, and adjustments to concepts such as data saturation. Future research agendas need to explore the impact of AI on knowledge construction, mitigating algorithmic bias, and protecting participant rights. Ultimately, the future of qualitative research lies in balancing technological innovation with humanistic values, ensuring that AI serves as a partner that enhances, rather than replaces, the role of human researchers (Brailas, 2025; Chakravorti et al., 2024; Hasan, 2025).

9. CONCLUSION

The integration of Large Language Models (LLMs) into qualitative research marks a transformative phase in the evolution of interpretive methodologies, where technological advancement intersects with long-standing epistemological and ethical commitments. While LLMs offer significant benefits in terms of efficiency, scalability, and analytical support, their adoption must be approached with careful consideration. Rather than replacing human interpretation, these tools should be viewed as complementary resources that enhance, but do not override, the researcher's critical judgment. A balanced approach one that neither rejects nor uncritically embraces AI ensures that innovation is aligned with the foundational principles of qualitative inquiry.

Despite their capabilities, LLMs present notable limitations, including a lack of contextual understanding, cultural sensitivity, and true interpretive depth. Their reliance on statistical patterns rather than lived experience raises concerns about bias, misrepresentation, and the potential marginalization of underrepresented voices. Additionally, the opacity of AI systems challenges the transparency and accountability required in rigorous qualitative research. Addressing these issues requires the adoption of hybrid methodological frameworks, where AI supports tasks such as data organization and preliminary coding, while human researchers retain authority over interpretation, validation, and theoretical development.

Looking ahead, the future of qualitative research depends on the responsible and ethical integration of LLMs, supported by researcher competency, institutional guidance, and evolving methodological standards. By fostering AI literacy, maintaining reflexivity, and ensuring transparency in AI-assisted processes, researchers can harness the strengths of LLMs without compromising research integrity. Ultimately, the goal is to preserve the human-centered essence of qualitative inquiry capturing meaning, context, and lived experience while leveraging technological tools to expand its scope and impact in an increasingly data-driven world.

REFERENCES

Asmawi, A., & Alam, S. (2024). *Qualitative research : Understanding its underlying philosophies. 2*(Minaee, S., Mikolov, T., Nikzad, N., Chenaghlu, M., Socher, R., Amatriain, X., Gao, J. (2025). Large Language Models : A Survey. Wosny, M., Hastings, J. (2024). Applying Large Language Models to Interpret Qualitative Interviews in Healthcare. *Studies*, •••, 1–14.

Bennis, I., & Mouwafaq, S. (2025). Advancing AI-driven thematic analysis in qualitative research: a comparative study of nine generative models on Cutaneous Leishmaniasis data. *BMC Medical Informatics and Decision Making, 25*(Minaee, S., Mikolov, T., Nikzad, N., Chenaghlu, M., Socher, R., Amatriain, X., Gao, J. (2025). Large Language Models : A Survey. Wosny, M., Hastings, J. (2024). Applying Large Language Models to Interpret Qualitative Interviews in Healthcare. *Studies, 124*(1), 124. Advance online publication. DOI: 10.1186/s12911-025-02961-5 PMID: 40065373

Brailas, A. (2025). *Artificial Intelligence in Qualitative Research : Beyond Outsourcing Data Analysis to the Machine. Minaee, S., Mikolov, T., Nikzad, N., Chenaghlu, M., Socher, R., Amatriain, X., Gao, J. (2025). Large Language Models : A Survey. Wosny, M., Hastings, J. (2024). Applying Large Language Models to Interpret Qualitative Interviews in Healthcare. Studies.*

Chakravorti, T., Wang, X., Venkit, P. N., Koneru, S., Munger, K., & Rajtmajer, S. (2024). *Social Scientists on the Role of AI in Research. Minaee, S., Mikolov, T., Nikzad, N., Chenaghlu, M., Socher, R., Amatriain, X., Gao, J. (2025). Large Language Models : A Survey. Wosny, M., Hastings, J. (2024).* Applying Large Language Models to Interpret Qualitative Interviews in Healthcare. Studies.

Dai, S., Xiong, A., & Ku, L. (2023).. . *Analysis*.

Erciyes, E. (2020). *Paradigms of Inquiry in the Qualitative Research. 16*(Minaee, S., Mikolov, T., Nikzad, N., Chenaghlu, M., Socher, R., Amatriain, X., Gao, J. (2025). Large Language Models : A Survey. Wosny, M., Hastings, J. (2024). Applying Large Language Models to Interpret Qualitative Interviews in Healthcare. *European Scientific Journal, 16*(7), 181–200. DOI: 10.19044/esj.2020.v16n7p181

Hasan, K. (2025). *Navigating the meta-crisis of generativity : adapting qualitative research quality criteria in the era of generative AI.*

Hitch, D. (2024). Artificial Intelligence Augmented Qualitative Analysis: The Way of the Future? *Qualitative Health Research, 34*(Minaee, S., Mikolov, T., Nikzad, N., Chenaghlu, M., Socher, R., Amatriain, X., Gao, J. (2025). Large Language Models : A Survey. Wosny, M., Hastings, J. (2024). Applying Large Language Models to Interpret Qualitative Interviews in Healthcare. *Studies, 34*(7), 595–606. DOI: 10.1177/10497323231217392 PMID: 38064244

Kempny, C., Yilmaz-Aslan, Y., & Brzoska, P. (2025). [Potentials and risks of using AI in qualitative research]. *Zeitschrift fur Evidenz, Fortbildung und Qualitat im Gesundheitswesen, Minaee, S., Mikolov, T., Nikzad, N., Chenaghlu, M., Socher, R., Amatriain, X., Gao, J. (2025). Large Language Models : A Survey. Wosny, M., Hastings, J. (2024). Applying Large Language Models to Interpret Qualitative Interviews in Healthcare. Studies.* Advance online publication. DOI: 10.1016/j.zefq.2025.10.004 PMID: 41241637

Lec, M. D. M., & Valenc, L. (2024). *Applications and Implications of Large Language Models in Qualitative Analysis : A New Frontier for Empirical Software Engineering. Minaee, S., Mikolov, T., Nikzad, N., Chenaghlu, M., Socher, R., Amatriain, X., Gao, J. (2025). Large Language Models : A Survey. Wosny, M., Hastings, J. (2024). Applying Large Language Models to Interpret Qualitative Interviews in Healthcare. Studies.*

Minaee, S., Mikolov, T., Nikzad, N., Chenaghlu, M., Socher, R., Amatriain, X., & Gao, J. (2025). *Large Language Models : A Survey. Minaee, S., Mikolov, T., Nikzad, N., Chenaghlu, M., Socher, R., Amatriain, X., Gao, J. (2025). Large Language Models : A Survey. Wosny, M., Hastings, J. (2024).* Applying Large Language Models to Interpret Qualitative Interviews in Healthcare. Studies.

Montes, C. M., Feldt, R., Martos, C. M., Ouhbi, S., & Premanandan, S. (2025). Large Language Models in Thematic Analysis : Prompt Engineering, Evaluation, and Guidelines for Qualitative Software Engineering Research. Minaee, S., Mikolov, T., Nikzad, N., Chenaghlu, M., Socher, R., Amatriain, X., Gao, J. (2025). Large Language Models : A Survey. Wosny, M., Hastings, J. (2024). Applying Large Language Models to Interpret Qualitative Interviews in Healthcare. *Studies*, •••, 1–17.

O'Kane, P., Ott, D. L., Smith, A. D., & Brown, T. C. (2022). Understanding Computer-Assisted Qualitative Data Analysis Software as a Tool to Enhance Systematic Literature Reviews in Human Resource Development. *Human Resource Development Review, 22*(Minaee, S., Mikolov, T., Nikzad, N., Chenaghlu, M., Socher, R., Amatriain, X., Gao, J. (2025). Large Language Models : A Survey. Wosny, M., Hastings, J. (2024). Applying Large Language Models to Interpret Qualitative Interviews in Healthcare. *Human Resource Development Review, 22*(2), 291–307. DOI: 10.1177/15344843221144668

Olla, P., Elliott, L., Abumeeiz, M., Mihelich, K., & Olson, J. (2024). Promptology: Enhancing Human–AI Interaction in Large Language Models. In *Information* (Vol. 15, Issue 10, p. 634). DOI: 10.3390/info15100634

Rashid, A., Jamil, N. A., & Ling, T. W. (2025). *Rethinking Artificial Intelligence (AI) in Qualitative Research. 8*(Minaee, S., Mikolov, T., Nikzad, N., Chenaghlu, M., Socher, R., Amatriain, X., Gao, J. (2025). Large Language Models : A Survey. Wosny, M., Hastings, J. (2024). Applying Large Language Models to Interpret Qualitative Interviews in Healthcare. *Studies, 211–214*(2), 211–214. Advance online publication. DOI: 10.31436/ijcs.v8i2.468

Roberts, J., Baker, M., & Andrew, J. (2024). Artificial intelligence and qualitative research: The promise and perils of large language model (LLM) 'assistance.' *Critical Perspectives on Accounting, 99*(Minaee, S., Mikolov, T., Nikzad, N., Chenaghlu, M., Socher, R., Amatriain, X., Gao, J. (2025). Large Language Models : A Survey. Wosny, M., Hastings, J. (2024). Applying Large Language Models to Interpret Qualitative Interviews in Healthcare. *Studies, 102722*, •••. https://doi.org/https://doi.org/10.1016/j.cpa.2024.102722

Wen, C., Clough, P., Paton, R., & Middleton, R. (2025). Leveraging large language models for thematic analysis : a case study in the charity sector. *AI & SOCIETY, Minaee, S., Mikolov, T., Nikzad, N., Chenaghlu, M., Socher, R., Amatriain, X., Gao, J. (2025). Large Language Models : A Survey. Wosny, M., Hastings, J. (2024). Applying Large Language Models to Interpret Qualitative Interviews in Healthcare. Studies*. Advance online publication. DOI: 10.1007/s00146-025-02487-4

Whitty, M. T. (2025). *How to Conduct AI-Assisted (Large Language Model-Assisted).* Content Analysis in Information Science and Cyber Security Research.

Wosny, M., & Hastings, J. (2024). Applying Large Language Models to Interpret Qualitative Interviews in Healthcare. *Studies in Health Technology and Informatics, 316*(Minaee, S., Mikolov, T., Nikzad, N., Chenaghlu, M., Socher, R., Amatriain, X., Gao, J. (2025). Large Language Models : A Survey. Wosny, M., Hastings, J. (2024). Applying Large Language Models to Interpret Qualitative Interviews in Healthcare. *Studies, 791–795*. Advance online publication. DOI: 10.3233/SHTI240530 PMID: 39176911

Zhao, Z., & Liu, Y. (2025). *A Confidence – Diversity Framework for Calibrating AI Judgement in Accessible Qualitative Coding Tasks*. 1–23.

KEY TERMS AND DEFINITIONS

AI Literacy: Researcher competency encompassing understanding of how LLMs function, their capabilities and limitations, effective prompt engineering, critical evaluation of outputs, and ethical implications of AI integration.

Algorithmic Bias: Systematic distortions in LLM outputs arising from biases embedded in training data, potentially perpetuating stereotypes or marginalizing non-dominant perspectives in research interpretation.

Black-Box Problem: The opacity of LLM decision-making processes, where the computational pathways producing specific outputs remain fundamentally inaccessible and unexplainable, challenging research transparency and auditability.

Hybrid Approaches: Methodological frameworks strategically combining human interpretive expertise with AI computational capabilities, maintaining human analytical authority while leveraging LLM efficiency for specific tasks.

Informed Consent: Ethical principle requiring that research participants understand how their data will be collected, processed, stored, and used, including any AI involvement, before voluntarily agreeing to participate.

Large Language Models (LLMs): Advanced artificial intelligence systems trained on vast text corpora to understand, generate, and manipulate human language through statistical pattern recognition and transformer architectures.

Qualitative Research: Interpretive methodology emphasizing contextual understanding, human experience, and meaning-making through in-depth analysis of non-numerical data such as interviews, observations, and documents.

Reflexivity: The ongoing critical examination of how researcher positionality, assumptions, and relationships with participants influence data collection, analysis, and interpretation throughout the research process.

Thematic Analysis: Qualitative analytical method involving identification, analysis, and reporting of patterns of meaning across datasets through iterative coding, theme development, and conceptual synthesis.

Trustworthiness Criteria: Lincoln and Guba's framework for assessing qualitative research rigor through credibility, transferability, dependability, and confirmability rather than traditional positivist validity measures.

Chapter 12
The Algorithm as a Cultural Informant:
Exploring AI-Mediated Realities Through Computational Ethnography

Rismawati Rismawati
http://orcid.org/0000-0001-8046-6764
Universitas Muhammadiyah Palopo, Indonesia

Usep Suhud
http://orcid.org/0000-0002-1728-0737
Universitas Negeri Jakarta, Indonesia

Eko Ganis Sukoharsono
Universitas Brawijaya, Indonesia

ABSTRACT

This chapter also treats algorithms as culture informants, extending an account of computational ethnography that attempts to link interpretive depth with computational scale. From a qualitative and constructivist perspective, algorithms are cast as epistemic agents, cultural artefacts encoding design values that serve to cultural actors to mediate meaning,norms and visibility. The chapter combines digital traces, human-AI interaction and reflexive recordings with more traditional ethnographic techniques such as interviews and fieldnotes. It introduces an analytical hybrid combining NLP, topic modeling, sentiment analysis and LLM facilitated reflexive coding under human governance. Transparency, bias auditing, and Human-Centered AI principles ensure the ethical and epistemic rigor of our approaches. The chapter

DOI: 10.4018/979-8-3373-7640-0.ch012

nt © 2026, IGI Global Scientific Publishing. Copying or distributing in print or electronic forms without written permission of IGI Global Scientific Publishing is prohibited. Use of this chapter to generative artificial intelligence (AI) technologies is expressly prohibited. The publisher reserves all rights to license its use for generative AI training and machine learning model development.

finishes by discussing the implications for social, organizational and CSR research in promoting reflexive, co-constructed human-AI inquiry that considers algorithms as situated cultural sources of knowledge.

1. INTRODUCTION

1.1 Context and Rationale

Modern social existence is progressively integrated inside digital environments influenced by platforms, data frameworks, and algorithmic mechanisms. Social connection, cultural engagement, and public debate now occur not only through direct human communication but also inside contexts where digital footprints, interface design, and automated categorization dictate visibility, prioritization, and dissemination. In this setting, platforms do not simply facilitate social interaction; they actively mediate the conditions under which meaning is generated, experienced, and perceived (Kozinets & Gretzel, 2024; Roumbanis, 2025; Srivastava, 2023).

This change has amplified the significance of algorithms in structuring visibility, relevance, and engagement. Ranking systems, recommendation algorithms, and platform-responsive feedback mechanisms determine which storylines acquire prominence, which individuals receive attention, and which types of involvement are incentivized. Consequently, algorithms increasingly serve as socio-technical mediators that shape cultural significance, legitimize specific communicative practices, and dictate user navigation inside digital environments (Gritsenko & Wood, 2022; Taylor & Choi, 2022; Obreja, 2024). In this regard, algorithmic systems are not impartial background instruments but rather active contributors to modern meaning-making.

These advancements reveal significant constraints in traditional ethnography when addressing extensive, platform-mediated data. Although conventional ethnographic methods are essential for providing contextual depth, reflexivity, and interpretive richness, they frequently lack the capacity to track scattered, asynchronous, and algorithmically filtered interactions in digital environments. Recent literature advocates for methodological strategies that maintain ethnographic sensitivity while addressing computational scale and infrastructural complexity (Brooker, 2022; Jensen et al., 2022). This necessity justifies computational ethnography as a reflexive and adaptive reaction to AI-mediated social realities.

1.2 Research Gap and Chapter Objectives

While ethnography continues to be a potent method for analyzing culture, meaning, and lived experience, its traditional forms face increasing constraints in AI-mediated digital contexts. Modern social environments are progressively organized by platforms, algorithmic classification, recommendation systems, and data-driven infrastructures that determine what is visible, searchable, and socially significant. These circumstances contest traditional ethnographic beliefs regarding immersion, defined field sites, and prolonged observational presence, especially when encounters are fragmented, asynchronous, and perpetually mediated by computational systems (Abramson et al., 2025).

A subsequent methodological gap pertains to the conflict between interpretive depth and analytical scale. Conventional qualitative methods are essential for contextual sensitivity, reflexivity, and meaning-focused analysis; however, they frequently lack the capacity to handle extensive digital traces, user-generated content, and swiftly evolving online discourse without significantly simplifying data complexity (Salganik, 2022). The incorporation of artificial intelligence into qualitative research remains epistemologically ambiguous. Although AI can facilitate pattern detection, clustering, and topic exploration, it may also pose risks of abstraction, decontextualization, and premature closure if not regulated by critical and reflective human judgment (Chatzichristos, 2025; Williams, 2024).

This chapter aims to achieve four objectives in response to these tensions. Initially, it delineates the methodological constraints of traditional ethnography in algorithmically mediated environments. Secondly, it advocates a hybrid qualitative-computational methodology that links scale with interpretive profundity. Third, it perceives algorithms as both ethnographic entities and cultural informants. Fourth, it promotes methodological and epistemological clarity for the nascent discipline of computational ethnography (Abramson et al., 2025; Punziano, 2025).

1.3 Chapter Structure

This chapter comprises six principal sections. The Background section delineates the theoretical and methodological framework of the study by examining the evolution from classical ethnography to digital and computational research methodologies. It posits that algorithms must be perceived not solely as technological tools, but as cultural and epistemic agents that influence the production and interpretation of social meaning in digitally mediated contexts.

The primary emphasis of the chapter analyzes the methodological difficulties that emerge when ethnographic research engages with platformized, large-scale, and algorithmically filtered data. Special emphasis is placed on the constraints of traditional qualitative methods in these circumstances and the necessity for a hybrid framework that integrates computational breadth with interpretive profundity.

The segment on Solutions and Recommendations delineates the suggested reflexive computational ethnography framework. This section examines the significance of human oversight, interpretative validation, auditability, and ethical precautions in AI-assisted qualitative research, while providing practical guidelines for researchers operating in digitally mediated social contexts.

The Future Research Directions section delineates nascent methodological potential for advancing computational ethnography, encompassing concerns of transparency, multidisciplinary training, and specialized applications in areas such as organizational studies, corporate social responsibility, and sustainability research.

The Conclusion consolidates the chapter's principal point by emphasizing the necessity of harmonizing computing proficiency with ethnographic reflexivity, thus establishing computational ethnography as an essential methodological approach to algorithmically influenced social realities.

2. BACKGROUND

2.1. Constructivist and Interpretivist Paradigms in the Age of AI

Classical ethnography emerged as an interpretive methodology rooted in immersion, participant observation, and detailed description, aimed at comprehending the production of meaning within living social contexts (Geertz, 1973). Within this paradigm, the field serves not just as a venue for data gathering but also as a relational and reflexive environment where social reality is analyzed through continuous interaction with participants, practices, and situations. Ethnographic knowledge relies not solely on observation, but also on the researcher's capacity to contextualize meaning within cultural and social contexts.

The migration of interaction, communication, and cultural participation into digital contexts has altered the conditions of anthropological research. Social life today transpires across platforms characterized by swift circulation, widespread participation, and continuous data collection. These venues generate narratives and exchanges, as well as metadata, interaction traces, and algorithmically structured visibility that influence what becomes prominent, searchable, and socially significant (Kozinets & Gretzel, 2024; Roumbanis, 2025). Consequently, ethnographers are increasingly faced with social domains that are hybrid, dynamic, and mediated by computation.

This transition led to the emergence of digital ethnography and netnography, both of which modified ethnographic principles for online and hybrid environments.

These methodologies maintained the significance of interpretive profundity while acknowledging that digital contexts necessitate novel modes of access, observation, and contextualization (Brooker, 2022; Kozinets & Gretzel, 2024). Recent advancements have propelled this trajectory towards computational ethnography, which utilizes computational tools to analyze extensive digital traces while maintaining reflexivity and contextual interpretation (Abramson et al., 2025; Jensen et al., 2022). Figure 1 depicts the methodological transition from traditional ethnography to digital and computational ethnography, emphasizing the growing significance of algorithmically mediated social contexts.

Figure 1. Evolution of the Ethno-Computational Paradigm

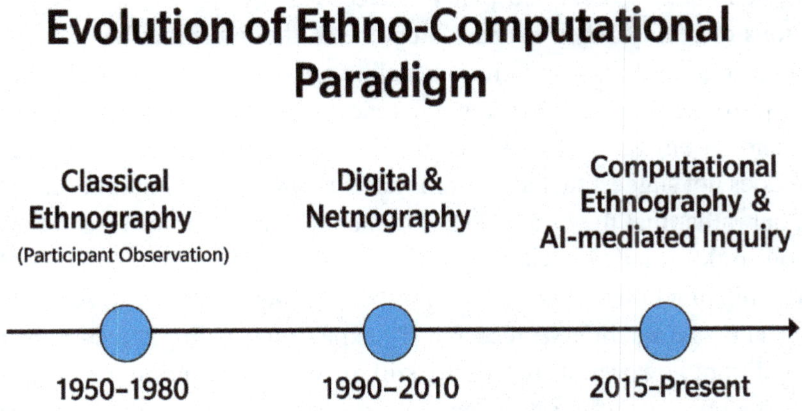

Source: Author adapted (2025) from Evolution of the Ethno-Computational Paradigm (2025)

Figure 1 The Intellectual Evolution of Ethnography, from classical origins to computational futures. The history is delineated into three paradigmatic epochs: Classical Ethnography (1950–1980), characterized by participant observation; Digital and Netnography (1990–2010), indicative of the adaptation of ethnographic methodologies to digital contexts; and Computational Ethnography and AI-mediated Inquiry (2015–Present), wherein artificial intelligence and algorithmic instruments are pivotal in data interpretation and cultural analysis. This evolution represents a steady epistemological shift from human-centered observation to hybrid human-machine co-production of knowledge, framed within reflexivity and algorithmic mediation in contemporary social research.

2.2 Algorithms as Cultural and Epistemic Actors

In digitally mediated contexts, algorithms may no longer be regarded solely as technological tools functioning in the periphery of social existence. They should be regarded as socio-technical constructs that engage in the creation, structuring, and dissemination of meaning. Algorithms, as cultural constructs, are influenced by certain institutional logics, design presuppositions, and normative evaluations of relevance, visibility, value, and permissible modes of engagement. They do not only process information; they establish the conditions that render information salient, interpretable, and socially significant.

This cultural embeddedness also imparts epistemic relevance to algorithms. In AI-mediated research environments, algorithmic outputs, including grouped topics, suggestions, sentiment scores, and pattern detections, do not serve as impartial representations of reality. They instead intervene in the interpretative process by emphasizing certain associations while concealing others. In this regard, algorithms function as epistemic agents: they shape inference, direct interpretative focus, and affect how scholars engage with and structure the field (Punziano, 2025; Williams, 2024). This does not suggest that algorithms have autonomous comprehension like to humans, but rather that they exercise epistemic influence within socio-technical configurations of knowledge creation.

The epistemic function of algorithms must be examined reflexively. AI-assisted analysis can enhance qualitative research by expediting coding, clustering, and pattern recognition; however, its results are still influenced by training data, model architecture, parameter configuration, and inherent biases. Consequently, algorithmic mediation need to be regarded not as a replacement for interpretation, but as a contextual stimulus for additional hermeneutic exploration. The importance of algorithms in computational ethnography resides in their dual role as both subjects of study and agents that influence knowledge production. Figure 2 demonstrates this idea by situating AI as an epistemic agent in the research process, rather than merely as a functional analytical tool.

Figure 2. Conceptualizing AI as an Epistemic Actor

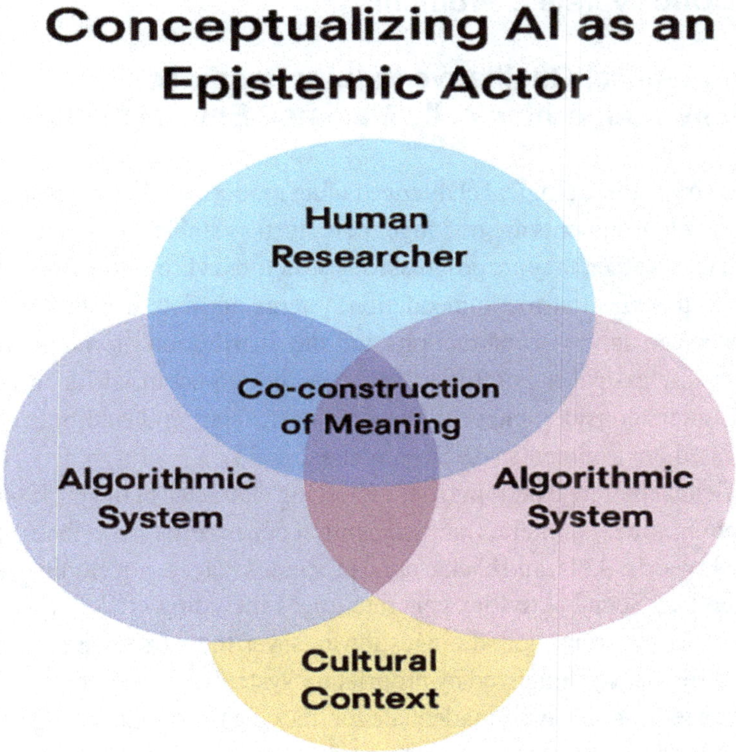

Source: Author adapted (2025)

Figure 2 illustrates that meaning in computational ethnography arises from the convergence of three domains: the human researcher, the algorithmic system, and the overarching cultural background. The intersection of these realms signifies the collaborative creation of meaning, when knowledge is not merely derived from facts but is negotiated through the interplay of human judgment and algorithmic facilitation. The picture further elucidates the chapter's principal assertion that algorithms ought not to be regarded as impartial arbiters of truth. Rather, they operate as contextual and debatable contributors to interpretive endeavors, rendering reflexivity, examination, and contextual validation essential. When examined in this manner, the algorithm transforms into not merely an analytical tool, but also a cultural and epistemic entity that necessitates interpretation within the socio-technical domain.

3. MAIN FOCUS OF THE CHAPTER: Issues, Controversies, Problems

3.1 Why Classical Qualitative Inquiry Becomes Insufficient in Algorithmically Mediated Environments

The primary methodological challenge facing modern qualitative research is that several social environments are no longer perceived as stable, confined, and readily visible domains. Rather, they are progressively structured via platform infrastructures, algorithmic categorization, recommendation systems, and data-driven feedback loops that influence the nature of interactions and the distribution of visibility. In these circumstances, classical qualitative research, although essential for interpretative depth, encounters considerable challenges in encapsulating the complete intricacy of AI-mediated environments (Abramson et al., 2025).

A significant challenge pertains to the evolving character of the profession itself. In digital environments, interaction frequently occurs across interfaces, comment threads, dashboards, APIs, and background processes that are not readily perceptible to the researcher. Social activities generate traces including clicks, modifications, ratings, timestamps, prompts, and suggestion logs; yet, these traces are inconsistently exposed and frequently organized by proprietary systems. Consequently, traditional participant observation may be inadequate for tracking the organization of meaning, engagement, and salience across platformed environments (Kozinets & Gretzel, 2024; Roumbanis, 2025).

A secondary issue pertains to analytical scale. Classical qualitative methods excel in providing thick descriptions, contextual interpretations, and intimate engagement with participants; however, they are inadequately equipped to handle large volumes of digital material, including social media logs, user comments, algorithmic meta-data, and platform-generated interaction histories, without significantly diminishing the empirical field. Consequently, researchers might be compelled to subsample, truncate, or selectively analyze data, thereby introducing bias and constraining the interpretive scope of the study (Salganik, 2022).

A third controversy pertains to the epistemological implications of algorithmic mediation. Researchers are now examining not only individuals and practices but also the surroundings where algorithmic systems actively influence relevance, attention, and communicative visibility. This indicates that the field is partially pre-structured prior to the researcher's entry. The inadequacy of traditional qualitative inquiry should not be interpreted as a methodological failure; rather, it reflects a transformed research landscape that necessitates hybrid methodologies integrating ethnographic reflexivity with computational acuity and infrastructural cognizance (Abramson et al., 2025).

3.2 The Core Methodological Tension: Scale versus Interpretive Depth

A primary debate in modern qualitative research revolves around the conflict between analytical breadth and interpretive profundity. Digital environments produce vast quantities of data, encompassing posts, comments, interactions, information, prompts, rankings, and suggestion traces. These materials surpass the feasible limits of manual observation and detailed analysis when approached exclusively through traditional qualitative methods. Conversely, the merit of qualitative inquiry is in its capacity to elucidate nuance, context, ambiguity, symbolism, and experiential significance—elements that are frequently diminished when analysis is predominantly governed by computational abstraction.

This tension engenders a methodological quandary. If researchers adhere solely to traditional close-reading techniques, they may neglect the overarching patterns, circulations, and infrastructure dynamics that influence digital social life on a large scale. However, an excessive dependence on computational tools may compromise the contextual sensitivity and reflexive judgment that render qualitative investigation analytically unique. In essence, scale devoid of interpretation can result in technically proficient yet socially superficial outcomes, whereas interpretation lacking adequate scale may generate comprehensive but incomplete narratives of algorithmically influenced contexts (Salganik, 2022).

The debate intensifies with the introduction of AI-assisted tools into the research process. Topic modeling, clustering, sentiment analysis, and big language models can identify patterns and structure extensive corpora; nevertheless, they do not inherently ensure accurate interpretation. Their outputs are influenced by model architecture, training data, prompt design, and inherent assumptions, indicating that computational scale should consistently undergo interpretive examination rather than be regarded as epistemic closure (Chatzichristos, 2025). The fundamental concern, consequently, is not whether scale or depth should dominate, but rather how both might be effectively integrated. The existing conflict serves as the primary justification for the hybrid framework introduced in the subsequent section, wherein Figure 3 delineates a methodological structure that integrates digital breadth with ethnographic depth.

3.3 Proposed Hybrid Computational Ethnography Framework

This chapter introduces a hybrid computational ethnography paradigm to address the limitations discussed before, aiming to integrate computational scale with anthropological interpretation. The paradigm posits that digital social realities cannot be comprehensively grasped using only manual qualitative research or exclusively automated computational methods. Methodological adequacy relies on an iterative relationship among extensive data processing, contextual interpretation, and reflective human judgment.

The approach allows researchers to analyze dispersed and voluminous digital traces, including comments, interaction logs, platform-generated text, and other types of digitally mediated conversation. Methods such natural language processing, topic modeling, thematic grouping, and AI-assisted coding can facilitate the identification of patterns, repeating themes, discursive anomalies, and emergent structures that may be challenging to discern from careful reading alone. Nonetheless, these operations do not, in and of themselves, represent interpretation. Their significance resides in their ability to expand the empirical scope of investigation and produce preliminary analytical insights for more profound qualitative exploration (Abramson et al., 2025; Jensen et al., 2022).

At the ethnographic level, the framework emphasizes the importance of reflexivity, contextualization, and interpretative accountability. Computational results should be interpreted in conjunction with fieldnotes, researcher memoranda, platform circumstances, and the contextual knowledge of participants and discursive environments. Thus, algorithms are included into the research process not as independent decision-makers, but as collaborative collaborators whose results necessitate examination, comparison, and verification. The paradigm posits that knowledge creation is co-constructed, albeit with unequal delegation; interpretive authority is anchored in the researcher's reflexive engagement with the field (Punziano, 2025).

Figure 3 illustrates the data architecture of the proposed hybrid computational ethnography framework, demonstrating the integration of digital traces, computational analysis, and interpretative validation within a reflexive research process.

Figure 3. Data Architecture in Computational Ethnography

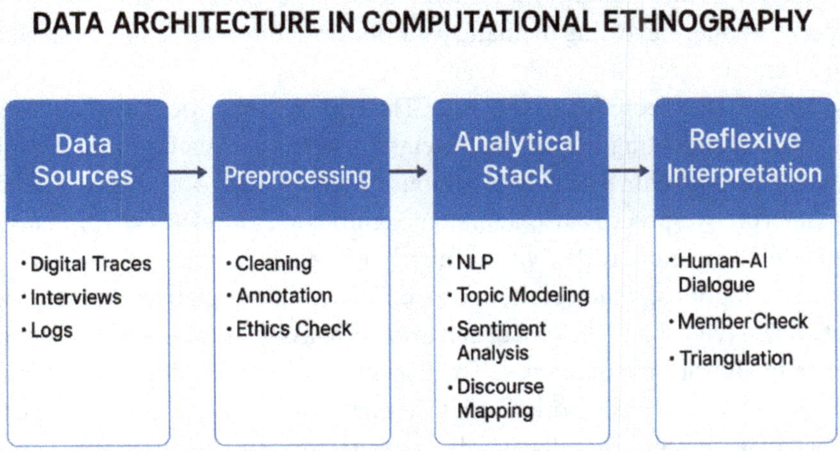

Source: Author adapted (2025).

Figure 3 illustrates that the suggested framework is organized as a layered methodical architecture rather than a straight progression of technical procedures. At the foundational level, digital footprints and platform-generated data compose the empirical domain. These resources undergo computational processes that facilitate organization, grouping, and the identification of patterns within extensive content volumes. However, the figure clearly indicates that computational processing does not conclude the analytical process. Outputs go into a phase of interpretive validation, wherein the researcher assesses themes, contrasts divergent examples, recontextualizes findings, and generates introspective notes. The architecture thus illustrates the chapter's principal argument: computational ethnography attains methodological credibility solely when scale is persistently linked to contextual depth and when algorithmic results are held accountable to human interpretation. The framework serves not merely as a method for managing increased data, but as a reflexive paradigm for analyzing social significance in algorithmically mediated contexts. Following the delineation of the proposed framework, the subsequent part addresses the practical and epistemological prerequisites necessary to implement human–AI co-analysis as an interpretative process.

3.4 Human–AI Co-Analysis as an Interpretive Procedure

Human–AI co-analysis should not be perceived as the assignment of qualitative interpretation to automated systems. It is more accurately understood as an interpretive method wherein computing results are included into the analytical process as tentative proposals, alternative interpretations, or organized stimuli for additional contemplation. In this framework, AI aids in the organizing of extensive corpora by suggesting codes, revealing thematic patterns, detecting anomalies, or producing alternate interpretations, while the researcher retains responsibility for contextualization, evaluation, and theoretical assessment. The analytical significance of co-analysis resides not in supplanting human interpretation, but in enhancing it by organized engagement with machine-generated options (Williams, 2024; Punziano, 2025).

This interpretive process also alters the significance of reflexivity. In classical qualitative investigation, reflexivity primarily examines how the researcher's perspective, assumptions, and social context influence interpretation. In AI-assisted research, reflexivity should be broadened to encompass the socio-technical contexts that generate algorithmic outcomes. The design of the model, the training dataset, prompting methodologies, and inference settings collectively determine what the system perceives, emphasizes, disregards, or misidentifies. Consequently, reflexivity serves as a co-audit of both self and system, necessitating that researchers examine their interpretative decisions as well as the mediating principles of the instruments they utilize.

Consequently, stringent human–AI co-analysis relies on clear procedural safeguards. This encompasses transparent prompt logs, version tracking, side-by-side comparisons of human and AI coding, discrepant-case analysis, iterative codebook refining, and reflexive memoing that records the rationale for accepting, revising, or rejecting certain outputs. These techniques guarantee that AI functions as a designated collaborator instead of an unchallenged analytical authority, maintaining interpretive complexity while facilitating broader empirical scope (Chatzichristos, 2025; Tracy, 2020; Williams, 2024). In this context, co-analysis transforms from a mere shortcut for interpretation into a systematic approach for maintaining qualitative rigor in computational ethnography. This interpretive framework demonstrates that human–AI collaboration is intrinsically linked to issues of transparency, trustworthiness, and ethical governance, which will be discussed in the subsequent section.

4. ETHICAL AND EPISTEMOLOGICAL CHALLENGES OF AI AS INFORMANT

4.1. Addressing the Black Box Dilemma and Algorithmic Bias

A significant problem in AI-assisted qualitative research is the black box dilemma. Modern AI systems, especially deep learning models and huge language models, frequently deliver results without offering transparent insight into the processes that led to those results. For qualitative researchers, this lack of clarity is not only a technical issue; it constitutes an epistemological challenge. If the pathway from data to output is unclear, the basis for assessing trustworthiness, confirmability, and interpretative reasoning also becomes challenging (Bender et al., 2021). In computational ethnography, this poses a significant risk: researchers may depend on themes, clusters, or summaries that seem cohesive while being unable to elucidate whose voices, assumptions, or omissions have influenced them.

Algorithmic bias is closely associated with opacity. Bias can infiltrate via training data, labeling methodologies, model tuning, prompting techniques, and deployment contexts. In practice, this indicates that AI systems can replicate prevailing narratives, inadequately represent disadvantaged dialects, and generate seemingly impartial portrayals that are, in reality, socially inequitable (Kordzadeh & Ghasemaghaei, 2022; Ukanwa, 2024). Consequently, eliminating bias should not be regarded as a mere technical modification; it must be recognized as a methodological and ethical imperative.

Figure 4 encapsulates the black box issue by depicting the emergence of outputs through concealed model layers, ingrained biases, and restricted visibility, while simultaneously highlighting transparency, auditing, and reflexivity as essential safeguards in AI-assisted research.

Figure 4. The "Black Box" Dilemma: Conceptual Scheme

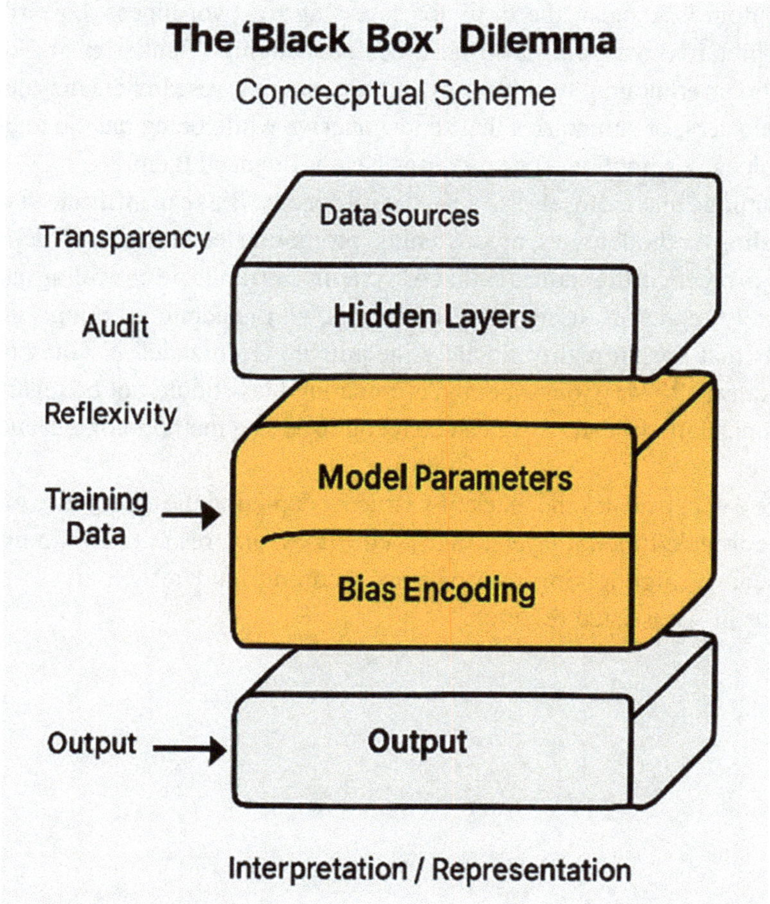

Source: Adapted by the author (2025).

Figure 4 elucidates that the issue extends beyond the mere complexity of algorithms; rather, such complexity can obfuscate accountability. The illustration depicts a progression from training data via hidden layers and bias encoding to interpretive output, concurrently highlighting transparency, auditing, and reflexivity as interventions that reestablish the connection between computational processes and methodological accountability. This necessitates that researchers meticulously document datasets, maintain timely and versioned logs, utilize audit trails, and regard AI-generated conclusions as preliminary rather than definitive. Thus, transparency transforms from a mere commitment to complete system openness into a systematic approach aimed at maintaining rigor, reflexivity, and interpretative accountability in computational ethnography (Gebru et al., 2021; Mökander & Floridi, 2023; Laine et al., 2024; Chuan et al., 2024).

4.2 Data Privacy, Consent, and Positionality

Computational ethnography inherently navigates the conflicts between the depth of platform-generated data and participants' privacy concerns. Although digital data may be publicly accessible, its aggregation, linking, and recontextualization might reveal sensitive characteristics or cause harm that beyond the initial context of data sharing. In this scenario, ethical risk emanates not solely from access, but from the researcher's ability to render traces freshly visible, searchable, and interpretable beyond their original contextual integrity. Recent qualitative research utilizing social media data indicates that numerous studies continue to underreport privacy protections, anonymization methods, and risk assessments, highlighting a continuous disparity between ethical norms and reporting practices.

These problems render consent a fundamental issue in computational ethnography. In platformized systems, consent must not be regarded as a singular procedural formality. Participants must be apprised of the types of data to be collected, as well as the utilization, storage, linkage, and potential reuse of metadata, inferred variables, and AI-assisted analytical tools. Innovative participant-centered models, such as data donation, present a viable alternative by centering access on defined participant rights and enhancing researcher-participant communication, while facilitating more transparent workflows for trace-data collection (Carrière et al., 2024; Hase et al., 2024). Nonetheless, data donation does not alleviate ethical responsibilities, as platform non-compliance, export restrictions, and interface obstacles may still influence participation and distort the sample. Consequently, layered permission, transparent documentation of the data lifecycle, and continuous chances for opting out are necessary.

Positionality is equally significant when integrating AI techniques into qualitative analysis. When big language models aid in summarizing interviews, aggregating topics, or composing memos, their outputs integrate into the ethnographic narrative. However, these outputs are never impartial. They embody the presuppositions inherent in training data, model architecture, classification frameworks, and prompting methodologies, indicating that AI-generated interpretations possess contextual viewpoints rather than universal verities (Suchman, 2023; Scheuerman et al., 2024; Marshall & Naff, 2024). Consequently, the onus of representation must unequivocally rest with the researcher. This necessitates detailed AI usage logs, disclosure of origins, human mediation for disputed interpretations, and member verification when applicable. Viewed in this manner, privacy, consent, and positionality are not ancillary issues, but essential prerequisites for ethically sound computational ethnography.

4.3 Towards an Ethical Framework for Computational Ethnography

An ethical framework for computational ethnography must commence with the idea that artificial intelligence should enhance, rather than supplant, human interpretation. Qualitative research necessitates a methodology consistent with Human-Centered AI (HCAI), wherein human agency, accountability, and interpretative judgment are paramount throughout the study process. From this viewpoint, AI-assisted technologies are beneficial solely when they function under substantial human oversight, incorporating written prompts, version tracking, transparent protocols, and explicit mechanisms for review and override (Shneiderman, 2022; Allen et al., 2024).

This framework must also incorporate interpretive governance. In computational ethnography, the primary inquiry is not merely if an AI system can discern themes, condense narratives, or categorize patterns, but rather if those outputs can be contextually substantiated and ethically defended. AI-generated outcomes should be corroborated via fieldnotes, interviews, participant comments, and researcher memos instead than being regarded as self-evident conclusions. Human-in-the-loop judgment is not merely an optional addition; it serves as a methodological safeguard against decontextualization, premature closure, and the perpetuation of structural bias (Abujaber et al., 2024; Ossa et al., 2024).

An further element of the system is risk-aware deployment. Researchers must evaluate if specific models disadvantage some groups, diminish language or cultural nuance, or generate outcomes with deceptive confidence. Ethical practice necessitates bias assessments, transparent disclosure of model limits, and meticulous communication of uncertainty when AI-generated outputs influence significant assertions. Moreover, explainability should be regarded as an integral component of responsibility. This encompasses the dissemination of accessible descriptions of AI applications, the preservation of audit trails, the documentation of interpretive choices, and the establishment of avenues for participants to inquire about or challenge AI-assisted representations (AI Now Institute, 2023; Singhal et al., 2024).

Collectively, these components indicate that an ethics of computational ethnography is not an ancillary aspect of methodology, but rather a prerequisite for methodological credibility itself. It integrates human-centered design, interpretative precision, and public accountability, guaranteeing that algorithmic support is recorded, contestable, and socially responsible within qualitative research.

5. FUTURE RESEARCH DIRECTIONS

5.1 Toward Reflexive and Auditable Qualitative AI

A primary future focus for computational ethnography is the advancement of reflexive and verifiable qualitative AI. With the increasing integration of AI-assisted tools in coding, memoing, clustering, and interpretive support, the primary methodological inquiry has shifted from whether artificial intelligence can aid qualitative research to how this assistance can be structured to ensure transparency, reflexivity, and accountability. Future qualitative AI should be developed not as a mechanism for automated interpretive replacement, but as a structured framework where computational outputs are subject to examination, modification, and contextual verification.

Reflexive qualitative AI acknowledges that model outputs are inherently biased. An AI system's identification of themes, patterns, or summaries is influenced by training data, model architecture, prompting tactics, and researchers' choices on data preparation and analytical application. Consequently, future methodological advancements must transcend the presumption that AI merely accelerates current qualitative processes. Researchers must clearly investigate how socio-technical factors influence what the system perceives, emphasizes, or disregards. In this context, reflexivity in the AI era necessitates a dual examination of both the individual scholar's interpretive assumptions and the operational frameworks of the tools utilized (Suchman, 2023).

Auditability is equally significant. Future qualitative AI must be underpinned by procedural infrastructures that render interpretative conclusions transparent and subject to review. This encompasses prompt logs, reasoning tracing, comparative analyses of human and AI coding, model and version information, discrepancy notes, and defined criteria for the acceptance, revision, or rejection of AI proposals (Allen et al., 2024; Dunivin et al., 2025). Such techniques guarantee that model outputs stay tentative and debatable instead of serving as obscure analytical conclusions. Auditability enhances methodological rigor by reinforcing credibility, reliability, and confirmability through the clear documenting of the processes used to develop and validate findings (Tracy, 2020).

In the future, reflexive and auditable qualitative AI is expected to provide a definitive standard for computational ethnography. Hybrid workflows will progressively integrate computational scale with interpretive depth; yet, their legitimacy will hinge on the preservation of human authorship, contextual sensitivity, and ethical responsibility. The future of qualitative AI resides not in complete automation, but in fostering transparent, contested, and reflexively governed interactions between human researchers and algorithmic systems.

5.2. Algorithmic Literacy and Interdisciplinary Research Training

A significant future path for computational ethnography is to the advancement of algorithmic literacy and interdisciplinary research training. As qualitative research increasingly incorporates AI-assisted tools, platform infrastructures, and extensive digital footprints, scholars can no longer depend exclusively on traditional interpretative skills. Although reflexivity, contextual sensitivity, and theoretical insight are essential, they must now be augmented by a practical comprehension of the functioning of algorithmic systems, the formation of model outputs, and the identification of their limitations. In this sense, algorithmic literacy encompasses more than mere technical proficiency; it signifies the capacity to critically analyze the socio-technical contexts that govern the generation, dissemination, and interpretation of computer outputs as evidence.

This literacy is essential as AI-assisted inquiry presents novel methodological risks. Researchers must recognize the impact of training data, prompting tactics, model versions, and interface design on coding recommendations, summaries, or theme clustering. In the absence of this competence, qualitative researchers may utilize computational outputs instrumentally, neglecting their inherent assumptions, failure modes, and representational biases. Consequently, next research training must encompass not only interpretative reasoning but also rapid design, model auditing, discrepancy analysis, provenance disclosure, and human override processes (Allen et al., 2024).

Interdisciplinary collaboration will be crucial to this endeavor. Computational ethnography functions at the convergence of qualitative methods, data science, platform studies, and artificial intelligence ethics. Consequently, successful training must transcend disciplinary boundaries. Future research environments should promote collaboration among ethnographers, communication academics, sociologists, computational researchers, and policy or design experts. This collaboration can facilitate the establishment of common norms for public reporting, interpretative governance, and method selection, while ensuring that computational support aligns with qualitative ideals instead of undermining them.

The future of computational ethnography relies not only on improved technologies but also on more adequately trained researchers. An effective collaborative pedagogy should instruct when to trust model outputs, when to scrutinize them, when to override them, and when to omit them from analysis entirely. In this regard, algorithmic literacy is emerging as a fundamental prerequisite for responsible, reflective, and methodologically sound qualitative research in AI-mediated contexts.

5.3. Expanding Applications in Organizational, CSR, and Sustainability Research

A significant potential trajectory for computational ethnography is its increasing utilization in organizational, corporate social responsibility, and sustainability studies. As firms increasingly engage via digitally mediated channels, the social dynamics of enterprises are no longer restricted to official reporting, managerial dialogue, or internal frameworks. Organizational meaning is now perpetually negotiated across websites, social media platforms, stakeholder forums, comment sections, and algorithmically managed visibility systems. This presents a substantial opportunity for computational ethnography to investigate how organizations are perceived, challenged, and legitimized in digital contexts.

This method is particularly promising within the realm of CSR and sustainability, as much of the current discourse on responsibility, transparency, and accountability occurs through fragmented digital traces. Corporate sustainability assertions, ESG narratives, diversity communications, and publicly stated ethical pledges are now commonly perceived, magnified, scrutinized, or opposed through platformed interactions. A computational ethnographic method enables the large-scale analysis of these dynamics while maintaining sensitivity to speech, symbolism, and cultural significance. Instead of viewing stakeholder responses as discrete reactions, researchers can analyze how these responses are structured, disseminated, and influenced by algorithmic visibility and platform dynamics.

This methodology facilitates deeper examination of topics such as greenwashing, symbolic legitimacy, and the politics of organizational representation. Computational ethnography can elucidate how corporate narratives on sustainability proliferate, where doubt arises, and how public trust is mediated through digital discourse. AI-assisted qualitative analysis might enhance the examination of corporate communication not solely as message transmission, but as a dynamic domain of contested interpretations.

Subsequent study in this domain may progress from static document analysis to more comprehensive investigations of the production, mediation, and interpretation of CSR and sustainability narratives within hybrid socio-technical contexts. Computational ethnography is especially beneficial for researchers focused on stakeholder engagement, platform governance, reputational dynamics, and the cultural politics of corporate accountability. This method integrates interpretive depth with digital complexity, so enhancing qualitative research into modern organizational dynamics.

5.4 Future Methodological Development

A prospective advancement in computational ethnography involves the enhancement of methodological frameworks that facilitate collaboration between human interpretation and computational support, ensuring that none is subsumed by the other. As the discipline evolves, the task will shift from merely assessing AI's capacity to assist qualitative research to determining how methodological frameworks can sustain that assistance in a manner that is reflexive, iterative, and attuned to context. This necessitates progressing beyond sporadic applications of AI to well-defined processes that delineate the timing of computational tool integration, the criteria for evaluating their outputs, and the circumstances under which these outputs are accepted, modified, or dismissed.

Future methodological advancements should concentrate on formalizing hybrid approaches for qualitative research. This may encompass organized cycles of human coding and AI-assisted grouping, iterative comparisons between model-generated themes and field-based interpretations, as well as defined processes for discrepancy analysis. Such advancements are significant as they enable researchers to leverage computational diversity while preserving interpretive responsibility. Future methodologies should not focus solely on automation, but rather on a structured interplay between computer recommendations and ethnographic discernment.

This approach facilitates the exploration of more intricate forms of inquiry. Computational ethnography is likely to broaden into multimodal research contexts that integrate text, images, video, and interface analysis; into cross-platform investigations that track narratives across various digital environments; and into longitudinal frameworks that monitor the evolution of discourse over time. As these opportunities expand, the necessity for transparent human–AI collaboration frameworks will become increasingly critical. In the absence of such models, methodological innovation may yield scale devoid of rigor or speed lacking interpretative depth. Figure 5 depicts a potential trajectory for this evolution by portraying reflexive human–AI collaboration as a cyclical process rather than a linear one..

Figure 5. Reflexive Human–AI Collaboration Model

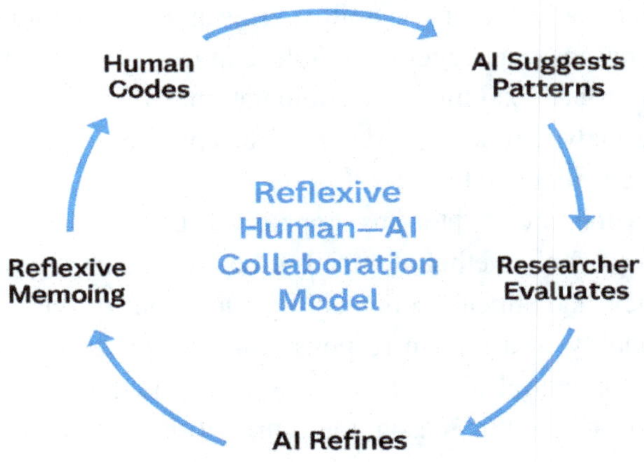

Source: Adapted by the author (2025).

Figure 5 illustrates that future computational ethnography is optimally comprehended as a cyclical process involving human coding, AI-facilitated pattern recognition, researcher assessment, model enhancement, and reflexive documentation. The figure is significant since it refutes the presumption that AI provides definitive interpretations. It conceptualizes collaboration as a cyclical process wherein computational outputs are continually vulnerable to human evaluation, contextual reinterpretation, and methodological modification. The future of computational ethnography, when considered in this manner, does not include supplanting qualitative interpretation, but rather in creating more resilient and transparent interactions between researchers and algorithmic systems.

CONCLUSION

This chapter has argued that the methodological conditions of qualitative inquiry are being fundamentally reshaped by digitally mediated and algorithmically organized social environments. In such contexts, classical ethnography remains indispensable for its interpretive depth, reflexivity, and contextual sensitivity, yet it also encounters important limitations when social life is dispersed across platforms, structured by recommendation systems, and recorded through large volumes of digital traces. These transformations require not the abandonment of ethnographic principles, but their methodological adaptation.

In response to this challenge, the chapter has proposed a hybrid computational ethnography framework that brings computational scale into dialogue with ethnographic interpretation. Rather than treating algorithms as neutral tools or autonomous interpretive agents, the chapter has conceptualized them as cultural and epistemic actors that participate in shaping visibility, relevance, and meaning within digital environments. This position makes it possible to move beyond a purely instrumental view of AI and instead examine how computational systems become part of the field, part of the analytic process, and part of the epistemological conditions under which knowledge is produced.

At the same time, the chapter has emphasized that the integration of AI into qualitative inquiry is only methodologically credible when it is governed by reflexivity, transparency, and ethical accountability. Questions of opacity, bias, privacy, consent, positionality, and human responsibility cannot be treated as secondary concerns. They are central to the legitimacy of computational ethnography as a qualitative approach. For this reason, the future of the field lies not in automated interpretation, but in the development of reflexive, auditable, and human-centered forms of collaboration between researchers and algorithmic systems.

Ultimately, computational ethnography should be understood as an evolving response to the complexity of contemporary social worlds. Its value lies in its ability to connect interpretive depth with digital complexity, thereby enabling qualitative researchers to remain attentive to meaning, power, and context in environments increasingly shaped by algorithmic mediation.

REFERENCES

Abramson, C. M., Li, Z., Prendergast, T., & Dohan, D. (2025). *Qualitative Research in an Era of AI: A Pragmatic Approach to Data Analysis, Workflow, and Computation*. arXiv preprint. https://doi.org//arXiv.2509.12503DOI: 10.48550

Abujaber, A. A., Abujaber, A., & Eltoukhy, M.. (2024). Ethical framework for artificial intelligence in healthcare: A scoping review. *BMC Medical Ethics*, *25*(1), 71. DOI: 10.1186/s12910-024-01161-2 PMID: 38898408

AI Now Institute. (2023). *Algorithmic accountability: Moving beyond audits*. AI Now Institute., DOI: 10.5281/zenodo.7866032

Allen, K., Reid, R., & El-Sayed, A. (2024). Human-centered AI as a framework guiding the development and implementation of AI in healthcare. *Discover Health Systems*, *3*, 62. DOI: 10.1016/j.dhsy.2024.100062

Bender, E. M., Gebru, T., McMillan-Major, A., & Shmitchell, S. (2021). On the dangers of stochastic parrots: Can language models be too big? *Proceedings of the 2021 ACM Conference on Fairness, Accountability, and Transparency*, 610–623. DOI: 10.1145/3442188.3445922

Brooker, P. (2022). Computational ethnography: A view from sociology. *Big Data & Society*, *9*(1), 1–6. DOI: 10.1177/20539517211069892

Carrière, T. C., Boeschoten, L., & van der Veen, E.. (2024). Best practices for studies using digital data donation. [REMOVED HYPERLINK FIELD]. *Quality & Quantity*, *59*(S1, Suppl 1), 389–412. DOI: 10.1007/s11135-024-01983-x PMID: 40191694

Chatzichristos, G. (2025). Qualitative Research in the Era of AI: A Return to Positivism. *International Journal of Qualitative Methods*, *24*, 16094069251337583. Advance online publication. DOI: 10.1177/16094069251337583

Chuan, C.-H., Surdeanu, M., & Markman, A. B. (2024). EXplainable artificial intelligence (XAI) for facilitating algorithmic awareness and fairness. *Computers in Industry*, *152*, 103067. DOI: 10.1016/j.compind.2024.103067

Dunivin, Z. O., Johnston, J., McFarland, D. A., & Nelson, L. K. (2025). Scaling hermeneutics: A guide to qualitative coding with large language models. *EPJ Data Science*, *14*, 28. DOI: 10.1140/epjds/s13688-025-00548-8

Gebru, T., Morgenstern, J., Vecchione, B., Vaughan, J. W., Wallach, H., Daumé, H.III, & Crawford, K. (2021). Datasheets for datasets. *Communications of the ACM*, *64*(12), 86–92. DOI: 10.1145/3458723

Geertz, C. (1973). Deep play: Notes on the Balinese cockfight. *Daedalus, 101*(1), 1–37. https://web.mit.edu/allanmc/www/geertz.pdf

Gritsenko, D., & Wood, M. (2022). Algorithmic governance: A modes of governance approach. *Regulation & Governance, 16*(1), 45–62. DOI: 10.1111/rego.12367

Hase, V., Ausloos, J., Boeschoten, L., Pfiffner, N., Janssen, H., Araujo, T., Carrière, T., de Vreese, C., Haßler, J., Loecherbach, F., Kmetty, Z., Möller, J., Ohme, J., Schmidbauer, E., Struminskaya, B., Trilling, D., Welbers, K., & Haim, M. (2024). Fulfilling data access obligations: How could (and should) platforms facilitate data donation studies? *Internet Policy Review, 13*(3). Advance online publication. DOI: 10.14763/2024.3.1793

Jensen, L. X., Bearman, M., Boud, D., & Konradsen, F. (2022). Digital ethnography in higher education teaching and learning—A methodological review. *Higher Education, 84*(5), 555–575. DOI: 10.1007/s10734-022-00838-4

Kordzadeh, N., & Ghasemaghaei, M. (2022). Algorithmic bias: Review, synthesis, and future research directions. [IDEAS/RePEc]. *The Information Society, 38*(3), 388–409. DOI: 10.1080/01972243.2021.1913749

Kozinets, R. V., & Gretzel, U. (2024). Netnography evolved: New contexts, scope, procedures and sensibilities. *Annals of Tourism Research, 104*, 103693. DOI: 10.1016/j.annals.2023.103693

Laine, J., Mäntymäki, M., & Islam, A. K. M. N. (2024). Ethics-based AI auditing: A systematic literature review. *Telecommunications Policy, 48*(6), 102782. DOI: 10.1016/j.telpol.2024.102782

Marshall, D. T., & Naff, D. B. (2024). The ethics of using artificial intelligence in qualitative research. *Journal of Empirical Research on Human Research Ethics; JERHRE, 19*(3), 92–102. DOI: 10.1177/15562646241262659 PMID: 38881315

Mökander, J., & Floridi, L. (2023). Auditing of AI: Legal, ethical and technical approaches. *Digital Society : Ethics, Socio-Legal and Governance of Digital Technology, 2*(3), 38. DOI: 10.1007/s44206-023-00074-y

Obreja, D.-M. (2024). Bridging awareness and resistance: Using algorithmic literacies to contest algorithmic governance. *Big Data & Society, 11*(2), 1–13. DOI: 10.1177/20539517241296046

Ossa, L. A., Salcedo, N., & Gómez, C. (2024). Integrating ethics in AI development: A qualitative study. *BMC Medical Ethics, 25*(1), 12. DOI: 10.1186/s12910-023-00975- PMID: 38297294

Punziano, G. (2025). Adaptive epistemology: Embracing generative AI as a (post-digital) paradigm in social science research. *Societies (Basel, Switzerland), 15*(7), 205. DOI: 10.3390/soc15070205

Roumbanis, L. (2025). On algorithmic mediations. *Social Sciences Information. Information Sur les Sciences Sociales, 64*(1), 3–20. DOI: 10.1177/13684310251319677

Salganik, M. J. (2022). *Bit by bit: Social research in the digital age* (Updated ed.). Princeton University Press.

Scheuerman, M. K., Sharma, A., & Sanches, P.. (2024). Products of positionality: How tech workers shape identity through design. *Proceedings of the ACM on Human-Computer Interaction.* DOI: 10.1145/3613904.3641890

Shneiderman, B. (2022). *Human-Centered AI.* Oxford University Press., DOI: 10.1093/oso/9780192845290.001.0001

Singhal, A., Kataria, S., & Mohanty, A.. (2024). Toward fairness, accountability, transparency, and ethics in AI for health information on social media platforms. *JMIR Medical Informatics, 12*, e50048. DOI: 10.2196/50048 PMID: 38568737

Srivastava, S. (2023). Algorithmic governance and the international politics of Big Tech. *Perspectives on Politics, 21*(3), 989–1000. DOI: 10.1017/S1537592721003145

Suchman, L. (2023). The uncontroversial “thingness” of AI. *Big Data & Society, 10*(2), 20539517231206794. Advance online publication. DOI: 10.1177/20539517231206794

Taylor, S. H., & Choi, M. (2022). An initial conceptualization of algorithm responsiveness. *Social Media + Society, 8*(4), 1–12. DOI: 10.1177/20563051221144322

Tracy, S. J. (2020). *Qualitative research methods: Collecting evidence, crafting analysis, communicating impact* (2nd ed.). Wiley-Blackwell.

Ukanwa, K. (2024). Algorithmic bias: Social science research integration and implications. *Journal of Behavioral and Experimental Economics, 110*, 102186. DOI: 10.1016/j.socec.2024.102186

Williams, A. (2024). Re-interpreting interpretivism: Artificial intelligence and the future of qualitative meaning. *Frontiers in Sociology, 9*, 145–160. DOI: 10.3389/fsoc.2024.123456

Section 3
Artificial Intelligence Literacy in Research

Chapter 13
Rethinking Qualitative Methodology Through the Lens of AI Literacy:
Navigating Trustworthiness, Reflexivity, and Human–AI Collaboration

Freda Awonakie Akwetey
http://orcid.org/0009-0002-4013-2668
University of South Africa, South Africa

Francis Arthur
http://orcid.org/0000-0001-7792-6563
University of Cape Coast, Ghana

ABSTRACT

This chapter advances the AI-Literate Qualitative Methodology (AILQM) framework as an integrative model for embedding artificial intelligence within qualitative research while safeguarding epistemological integrity, methodological rigour, and ethical responsibility. It articulates six interrelated literacy dimensions: Cognitive and Data Literacy, Ethical and Equity Literacy, Methodological Literacy, Collaborative Literacy, Trustworthiness Literacy, and Policy and Governance Literacy. Together, these competencies enable researchers to critically interpret AI outputs, document human-AI decision-making, and uphold credibility, dependability, confirmability, and transferability. By reframing AI literacy as a methodological orientation rather

DOI: 10.4018/979-8-3373-7640-0.ch013

ht © 2026, IGI Global Scientific Publishing. Copying or distributing in print or electronic forms without written permission of IGI Global Scientific Publishing is prohibited. Use of this chapter to generative artificial intelligence (AI) technologies is expressly prohibited. The publisher reserves all rights to license its use for generative AI training and machine learning model development.

than a technical skill, the AILQM framework provides a principled pathway for conducting reflexive, transparent, and human-centred qualitative research in the evolving AI landscape.

1. INTRODUCTION

The accelerating integration of Artificial Intelligence (AI) into academic research is producing a profound methodological shift across disciplines. Within qualitative inquiry, the emergence of large language models (LLMs), natural language processing (NLP), and generative AI systems is transforming how researchers transcribe, code, interpret, and represent qualitative data (Bandi et al., 2023). These technologies promise increased efficiency and analytic augmentation; however, they simultaneously challenge long-standing assumptions about authorship, interpretation, reflexivity, and methodological trustworthiness within interpretivist and constructivist research traditions (Kalaitzidis, 2025).

Qualitative research has historically positioned the researcher as the primary instrument of analysis, an interpretive agent whose positionality, reflexivity, and contextual awareness shape the process of knowledge construction (Marshall & Naff, 2024). The introduction of AI into analytic processes complicates this assumption. AI systems can now participate in tasks once considered inherently human, including generating preliminary codes, clustering themes, summarising narratives, translating interviews, and drafting analytic memos (Christou, 2023). While these capabilities expand the analytical toolkit available to researchers, they also raise critical methodological questions: How does algorithmic mediation reshape interpretive authority? What becomes of reflexivity when elements of analysis are partially automated? And how can the principles of qualitative trustworthiness be preserved within AI-assisted research processes?

Current scholarship addressing AI in research has focused predominantly on issues of ethics, academic integrity, and regulatory governance (Pattyn, 2025). While these concerns are essential, they represent only part of the methodological challenge. Emerging discussions increasingly highlight the need for researchers to develop forms of AI literacy capable of supporting critical engagement with algorithmic systems (Ikonen, 2025). However, existing conceptualisations of AI literacy often frame it narrowly as technical proficiency or digital skill. Such interpretations are insufficient for qualitative inquiry, where the central methodological concern lies not merely in using AI tools but in understanding how these systems influence interpretation, reflexivity, and epistemological accountability.

A key limitation in the current literature is therefore the absence of a coherent methodological framework that situates AI literacy within the epistemological foundations of qualitative research. Although studies have acknowledged risks such as algorithmic bias, hallucinated outputs, and the potential outsourcing of analytic judgement (Keogh, 2025), fewer contributions have articulated how qualitative researchers should systematically integrate AI while preserving interpretive integrity. This gap reflects a broader disconnect between the rapid adoption of AI tools in research practice and the slower development of methodological frameworks capable of guiding their responsible use (Perkins & Roe, 2025).

Moreover, other studies (e.g., Nguyen-Trung, 2025; Zhang et al., 2025) increasingly explore how artificial intelligence can support qualitative research processes, particularly through large language models and AI-assisted analytic tools. Evidence indicates that AI can enhance efficiency in qualitative workflows by assisting with tasks such as transcription, coding, thematic organisation, and data summarisation (Nguyen-Trung, 2025; Zhang et al., 2025). Studies further show that well-designed prompts and improved understanding of AI capabilities can strengthen researchers' trust in AI-generated outputs and improve transparency in human-AI interaction (Zhang et al., 2025). Comparative investigations between human-only and machine-assisted analysis also indicate that AI can substantially reduce analytic time and identify patterns within large qualitative datasets, although interpretive nuances such as ambiguous or context-dependent meanings remain difficult for AI systems to capture (Martin et al., 2026). Other research highlights the value of hybrid approaches in which AI supports analytic processes while human researchers retain interpretive oversight, particularly when prompts and analytic procedures are carefully designed (Akman et al., 2025).

Beyond data analysis, recent studies also examine the role of AI across broader qualitative research workflows, including data collection contexts such as virtual focus groups. In such settings, AI-generated summaries can improve organisational efficiency, although human moderation remains necessary to preserve contextual depth and emotional richness in discussions (Chen et al., 2025). Emerging conceptual work further frames AI as a collaborative research partner capable of contributing to distributed cognition within qualitative inquiry, while still requiring researchers to maintain reflexive responsibility for interpretation and knowledge production (Costa et al., 2025). Despite these advances, the literature remains fragmented and often focuses on specific applications of AI such as prompt engineering, thematic analysis support, or human–AI comparison experiments (Akman et al., 2025; Nguyen-Trung, 2025; Zhang et al., 2025). Consequently, there is limited theoretical work that synthesises these developments into a coherent methodological framework capable of guiding the responsible and reflexive integration of AI within qualitative research practice. This gap highlights the need for conceptual approaches that move beyond tool-specific applications to articulate the competencies and methodological principles necessary for AI-literate qualitative inquiry.

This chapter addresses that gap by proposing the AI-Literate Qualitative Methodology (AILQM) framework, a conceptual model that positions AI literacy as a multidimensional methodological competency encompassing cognitive, ethical, methodological, collaborative, trustworthiness, and governance dimensions. Rather than viewing AI as either a threat to qualitative authenticity or a neutral productivity tool, the framework conceptualises AI as an augmentative collaborator embedded within a human-AI analytical partnership whose integration must be governed by reflexive awareness, transparency, and epistemic accountability. By situating AI literacy within broader methodological debates surrounding qualitative trustworthiness, reflexivity, and interpretive authority, the chapter contributes to the evolving discourse on AI-assisted research. In doing so, it argues that AI literacy should be understood not as an optional digital skill but as a necessary condition for sustaining methodological rigour and interpretive credibility in an increasingly AI-augmented research landscape.

2. THE RISE OF AI IN QUALITATIVE RESEARCH PRACTICE

The rapid advancement of artificial intelligence (AI), particularly large language models (LLMs) and natural language processing (NLP) systems, has significantly transformed qualitative research practice. AI is no longer confined to peripheral administrative functions but increasingly participates in analytic processes traditionally reserved for human researchers (Bender et al., 2021; Dwivedi et al., 2023).

One of the earliest and most widespread applications is automated transcription. Deep learning-based speech recognition systems have improved the speed and affordability of converting audio interviews and focus group recordings into text (Kanbach et al., 2024). However, concerns remain regarding accuracy across diverse accents, dialects, and multilingual contexts, especially in underrepresented linguistic communities (Adarkwah, 2024; Buolamwini & Gebru, 2018). Transcription errors are not merely technical issues; they can shape interpretation and influence coding outcomes.

AI is also increasingly used for preliminary thematic coding and data clustering. NLP models can identify co-occurrence patterns, semantic similarity, and latent topic structures across large datasets. While this enhances efficiency, algorithmic coding is fundamentally probabilistic rather than interpretive. It identifies patterns based on statistical association rather than lived meaning or contextual nuance (Bender et al., 2021; Floridi & Chiriatti, 2020). The risk lies in conflating computational pattern recognition with qualitative understanding.

Similarly, sentiment and discourse analysis tools classify emotional tone or rhetorical positioning within textual corpora. These techniques are particularly common in digital ethnography and social media research. However, scholars caution that computational sentiment models often reduce complex affective states into simplified positive-negative binaries, overlooking ambiguity and cultural context (Crawford, 2021; Noble, 2018).

The rise of generative AI has expanded applications to text summarisation and data reduction. LLMs can synthesise lengthy transcripts into condensed thematic overviews within seconds. Although such summarisation increases productivity, it involves algorithmic filtering shaped by training data and model architecture. These systems may foreground dominant linguistic patterns while marginalising subtle, culturally embedded, or contradictory meanings (Bender et al., 2021; Dwivedi et al., 2023).

In multilingual research settings, AI-enabled translation of qualitative data offers practical advantages. Neural machine translation can provide initial translations of interviews conducted in diverse languages. Yet translation is an interpretive act, embedded in cultural and contextual meaning-making. Over-reliance on automated systems may distort idiomatic expressions or culturally specific concepts (Adarkwah, 2024; Floridi & Chiriatti, 2020).

More recently, researchers have begun using generative AI for drafting analytic memos and reports, organising themes, and refining academic prose. While such tools may augment productivity, they introduce new questions regarding epistemic ownership, intellectual accountability, and the boundaries of authorship (Dwivedi et al., 2023; Tlili et al., 2023).

Collectively, these applications introduce significant efficiency gains but also epistemological tension. AI systems can process vast textual corpora rapidly and generate human-like outputs; however, they lack contextual consciousness, embodiment, and reflexive awareness. As emerging scholarship on GenAI in education and research demonstrates, AI outputs may appear authoritative yet require sustained human oversight to prevent misinterpretation, bias amplification, or fabrication (Adarkwah, 2024; Tlili et al., 2023).

For qualitative researchers grounded in interpretivist and constructivist paradigms, the central challenge is not whether AI can assist analytic processes, but how its integration reshapes interpretation, reflexivity, and trustworthiness. The rise of AI in qualitative research therefore demands critical AI literacy rather than uncritical adoption or categorical rejection.

3. DEFINING AI LITERACY IN QUALITATIVE INQUIRY

AI literacy in qualitative research extends beyond technical proficiency with digital tools. It refers to a researcher's critical capacity to understand, interrogate, and responsibly integrate AI systems within interpretive inquiry. Recent scholarship conceptualises AI literacy as encompassing technical understanding, ethical awareness, and critical engagement with socio-technical systems (Arkorful et al., 2025; Long & Magerko, 2020; Ng et al., 2021; UNESCO, 2023). Within qualitative inquiry, however, AI literacy assumes a distinctly epistemological dimension.

First, AI literacy requires the ability to understand how AI systems generate outputs (Abreh et al., 2025). Large language models (LLMs) and NLP systems operate through probabilistic pattern prediction trained on vast corpora rather than through contextual understanding or intentional reasoning (Bender et al., 2021; Bommasani et al., 2021). Researchers must therefore recognise that AI-generated codes, summaries, and interpretations are statistical approximations rather than situated meaning-making.

Second, AI literacy involves the capacity to critically evaluate AI-generated codes and analytic summaries. Algorithmic outputs may appear coherent and authoritative, yet they are shaped by training data distributions and optimisation objectives (Dwivedi et al., 2023; Floridi & Chiriatti, 2020). Without critical scrutiny, researchers risk mistaking fluency for validity. AI-literate researchers interrogate the alignment between computational pattern detection and theoretically grounded interpretation.

Third, AI literacy includes the competence to detect and mitigate algorithmic bias. Bias may emerge from skewed training datasets, underrepresentation of linguistic or cultural groups, or embedded normative assumptions (Crawford, 2021; Noble, 2018). In qualitative research, such bias can distort thematic emphasis, silence marginal voices, or reinforce dominant narratives. Recognising these structural limitations is central to maintaining epistemic justice.

Fourth, AI literacy demands transparency in AI usage. Emerging research ethics guidelines emphasise disclosure of AI involvement in data analysis, documentation of prompts, and clear delineation between human and AI-generated contributions (European Commission, 2023; UNESCO, 2023). Transparency strengthens credibility and allows peer scrutiny of analytic processes.

Finally, AI literacy safeguards methodological trustworthiness. In the tradition of Lincoln and Guba's (1985) criteria of credibility, dependability, confirmability, and transferability, AI integration must not erode reflexivity or analytic rigour. Instead, AI should function as an augmentative tool within a reflexively grounded interpretive framework.

Thus, AI literacy becomes an epistemic safeguard - ensuring that AI operates as an assistive collaborator rather than a substitute for human interpretive authority. In qualitative inquiry, where meaning is co-constructed and context-bound, AI literacy is not optional technical knowledge; it is a methodological necessity.

4. THEORETICAL UNDERPINNINGS

4.1. From Positivist Grounded Theory to Constructivist Reflexivity in the Age of AI

The epistemological trajectory of qualitative research provides critical context for understanding AI's contemporary methodological implications. Early grounded theory (GT), introduced by Glaser and Strauss (1967), emerged during a period dominated by post-war positivism. At that time, social science research largely prioritised the testing of pre-existing theories over theory generation (Glaser & Strauss, 1967; Timonen et al., 2018). Classical GT sought to restore theory-building to empirical inquiry through systematic and rigorous procedures. Its philosophical orientation assumed that theory could be inductively discovered from data, reflecting an objectivist belief that social patterns "exist" independently of the researcher (Annells, 1996; Charmaz, 2006). While this formalisation strengthened qualitative research's legitimacy within positivist academic traditions, it retained a commitment to methodological structure and analytic neutrality.

The first significant epistemological shift occurred with the development of Straussian GT. Strauss and Corbin (1998) reoriented GT away from strict positivism toward a more interpretive stance, acknowledging that researchers inevitably bring theoretical sensitivity and conceptual frameworks into analysis (Charmaz, 2000; Strauss & Corbin, 1998). Rather than discovering theory in a vacuum, researchers construct meaning through interaction with data. This shift softened the classical insistence on entering the field without engagement with prior literature and introduced a more reflexive understanding of analytic practice (Heath & Cowley, 2004; Jones & Alony, 2011). Although Straussian GT maintained procedural rigour, it was philosophically grounded in realism and symbolic interactionism rather than strict objectivism (Kenny & Fourie, 2015).

Constructivist grounded theory, as articulated by Charmaz (2006), further deepened this interpretive turn. Here, theory is not discovered but co-constructed through the researcher's situated engagement with participants, contexts, and prior knowledge (Charmaz, 2006, 2014). Constructivist GT reframed methodology as heuristic rather than prescriptive, positioning reflexivity as central to epistemological credibility (Bryant, 2007). This development marked qualitative research's broader movement from positivism toward constructivism, privileging context, interaction, and interpretive depth.

However, the growing integration of artificial intelligence into qualitative inquiry introduces a potential counter-movement. AI systems particularly machine learning (ML) and natural language processing (NLP) models such as large language models (LLMs) operate through pattern detection, probabilistic prediction, and structured data processing. These logics mirror the positivist emphasis on regularity, repeatability, and observable pattern extraction characteristic of classical GT. Contemporary scholarship notes AI's capacity to enhance transcription efficiency, automate coding, and identify descriptive themes at scale (Hamilton et al., 2023; Parker et al., 2023; Tai et al., 2024). AI tools can sometimes detect patterns overlooked by human analysts, thereby complementing manual analysis (Morgan, 2023).

Yet important epistemological tensions remain. While AI demonstrates strength in identifying surface-level and descriptive regularities, it struggles with deeply contextual, interpretive, and meaning-laden themes that require situated human judgement (Morgan, 2023). Concerns have also been raised regarding hallucinated outputs, inaccuracies, prompt dependency, and ethical risks, particularly when researchers lack critical AI literacy (Chubb, 2023; Liu et al., 2023). Moreover, AI systems may embed or reproduce structural biases present in their training data (Varsha, 2023), reinforcing power asymmetries rather than challenging them.

While grounded theory provides a useful illustration of the epistemological evolution of qualitative methodology, other qualitative traditions such as ethnography and phenomenology offer additional insight into the methodological tensions created by artificial intelligence. These approaches place even greater emphasis on embodiment, lived experience, and contextual immersion, raising distinct questions regarding the role of computational tools in qualitative research.

Ethnography, rooted in anthropology and interpretive sociology, is fundamentally concerned with understanding social practices through prolonged engagement with communities and cultural contexts (Hammersley & Atkinson, 2019). Ethnographic knowledge is produced through participant observation, informal interaction, field notes, and the researcher's embodied presence within the social environment being studied. The ethnographer's sensory experiences, positionality, and interpretive judgement therefore become integral components of knowledge production (Pink et al., 2022; Pink et al., 2020).

The integration of AI into ethnographic research presents both opportunities and limitations. On one hand, AI-powered tools can support the management of large volumes of ethnographic data, including field notes, interview transcripts, and audiovisual materials. Natural language processing and machine learning algorithms can assist researchers in identifying recurring linguistic patterns, thematic clusters, or interactional structures within large qualitative datasets (Hamilton et al., 2023). Such tools may therefore enhance analytic efficiency and enable researchers to explore patterns that might otherwise remain unnoticed.

However, ethnography's epistemological commitment to situated and embodied knowledge creates important tensions with AI-based analytical systems. AI operates primarily through computational abstraction and statistical modelling, whereas ethnographic insight often emerges from subtle contextual cues such as tone, gestures, spatial arrangements, or power dynamics within social interactions (Pink et al., 2016). These forms of tacit and experiential knowledge are difficult to translate into machine-readable data structures. Consequently, while AI may assist with descriptive categorisation, it cannot substitute for the ethnographer's interpretive immersion in the field.

Phenomenology presents an even more pronounced epistemological contrast. Originating in the philosophical work of Husserl (1960; 1970) and later expanded by scholars such as Heidegger and Merleau-Ponty, phenomenology seeks to understand how individuals experience and interpret phenomena in their lived reality (Heidegger, 2005; Merleau-Ponty et al., 2013; Smith, 2018; Smith et al., 2009). Phenomenological research therefore focuses on subjective meaning-making, intentionality, and the structures of lived experience (Zahavi, 2019). Data analysis typically involves deep engagement with participants' narratives in order to uncover essential meanings underlying those experiences.

Within this methodological tradition, AI-assisted analysis raises complex philosophical questions. Because phenomenology prioritises the first-person perspective and embodied consciousness, computational systems that rely on probabilistic pattern recognition may struggle to capture the richness and nuance of lived experience. AI may assist researchers in organising narrative data or identifying linguistic patterns within transcripts, but the interpretive act of discerning the essence of experience remains inherently human. As Morgan (2023) argues, algorithmic pattern detection may reveal recurring textual structures but cannot easily grasp the existential meaning embedded in human narratives.

Nevertheless, AI may still play a supportive role in phenomenological research when used cautiously and reflexively. For example, AI tools may assist researchers in exploring alternative thematic interpretations, generating analytic prompts, or organising large qualitative datasets. When embedded within a reflexive research process, such tools can function as analytic aids rather than epistemological substitutes. In this sense, AI may expand the researcher's interpretive horizon while leaving the ultimate responsibility for meaning-making with the human analyst.

Generally, ethnography and phenomenology demonstrate that the implications of AI in qualitative research extend beyond grounded theory. Methods that rely heavily on embodiment, immersion, and lived experience highlight the limitations of computational analysis and reinforce the need for human interpretive authority in qualitative inquiry. Recent methodological innovations such as Computational Grounded Theory (CGT) attempt to reconcile these tensions by integrating computational analysis with human interpretive oversight (Nelson, 2020). Rather than replacing constructivist commitments, such hybrid approaches position AI as a structured analytic assistant within a reflexive interpretive framework. Nevertheless, scholars caution that AI's emphasis on prediction, regularity, and data-driven inference may inadvertently revive positivist orientations within qualitative inquiry (Christou, 2023). Table 1 presents the epistemological evolution of grounded theory and AI integration in qualitative research.

Table 1. Epistemological Evolution of Grounded Theory and AI Integration in Qualitative Research

Epistemological Phase	Core Assumptions About Knowledge	Role of Researcher	Methodological Orientation	AI Alignment or Tension	Key Scholarly Anchors
Classical Grounded Theory	Theory exists in data and can be inductively discovered; emphasis on objectivity and empirical regularities	Neutral observer extracting patterns from data	Structured, systematic coding procedures	Strong alignment with AI's data-driven pattern detection and repeatability	"(Glaser & Strauss, 1967; Annells, 1996; Charmaz, 2006)"
Straussian Grounded Theory	Theory is systematically constructed through interaction between data and researcher	Interpretive analyst with theoretical sensitivity	Inductive–deductive interplay; structured but reflexive coding	Partial alignment; AI can assist structure but may underrepresent interpretive nuance	"(Strauss & Corbin, 1998; Heath & Cowley, 2004; Kenny & Fourie, 2015)"
Constructivist Grounded Theory	Knowledge is co-constructed through situated engagement; meaning is context-bound	Reflexive co-constructor of meaning	Heuristic, flexible, interpretive methodology	Epistemological tension; AI struggles with contextual and interpretive depth	"(Charmaz, 2006, 2014; Bryant, 2007)"
Ethnographic Research Traditions	Social reality is understood through cultural immersion, lived practices, and contextual interpretation	Embedded observer engaging with participants and environments	Participant observation, fieldwork, and thick description	AI can assist with data organisation and pattern detection but struggles with embodied, situational understanding	"(Hammersley & Atkinson, 2019; Pink et al., 2022)"
Phenomenological Inquiry	Knowledge emerges from individuals' lived experiences and subjective meaning-making	Interpretive analyst uncovering the essence of lived experience	Narrative interpretation and phenomenological reduction	AI may assist with textual organisation but cannot fully capture experiential meaning and consciousness	"(Smith et al., 2009; van Manen, 2016)"

continued on following page

Table 1. Continued

Epistemological Phase	Core Assumptions About Knowledge	Role of Researcher	Methodological Orientation	AI Alignment or Tension	Key Scholarly Anchors
AI-Enhanced Qualitative Inquiry	Patterns identified through computational regularity; probabilistic prediction informs analysis	Human-AI collaborative analyst; requires oversight	Hybrid computational-interpretive approaches	Risk of positivist resurgence; requires reflexive governance	"(Hamilton et al., 2023; Morgan, 2023; Nelson, 2020; Christou, 2023)"
Post-Qualitative / Sociomaterial Perspectives	Knowledge emerges through entanglements between human and non-human actors, including technologies, algorithms, and data infrastructures	Researcher becomes part of a distributed network of human and technological actors shaping inquiry	Experimental, relational, and non-representational approaches that question traditional methodological boundaries	AI is viewed not only as a tool but as a methodological actor influencing how patterns, interpretations, and analytic possibilities emerge	"(St. Pierre, 2019; Aagaard, 2022; Leonelli, 2023)"

Beyond these debates, emerging scholarship in post-qualitative inquiry further complicates the relationship between artificial intelligence and qualitative research. Post-qualitative approaches challenge conventional human-centred epistemologies by recognising the agency of non-human actors, including technologies, algorithms, and data infrastructures, in the production of knowledge (Aagaard, 2022; St. Pierre, 2019; 2021; 2024). From this perspective, AI is not merely an analytic tool used by researchers but part of a broader sociotechnical assemblage that participates in shaping methodological processes and interpretive possibilities. Scholars argue that computational systems, prompts, datasets, and algorithmic outputs can influence how qualitative patterns emerge and how interpretations are constructed, thereby redistributing epistemic authority within the research process (Leonelli, 2023). Rather than positioning AI solely as an efficiency-enhancing instrument, post-qualitative perspectives invite researchers to critically examine how human and non-human actors collaboratively produce knowledge. This perspective expands the methodological debate by suggesting that AI may function as a methodological partner within research assemblages, raising new questions about reflexivity, responsibility, and the governance of algorithmically mediated qualitative analysis.

4.2. Implications for AILQM

Within this epistemological landscape, the AI-Literate Qualitative Methodology (AILQM) Framework responds directly to the risk of epistemic regression. By embedding AI integration within constructivist epistemology, critical digital literacy, and human-centred AI principles, AILQM safeguards interpretive authority while leveraging computational affordances. It acknowledges that AI's strengths lie in pattern detection and organisational efficiency, but maintains that meaning remains co-constructed through reflexive human judgement. Thus, AILQM does not reject AI's analytic potential; rather, it situates AI within a constructivist governance architecture that prioritises reflexivity, ethical vigilance, contextual sensitivity, and epistemological transparency.

Drawing on interpretivist and constructivist traditions, the proposed AI-Literate Qualitative Methodology (AILQM) Framework conceptualises AI literacy as a multidimensional and integrative construct anchored in a Human–AI Collaborative Process. Rather than reducing AI literacy to a limited set of competencies, AILQM advances six interrelated literacy domains that collectively sustain epistemological integrity, methodological rigour, and ethical responsibility in AI-mediated qualitative inquiry. These domains include:

1. **Cognitive and Data Literacy** – Developing critical understanding of algorithmic architectures, probabilistic reasoning, data-processing logics, and the epistemic limitations of AI systems, alongside the capacity to interrogate data provenance, quality, and embedded bias.
2. **Ethical and Equity Literacy** – Ensuring responsible data governance, informed consent, privacy protection, and reflexive engagement with algorithmic bias, structural inequities, and power asymmetries.
3. **Methodological Literacy** – Integrating AI tools within qualitative workflows in ways that preserve reflexivity, transparency, auditability, and interpretive authority.
4. **Collaborative Literacy** – Cultivating the capacity to engage AI as a dialogic analytic partner, facilitating iterative co-construction of meaning while maintaining human interpretive centrality.
5. **Trustworthiness Literacy** – Safeguarding credibility, dependability, confirmability, and transferability through systematic documentation and critical evaluation of AI's role across analytic stages.
6. **Policy and Governance Literacy** – Situating AI integration within legal, cultural, and institutional frameworks that regulate data use, privacy, and research ethics.

AILQM thus functions as a comprehensive heuristic and governance-oriented scaffold for embedding AI responsibly into qualitative research, ensuring that technological innovation remains aligned with the foundational commitments of interpretivist and constructivist inquiry.

5. ETHICAL DILEMMAS AND REFLEXIVITY IN AI-ASSISTED INQUIRY

The integration of AI into qualitative research introduces a range of ethical dilemmas that extend beyond conventional concerns of confidentiality and consent. In interpretive inquiry, ethics is not merely procedural but relational and reflexive (Finlay, 2017; Tracy, 2010). AI-assisted analysis complicates this relational ethic by introducing non-human systems into processes of meaning-making.

One central dilemma concerns informed consent and secondary data use. When qualitative data are processed through AI systems - particularly cloud-based or proprietary platforms - researchers may inadvertently expose participant data to third-party infrastructures. Recent scholarship highlights the opacity of data processing pipelines in AI ecosystems and the difficulty of ensuring full transparency regarding where and how data are stored or reused (Kitchin, 2021; Mittelstadt, 2019). In contexts involving vulnerable populations, this raises significant ethical questions about data sovereignty and participant autonomy.

A second dilemma involves epistemic authority and interpretive displacement. Qualitative research traditionally centres the researcher's reflexive engagement with data. However, when AI tools generate codes, themes, or summaries, there is a risk that researchers may defer interpretive authority to algorithmic outputs. Zuboff (2019) argues that algorithmic systems can subtly reshape human judgement by structuring what is visible and actionable. Within qualitative inquiry, this may lead to the normalisation of computational framings that marginalise alternative or counter-hegemonic interpretations.

Third, AI-assisted inquiry raises concerns about reproduction of structural bias and epistemic injustice. Scholars of digital inequality have shown that algorithmic systems often embed dominant cultural narratives and underrepresent marginal voices (Benjamin, 2019; Eubanks, 2018). When applied to qualitative data, such systems may inadvertently prioritise dominant discursive patterns while attenuating culturally specific or minoritised expressions. This is particularly consequential in postcolonial or Global South research contexts, where linguistic nuance and socio-historical positionality are central to interpretation.

Reflexivity therefore becomes more complex in AI-mediated research. Finlay (2017) conceptualises reflexivity as ongoing critical self-awareness regarding positionality, power, and interpretive influence. In AI-assisted inquiry, reflexivity must extend to the researcher's technological choices: Why was a particular AI tool selected? How were prompts framed? What assumptions were embedded in algorithmic suggestions? How might the tool's training corpus shape analytic outputs? This expanded reflexivity aligns with calls for "algorithmic reflexivity" in digital research (Pink et al., 2018).

Finally, AI-assisted inquiry challenges notions of accountability. If analytic interpretations emerge from human–AI collaboration, who bears responsibility for error, bias, or misrepresentation? Responsible innovation scholarship argues that accountability must remain human-centred, even when technological systems are involved (Macnaghten, Owen & Stilgoe, 2013). Researchers must therefore retain interpretive ownership and clearly document the role AI played in analytic processes.

Ethical dilemmas in AI-assisted qualitative research are not peripheral concerns - they strike at the heart of interpretivist methodology. Rather than diminishing reflexivity, AI integration demands its expansion. Researchers must critically examine not only their own positionality but also the socio-technical infrastructures that co-shape meaning-making.

6. METHODOLOGICAL TRANSFORMATIONS AND THE "BLACK-BOX" PROBLEM

The incorporation of artificial intelligence into qualitative research practice is reshaping established methodological conventions, particularly in relation to data analysis, interpretation, and standards of rigour (Costa et al., 2025; Dellafiore et al., 2025). AI-assisted tools for transcription, coding, pattern recognition, and summarisation are increasingly embedded within qualitative workflows, altering how researchers engage with data and make analytic decisions (Bryda & Sadowski, 2024). While these developments offer opportunities for efficiency and scalability, they also introduce methodological challenges that require careful scrutiny, especially in relation to the "black-box" nature of many AI systems.

One significant methodological transformation concerns the shifting locus of analytic labour. Traditionally, qualitative analysis has relied on prolonged immersion in data, iterative coding, and interpretive sense-making by the researcher (Costa et al., 2025). AI tools can accelerate these processes by generating preliminary codes, clustering themes, or identifying linguistic patterns across large datasets. However, such automation risks reconfiguring qualitative analysis into a process that privileges speed over depth if not critically managed. AI literacy becomes essential in ensuring that automated outputs are treated as analytic aids rather than substitutes for interpretive judgement. Methodologically robust qualitative research in an AI-enhanced context requires researchers to remain actively engaged in validating, refining, and contextualising AI-generated insights.

The "black-box" problem presents a central methodological concern in AI-assisted qualitative inquiry. Many AI systems, particularly proprietary large language models and machine learning algorithms, operate with limited transparency regarding how inputs are transformed into outputs. This opacity challenges core qualitative principles such as transparency, dependability, and confirmability (Jones, 2025; Pappachan et al., 2024). When researchers cannot fully explain how an AI tool arrived at a particular coding suggestion or thematic structure, the auditability of the analytic process is compromised. As a result, methodological trustworthiness increasingly depends on researchers' ability to document AI use, articulate its role in analysis, and clarify where human interpretation intervenes (Friese, 2026).

Methodological transformations are also evident in the evolving relationship between inductive and deductive reasoning in qualitative research. AI-assisted analysis often relies on probabilistic pattern recognition informed by prior training data, which may subtly introduce deductive or quasi-deductive tendencies into ostensibly inductive designs such as grounded theory or phenomenology (Papageorgiou et al., 2025). Without reflexive oversight, this can lead to premature thematic closure or the privileging of dominant discourses embedded in training corpora (Bender et al., 2021). Researchers must therefore critically assess how AI tools shape analytic trajectories and remain attentive to the risk of constraining emergent meanings.

Addressing the black-box problem requires a methodological reorientation rather than outright rejection of AI tools. Emerging scholarship advocates for human-in-the-loop approaches, in which AI systems are deliberately embedded within iterative cycles of human interpretation, verification, and reflexive critique (Shneiderman, 2020). In qualitative research, this involves treating AI outputs as provisional prompts that stimulate analytic reflection rather than definitive representations of meaning. Methodological transparency can be further enhanced through detailed reporting of prompt design, analytic stages, and decision points, allowing readers to assess the credibility of AI-assisted analyses.

Ultimately, the methodological transformations brought about by AI demand an expansion of qualitative rigour to account for algorithmic mediation. The black-box problem underscores the necessity of AI literacy as a methodological competence that enables researchers to navigate opacity, maintain interpretive control, and uphold the standards of qualitative trustworthiness. By embedding AI within reflexive, transparent, and human-led analytic frameworks, qualitative researchers can leverage technological innovation without undermining the epistemological foundations of their work.

7. THE AI-LITERATE QUALITATIVE METHODOLOGY

7.1. The AI-Literate Qualitative Methodology (AILQM) Framework

The AI-Literate Qualitative Methodology (AILQM) Framework (see Figure 1) presents an integrative and structured model for embedding artificial intelligence within qualitative research while maintaining epistemological integrity, methodological rigour, and ethical responsibility. Central to the framework is the Human-AI Collaborative Process, which positions AI not as an autonomous analytic authority but as a reflexive partner in knowledge production. This central hub recognises that AI outputs must be interpreted, validated, and integrated through human judgement to preserve the interpretive and contextual depth that defines qualitative inquiry.

Surrounding the central hub are six interrelated literacy dimensions, each representing a critical competency for AI-literate qualitative research. Cognitive and Data Literacy focuses on researchers' understanding of AI systems, including their mechanisms, limitations, and data-processing characteristics. This dimension also emphasises the evaluation of data quality, provenance, and bias, ensuring that AI-generated insights are interpreted as probabilistic and context-dependent rather than objective or definitive. Through this literacy, researchers are equipped to critically appraise AI outputs and integrate them meaningfully into their analyses.

Ethical and Equity Literacy foregrounds researchers' obligations to uphold participant rights, ensure informed consent, and protect sensitive data (Badawy et al., 2025). This dimension extends ethical practice by incorporating reflexive engagement with algorithmic bias, social inequities, and power asymmetries embedded in AI tools. By operationalising ethical literacy as a continuous reflexive process rather than a procedural requirement, the framework promotes research practices that are socially responsible, inclusive, and equitable. A critical component embedded within the Ethical and Equity Literacy dimension of the AILQM framework is the concept of epistemic justice. Epistemic justice refers to the fair recognition and representation of diverse knowledge systems and voices within processes of knowledge production.

In AI-assisted qualitative research, epistemic injustice may arise when algorithmic systems reproduce biases embedded in their training datasets. Such biases may privilege dominant linguistic or cultural perspectives while marginalising alternative ways of knowing, particularly those associated with indigenous communities or historically underrepresented groups. To address these concerns, Ethical and Equity Literacy within the AILQM framework encourages researchers to engage in reflexive practices that safeguard epistemic justice by:

- critically evaluating AI-generated interpretations for cultural or contextual bias
- ensuring that participant voices remain central to analysis
- validating AI-assisted findings through contextual knowledge and researcher judgement
- recognising forms of local or indigenous knowledge that AI systems may fail to interpret accurately

Methodological Literacy guides the practical integration of AI tools within qualitative workflows. It emphasises the deliberate positioning of AI as a supportive instrument for data organisation, coding, and pattern recognition while ensuring that human interpretation remains central. Methodological literacy also includes rigorous documentation of human–AI decision-making processes, which enhances transparency and facilitates auditability. This dimension aligns with human-centred AI principles and reinforces the interpretive rigour that is foundational to qualitative methodology.

Collaborative Literacy reflects the iterative and relational aspects of human-AI engagement (Sidra & Mason, 2025). It equips researchers to actively co-construct knowledge with AI systems, interpreting outputs reflexively, and adapting analytic strategies in response to AI suggestions. This dimension highlights that AI-mediated qualitative research is inherently collaborative, requiring constant negotiation between human expertise and machine-generated insights. Also, the collaborative literacy dimension of the AILQM framework can be further understood through the concept of dialogic human-AI interaction. Unlike traditional analytic software that performs predetermined computational tasks, large language models (LLMs) operate through interactive dialogue in which researchers iteratively refine prompts, interpret responses, and guide analytic exploration.

This dialogic process reflects principles of reflexive knowledge construction, where meaning emerges through an ongoing exchange between human judgement and algorithmic suggestion. In qualitative research workflows, researchers may engage in multiple conversational cycles with AI systems during stages such as code generation, thematic synthesis, theoretical memoing, and methodological reflection. For example, during preliminary coding, a researcher might request AI-generated suggestions for possible categories based on interview transcripts. The AI output may highlight recurring linguistic patterns or potential thematic groupings. However, rather than accepting these outputs uncritically, the researcher engages in a dialogic process of verification, modification, and reinterpretation, comparing AI suggestions against contextual understanding of the data. Through this iterative interaction, AI becomes a cognitive scaffold that stimulates analytic thinking rather than replacing interpretive reasoning.

Dialogic engagement with LLMs therefore requires several competencies. Researchers must be able to formulate methodologically informed prompts, critically evaluate generated outputs, and recognise the probabilistic nature of AI-generated insights. Furthermore, reflexive documentation of these interactions is necessary to maintain transparency and analytic credibility. This dialogic model aligns with emerging perspectives on human-centred artificial intelligence, which emphasise collaborative rather than autonomous machine intelligence (Shneiderman, 2020). Within the AILQM framework, collaborative literacy therefore represents the capacity to navigate and govern this dialogic relationship, ensuring that AI-mediated interactions enhance analytic insight while preserving the interpretive agency of the researcher.

Trustworthiness Literacy explicitly addresses the qualitative criteria of credibility, dependability, confirmability, and transferability (Ahmed, 2024). By systematically evaluating and documenting the role of AI at each analytic stage, researchers can maintain methodological transparency and enhance the legitimacy of their findings. This dimension ensures that AI is integrated in a manner that strengthens rather than diminishes traditional standards of qualitative rigor.

Finally, Policy and Governance Literacy situates AI engagement within broader structural and institutional contexts (Mukherjee, 2025). This dimension emphasises adherence to legal, cultural, and institutional frameworks that govern data use, privacy, and research ethics. By incorporating policy literacy, the framework recognises that responsible AI integration is contingent not only on individual competence but also on supportive regulatory and organisational environments.

The AILQM framework represents a comprehensive and flexible model for navigating the complexities of AI-assisted qualitative research. By combining cognitive, ethical, methodological, collaborative, and trustworthiness literacies within a contextually informed policy framework, the model provides researchers with a roadmap for leveraging AI's analytical capabilities while safeguarding the humanistic and interpretive commitments central to qualitative inquiry. The framework thus advances AI literacy from a technical skill to a foundational methodological orientation, promoting responsible, reflexive, and rigorous qualitative research in the AI era.

Figure 1. AI-Literate Qualitative Methodology (AILQM) Framework

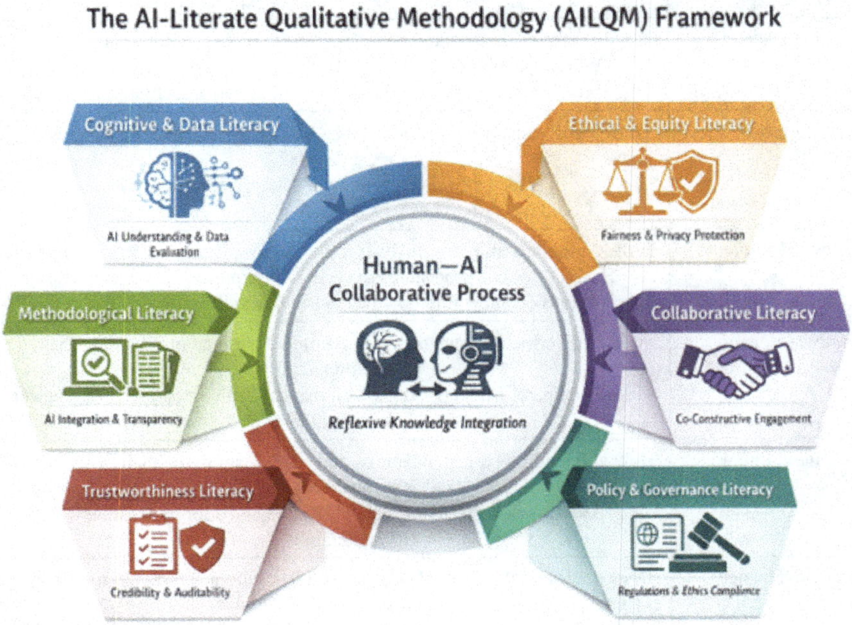

Source: Authors' own work

7.2. Practical Guidance for Applying the AILQM Framework

To enhance the practical applicability of the AI-Literate Qualitative Methodology (AILQM) framework, the framework may be operationalised through structured reflective tools that assist researchers in determining whether and how artificial intelligence should be integrated into qualitative research workflows. Such tools support methodological reflexivity and ensure that AI integration aligns with the epistemological and ethical commitments of qualitative inquiry. One practical mechanism is a self-assessment checklist (see Table 2) that allows researchers to evaluate their preparedness to use AI tools responsibly. This checklist encourages critical reflexivity, ensuring that the researcher remains the primary interpretive authority within AI-assisted qualitative inquiry.

Table 2. Self-Assessment Checklist for AI-Assisted Qualitative Research

Literacy Dimension	Reflective Questions for Researchers
Cognitive and Data Literacy	Do I understand how the AI system generates outputs? What type of data or training information influences its responses? Have I examined potential data biases that may affect interpretation?
Ethical and Equity Literacy	Have participants been informed that AI tools may assist in data processing or analysis? Does the tool comply with data privacy standards? Could the system reproduce social, linguistic, or cultural biases?
Methodological Literacy	Is the AI tool assisting the analysis rather than replacing human interpretation? Have I designed procedures to verify AI outputs against the raw data? Am I documenting how AI contributes to the analytic process?
Collaborative Literacy	Am I critically engaging with AI outputs instead of accepting them automatically? Have I compared AI suggestions with manual coding, theoretical frameworks, or prior literature?
Trustworthiness Literacy	How will credibility, dependability, and confirmability be maintained when AI is involved in analysis? Are AI-assisted decisions clearly documented for transparency and auditability?
Policy and Governance Literacy	Does the use of this AI tool comply with institutional ethics approval and data governance regulations? Are there national or organisational policies guiding AI use in research?

7.3. Human-AI Analytical Decision-Making Framework for AI Use in Qualitative Research

In addition to reflective checklists, researchers may benefit from a structured decision-making framework (see Table 3) that guides the responsible adoption of AI tools during different stages of qualitative research. This framework translates the conceptual principles of AILQM into practical methodological steps, enabling researchers to incorporate AI tools while maintaining analytic transparency.

Table 3. Decision-Making Flow for the Use of AI in Qualitative Research

Stage	Key Question	Researcher Action
1. Define the Analytical Task	Is the task organisational or interpretive in nature?	Use AI for organisational tasks such as transcript sorting, indexing, or preliminary coding suggestions.
2. Evaluate AI Suitability	Can AI assist efficiency without compromising interpretive depth?	Determine whether AI is appropriate for the task or whether manual analysis is preferable.
3. Assess Ethical and Data Considerations	Does the AI tool meet privacy and ethical standards?	Ensure compliance with institutional review board requirements and data protection policies.
4. Implement Human-AI Collaboration	How will AI support the analysis process?	Use AI to generate coding suggestions, keyword clusters, or thematic groupings.
5. Conduct Human Verification	Are AI outputs supported by the original qualitative data?	Compare AI outputs with transcripts, field notes, and theoretical frameworks.
6. Document Analytical Decisions	How will the role of AI be reported?	Maintain a methodological audit trail documenting how AI contributed to analysis.

Figure 2. Human-AI Analytical Decision Cycle

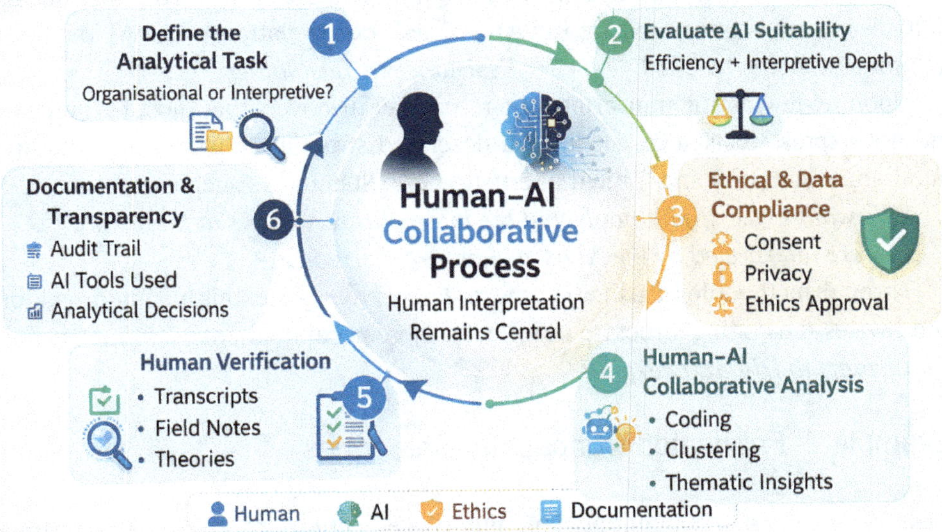

Source: Authors' own work

The Human-AI Analytical Decision Cycle (see Figure 2) illustrates the iterative process through which qualitative researchers integrate artificial intelligence tools into analytical workflows. The cycle begins with defining the analytical task and evaluating the suitability of AI for the research purpose. It then progresses through ethical and data compliance checks before entering the human-AI collaborative analysis stage. Researchers subsequently verify AI-generated outputs through direct comparison with transcripts, field notes, and theoretical frameworks. The process concludes with documentation and transparent reporting of methodological decisions. This cyclical framework emphasises that while AI can enhance analytical efficiency, human interpretation remains central to qualitative inquiry.

7.4 Illustrative Examples of Methodological Literacy in Practice

To demonstrate how the AILQM framework functions in practice, the following hypothetical examples illustrate how methodological literacy enables researchers to critically engage with AI-generated outputs and identify potential analytical inaccuracies. These examples demonstrate how methodological literacy functions as a safeguard against AI hallucinations and interpretive distortions, ensuring that qualitative analysis remains grounded in empirical data rather than algorithmic inference.

Example 1: Misinterpreted Participant Sentiment

A qualitative researcher investigating teachers' experiences with digital learning uploads interview transcripts into an AI-assisted coding platform. The AI identifies a dominant theme labelled *"Teacher Resistance to Technology."*

Upon reviewing the transcripts, the researcher discovers that most participants did not express resistance but instead described structural challenges, including unreliable electricity and limited internet access. Statements such as:

"We want to use digital tools, but the internet rarely works at the school."

... were interpreted by the AI as evidence of resistance.

Through methodological literacy, the researcher recognises this misinterpretation and revises the coding structure to reflect a more accurate theme: *"Infrastructure Constraints to Digital Learning."*

Example 2: Fabricated Conceptual Categories

In another study examining community perceptions of sustainable fashion practices, a researcher uses an AI tool to generate initial thematic clusters from focus group transcripts. The AI produces a theme labelled *"Traditional Textile Sustainability Practices."*

However, when the researcher cross-checks the transcripts, it becomes evident that participants never explicitly discussed sustainability concepts. Instead, participants described traditional textile practices in relation to economic survival and cultural heritage.

The AI tool had generated an interpretive category based on patterns in its training data rather than the empirical data itself. Recognising this hallucination, the researcher reframes the theme as *"Cultural and Economic Motivations for Traditional Textile Production."*

Example 3: AI-Generated Quotations

A researcher analysing student reflections on online learning uses an AI summarisation tool to produce thematic summaries. The tool generates a quotation that appears highly representative of the dataset:

"Online learning has improved my independence as a learner."

However, when the researcher returns to the transcripts, he/she discovers that *no participant used this exact phrase*. The statement was generated by the AI as a paraphrased summary rather than a direct quotation.

Through reflexive methodological practice, the researcher removes the fabricated quotation and instead uses authentic participant statements to represent the theme.

Example 4: Misclassification of Emotional Tone

In a study exploring nurses' experiences during a public health crisis, an AI-based sentiment analysis tool categorises many statements as *negative emotional responses*.

Closer examination by the researcher reveals that participants were describing *professional responsibility and resilience,* not negativity. For instance, the statement:

"The work was extremely difficult, but we remained committed to caring for our patients."

was classified as negative sentiment because of the word "difficult."

The researcher therefore revises the analysis to reflect a more appropriate theme: *"Professional Commitment in Challenging Conditions."*

7.5. Comparison of AI Tools for Qualitative Research

The growing availability of AI-enabled qualitative research platforms demonstrates the increasing integration of computational tools into qualitative analysis workflows. Traditional computer-assisted qualitative data analysis software (CAQDAS) such as NVivo, ATLAS.ti, and MAXQDA (see Table 4) have incorporated AI capabilities including automated coding, sentiment analysis, and pattern detection to support researchers in analysing large datasets more efficiently. These tools typically combine machine learning and natural language processing techniques to identify recurring linguistic patterns, cluster related concepts, and generate summaries that accelerate analytic processes.

In contrast, general large language model platforms such as ChatGPT provide flexible dialogic interaction that can support conceptual development, reflexive memo writing, and exploratory coding, although they lack the structured project management and traceability features of traditional CAQDAS environments.

Meanwhile, AI transcription platforms such as Otter.ai play a crucial role in the early stages of qualitative research by automating the conversion of audio recordings into textual transcripts, thereby significantly reducing the time required for data preparation prior to analysis.

In all, these platforms illustrate that AI technologies are increasingly embedded across multiple stages of qualitative research workflows, including transcription, coding, thematic development, and analytic synthesis. However, despite their efficiency advantages, researchers must remain aware that AI-generated outputs require careful human interpretation to ensure that contextual meaning, cultural nuance, and theoretical significance are preserved within qualitative analysis.

Table 4. Comparison of AI Tools/Platforms for Qualitative Research

AI Platform	Primary Function in Qualitative Research	Key AI Capabilities	Typical Research Workflow Stage	Strengths	Limitations
NVivo AI Assistant	Computer-Assisted Qualitative Data Analysis Software (CAQDAS)	AI-assisted coding, automated insights, sentiment detection, pattern identification	Coding, thematic development, data visualization	Designed specifically for academic qualitative research; supports structured coding frameworks and traceable analytic decisions	AI suggestions still require human validation; software licensing cost may limit accessibility
ATLAS.ti AI Coding	CAQDAS with integrated AI analysis	Automatic open coding, document summarization, AI-assisted thematic clustering	Coding, thematic analysis, conceptual mapping	Allows researchers to "chat" with documents and automatically generate codes and insights	AI coding may oversimplify context or generate superficial categories
MAXQDA AI Assist	Qualitative and mixed-methods data analysis software	AI-generated summaries, paraphrasing, coding suggestions, multilingual support	Data preparation, coding, summarization	Strong integration with mixed-methods research and multilingual datasets	Requires manual interpretation of AI outputs; limited contextual understanding

continued on following page

Table 4. Continued

AI Platform	Primary Function in Qualitative Research	Key AI Capabilities	Typical Research Workflow Stage	Strengths	Limitations
ChatGPT / LLM Platforms	General large language model for text analysis	Thematic summarization, conceptual synthesis, idea generation, memo writing	Literature review, memoing, conceptual development, exploratory coding	Highly flexible; supports dialogic interaction for reflexive analysis	Not a dedicated qualitative research tool; outputs may be prompt-dependent and require verification
Otter.ai	AI transcription platform	Automatic speech recognition, meeting transcription, keyword extraction	Data transcription and preparation	Rapid transcription of interviews and focus groups; integrates with other research tools	Limited analytic capabilities beyond transcription
Dovetail AI	Cloud-based qualitative analysis platform	Automated tagging, theme clustering, AI insight extraction	Coding, collaborative analysis	Strong collaboration features for research teams and UX researchers	Less established in academic research compared with CAQDAS tools
Delve AI	AI-assisted qualitative coding platform	AI-supported thematic analysis, coding suggestions, research workflow automation	Coding, theme identification	Simplifies qualitative coding for beginners	Less advanced visualization and theoretical modelling capabilities

7.6. AI Integration Across the Qualitative Research Workflow

The integration of artificial intelligence into qualitative research workflows is increasingly evident across multiple stages of the research process. Rather than functioning as a replacement for qualitative interpretation, AI tools primarily support data management, pattern detection, and analytic organisation, thereby enhancing the efficiency of qualitative analysis. During the early stages of research, AI platforms can assist researchers in refining research questions, developing interview guides, and identifying relevant literature. However, the conceptual framing of qualitative studies remains dependent on the researcher's theoretical orientation and methodological expertise.

In the data preparation stage, AI-driven transcription technologies significantly reduce the time required to convert interview recordings into textual transcripts. While these tools enhance efficiency, researchers must still verify transcription accuracy and ensure that contextual meaning and participant confidentiality are preserved.

AI-assisted tools become more prominent during coding and thematic analysis, where natural language processing techniques can identify recurring linguistic patterns and suggest preliminary codes. Nevertheless, the interpretive nature of qualitative analysis requires researchers to critically evaluate these suggestions and ensure that coding decisions reflect the theoretical and contextual dimensions of the data.

In later stages such as memoing, validation, and theory development, AI tools may support analytic reflection and conceptual synthesis. However, the interpretive responsibility for connecting empirical insights to theoretical frameworks remains fundamentally human. This workflow illustrates that AI functions most effectively as an analytical augmentation tool within a human-centred qualitative methodology. By integrating AI tools within a reflexive research process, scholars can leverage computational efficiencies while preserving the interpretive depth that characterises qualitative inquiry. Table 5 shows AI integration across the qualitative research workflow.

Table 5. AI Integration Across the Qualitative Research Workflow

Research Stage	Key Research Activities	Role of AI	Example AI Tools/Platforms	Human Researcher Responsibilities
1. Research Design & Data Collection	Formulating research questions, designing interview guides, planning fieldwork, conducting interviews and observations	AI may assist in drafting interview questions, generating prompts, or identifying relevant literature to refine research design	ChatGPT, Claude, Elicit, Scite	Ensure research questions align with theoretical framework; maintain methodological coherence and ethical considerations
2. Data Transcription & Preparation	Converting interview recordings into text, cleaning transcripts, anonymising participant information	Automatic speech recognition converts audio into transcripts; AI may also identify keywords and speaker segmentation	Otter.ai, Trint, Sonix, Whisper	Review and correct transcription errors; ensure contextual accuracy and confidentiality of participant data
3. Data Familiarisation & Exploration	Reading transcripts, identifying preliminary patterns, summarising large datasets	AI can generate summaries, highlight recurring terms, and assist with initial descriptive categorisation	ChatGPT, NVivo AI Assistant, MAXQDA AI Assist	Critically interpret summaries and remain attentive to contextual meaning within participant narratives

continued on following page

Table 5. Continued

Research Stage	Key Research Activities	Role of AI	Example AI Tools/Platforms	Human Researcher Responsibilities
4. Initial Coding	Assigning codes to segments of qualitative data	AI-assisted coding may generate suggested codes or cluster similar textual segments	ATLAS.ti AI Coding, NVivo AI Assistant, Delve AI	Validate or refine AI-generated codes; ensure coding reflects theoretical sensitivity and contextual understanding
5. Thematic Development	Identifying patterns across codes, grouping categories into themes	AI can assist with clustering related codes and identifying semantic relationships between concepts	MAXQDA AI Assist, Dovetail AI, ChatGPT	Interpret relationships between themes; ensure analytic depth and theoretical relevance
6. Reflexive Memoing & Interpretation	Writing analytic memos, reflecting on emerging theoretical insights, linking findings to literature	AI can assist with summarising analytic notes or generating conceptual connections between ideas	ChatGPT, Claude	Maintain reflexive interpretation; critically evaluate AI suggestions against raw data
7. Validation & Trustworthiness Checks	Triangulation, member checking, peer debriefing, audit trail documentation	AI may assist in organising coded data, identifying inconsistencies, and summarising validation results	NVivo AI Assistant, ATLAS.ti AI	Ensure credibility, dependability, confirmability, and transferability of findings
8. Theory Development & Reporting	Integrating findings into conceptual models or theoretical frameworks	AI may assist in synthesising themes into conceptual diagrams or summarising findings	ChatGPT, Scite, Elicit	Ensure theoretical coherence, contextual interpretation, and scholarly rigor in reporting

7.7 Cross-Cultural Considerations in AI-Assisted

The integration of AI into qualitative research also raises important cross-cultural considerations, particularly when analysing data generated within diverse social and linguistic contexts. Many AI systems are trained primarily on datasets derived from Western linguistic and cultural environments, which may introduce interpretive biases when analysing narratives from other cultural contexts (Aljanada et al., 2025; Voutyrakou & Skordoulis, 2025).

In cross-cultural qualitative research, meanings are often embedded in local idioms, metaphors, and culturally specific expressions that may not be accurately interpreted by AI systems. For example, expressions used in African, Asian, or Indigenous communities may carry symbolic meanings that extend beyond their literal textual form. Researchers therefore must exercise caution when relying on AI-generated interpretations of culturally situated narratives. Within the AILQM framework, cognitive and ethical literacy includes the ability to recognise these limitations and ensure that culturally grounded interpretations remain central to analysis.

Cross-cultural reflexivity also requires researchers to consider how AI systems may reproduce structural biases embedded in their training data, potentially marginalising non-dominant voices or reinforcing existing epistemic hierarchies.

7.8 Limitations of AI tools in Qualitative Research

While AI technologies offer significant opportunities for enhancing qualitative analysis, several limitations must be acknowledged. First, AI systems are prone to hallucinations, generating plausible but inaccurate outputs that may misrepresent underlying data. Second, AI-generated analyses are prompt-dependent, meaning that results may vary significantly based on how questions or instructions are formulated.

Additionally, widely used AI tools vary in their methodological suitability. For example, transcription tools prioritise efficiency but may struggle with accent variation, background noise, or multilingual dialogue. Coding assistance tools may detect lexical patterns but cannot independently interpret contextual meaning or theoretical significance.

Large language models are also limited by training data constraints, which may embed cultural biases or incomplete representations of social experiences. Consequently, AI-generated insights should always be treated as probabilistic suggestions rather than authoritative analytic conclusions. Recognising these limitations reinforces the central premise of the AILQM framework: AI should function as a reflexive analytic assistant rather than a substitute for human interpretation.

8. IMPLICATIONS FOR TRAINING, POLICY, AND RESEARCH ETHICS

The AILQM framework carries significant implications for researcher preparation, institutional governance, and ethical oversight. As artificial intelligence becomes embedded within qualitative workflows, competence in AI engagement can no longer be optional or peripheral. Instead, it must be systematically integrated into research training, policy development, and ethical review structures. To ensure practical applicability, the key implications are summarised in Table 6.

Table 6. Practical Implications of the AILQM Framework for Training, Policy, and Research Ethics

Domain	Key Implications	Practical Actions	Responsible Actors
Researcher Training	AI literacy should be embedded within postgraduate research training and professional development.	Introduce mandatory modules on AI fundamentals, algorithmic bias, prompt design, data governance, and reflexive AI use. Incorporate supervised human–AI analytic exercises. Develop case-based ethical simulations.	Universities, Graduate Schools, Research Supervisors
Methodological Capacity Building	Qualitative researchers require structured guidance on integrating AI without compromising interpretive depth.	Provide methodological workshops on documenting human–AI decision trails, validating AI-assisted coding, and triangulating machine outputs with human interpretation.	Faculties, Methodology Centres, Research Associations
Ethical Review Processes	Existing research ethics protocols may not sufficiently address AI-specific risks.	Revise ethics application templates to include AI usage disclosures, data handling procedures, bias mitigation strategies, and AI tool transparency. Require explicit justification for AI-assisted analysis.	Institutional Review Boards, Ethics Committees
Data Governance and Privacy	AI tools may introduce cross-border data processing and privacy risks.	Develop institutional guidelines on approved AI platforms, data storage, anonymisation procedures, and compliance with national data protection laws.	University Management, Data Protection Officers
Equity and Inclusion	AI systems may reproduce social and cultural biases.	Establish bias auditing procedures for AI outputs. Encourage participatory validation involving diverse stakeholders. Promote culturally responsive AI usage.	Research Teams, Ethics Boards
Policy Development	Responsible AI integration requires institutional and national regulatory alignment.	Develop AI research policies outlining acceptable use, accountability structures, intellectual property considerations, and publication transparency requirements.	Universities, National Research Councils, Ministries of Education
Scholarly Reporting Standards	Transparency in AI use is essential for maintaining trustworthiness.	Require explicit reporting of AI tools used, version numbers, prompts, validation procedures, and the extent of AI contribution in published studies.	Journal Editors, Publishers, Research Institutions

Collectively, these implications reinforce that AI literacy must be institutionalised rather than individualised. Training should cultivate reflexive competence, policy should ensure structural accountability, and ethical oversight should evolve to address algorithmic complexity. Without coordinated institutional action, the integration of AI into qualitative research risks being inconsistent, opaque, and ethically vulnerable. Importantly, the AILQM framework positions ethics not as a compliance exercise but as a continuous reflective process. Researchers must remain vigilant regarding data sovereignty, participant autonomy, and epistemic justice. Ethical reflexivity becomes particularly crucial in contexts where AI systems are trained predominantly on data from the Global North, potentially marginalising local knowledge systems. Institutions, especially within developing research contexts, should therefore invest in capacity building that strengthens both technological competence and ethical discernment.

9. CONCLUSION AND FUTURE RESEARCH DIRECTIONS

The framework contributes theoretically by extending conversations on AI integration beyond tool usage towards epistemological positioning. It underscores that qualitative research remains inherently interpretive and context-bound, even when assisted by advanced computational systems. Practically, it offers a structured roadmap for embedding AI within research processes without eroding credibility, dependability, confirmability, or transferability. The integration of AI into qualitative methodology demands neither uncritical enthusiasm nor categorical resistance. Instead, it calls for principled adaptation grounded in literacy, reflexivity, and ethical responsibility. The AILQM framework provides a structured pathway for navigating this transformation, ensuring that qualitative inquiry remains human-centred, contextually sensitive, and methodologically robust in the evolving AI landscape.

Future research should pursue several directions. First, empirical validation of the AILQM framework is needed across disciplinary contexts to examine how the literacy dimensions interact and predict research quality outcomes. Quantitative and mixed-method studies could develop and validate measurement scales for AI literacy in qualitative research settings. Also, comparative studies should explore variations in AI adoption across geographic, institutional, and cultural contexts. Such research would illuminate structural disparities in access, governance capacity, and ethical preparedness. Additionally, longitudinal investigations are required to assess how sustained AI engagement reshapes researcher identity, reflexivity, and methodological decision-making over time. Moreover, future scholarship should examine the implications of emerging generative and multimodal AI systems for narrative analysis, discourse studies, and ethnographic interpretation. This includes investigating how transparency, explainability, and reproducibility standards can be strengthened in increasingly complex AI environments. Finally, policy-oriented research should evaluate the effectiveness of institutional AI governance frameworks and their impact on research integrity. Developing international reporting standards for AI-assisted qualitative research would further enhance scholarly accountability.

REFERENCES

Aagaard, J. (2022). Troubling the troublemakers: Three challenges to post-qualitative inquiry. *International Journal of Qualitative Research, 15*(3), 311–325. DOI: 10.1177/19408447211052668

Abreh, M. K., Arthur, F., Akwetey, F. A., & Nortey, S. A. (2025). Modelling STEM students' intention to learn artificial intelligence (AI) in Ghana: A PLS-SEM and fsQCA approach. *Discover Artificial Intelligence, 5*(1), 223. DOI: 10.1007/s44163-025-00466-8

Adarkwah, M. A. (2024). Generative artificial intelligence and education: Ethical considerations and implications for research practice. *Education and Information Technologies.*

Ahmed, S. K. (2024). The pillars of trustworthiness in qualitative research. *Journal of Medicine, Surgery, and Public Health, 2*, 100051. DOI: 10.1016/j.glmedi.2024.100051

Akman, N., Uzun, E., & Arslan, P. (2025). Human researcher vs. AI-supported qualitative data analysis: Hybrid prompt design for constant comparison analysis. *Education Mind, 4*(1), 46–65. DOI: 10.58583/EM.4.1.5

Aljanada, R. A., Alfaisal, A. M., AlHamad, A. Q. M., Himdi, H., Alfaisal, R., & Salloum, S. A. (2025). Cultural Implications in Generative AI Using Text Clustering and Sentiment Analysis. In *Generative AI in Creative Industries* (pp. 171–185). Springer Nature Switzerland. DOI: 10.1007/978-3-031-89175-5_11

Annells, M. (1996). Grounded theory method: Philosophical per spectives, paradigm of inquiry, and postmodernism. *Qualitative Health Research, 6*(3), 379–393. DOI: 10.1177/104973239600600306

Arkorful, V., Arthur, F., Boateng, E., Ofosu-Koranteng, M., Salifu, I., Attom, E. R., Tetteh, S. A., Quayson, E., Asare-Bediako, S., & Nortey, S. A. (2025). Exploring artificial intelligence literacy among basic school teachers in Ghana. *Discover Education, 4*(1), 250. DOI: 10.1007/s44217-025-00630-3

Badawy, W., Zinhom, H., & Shaban, M. (2025). Navigating ethical considerations in the use of artificial intelligence for patient care: A systematic review. *International Nursing Review, 72*(3), e13059. DOI: 10.1111/inr.13059 PMID: 39545614

Bandi, A., Adapa, P. V. S. R., & Kuchi, Y. E. V. P. K. (2023). The power of generative ai: A review of requirements, models, input–output formats, evaluation metrics, and challenges. *Future Internet, 15*(8), 260. DOI: 10.3390/fi15080260

Bender, E. M., Gebru, T., McMillan-Major, A., & Shmitchell, S. (2021). On the dangers of stochastic parrots: Can language models be too big? In *Proceedings of the 2021 ACM Conference on Fairness, Accountability, and Transparency* (pp. 610–623). ACM. DOI: 10.1145/3442188.3445922

Bender, E. M., Gebru, T., McMillan-Major, A., & Shmitchell, S. (2021). On the dangers of stochastic parrots: Can language models be too big? In *Proceedings of the 2021 ACM conference on fairness, accountability, and transparency* (pp. 610-623). DOI: 10.1145/3442188.3445922

Benjamin, R. (2019). Race after technology: Abolitionist tools for the new Jim code. Polity Press. https://www.politybooks.com/bookdetail?book_slug=race-after-technology-abolitionist-tools-for-the-new-jim-code--9781509526390&utm

Bommasani, R., Hudson, D. A., Adeli, E., Altman, R., Arora, S., von Arx, S., Bernstein, M. S., Bohg, J., Bosselut, A., Brunskill, E., Brynjolfsson, E., Buch, S., Card, D., Castellon, R., Chatterji, N. S., Chen, A., Creel, K., Davis, J. Q., Demszky, D., . . . Liang, P. (2021). On the opportunities and risks of foundation models. arXiv. https://doi.org//arXiv.2108.07258DOI: 10.48550

Bryant, A. (2007). A constructive/ist response to Glaser's 'Constructivist Grounded Theory?' Historical Social Research/Historische So zialforschung. Supplement,no.19 (pp. 106–13). JSTOR. https:// www.jstor.org/stable/40981072 (Accessed25Apr.2025).

Bryda, G., & Sadowski, D. (2024). From words to themes: AI-powered qualitative data coding and analysis. In *World conference on qualitative research* (pp. 309-345). Cham: Springer Nature Switzerland. DOI: 10.1007/978-3-031-65735-1_19

Buolamwini, J., & Gebru, T. (2018). Gender shades: Intersectional accuracy disparities in commercial gender classification. Proceedings of Machine Learning Research, 81, 1–15. http://proceedings.mlr.press/v81/buolamwini18a.html

Charmaz, K. (2000). Grounded theory: Objectivist and constructivist methods. In N. Denzin & Y. Lincoln (Eds.), Handbook of qualitative research 2 (pp. 509–535). Sage.

Charmaz, K. (2006). Constructing grounded theory: A practical guide through qualitative analysis. *Sage (Atlanta, Ga.).*

Charmaz, K. (2014). *Constructing grounded theory* (2nd ed.). Sage., https://books.google.com/books/about/Constructing_Grounded_Theory.html?id=v_GGAwAAQBAJ&utm

Charmaz, K. (2014). Constructing grounded theory. *Sage (Atlanta, Ga.).*

Chen, H., Wang, Y., & Chang, C. (2025). Enhancing qualitative inquiry: AI-assisted focus group data collection. Qualitative Research Journal. https://doi.org/, P. (2023). How to use Artificial Intelligence (AI) as a resource, methodological and analysis tool in qualitative research? The Qualitative Report, 28(7), 1968–1980. DOI: 10.1108/QRJ-04-2025-0145Christou

Chubb, L. A. (2023). Me and the machines: Possibilities and pitfalls of using artificial intelligence for qualitative data analysis. *International Journal of Qualitative Methods, 22*, 16094069231193593. Advance online publication. DOI: 10.1177/16094069231193593

Costa, A. P., Bryda, G., Christou, P. A., & Kasperiuniene, J. (2025). AI as a Co-researcher in the Qualitative Research Workflow: Transforming Human-AI Collaboration. *International Journal of Qualitative Methods, 24*, 16094069251383739. DOI: 10.1177/16094069251383739

Costa, A. P., Bryda, G., Christou, P. A., & Kasperiuniene, J. (2025). AI as a co-researcher in the qualitative research workflow: Transforming human–AI collaboration. *International Journal of Qualitative Methods, 24*, 16094069251383739. Advance online publication. DOI: 10.1177/16094069251383739

Crawford, K. (2021). *Atlas of AI: Power, politics, and the planetary costs of artificial intelligence*. Yale University Press.

Dellafiore, F., Saba, A., Collaro, C., & Artioli, G. (2025). Artificial Intelligence in Qualitative Research: Insights From Experts via Reflexive Thematic Analysis. *Qualitative Health Research*, •••, 10497323251389800. PMID: 41351219

Dwivedi, Y. K., Kshetri, N., Hughes, L., Slade, E. L., Jeyaraj, A., Kar, A. K., Baabdullah, A. M., Koohang, A., Raghavan, V., Ahuja, M., Albanna, H., Albashrawi, M., Alalwan, A. A., Al-Araj, R., Al-Debei, M. M., Al-Khowaiter, W. A., Al-Somali, S. A., Alweshah, G. A., Alzoubi, H. M., & Wright, R. (2023). So what if ChatGPT wrote it? Multidisciplinary perspectives on generative AI. *International Journal of Information Management, 71*, 102642. DOI: 10.1016/j.ijinfomgt.2023.102642

Eubanks, V. (2018). Automating inequality: How high-tech tools profile, police, and punish the poor. St. Martin's Press. https://us.macmillan.com/books/9781250074317/automatinginequality

European Commission. (2023). Ethics guidelines for trustworthy AI (updated framework). https://digital-strategy.ec.europa.eu/en/library/ethics-guidelines-trustworthy-ai

Finlay, L. (2017). Championing "reflexivities.". *Qualitative Psychology*, *4*(2), 120–125. DOI: 10.1037/qup0000075

Floridi, L., & Chiriatti, M. (2020). GPT-3: Its nature, scope, limits, and consequences. *Minds and Machines*, *30*(4), 681–694. DOI: 10.1007/s11023-020-09548-1

Friese, S. (2026). From Coding to Conversation: A New Methodological Framework for AI-Assisted Qualitative Analysis. *Qualitative Inquiry*, 10778004251412871. DOI: 10.1177/10778004251412871

Glaser, B., & Strauss, A. (1967). *The discovery of grounded theory: Strategies for qualitative research*. Aldine.

Hamilton, L., Elliott, D., Quick, A., Smith, S., & Choplin, V. (2023). Exploring the use of AI in qualitative analysis: A comparative study of guaranteed income data. *International Journal of Qualitative Methods*, *22*, 16094069231201504. Advance online publication. DOI: 10.1177/16094069231201504

Hammersley, M., & Atkinson, P. (2019). *Ethnography: Principles in practice*. Routledge. DOI: 10.4324/9781315146027

Heath, H., & Cowley, S. (2004). Developing a grounded theory approach: A comparison of Glaser and Strauss. *International Journal of Nursing Studies*, *41*(2), 141–150. DOI: 10.1016/S0020-7489(03)00113-5 PMID: 14725778

Heidegger, M. (2005). *Introduction to phenomenological research*. Indiana university press. DOI: 10.2307/j.ctvt1sgpb

Husserl, E. (1960). *Cartesian meditations: An introduction to phenomenology* (Cairns, D., Trans.). Martinus Nijhoff. (Original work published 1931) DOI: 10.1007/978-94-017-4952-7

Husserl, E. (1970). *The crisis of European sciences and transcendental phenomenology: An introduction to phenomenological philosophy* (Carr, D., Trans.). Northwestern University Press. (Original work published 1936)

Ikonen, S. (2025). Generative AI in strategy work.

Jones, K. M. (2025). Generative AI in Qualitative Research and Related Transparency Problems: A Novel Heuristic for Disclosing Uses of AI. *International Journal of Qualitative Methods*, *24*, 16094069251404329. DOI: 10.1177/16094069251404329

Jones, M., & Alony, I. (2011). Guiding the use of grounded theory in doctoral studies– an example from the Australian film industry. *International Journal of Doctoral Studies*, *6*, 95–114. DOI: 10.28945/1429

Kanbach, D. K., Heiduk, L., Blueher, G., Schreiter, M., & Lahmann, A. (2024). The GenAI is out of the bottle: Generative artificial intelligence from a business model innovation perspective. *Review of Managerial Science*, *18*(4), 1189–1220. DOI: 10.1007/s11846-023-00696-z

Kenny, M., & Fourie, R. (2015). Contrasting classic, Straussian, and constructivist grounded theory: Methodological and philo sophical conflicts. *The Qualitative Report*, *20*(8), 1270–1289. DOI: 10.46743/2160-3715/2015.2251

Keogh, J. (2025). Frankenstein, thematic analysis and generative artificial intelligence: Quality appraisal methods and considerations for qualitative research. *PLoS One*, *20*(9), e0330217. DOI: 10.1371/journal.pone.0330217 PMID: 40911617

Kitchin, R. (2021). *The data revolution: A critical analysis of big data, open data and data infrastructures* (2nd ed.). Sage., DOI: 10.4135/9781526469403

Leonelli, S. (2023). *Philosophy of open science*. Cambridge University Press. DOI: 10.1017/9781009416368

Lincoln, Y. S., & Guba, E. G. (1985). Naturalistic inquiry. *Sage (Atlanta, Ga.).*

Liu, H., Ning, R., Teng, Z., Liu, J., Zhou, Q., & Zhang, Y. (2023). *Evaluating the logical reasoning ability of ChatGPTand GPT-4*. Cornell University.

Long, D., & Magerko, B. (2020). What is AI literacy? Competencies and design considerations. In Proceedings of the 2020 CHI Conference on Human Factors in Computing Systems (Paper 574). ACM. DOI: 10.1145/3313831.3376727

Marshall, D. T., & Naff, D. B. (2024). The ethics of using artificial intelligence in qualitative research. *Journal of Empirical Research on Human Research Ethics; JERHRE*, *19*(3), 92–102. DOI: 10.1177/15562646241262659 PMID: 38881315

Martin, S., Beecham, E., Kursumovic, E., Armstrong, R. A., Cook, T. M., Déom, N., Kane, A. D., Moniz, S., Soar, J., & Vindrola-Padros, C. (2026). Comparing human vs. machine-assisted analysis to develop a new approach for big qualitative data analysis. *PLOS Digital Health*, *5*(2), e0000576. DOI: 10.1371/journal.pdig.0000576 PMID: 41739812

Merleau-Ponty, M., Landes, D., Carman, T., & Lefort, C. (2013). *Phenomenology of perception*. Routledge. DOI: 10.4324/9780203720714

Mittelstadt, B. (2019). Principles alone cannot guarantee ethical AI. *Nature Machine Intelligence*, *1*(11), 501–507. DOI: 10.1038/s42256-019-0114-4

Morgan, D. L. (2023). Exploring the use of artificial intelligence for qualitative data analysis: The case of ChatGPT. *International Journal of Qualitative Methods*, *22*, 16094069231211248. Advance online publication. DOI: 10.1177/16094069231211248

Morgan, D. L. (2023). Exploring the use of artificial intelligence for qualitative data analysis: The case of ChatGPT. *International Journal of Qualitative Methods*, *22*, 16094069231211248. DOI: 10.1177/16094069231211248

Mukherjee, B. N. (2025). Navigating AI Governance: National and International Legal and Regulatory Frameworks. In *Navigating the Intersection of AI Policy, Technology, and Governance* (pp. 201-224). IGI Global Scientific Publishing.

Nelson, L. K. (2020). Computational grounded theory: A method ological framework. *Sociological Methods & Research*, *49*(1), 3–42. DOI: 10.1177/0049124117729703

Ng, D. T. K., Leung, J. K. L., Chu, S. K. W., & Qiao, M. S. (2021). Conceptualizing AI literacy: An exploratory review. *Computers and Education: Artificial Intelligence*, *2*, 100041. DOI: 10.1016/j.caeai.2021.100041

Nguyen-Trung, K. (2025). ChatGPT in thematic analysis: Can AI become a research assistant in qualitative research? *Quality & Quantity*, *59*(6), 4945–4978. DOI: 10.1007/s11135-025-02165-z

Noble, S. U. (2018). *Algorithms of oppression: How search engines reinforce racism.* NYU Press., DOI: 10.18574/nyu/9781479833641.001.0001

Papageorgiou, G., Skamnia, E., & Economou, P. (2025). AI-Assisted Literature Review: Integrating Visualization and Geometric Features for Insightful Analysis. *Wiley Interdisciplinary Reviews. Data Mining and Knowledge Discovery*, *15*(2), e70016. DOI: 10.1002/widm.70016

Pappachan, P., Moslehpour, M., Bansal, R., & Rahaman, M. (2024). Transparency and accountability. In *Challenges in large language model development and AI ethics* (pp. 178–211). IGI Global Scientific Publishing. DOI: 10.4018/979-8-3693-3860-5.ch006

Parker, R. D., Mancini, K., & Abram, M. D. (2023). Natural Language processing enhanced qualitative methods: An opportunity to improve health outcomes. *International Journal of Qualitative Methods*, *22*, 16094069231214144. Advance online publication. DOI: 10.1177/16094069231214144

Pattyn, F. (2025). The value of generative AI for qualitative research: A pilot study. *Journal of Data Science and Intelligent Systems*, *3*(3), 184–191. DOI: 10.47852/bonviewJDSIS42022964

Pink, S., Fors, V., Lanzeni, D., Duque, M., Sumartojo, S., & Strengers, Y. (2022). *Design ethnography: Research, responsibilities, and futures*. Routledge. DOI: 10.4324/9781003083665

Pink, S., Mackley, K. L., Morosanu, R., Mitchell, V., & Bhamra, T. (2020). *Making homes: Ethnography and design*. Routledge. DOI: 10.4324/9781003085942

Pink, S., Sinanan, J., Hjorth, L., & Horst, H. (2016). Tactile digital ethnography: Researching mobile media through the hand. *Mobile Media & Communication*, *4*(2), 237–251. DOI: 10.1177/2050157915619958

Pink, S., Sumartojo, S., Lupton, D., & Heyes La Bond, C. (2018). Mundane data: The routines, contingencies and accomplishments of digital living. *Big Data & Society*, *4*(1), 2053951717700924. Advance online publication. DOI: 10.1177/2053951717700924

Shneiderman, B. (2020). Human-centered artificial intelligence: Reliable, safe & trustworthy. *International Journal of Human-Computer Interaction*, *36*(6), 495–504. DOI: 10.1080/10447318.2020.1741118

Sidra, S., & Mason, C. (2025). Generative AI in human-AI collaboration: Validation of the collaborative AI literacy and collaborative AI metacognition scales for effective use. *International Journal of Human-Computer Interaction*, •••, 1–25.

Smith, J. A. (2018). "Yes it is phenomenological": A reply to Max Van Manen's critique of interpretative phenomenological analysis. *Qualitative Health Research*, *28*(12), 1955–1958. DOI: 10.1177/1049732318799577 PMID: 30241456

Smith, J. A., Flowers, P., & Larkin, M. (2009). *Interpretative phenomenological analysis: Theory, method and research*. Sage.

St. Pierre, E. A. (2019). Post qualitative inquiry in an ontology of immanence. *Qualitative Inquiry*, *25*(1), 3–16. DOI: 10.1177/1077800418772634

St. Pierre, E. A. (2021). Why post qualitative inquiry? *Qualitative Inquiry*, *27*(2), 163–166. DOI: 10.1177/1077800420931142

St. Pierre, E. A. (2024). Reading for post qualitative inquiry. *International Journal of Qualitative Studies in Education : QSE*, *37*(6), 1600–1610. DOI: 10.1080/09518398.2024.2340041

Stilgoe, J., Owen, R., & Macnaghten, P. (2013). Developing a framework for responsible innovation. *Research Policy*, *42*(9), 1568–1580. DOI: 10.1016/j.respol.2013.05.008

Strauss, A., & Corbin, J. (1998). *Basics of qualitative research: Techniques and procedures for developing grounded theory*. Sage Publications, Inc.

Tai, R. H., Bentley, L. R., Xia, X., Sitt, J. M., Fankhauser, S. C., Chicas-Mosier, A. M., & Monteith, B. G. (2024). An examination of the use of large language models to aid analysis of textual data. *International Journal of Qualitative Methods, 23*, 16094069241231168. Advance online publication. DOI: 10.1177/16094069241231168

Timonen, V., Foley, G., & Conlon, C. (2018). Challenges when using grounded theory: A pragmatic introduction to doing GT research. *International Journal of Qualitative Methods, 17*(1), 1–10. DOI: 10.1177/1609406918758086

Tlili, A., Shehata, B., Adarkwah, M. A., Bozkurt, A., Hickey, D. T., Huang, R., & Agyemang, B. (2023). What if the devil is my guardian angel? ChatGPT as a case study of generative AI in education. *Smart Learning Environments, 10*(1), 15. Advance online publication. DOI: 10.1186/s40561-023-00237-x

Tracy, S. J. (2010). Qualitative quality: Eight "big-tent" criteria for excellent qualitative research. *Qualitative Inquiry, 16*(10), 837–851. DOI: 10.1177/1077800410383121

UNESCO. (2023). Guidance for generative AI in education and research. https://unesdoc.unesco.org/ark:/48223/pf0000386693

Varsha, P. S. (2023). How can we manage biases in artificial in telligence systems– a systematic literature review. *International Journal of Information Management Data Insights, 3*(1), 100165. DOI: 10.1016/j.jjimei.2023.100165

Voutyrakou, D. A., & Skordoulis, C. (2025). Exploring gender bias in AI-generated definitions of role models: A cross-linguistic perspective. *AI and Ethics, 5*(5), 5485–5500. DOI: 10.1007/s43681-025-00792-8

Zahavi, D. (2019). Getting it quite wrong: Van Manen and Smith on phenomenology. *Qualitative Health Research, 29*(6), 900–907. DOI: 10.1177/1049732318817547 PMID: 30565516

Zhang, H. E., Wu, C., Xie, J., Lyu, Y., Cai, J., & Carroll, J. M. (2025). Harnessing the power of AI in qualitative research: Exploring, using and redesigning ChatGPT. *Computers in Human Behavior: Artificial Humans, 4*, 100144. DOI: 10.1016/j.chbah.2025.100144

Zuboff, S. (2019). *The age of surveillance capitalism: The fight for a human future at the new frontier of power*. PublicAffairs.w.

Chapter 14
AI Literacy

Apoorav Sharma
http://orcid.org/0009-0002-9966-7691
Lovely Professional University, India

Lovleen Marwaha
Lovely Professional University, India

ABSTRACT

Drawing on philosophical, epistemological, methodological, ethical, and pedagogical perspectives, the chapter examines how AI literacy supports reflexive engagement, preserves interpretive authority, and enhances methodological rigor in artificial intelligence–augmented qualitative research. It explores the implications of artificial intelligence for knowledge production, researcher agency, ethics, and education, while addressing challenges such as bias, opacity, automation bias, and unequal access. The chapter ultimately argues that AI literacy is not a fixed technical skill but a dynamic, critical, and reflexive practice. By embedding AI literacy within qualitative methodology, researchers and educators can ensure that artificial intelligence enhances rather than undermines the core values of qualitative inquiry, including contextual depth, ethical responsibility, and epistemic accountability.

1. INTRODUCTION: WHY AI LITERACY MATTERS IN QUALITATIVE RESEARCH

Artificial intelligence is no longer a specialty field of computational science, but is now an ubiquitous socio-technological phenomenon that influences modern knowledge production, decision-making, and social intercourse (Chang & Choi, 2025). Artificial intelligence systems have become a standard feature of data creation, processing, translation, and sharing in practice across different fields, including

DOI: 10.4018/979-8-3373-7640-0.ch014

ht © 2026, IGI Global Scientific Publishing. Copying or distributing in print or electronic forms without written permission of IGI Global Scientific Publishing is prohibited. Use of this chapter to generative artificial intelligence (AI) technologies is expressly prohibited. The publisher reserves all rights to license its use for generative AI training and machine learning model development.

healthcare, education, governance, economics, media studies, and social sciences (Chatzichristos et al., 2025). This growth has been precipitated by the developments in machine learning and natural language processing, as well as generative artificial intelligence, that allow identifying patterns automatically, text generation, predictive modelling, and the synthesis of large volumes of data (Barrera et al., 2025). This leads to the fact that artificial intelligence no longer functions as an infrastructure situation but increasingly takes part in the formulation of epistemic processes themselves (Kaplan & Meylani, 2025).

In the field of research, the artificial intelligence technologies are affecting the way the questions are framed, the way the data are gathered and analyzed, as well as the ways the data are interpreted and reported (Asrifan et al., 2025). These changes have significant repercussions on qualitative research that has always focused on the depth in situations, interpretive rigor, reflexivity, and meaning-making. Whereas quantitative disciplines have addressed extensively the opportunities of artificial intelligence by computational modeling and statistical automation, qualitative inquiry has a different range of opportunities and challenges. Artificial intelligence is now used to help with transcription, encoding, thematic clustering, discourse analysis, and reflexive memo writing, thus penetrating into the space historically anchored in human judgment and interpretive sensibility.

In spite of this increasing incorporation, most scholars approach artificial intelligence technologies in an instrumental and non-critical way and see it as a neutral tool, instead of a socio-technical system with its assumptions, values, and power. This role of disuse and a lack of knowledge indicates how desperately artificial intelligence literacy is necessary. Artificial intelligence literacy has also recently taken root in the new scholarship of a multidimensional competence, beyond technical awareness, of critical, ethical, and socio-cultural awareness of artificial intelligence systems and their impacts to human practices. Such literacy is not a choice in the context of qualitative research, but methodological integrity and epistemic responsibility cannot exist without it.

1.2 Digital Literacy to AI Literacy: Conceptual Development.

The idea of artificial intelligence literacy is a continuation of previously developed frames of digital literacy and data literacy, which focused on being able to access, judge and act responsibly upon digital information and technology (Selwyn, 2019). Digital literacy is the result of the spread of information and communication technologies and is characterized by the skills of navigation, evaluation, and involvement in digital space. The concept of data literacy broadened the scope of this to encompass data-structure, data-source, data-limit, and data-use competencies

(Tran & Tran, 2023). Artificial intelligence however opens a qualitatively different paradigm that transcends the territories of these previous literacies.

Artificial intelligence systems are not tools of accessing information and processing data; they are proactive producers and can deduce patterns and mediate human actions. Artificial intelligence literacy must, therefore, be conceptualized further, to incorporate technical knowledge and socio-technical conscious. According to recent studies, the conceptualization of artificial intelligence literacy is a human competence, which integrates explicit knowledge of artificial intelligence processes with implicit experience in the interaction with artificial intelligence systems, and the inseparable nature of the social and technical aspects. This change is indicative of an increasing awareness that artificial intelligence is redefining practices and the way of thinking, all the way to knowledge, agency, and authority.

Moreover, artificial intelligence literacy reacts to the democratisation of artificial intelligence technologies. Non-technical users interact more and more with complex algorithmic systems without formal training, due to the proliferation of leveraged generative artificial intelligence systems. Research investigating the beliefs and opinions of non-technical publics about artificial intelligence show that people tend to create their own informal folk theories to explain artificial intelligence, and such theories can mask prejudices, constraints, and power dynamics. In the context of qualitative researchers, whose activity is usually based on the creation of meaning and interpretation, unquestioned assumptions regarding artificial intelligence may become a serious methodological threat. The artificial intelligence literacy, therefore, is the key development of research competence, which allows scholars to interact with intelligent systems critically and reflexively.

1.3 AI as a Socio-Technical System, Not Just a Tool

One of the core assumptions of artificial intelligence literacy is the acknowledgment of the artificial intelligence as a socio-technical system and not a tool of neutrality or autonomy (Kudina & van de Poel, 2024). The artificial intelligence systems are developed in intricate aggregations of information, formulas, organizational interests, human workers, and cultural suppositions. They also represent the situations where they are created and implemented, such as past injustices, values, and epistemological priorities. Viewing artificial intelligence as a technical object of study blurs these underlying dimensions and threatens the recreation of prejudices and omissions in research.

The socio-technical views underline that artificial intelligence technologies mediate human activity and do not substitute it (Cha & Wong, 2025). In qualitative research, the mediation influences voice representation, meaning classification, and narrations. To illustrate, automatic text analyzers can give more weight to frequency than nuance, coherence than contradiction, or established forms of language than marginalized ones. In the absence of a sufficient level of literacy in artificial intelligence, researchers can unintentionally give interpretive power to algorithms, thereby impairing the commitments to reflexivity and contextualism of qualitative research.

The artificial intelligence literacy will thus involve the knowledge of how artificial intelligence systems are trained, what they are based on, how they produce outputs, and where they become uncertain and erroneous. It also presupposes the knowledge of the overall social impact of using artificial intelligence, such as the problem of surveillance, data ownership, and power asymmetry. This kind of literacy in line with interdisciplinary socio-technical concepts that put artificial intelligence in a co-constitutive position with social practice as opposed to an external one.

1.4 The Relevance of AI Literacy for Qualitative Methodologists

For qualitative methodologists, artificial intelligence literacy is directly linked to methodological rigor, ethical responsibility, and epistemic transparency. Qualitative research relies on interpretive judgment, reflexivity, and contextual sensitivity, all of which are challenged when artificial intelligence tools are introduced without critical engagement. Artificial intelligence literacy equips researchers to evaluate when and how artificial intelligence can support qualitative inquiry without compromising its core principles (Gazos et al., 2025).

Moreover, artificial intelligence literacy enables qualitative researchers to articulate and justify their methodological choices in artificial intelligence–augmented studies. As artificial intelligence tools increasingly shape research workflows, transparency regarding their use becomes essential for credibility and reproducibility. Understanding the limitations and affordances of artificial intelligence systems allows researchers to maintain interpretive authority while leveraging computational support in a responsible and reflexive manner. In this sense, artificial intelligence literacy is not about technical mastery but about informed engagement with intelligent systems as methodological actors within qualitative research practice.

2. CONCEPTUAL FOUNDATIONS OF AI LITERACY

2.1 Defining Artificial Intelligence: Narrow AI, General AI, and Generative AI

Artificial intelligence is a broad and evolving concept that encompasses a range of computational systems designed to perform tasks that typically require human intelligence (Ronge et al., 2025). Contemporary scholarly discourse commonly distinguishes among narrow artificial intelligence, general artificial intelligence, and generative artificial intelligence, each of which has distinct implications for research practices and literacy requirements.

Narrow artificial intelligence, also referred to as weak artificial intelligence, denotes systems designed to perform specific, well-defined tasks within constrained domains. Examples include recommendation algorithms, speech recognition systems, automated transcription tools, and machine learning models for sentiment analysis. These systems do not possess understanding or consciousness; rather, they operate through statistical pattern recognition and optimization based on predefined objective. Narrow artificial intelligence currently dominates real-world applications and is the primary form encountered in qualitative research workflows, particularly in tasks such as transcription, coding assistance, and text analytics (Kalota, 2024).

General artificial intelligence, in contrast, refers to a hypothetical form of intelligence capable of understanding, learning, and applying knowledge across domains in a manner comparable to human cognitive flexibility. While general artificial intelligence remains largely theoretical, its conceptual presence shapes public narratives, ethical debates, and popular imaginaries surrounding artificial intelligence. For qualitative researchers, distinguishing between narrow and general artificial intelligence is essential to avoid anthropomorphizing current systems or overestimating their epistemic capacities.

Generative artificial intelligence represents a more recent and transformative development within the landscape of narrow artificial intelligence. Generative systems, including large language models and multimodal generators, are capable of producing novel text, images, audio, and code based on probabilistic modeling of large-scale datasets (Corchado et al., 2023). Unlike earlier artificial intelligence tools that primarily classified or predicted, generative artificial intelligence actively participates in content creation, thereby blurring boundaries between analysis, interpretation, and authorship. This shift has profound implications for qualitative inquiry, as generative systems increasingly mediate processes traditionally understood as interpretive and creative rather than computational.

Understanding these distinctions is a foundational component of artificial intelligence literacy. Without conceptual clarity regarding the type of artificial intelligence involved, researchers risk mischaracterizing system capabilities, attributing unwarranted agency, or misunderstanding the methodological and ethical stakes of artificial intelligence integration.

2.2 What Is AI Literacy? Key Definitions and Models

Artificial intelligence literacy has emerged as a multidimensional construct that extends beyond technical proficiency to encompass critical, ethical, and socio-technical understanding. Recent scholarship converges on the view that artificial intelligence literacy involves the capacity to meaningfully interpret, evaluate, and responsibly engage with artificial intelligence systems in context (Gutierrez et al., 2023).

One influential line of research defines artificial intelligence literacy as a socio-technical competence that integrates explicit knowledge of artificial intelligence processes with experiential understanding of human–artificial intelligence interaction. This definition emphasizes that artificial intelligence literacy is not reducible to knowing how algorithms function; it also involves understanding how artificial intelligence systems shape and are shaped by human practices, institutional settings, and social values.

Educational frameworks developed in the mid-2020s further conceptualize artificial intelligence literacy as an interdisciplinary construct composed of technical understanding, interactional skills, ethical reasoning, and societal awareness. These models explicitly reject purely computational definitions of literacy and instead situate artificial intelligence literacy within broader traditions of critical digital education. From this perspective, being literate in artificial intelligence means being able to question outputs, recognize limitations, identify potential harms, and reflect on power relations embedded in artificial intelligence systems.

Scoping reviews conducted between 2022 and 2025 highlight persistent inconsistencies in how artificial intelligence literacy is defined and operationalized across disciplines, yet they also identify growing consensus around its multidimensional nature (Gangwal et al., 2024). Importantly for qualitative research, these models foreground interpretation, reflexivity, and ethical judgment as core components, aligning artificial intelligence literacy with established qualitative epistemologies. While general AI literacy models focus on technical 'know-how,' this chapter uniquely contributes a **Qualitative AI Literacy** framework that centers on 'epistemic accountability' the researcher's duty to remain the primary 'knower' in a hybrid human-AI assemblage.

This chapter makes three key contributions to existing AI literacy scholarship:

1. It reframes AI literacy as a *methodological competency* specific to qualitative inquiry rather than a general digital skill.
2. It introduces the concept of *epistemic accountability*, emphasizing that researchers must remain the primary interpreters in AI-mediated analysis.
3. It operationalizes AI literacy across the qualitative research lifecycle, linking abstract competencies to concrete research practices, risks, and safeguards.

2.3 Core Dimensions of AI Literacy

Artificial intelligence literacy is best understood as a constellation of interrelated dimensions rather than a single skill set. The following dimensions are particularly salient for qualitative researchers.

2.3.1 Technical Awareness

Technical awareness refers to a foundational understanding of how artificial intelligence systems are developed, trained, and deployed. This includes basic knowledge of machine learning, model training, inputs and outputs, and system limitations (Kirova et al., 2023). For qualitative researchers, technical awareness does not imply programming expertise but rather the ability to critically interpret how artificial intelligence tools process qualitative data, generate patterns, or produce text. Such awareness is essential to avoid misinterpreting artificial intelligence outputs as neutral or objective representations of social reality.

2.3.2 Data Literacy

Data literacy constitutes a central pillar of artificial intelligence literacy, given that artificial intelligence systems are fundamentally data-driven. This dimension involves understanding how data are collected, selected, labeled, and structured, as well as how biases and exclusions within data influence artificial intelligence outputs. In qualitative research, where data are often context-rich, relational, and situated, data literacy enables researchers to assess whether artificial intelligence tools adequately capture nuance, voice, and meaning or whether they flatten complexity through abstraction.

2.3.3 Algorithmic Thinking

Algorithmic thinking refers to the capacity to reason about processes, rules, and decision pathways underlying artificial intelligence systems. It involves understanding that artificial intelligence outputs are generated through probabilistic inference rather than comprehension or intention. For qualitative methodologists, algorithmic thinking supports critical evaluation of automated coding, clustering, or thematic suggestions, reinforcing the distinction between computational pattern detection and interpretive meaning-making (Radanliev, 2025).

2.3.4 Critical and Ethical Reasoning

Critical and ethical reasoning is a defining dimension of artificial intelligence literacy. It encompasses the ability to identify ethical risks, such as bias, surveillance, misuse of data, and erosion of consent, as well as to situate artificial intelligence use within broader power structures. Scholarship from 2022 onward consistently emphasizes that artificial intelligence literacy must include normative judgment and ethical reflexivity, particularly in research contexts involving vulnerable populations or sensitive data.

2.3.5 Human–AI Interaction Literacy

Human–artificial intelligence interaction literacy focuses on how individuals engage with artificial intelligence systems in practice. This includes understanding prompt construction, interpretation of outputs, and the relational dynamics that emerge when artificial intelligence systems are treated as collaborators or advisors. Studies of non-technical users demonstrate that people often anthropomorphize artificial intelligence, attributing understanding or intentionality where none exists (West et al., 2023). For qualitative researchers, interaction literacy supports reflexivity regarding how artificial intelligence shapes analytic decisions and interpretive trajectories.

2.4 AI Literacy vs. Data Literacy vs. Digital Literacy

While artificial intelligence literacy builds upon digital and data literacy, it is conceptually distinct. Digital literacy primarily concerns navigation and participation in digital environments, whereas data literacy emphasizes understanding and ethical use of data. Artificial intelligence literacy, by contrast, addresses systems that actively mediate knowledge production and decision-making. It requires engagement with inference, automation, and generativity, as well as with the socio-technical consequences of delegating interpretive tasks to machines (Kreinsen & Schulz, 2023).

Importantly, artificial intelligence literacy integrates but also transcends these earlier literacies by foregrounding issues of agency, accountability, and epistemic authority. In qualitative research, this distinction is critical, as artificial intelligence tools increasingly shape analytic processes rather than merely supporting them.

2.5 AI Literacy as a Continuum Rather Than a Fixed Skill

Contemporary scholarship increasingly conceptualizes artificial intelligence literacy as a continuum rather than a fixed or binary attribute (Van Audenhove et al., 2024). Individuals and researchers may possess varying levels of artificial intelligence literacy depending on context, task, and experience. This perspective acknowledges that artificial intelligence technologies evolve rapidly and that literacy must be continually updated through practice, reflection, and critical engagement.

Viewing artificial intelligence literacy as dynamic aligns with qualitative traditions that emphasize learning as situated and reflexive. Rather than aiming for complete mastery, qualitative researchers benefit from cultivating adaptive literacy that supports ongoing evaluation of artificial intelligence tools and their methodological implications. Such an approach reinforces the role of the researcher as an accountable and reflexive agent within an artificial intelligence–mediated research landscape. Figure 1 presents a multidimensional framework that situates AI literacy within the socio-technical and epistemological foundations of qualitative research.

Figure 1. This figure conceptualizes AI literacy as a reflexive and multidimensional competency embedded within qualitative research practice. It illustrates the interaction between core AI literacy dimensions—technical awareness, data literacy, algorithmic thinking, human–AI interaction literacy, and critical-ethical reasoning—and foundational qualitative principles such as contextual depth, epistemic accountability, and knowledge co-construction. The outer layer highlights the broader socio-technical environment, including power relations, institutional constraints, and knowledge production regimes that shape AI-mediated qualitative inquiry.

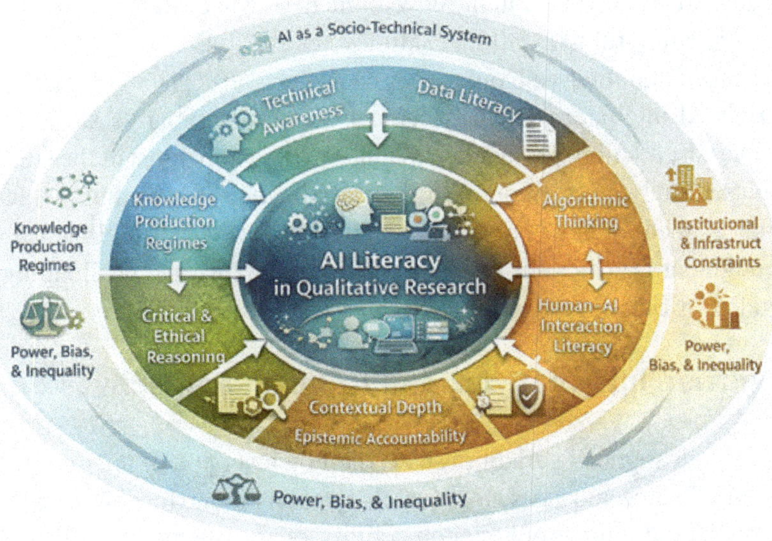

3. PHILOSOPHICAL AND EPISTEMOLOGICAL IMPLICATIONS OF AI LITERACY

3.1 AI and Knowledge Production

The field of artificial intelligence is progressively involved in the knowledge production processes, transforming the knowledge generation, structuring, authentication, and distribution. In qualitative research that relies on a more traditional approach, the creation of knowledge is perceived as an interpretive and relational process that is based on context, reflexivity, and the creation of meanings (Lin & Dai, 2025). The fact that the process of the emergence of knowledge is now admitted to be effected by the introduction of the artificial intelligence systems also questions some of the basic assumptions about the origin of knowledge sources and the parties and objects that contribute to the creation of knowledge.

Artificial intelligence systems aid in the knowledge creation by identifying trends, creating descriptions, and suggesting meanings relying on bulk information. Such contributions are however not epistemically neutral. The artificial intelligence systems are based on historically situated data, algorithm design decisions, and optimization objectives that are based on specific worldviews and institutional interests. Recent literature underlines the fact that the output of artificial intelligence is to be interpreted as a recreation of the original truth that is not a discovery. In the context of artificial intelligence literacy, researchers should be aware that artificial intelligence systems are actively involved in the development of what can be considered salient, relevant or meaningful to a dataset (Salimova et al., 2025).

In the case of qualitative inquiry, this causes serious epistemological concerns. When artificial intelligence helps to detect themes, summarize stories, or create analytic memos, it will be included in the interpretive apparatus of the study instead of an unbiased analytical support. Artificial intelligence literacy thus involves a clear realization of mediation of epistemic practices by artificial intelligence and how human judgment is critical in validating, contextualizing and challenging insights informed by artificial intelligence.

3.2 Positivist, Interpretivist, and Critical Perspectives on AI

Various philosophical paradigms have their own version of the interpretation of artificial intelligence and its use in research. Positivists tend to view artificial intelligence as one of the sources of better objectivity, efficiency and reliability. Artificial intelligence tools are appreciated because of their ability to compute high amounts of data regularly and reduce human bias by means of automation (Mahbubi, 2025).

In this paradigm, artificial intelligence literacy can be scaled to technical competence, the appropriate use of tools, and optimization of the efficiency of analytics.

The interpretivist views, on which much qualitative research is based, provide an opposite picture. Interpretivism underlines that social reality is created in the course of contextual, situated, and subjective meaning-making. In this viewpoint, artificial intelligence cannot be considered as a disinterested observer because its results are influenced by previous human choices ingrained in the choice of data, its labeling and the design of the algorithm. Artificial intelligence literacy in an interpretivist sense thus places more emphasis on critical thinking, contextual interpretation, and reflexive consciousness of the influence of artificial intelligence in making analytic decisions and taking interpretive paths.

The critical perspectives are an extension of this analysis but they preempt power, inequality, and ideology. According to the critical scholars, artificial intelligence systems tend to replicate the prevailing social patterns, favoring some voices over others. Research on non-technical interaction with artificial intelligence shows that opaque systems may strengthen knowledge and power asymmetries, especially in situations where the outputs of algorithms are viewed as objective or authoritative. A critical view of artificial intelligence literacy is the ability to question whose knowledge is augmented, whose experiences are mediated and how artificial intelligence is involved in greater regimes of control, surveillance, and governance **Yim, 2024**).

3.3 AI Literacy Through a Constructivist Lens

A qualitative research on artificial intelligence literacy has a very fruitful approach to constructivist epistemology. Constructivism believes that knowledge can be constructed in the process of interaction between individuals, tools, and social contexts, as opposed to being found as an objective external reality. In this regard, artificial intelligence systems cannot be considered external tools but active agents in co-construction of knowledge.

Artificial intelligence literacy can take a constructivist approach to describe the way in which human researchers and artificial intelligence systems are co-constructive in the analytic outcomes produced. The use of artificial intelligence tools also shapes the way researchers perceive data, patterns that are present, and how data are predetermined to be interpreted. Meanwhile, researchers determine the outputs of artificial intelligence by selecting data, designing prompts, choosing parameters, as well as framing interpretation. This mutual interaction between the human and artificial intelligence agents has recently become explicit in socio-technical models of artificial intelligence literacy.

To qualitative researchers, the constructivist artificial intelligence literacy supports the significance of methodological transparency and reflexivity. Instead of hiding the role of artificial intelligence, scholars are now urged to clearly report the role of artificial intelligence-based systems in analytic decisions and how human interpretation influenced the ultimate creation of meaning. This level of transparency enhances epistemic responsibility and makes the application of artificial intelligence agreeable with the ethics of qualitative research.

3.4 Posthumanism, Sociomateriality, and AI

The sociomaterial and posthumanist theories also make the traditional human-based epistemologies more perplexing by questioning the distinction between human and non-human agents in the knowledge production (Kim et al., 2025). According to these views, agency is dispersed in networks of human beings, technologies, discourses and material situation instead of being found in individual actors.

In this paradigm, the idea of artificial intelligence systems is not simply an instrument, but an actant engaged in research practice and results. The posthumanist understanding of artificial intelligence literacy must recognize the implication of entanglement between the human-cognitive, algorithmic and material infrastructures. Artificial intelligence systems establish new research time frames, define new work analytically, and change the researcher-data relations. The sociomaterial views point to the fact that artificial intelligence does not merely facilitate research, but actually restructures research and the process of conducting it.

More recent interdisciplinary research on artificial intelligence literacy has underlined the importance of a wider range of ontologies to develop responsible relationships with artificial intelligence, with the perspective that responsible interactions with artificial intelligence should not be restricted to functionality, but rather to the wider socio-material implications of artificial intelligence. In the qualitative inquiry, it means that authorship, agency, and responsibility in the context of artificial intelligence mediation of research may be reconsidered.

3.5 Reflexivity in the Age of Intelligent Systems

Reflexivity has long been a cornerstone of qualitative methodology, requiring researchers to critically examine their positionality, assumptions, and influence on the research process. The integration of artificial intelligence introduces new layers of reflexivity, as researchers must now account not only for their own interpretive role but also for the mediating influence of intelligent systems.

Artificial intelligence literacy enhances reflexivity by enabling researchers to ask critical questions about how artificial intelligence tools shape analytic pathways. These questions include how prompts frame responses, how algorithms prioritize certain patterns, and how automation may obscure uncertainty or ambiguity (Dai,

2025). Without artificial intelligence literacy, researchers risk uncritically accepting artificial intelligence outputs, thereby diminishing reflexive engagement and methodological rigor.

Emerging research emphasizes that reflexive artificial intelligence use involves ongoing evaluation rather than one-time assessment. This aligns with qualitative traditions that view reflexivity as a continuous process embedded throughout the research lifecycle. Artificial intelligence literacy thus supports reflexive vigilance, ensuring that human judgment remains central to qualitative interpretation even in artificial intelligence–augmented research environments.

3.6 Who Is the "Knower" When AI Is Involved?

The involvement of artificial intelligence in qualitative research raises fundamental questions about epistemic authority and the nature of the knower. Traditionally, qualitative research positions the researcher as the primary knower, responsible for interpreting data in dialogue with participants and theory (Richter et al., 2025). Artificial intelligence complicates this positioning by introducing systems that generate interpretations, summaries, and analytic suggestions.

Artificial intelligence literacy helps clarify that artificial intelligence systems do not possess understanding, intentionality, or epistemic responsibility. They do not "know" in a human sense but generate outputs based on statistical inference from prior data. However, their influence on research outcomes is undeniable. As such, the knower in artificial intelligence–mediated qualitative research is best understood as a hybrid assemblage, in which human researchers retain epistemic responsibility while artificial intelligence functions as a mediating actor rather than an epistemic subject.

Recognizing this distinction is crucial for maintaining ethical and methodological accountability. Artificial intelligence literacy enables researchers to resist epistemic displacement, ensuring that interpretive authority remains grounded in human judgment, theoretical sensitivity, and contextual understanding. In doing so, artificial intelligence literacy supports a rearticulation of qualitative epistemology that acknowledges technological mediation without surrendering the human core of qualitative inquiry.

4. AI LITERACY AND QUALITATIVE RESEARCH DESIGN

4.1 AI Across the Qualitative Research Lifecycle

Artificial intelligence is becoming deeply integrated throughout the whole qualitative research lifecycle and affects the stages beyond the formulation of research questions to the collection of data, its analysis and the publication of results. Conventionally, qualitative research design has been based on flexibility, emergent and responsiveness to the context. The introduction of artificial intelligence into this lifecycle does not kill these characteristics but changes the manner in which they become enacted. Artificial intelligence systems have become common in initial phases of the activity like literature scan, concept mapping, and thematic gaps. At a more advanced level, artificial intelligence will aid in data management, transcription, and coding, pattern identification, and analytic summary (Su et al., 2023). At the dissemination phase, generative artificial intelligence can be used to assist in the drafting of qualitative results, visualization of findings, and translation of findings. The artificial intelligence literacy allows the researchers to explain such interventions as methodological choices as opposed to technical convenience, so the utilization of artificial intelligence will be non-incidental, transparent, and epistemologically sound (Table 1).

Table 1. Integration of artificial intelligence across the qualitative research lifecycle, outlining key applications, associated methodological and ethical risks, and corresponding AI literacy practices required to ensure reflexivity, transparency, and epistemic accountability.

Stage	AI Use	Risk	AI Literacy Practice
Design	Literature summarization	Bias toward dominant narratives	Critical interrogation
Data Collection	AI interviews/ transcription	Loss of nuance	Reflexive validation
Analysis	Coding, clustering	Automation bias	Triangulation
Interpretation	Summarization	Epistemic overreach	Human interpretive authority
Reporting	Writing support	Authorship ambiguity	Disclosure

Significantly, the use of artificial intelligence throughout the lifecycle of qualitative research is neither standard nor predetermined. Artificial intelligence literacy prepares the researcher to react selectively and reflexively to the potential application of artificial intelligence assistance in the sites of application and the places where the exclusive use of human interpretation is methodologically required.

4.2 AI-Supported Research Question Development

The phase of formulating research questions is very crucial and intellectually creative in qualitative research. The support of this phase is gradually provided by artificial intelligence technologies that summarise already available scholarship, outline common patterns of thought and demonstrate unexplored aspects of large literature collection. Reflexively, artificial intelligence will also increase the consciousness of researchers on interdisciplinary relationships and emerging debates.

Nonetheless, there are epistemic risks associated with the use of artificial intelligence-assisted research questions development. The artificial intelligence systems are inclined to prioritize major discourses, frequently used motifs, and linguistically salient ideas, which may enforce the current paradigms, but not to question them. In the absence of artificial intelligence literacy, researchers can easily choose the questions defined by the success of an algorithm instead of their theoretical importance or social importance.

Artificial intelligence literacy urges researchers to view suggestions generated by artificial intelligence as heuristic suggestions, but not as guideposts. Although the research questions may be based on theory, context, and ethical concern, human judgment is still vital. In this regard, artificial intelligence serves as a mental scaffold that can help, but not eliminate, scholarly investigation (Salhab, 2024).

4.3 Sampling, Recruitment, and AI Mediation

Qualitative research design focuses on sampling and recruitment because it defines who will be represented and the representation of experiences. There is an increase in the mediation of such processes by artificial intelligence as algorithmic recruitment tools, digital platforms, and data-driven sampling strategies. By improving the reach of recruitment in the framework of online resources, artificial intelligence systems can potentially help identify potential subjects, divide populations, or segment them.

Although tools of this nature may help to increase the efficiency and the distance, they also come with some types of mediation that need a closer examination. Digital systems can create the systematic bias in sampling; algorithm sampling can discriminate against or overrepresent a specific group, based on the biases existing in digital infrastructures or past data. Artificial intelligence literacy gives a researcher the ability to critically assess how the recruitment algorithms work, what data input they are based on, and how the inclusion and exclusion criteria are built.

The ethical concerns surrounding recruitment through artificial intelligence are also associated with transparency and informed consent. Artificial intelligence systems might have led to the participation of participants without their knowledge. Artificially intelligent informative qualitative research design focuses on the ethical clarity aspect, which implies that the participants are aware of how technologies influence research interactions and that the power asymmetry will be transparent.

4.4 AI Tools in Qualitative Data Collection

Qualitative data collection now employs artificial intelligence instruments to gather data in certain types of digital, large-scale, and multimodal studies. The literacy of artificial intelligence is needed to comprehend how these tools influence the nature, scope and quality of the data gathered.

4.4.1 Interviews and Conversational Agents

Artificial intelligence-powered conversational agents are now used to conduct or supplement qualitative interviews. These systems can deliver standardized prompts, manage turn-taking, and enable scalable data collection across diverse contexts. In certain settings, conversational agents may reduce interviewer bias or enhance participant anonymity.

However, artificial intelligence-mediated interviews lack the adaptive empathy, contextual sensitivity, and relational attunement of human interviewers (Shiri, 2024). Artificial intelligence literacy enables researchers to critically assess how conversational agents frame questions, respond to participant input, and shape interactional dynamics. In qualitative research, artificial intelligence is therefore best positioned as a complementary tool rather than a replacement for human-led interviewing.

4.4.2 Digital Ethnography and AI

Digital ethnography increasingly relies on artificial intelligence to navigate and analyze vast online environments. Artificial intelligence systems can track interaction patterns, identify thematic clusters, and surface emergent practices within digital communities. These capabilities allow ethnographers to engage with scale and complexity that would otherwise be unmanageable. At the same time, artificial intelligence-mediated observation risks decontextualizing social practices by prioritizing measurable behavior over lived meaning. Artificial intelligence literacy supports ethnographers in situating algorithmic outputs within thick description and interpretive analysis, ensuring that computational insights do not substitute for contextual understanding.

4.4.3 Observational and Multimodal Data

Qualitative research increasingly incorporates multimodal data, including images, video, audio, and embodied interaction. Artificial intelligence tools capable of facial recognition, gesture analysis, and audio classification offer new analytic possibilities for such data. However, automated interpretation of affect or intention risks oversimplifying complex human expressions. Artificial intelligence literacy enables researchers to critically evaluate the assumptions embedded in multimodal analysis tools and to determine when automated outputs require reinterpretation or rejection. Human interpretation remains essential for preserving ambiguity, context, and meaning in qualitative multimodal research (Kong et al., 2024).

Vignette 1: AI-Assisted Transcription in Multilingual Contexts

Context: A qualitative researcher conducting interviews in regional dialects uses an AI transcription tool.
AI Role: Automated speech-to-text conversion.
Risk: Misinterpretation of culturally specific expressions and erasure of linguistic nuance.
Impact: Distortion of participant voice → threatens credibility.
Mitigation: Researcher conducts manual verification of 15–20% transcripts and maintains an audit log of corrections.
AI Literacy Practice: Reflexive validation + linguistic sensitivity.

4.5 Researcher Agency and Decision-Making in AI-Augmented Studies

The integration of artificial intelligence into qualitative research design fundamentally reshapes researcher agency. Artificial intelligence tools increasingly offer recommendations, summaries, and analytic pathways that may influence decision-making throughout the research process. Without artificial intelligence literacy, researchers risk deferring to algorithmic authority, thereby diminishing their role as reflexive and accountable knowledge producers. Artificial intelligence literacy reinforces researcher agency by positioning artificial intelligence as a support system rather than a decision-maker. Researchers remain responsible for methodological choices, interpretive judgments, and ethical accountability, including decisions about when to rely on artificial intelligence, when to override its outputs, and how to document its role in the research process. In artificial intelligence–augmented qualitative studies, agency is exercised through critical engagement, transparency, and reflexivity. By cultivating artificial intelligence literacy, qualitative researchers can harness technological affordances while preserving the epistemological and ethical foundations of qualitative inquiry (Saddhono et al., 2024). As shown in Figure 2, AI mediates multiple stages of the qualitative research process; however, interpretive authority and ethical responsibility remain fundamentally human obligations. Ethical and governance considerations for AI-augmented qualitative research include the following:

IRB/Policy: Researchers should report the use of AI tools to Institutional Review Boards or Ethics Committees, particularly when cloud-based AI systems are used for storing, processing, or analysing participant data.
Data Governance: Researchers should address data ownership, access control, and the power asymmetry between technology providers and researchers when sensitive qualitative data are processed through AI platforms.

Figure 2. This figure illustrates the role of artificial intelligence across key stages of the qualitative research lifecycle, from research design and sampling to data analysis, interpretation, and dissemination. It highlights AI-supported functions alongside associated risks, ethical considerations, and required AI literacy practices at each stage. The figure emphasizes that, despite increasing AI mediation, human interpretive authority, reflexive judgment, and epistemic accountability must be retained throughout the research process.

5. AI LITERACY IN QUALITATIVE DATA ANALYSIS

5.1 The Shift from Manual Coding to AI-Assisted Analysis

The traditional features of qualitative data analysis include long and intensive interaction with data by reading, coding, and writing memos, as well as synthesizing the information intuitively. The importance of manual coding has been traditionally based on the need to have profound familiarity with data and to lead to reflexive interpretation. However, in recent years, the artificially intelligent has become a more significant part of this realm of analysis, making it possible to provide computational assistance to qualitative analysis in new ways. Such a change in the sole manual coding to AI-supported analysis is a major methodological change that will require a high level of artistic intelligence.

The qualitative analysis tools that use AI to aid qualitative studies can now assist with automated suggestions on coding, clustering of textual extracts, suggesting common linguistic patterns and generating analytic summaries (Gao et al., 2023). They can significantly shorten the time it takes to handle large qualitative data, and may be able to reveal patterns not easily visible with the researcher merely using these tools. Nevertheless, the analytical work that is a core of qualitative inquiry is not removed by the AI- assisted analysis. Instead, it reassigns analytic effort, making artificial intelligence a detection pattern system and not an interpretation agent.

By being able to generate artificial intelligence literacy researchers can critically evaluate how automated coding recommendations are created, what the assumptions of algorithmic categorizations are, and how the processes can influence the analytic results. In the absence of such literacy, the risk of efficiency gains at the cost of interpretive richness and epistemic responsibility is becoming a reality.

5.2 Natural Language Processing in Qualitative Research

The idea of natural language processing (NLP) has turned into one of the most powerful applications of artificial intelligence to qualitative data analysis. NLP methods can be used to compute human language on a large scale, which can be useful in performing tasks like tokenization, sentiment analysis, topic modeling, and semantic similarity evaluation. NLP is widely used in qualitative research, mostly to analyse interview transcripts, open-ended survey responses, policy documents, discourse on the internet, etc.

In spite of its power of analysis, NLP raises significant methodological issues. Language is contextual, culturally situated, and frequently ambiguous, and NLP systems are based on statistical regularities which are computed based on large corpora (Murali et al., 2024). Consequently, irony, metaphor, silence, and culturally relevant meanings can be misunderstood or destroyed. To literally interpret the NLP outputs, artificial intelligence literacy is therefore needed to disregard them and instead perceive them as a clear meaning.

In qualitative terms, NLP outcome must be perceived as points of entry to analysis as opposed to final evidence. The artificial intelligence literacy can assist a researcher to frame the patterns created by NLP in larger interpretive contexts that are based on theory, context, and reflexive interaction with data.

5.3 Machine Learning for Pattern Detection and Thematic Analysis

Substitution of qualitative research with machine learning and pattern detection and thematic analysis is increasingly being employed. Such methods are able to group similar text pieces, uncover subtextual thematic relationships, and uncover relational pattern in huge data. Machine learning provides scalable analysis opportunities to researchers dealing with large corpora (longitudinal interviews or digital ethnographic data etc.).

Nevertheless, the pattern detection performed by machine learning is quite different to what humans can perceive thematically. This is because machine learning systems are sensitive to statistical similarities, as opposed to conceptual purity or theoretical applicability (Vijayvergiya et al., 2024). In order to prevent the mixing of algorithmic clustering and interpretive theme development, artificial intelligence literacy allows researchers to identify this difference.

Unlike in automated phase models of thematic analysis, human researchers are left with the role to assess relevancy, coherence, and meaning concerning research questions and theoretical paradigms in AI-assisted thematic analysis. Artificial intelligence literacy in this way supports the work of a researcher as an interpretive agent who either validates, refines or dismisses machine-generated patterns instead of accepting them unquestioningly.

5.4 Risks of Automation Bias and Over-Reliance

Among the greatest dangers of qualitative analysis helped by AI lies automation bias the disposition to trust the results of an algorithm more than to doubt them. Automation bias is especially problematic when there is no labeling of data, and the central features of analytic validity are meaning, context, and nuance in qualitative research. Whenever a system is seen as objective or technically advanced, researchers can be enticed to accept AI-generated codes, summaries, or thematic suggestions. To identify and alleviate this bias, artificial intelligence literacy is necessary to anticipate the fallibility, context-dependence and value-biased nature of AI systems by foregrounding. Possible strategies to mitigate automation bias are triangulation with manual analysis, repeated validation of AI outputs, and documentation of analytic decision-making processes. Artificial intelligence literacy enables researchers to have a healthy distance toward automated tools and use their affordances of analysis responsibly.

5.5 Interpretive Authority: Human vs. AI

The increasing use of artificial intelligence in the analysis of qualitative data provokes basic questions of the interpretive authority. Conventionally, a researcher who is reflexively involved in data communication with both theory and participants holds interpretive power. This arrangement is complicated by artificial intelligence that produces suggestions in analysis that can affect interpretation. Artificial intelligence literacy explains that AI systems lack knowledge, will, and interpretational judgment. Their productions are the results of probabilistic inferences, but not the meaning making processes. Although AI can draw out patterns or create summaries, it is not possible to assess the importance, theoretical contribution, and contextual meaning. The interpretive authority is thus left to human responsibility. In qualitative analysis with AI assistance, power is enforced with the critical interaction of the algorithmic results, making sure that ultimate interpretations will be based on the contextual knowledge, reflexivity, and moral responsibility (Bryda & Sadowski, 2024).

5.6 Transparency, Explainability, and Auditability

The key methodological issues of AI-assisted qualitative analysis are transparency and explainability. Most AI systems are not transparent and act as blind boxes, so it is hard to follow the processes of generating particular outputs. The lack of transparency is a fundamental qualitative issue because it questions the principle of transparency and trustworthiness. Figurative literacy Artificial intelligence literacy allows researchers to approach the problem of explainability critically and use tools that facilitate auditability where feasible. Auditability is the process of recording the usage of AI tools, parameters, and outputs and how they were applied to analytic decisions. Such reporting increases the credibility and enables readers to determine the methodological quality of AI-enhanced qualitative research. Artificial intelligence literacy can promote methodological accountability through foregrounding transparency as well as aligning AI usage with the already accepted qualitative rigor standards.

AI Audit Checklist for Qualitative Research:

1. What AI tools were used and for what purpose?
2. What data were shared with AI systems?
3. Were outputs verified against raw data?
4. How were biases identified and addressed?
5. What role did AI play vs. human interpretation?
6. How is AI use disclosed in reporting?
7. Are there risks related to privacy or data ownership?

5.7 Maintaining Thick Description and Contextual Depth

Thick description and contextual depth are hallmarks of qualitative research, enabling rich understanding of social phenomena. AI-assisted analysis, with its emphasis on abstraction and pattern detection, risks eroding these qualities if not used judiciously.

Artificial intelligence literacy supports a balanced analytic approach in which computational tools are integrated without displacing close reading, narrative engagement, and contextual interpretation. Researchers must actively return to original data, attend to participant voices, and situate findings within social, cultural, and historical contexts (Kazemitabaar et al., 2024). Maintaining thick description in AI-assisted analysis requires intentional methodological design and reflexive practice. Artificial intelligence literacy empowers qualitative researchers to leverage computational support while preserving the depth, ambiguity, and richness that define qualitative inquiry. AI literacy can be directly aligned with established criteria of qualitative rigor. Credibility is strengthened through systematic validation of AI-generated outputs against raw data. Dependability is enhanced through detailed audit trails documenting AI tool usage and analytic decisions. Confirmability is supported by transparency in reporting how AI influenced interpretations. Reflexivity is expanded to include critical awareness of human–AI interaction and its epistemic implications. By integrating AI literacy with these foundational principles, qualitative researchers can maintain methodological integrity in AI-augmented contexts.

6. ETHICAL DIMENSIONS OF AI LITERACY

6.1 Ethical Frameworks for AI Use in Qualitative Research

Ethical considerations have always been central to qualitative research due to its close engagement with human experiences, narratives, and social contexts. The integration of artificial intelligence into qualitative research intensifies existing ethical concerns while introducing new challenges related to algorithmic mediation, data governance, and distributed responsibility. Artificial intelligence literacy plays a critical role in enabling researchers to navigate these complexities with methodological rigor and ethical sensitivity.

Traditional research ethics frameworks emphasize principles such as respect for persons, beneficence, non-maleficence, and justice. While these principles remain foundational, AI-mediated research requires their reinterpretation within socio-technical systems where agency is distributed across researchers, data infrastructures, and algorithmic processes. Artificial intelligence literacy supports this reinterpretation by helping researchers recognize that ethical responsibility cannot be delegated to technological systems, even when analytic processes are partially automated.

In qualitative inquiry, ethical artificial intelligence use aligns with principles of reflexivity, transparency, and contextual sensitivity. Rather than treating ethics as procedural compliance, artificial intelligence literacy encourages ongoing ethical reflection across the research lifecycle, from design and data collection to analysis and dissemination.

6.2 Bias, Fairness, and Representation in AI Systems

Bias is one of the most extensively documented ethical challenges associated with artificial intelligence systems. Because AI models are trained on historically situated data, they often reproduce or amplify existing social, cultural, and linguistic inequalities (Gao et al., 2023). In qualitative research, where representation, voice, and meaning are central concerns, such biases carry particularly serious epistemic and ethical implications.

AI-assisted qualitative analysis tools may privilege dominant linguistic patterns, cultural norms, or frequently represented groups while marginalizing less visible or non-standard forms of expression. Without artificial intelligence literacy, researchers may inadvertently incorporate these distortions into their interpretations, thereby undermining commitments to inclusivity and justice. Artificial intelligence literacy enables researchers to recognize bias as a socio-ethical issue rather than merely a technical flaw. Ethical engagement requires active human intervention, including validation of AI outputs against contextual knowledge, triangulation with manual analysis, and reflexive attention to whose voices are amplified or silenced. Such practices reinforce qualitative research's commitment to equitable knowledge production.

6.3 Informed Consent in AI-Mediated Research

Informed consent is a cornerstone of ethical qualitative research, grounded in respect for participant autonomy and transparency. Artificial intelligence-mediated research complicates informed consent by introducing technologies that may not be fully understood by participants and, in some cases, by researchers themselves. When AI systems are used for data collection, transcription, analysis, or dissemination, participants may be unaware of how their data are processed, transformed, or retained. Artificial intelligence literacy is essential for enabling researchers to communicate clearly about the role of AI in their studies, including potential risks, uncertainties, and limitations. Ethical AI literacy encourages a relational and ongoing approach to consent, particularly in longitudinal or adaptive research designs.

Rather than treating consent as a one-time procedural requirement, researchers are encouraged to revisit consent as AI tools evolve or are applied in new ways. This approach aligns with qualitative traditions that emphasize trust, dialogue, and participant agency (Borger et al., 2023).

6.4 Privacy, Surveillance, and Data Ownership

Artificial intelligence systems often rely on large-scale data processing and cloud-based infrastructures, raising significant concerns related to privacy, surveillance, and data ownership. Qualitative research frequently involves sensitive personal information, making these concerns especially acute. AI tools used for transcription, storage, or analysis may involve third-party platforms with opaque data retention policies or secondary data use practices. Artificial intelligence literacy equips researchers to critically assess where data are stored, who has access to them, and how long they are retained. Without such literacy, researchers risk violating participant confidentiality and undermining trust. Questions of data ownership become further complicated when AI systems generate derivative outputs, such as summaries, embeddings, or synthetic representations. Ethical artificial intelligence literacy encourages precautionary approaches that prioritize participant protection and recognize qualitative data as relational rather than extractive resources.

6.5 Power, Inequality, and Algorithmic Harm

Artificial intelligence systems are embedded within broader political, economic, and institutional power structures that shape who designs them, who has access to them, and whose interests they serve. In qualitative research, AI-mediated practices may unintentionally reinforce existing inequalities by privileging well-resourced institutions or researchers with access to advanced technologies. Algorithmic harm can occur when AI systems misclassify, misrepresent, or erase certain social groups, leading to epistemic injustice. Artificial intelligence literacy enables qualitative researchers to recognize these harms and to critically examine how AI use intersects with race, gender, class, language, and geography. From an ethical standpoint, artificial intelligence literacy involves situating methodological choices within broader structures of power and responsibility. Researchers are encouraged to resist technological determinism and to prioritize human values, contextual understanding, and social accountability in AI-augmented research practices (Cevik et al., 2025).

6.6 Responsible and Reflexive AI Literacy Practices

Responsible artificial intelligence literacy is not a static checklist but an ongoing practice of ethical reflection, methodological care, and reflexive engagement. It involves continuous questioning of how AI tools shape research decisions, relationships, and outcomes across the research lifecycle.

Reflexive AI literacy encourages researchers to document and critically reflect on their use of artificial intelligence, including the rationale for tool selection, interpretation of outputs, and limitations encountered. Such documentation enhances transparency and accountability while supporting ethical scrutiny by peers and reviewers. Ultimately, ethical artificial intelligence literacy supports a vision of qualitative research in which technological innovation is guided by human values, social responsibility, and epistemic humility. By cultivating responsible and reflexive practices, qualitative researchers can engage with artificial intelligence in ways that enhance, rather than compromise, the ethical foundations of qualitative inquiry. The integration of AI into qualitative research necessitates expanded governance frameworks. Institutional Review Boards (IRBs) must increasingly evaluate not only human subject protections but also the role of AI systems in data handling, analysis, and storage. Researchers are ethically obligated to disclose AI usage, particularly when using cloud-based systems that may process sensitive qualitative data. Issues of data ownership and third-party access introduce new risks, requiring researchers to critically assess platform policies and ensure compliance with institutional and legal standards. AI literacy thus extends into governance awareness, equipping researchers to navigate emerging regulatory and ethical expectations.

7. TEACHING AND LEARNING AI LITERACY IN QUALITATIVE METHODOLOGY

7.1 The Pedagogical Imperative for AI Literacy

The rapid integration of artificial intelligence into research practices has created a clear pedagogical imperative to equip emerging scholars with AI literacy. In qualitative methodology education, this imperative is especially pronounced, as AI increasingly mediates processes of data collection, analysis, and interpretation that have traditionally relied on human judgment and reflexive engagement. Teaching AI literacy is therefore not an optional curricular enhancement but a necessary response to changing research ecologies. AI literacy education in qualitative methods serves multiple purposes: it prepares students to engage responsibly with AI tools, fosters critical awareness of technological mediation, and strengthens ethical and epistemological reasoning. Without formal instruction, students may adopt AI tools informally or uncritically, relying on opaque systems without understanding their methodological implications (Yim, 2024). The pedagogical goal is not technical specialization but the cultivation of reflective qualitative researchers capable of informed judgment in AI-augmented research contexts.

7.2 Integrating AI Literacy into Qualitative Methods Curricula

Integrating AI literacy into qualitative research methods curricula requires alignment with core methodological principles rather than the creation of isolated technical modules. AI literacy can be embedded across instruction on research design, data analysis, reflexivity, and ethics, reinforcing the idea that AI is increasingly constitutive of qualitative practice. Curricular integration may involve revising learning outcomes to include critical evaluation of AI tools, transparency in AI use, and ethical reasoning related to algorithmic mediation. Comparative exercises that contrast manual and AI-assisted analysis can help students identify strengths, limitations, and biases of different approaches (Meylani, 2024). By situating AI literacy within established qualitative frameworks, educators avoid framing AI as a neutral add-on and instead emphasize its methodological significance.

7.3 Experiential and Practice-Based Learning Approaches

Experiential and practice-based learning approaches are particularly effective for teaching AI literacy in qualitative methodology. AI literacy develops most robustly through hands-on engagement with tools, coupled with structured reflection and critical discussion. Practice-based activities allow students to observe firsthand how AI shapes analytic processes and interpretive outcomes. Guided exercises may involve using AI tools for transcription, coding assistance, or thematic exploration, followed by reflexive analysis of how these tools influenced interpretation. Comparing AI-generated outputs with manual analysis encourages critical awareness of automation bias, epistemic limits, and interpretive responsibility (Mittelstadt et al., 2016). Such pedagogical strategies reinforce that AI literacy is not about efficiency alone but about reflective engagement with technology in context.

7.4 Challenges in Teaching AI Literacy

Teaching AI literacy in qualitative methodology presents several challenges. Students enter classrooms with uneven exposure to AI technologies, ranging from extensive familiarity to apprehension or skepticism. Educators must therefore design inclusive learning environments that support diverse entry points (Alamäki et al.,

2024). A further challenge is the rapid pace of technological change, which makes curricula tied to specific tools quickly obsolete. This underscores the importance of emphasizing conceptual understanding, ethical reasoning, and reflexive judgment over transient technical skills (Crawford, 2021). Institutional constraints, including limited resources and evolving policies on AI use, may also complicate pedagogical efforts. Addressing these challenges requires curricular flexibility, ongoing professional development, and institutional support for responsible AI education.

8. CHALLENGES, LIMITATIONS, AND FUTURE DIRECTIONS

The integration of artificial intelligence into qualitative research offers transformative possibilities, yet it also introduces persistent challenges and limitations that require critical attention. One major concern is the risk of methodological dilution. AI tools optimized for efficiency and scale may encourage surface-level analysis or reductive interpretation, potentially undermining the depth and contextual sensitivity that define qualitative inquiry.

Opacity remains a significant limitation. Many AI systems operate as black boxes, making it difficult to explain or audit analytic processes an issue that conflicts with qualitative commitments to transparency and trustworthiness. While AI literacy enhances critical engagement with such systems, structural constraints in tool design often limit explainability. Addressing this limitation will require collaboration between qualitative researchers and AI developers to prioritize transparency and auditability.

Bias and representational harm also persist. AI systems trained on historically uneven data may privilege dominant narratives while marginalizing less-represented voices. Although AI literacy enables researchers to recognize and challenge such biases, complete neutrality remains unattainable. Future methodological innovation should therefore focus on strategies that actively counteract algorithmic bias rather than inadvertently reproducing it.

Unequal access to AI tools presents another challenge. Researchers in under-resourced institutions or regions may be disadvantaged in increasingly AI-mediated research environments, while others may feel pressured to adopt AI regardless of methodological fit. AI literacy must therefore include resistance to technological determinism and attention to equity, open tools, and capacity-building initiatives. Looking forward, future directions include the development of explainable AI tools tailored for qualitative analysis, participatory approaches to AI design that incorporate qualitative insights, and longitudinal studies examining how AI literacy evolves over time. Interdisciplinary collaboration will be essential for aligning technological innovation with qualitative values.

CONCLUSION

In this chapter, I have contended that AI literacy is a cornerstone pillar in the foundation of qualitative methodology and not merely a nicety. In fact, as the penetration of AI into qualitative methodologies increases, the ways in which knowledge is not only created but also validated and legitimized come into play. Thus, the broader scope of AI literacy, as discussed in this chapter, moves far beyond technical competencies and includes critical, ethical, epistemological, and reflexive knowledge. Through an examination of the theoretical basis of AI literacy, the current chapter has demonstrated how it is, by necessity, a socio-technical practice for researchers to grasp. The philosophical and epistemological study demonstrated how AI challenges dominant understandings of objectivity and human agency, and it is within this context—of a reflexive and interpretative ethics—that AI Literacy can be seen as a tool to enable qualitative researchers to engage with AI while maintaining a commitment to methodological precision and contextual depth. Ultimately, this chapter positions AI literacy as a dynamic and reflexive practice rather than a fixed endpoint. As AI technologies continue to evolve, so too must the interpretive, ethical, and methodological capacities of qualitative researchers. Embracing AI literacy allows qualitative inquiry to engage productively with technological innovation while remaining grounded in its core commitments to meaning, context, and human experience.

REFERENCES

Alamäki, A., Nyberg, C., Kimberley, A., & Salonen, A. O. (2024, March). Artificial intelligence literacy in sustainable development: A learning experiment in higher education. [). Frontiers Media SA.]. *Frontiers in Education, 9*, 1343406. DOI: 10.3389/feduc.2024.1343406

Asrifan, A., Said, U. M. R., Jakob, J. C., & Wanci, R. (2025). AI literacy: foundations, importance, and broader implications. In *Transforming vocational education and training using AI* (pp. 17–48). IGI Global Scientific Publishing.

Barrera, B., Feliu, A., Espina, C., Brand, T., Zeeb, H., & Ahmed, F. (2025). Leveraging AI to enhance qualitative research: Experiences and recommendations from case studies in cancer prevention literacy across the European Union. *International Journal of Qualitative Methods, 24*, 16094069251365766. DOI: 10.1177/16094069251365766

Borger, J. G., Ng, A. P., Anderton, H., Ashdown, G. W., Auld, M., Blewitt, M. E., Brown, D. V., Call, M. J., Collins, P., Freytag, S., Harrison, L. C., Hesping, E., Hoysted, J., Johnston, A., McInneny, A., Tang, P., Whitehead, L., Jex, A., & Naik, S. H. (2023). Artificial intelligence takes center stage: Exploring the capabilities and implications of ChatGPT and other AI-assisted technologies in scientific research and education. *Immunology and Cell Biology, 101*(10), 923–935. DOI: 10.1111/imcb.12689 PMID: 37721869

Bryda, G., & Sadowski, D. (2024, January). From words to themes: AI-powered qualitative data coding and analysis. In *World conference on qualitative research* (pp. 309-345). Cham: Springer Nature Switzerland. DOI: 10.1007/978-3-031-65735-1_19

Cevik, A. A., Abu-Zidan, F. M., & Cevik, A. A. (2025). Utilizing AI-powered thematic analysis: Methodology, implementation, and lessons learned. *Cureus, 17*(6). Advance online publication. DOI: 10.7759/cureus.85338 PMID: 40621288

Cha, I., & Wong, R. Y. (2025, April). Understanding Socio-technical Factors Configuring AI Non-Use in UX Work Practices. In *Proceedings of the 2025 CHI Conference on Human Factors in Computing Systems* (pp. 1-17). DOI: 10.1145/3706598.3713140

Chang, C. I., & Choi, W. C. (2025). Exploring Challenges and Opportunities in Artificial Intelligence (AI) Literacy and Educational AI Development: A Qualitative Study of Teachers and Researchers' Perspectives. *Perspectives.*

Chatzichristos, C., Chatzichristos, G., Borremans, I., Gruyaert, S., De Vos, I., De Vos, M., & De Backere, F. (2025). Bridging the AI-literacy gap in health care: Qualitative analysis of the Flanders case study. *Journal of Medical Internet Research, 27*, e76709. DOI: 10.2196/76709 PMID: 41359863

Corchado, J. M., López, S., Garcia, R., & Chamoso, P. (2023). Generative artificial intelligence: Fundamentals. *ADCAIJ: advances in distributed computing and artificial intelligence journal, 12*, e31704-e31704.

Dai, Y. (2025). Integrating unplugged and plugged activities for holistic AI education: An embodied constructionist pedagogical approach. *Education and Information Technologies, 30*(5), 6741–6764. DOI: 10.1007/s10639-024-13043-w

Gangwal, A., Ansari, A., Ahmad, I., Azad, A. K., Kumarasamy, V., Subramaniyan, V., & Wong, L. S. (2024). Generative artificial intelligence in drug discovery: Basic framework, recent advances, challenges, and opportunities. *Frontiers in Pharmacology, 15*, 1331062. DOI: 10.3389/fphar.2024.1331062 PMID: 38384298

Gao, J., Cao, J., Yeo, S., Choo, K. T. W., Zhang, Z., Li, T. J. J., . . . Perrault, S. T. (2023). Impact of Human-AI Interaction on User Trust and Reliance in AI-Assisted Qualitative Coding. *arXiv preprint arXiv:2309.13858.*

Gao, J., Choo, K. T. W., Cao, J., Lee, R. K. W., & Perrault, S. (2023). CoAIcoder: Examining the effectiveness of AI-assisted human-to-human collaboration in qualitative analysis. *ACM Transactions on Computer-Human Interaction, 31*(1), 1–38. DOI: 10.1145/3617362

Gazos, A., Kahn, J., Kusche, I., Büscher, C., & Götz, M. (2025). Organising AI for safety: Identifying structural vulnerabilities to guide the design of AI-enhanced socio-technical systems. *Safety Science, 184*, 106731. DOI: 10.1016/j.ssci.2024.106731

Gutierrez, C. I., Aguirre, A., Uuk, R., Boine, C. C., & Franklin, M. (2023). A proposal for a definition of general purpose artificial intelligence systems. *Digital Society : Ethics, Socio-Legal and Governance of Digital Technology, 2*(3), 36. DOI: 10.1007/s44206-023-00068-w

Kalota, F. (2024). A primer on generative artificial intelligence. *Education Sciences, 14*(2), 172. DOI: 10.3390/educsci14020172

Kaplan, R., & Meylani, R. (2025). Dimensions of artificial intelligence literacy: A qualitative synthesis of contemporary research literature. *Journal of Computer and Education Research, 13*(26), 790–825. DOI: 10.18009/jcer.1648380

Kazemitabaar, M., Williams, J., Drosos, I., Grossman, T., Henley, A. Z., Negreanu, C., & Sarkar, A. (2024, October). Improving steering and verification in AI-assisted data analysis with interactive task decomposition. In *Proceedings of the 37th Annual ACM Symposium on User Interface Software and Technology* (pp. 1-19). DOI: 10.1145/3654777.3676345

Kim, S. K., Kim, T. Y., & Kim, K. (2025). Development and effectiveness verification of AI education data sets based on constructivist learning principles for enhancing AI literacy. *Scientific Reports, 15*(1), 10725. DOI: 10.1038/s41598-025-95802-4 PMID: 40155549

Kirova, V. D., Ku, C. S., Laracy, J. R., & Marlowe, T. J. (2023). The ethics of artificial intelligence in the era of generative AI. *Journal of Systemics, Cybernetics and Informatics, 21*(4), 42–50. DOI: 10.54808/JSCI.21.04.42

Kong, S. C., Cheung, M. Y. W., & Tsang, O. (2024). Developing an artificial intelligence literacy framework: Evaluation of a literacy course for senior secondary students using a project-based learning approach. *Computers and Education: Artificial Intelligence, 6*, 100214. DOI: 10.1016/j.caeai.2024.100214

Kreinsen, M., & Schulz, S. (2023). Towards the triad of digital literacy, data literacy and AI literacy in teacher education–A discussion in light of the accessibility of novel generative AI.

Kudina, O., & van de Poel, I. (2024). A sociotechnical system perspective on AI. *Minds and Machines, 34*(3), 21. DOI: 10.1007/s11023-024-09680-2

Lin, Z., & Dai, Y. (2025, April). Fostering Epistemic Insights into AI Ethics through a Constructionist Pedagogy: An Interdisciplinary Approach to AI Literacy. *Proceedings of the AAAI Conference on Artificial Intelligence, 39*(28), 29171–29177. DOI: 10.1609/aaai.v39i28.35190

Mahbubi, M. (2025). Digital Epistemology: Evaluating The Credibility Of Knowledge Generated By Ai. *YUDHISTIRA: Journal of Philoshopy, 1*(1), 8–18.

Meylani, R. (2024). Artificial intelligence in the education of teachers: A qualitative synthesis of the cutting-edge research literature. *Journal of Computer and Education Research, 12*(24), 600–637. DOI: 10.18009/jcer.1477709

Murali, V., Maddila, C., Ahmad, I., Bolin, M., Cheng, D., Ghorbani, N., & Rigby, P. C. (2024). AI-assisted Code Authoring at Scale: Fine-tuning, deploying, and mixed methods evaluation. *Proceedings of the ACM on Software Engineering, 1*(FSE), 1066-1085. DOI: 10.1145/3643774

Radanliev, P. (2025). Artificial intelligence: Reflecting on the past and looking towards the next paradigm shift. *Journal of Experimental & Theoretical Artificial Intelligence, 37*(7), 1045–1062. DOI: 10.1080/0952813X.2024.2323042

Richter, S., Giroux, M., Piven, I., Sima, H., & Dodd, P. (2025). A constructivist approach to integrating AI in marketing education: Bridging theory and practice. *Journal of Marketing Education, 47*(2), 94–111. DOI: 10.1177/02734753241288876

Ronge, R., Maier, M., & Rathgeber, B. (2025). Towards a definition of Generative artificial intelligence. *Philosophy & Technology*, *38*(1), 31. DOI: 10.1007/s13347-025-00863-y

Saddhono, K., Suhita, R., Rakhmawati, A., Rohmadi, M., & Sukmono, I. K. (2024, November). AI and Literacy: Developing Critical Thinking and Analytical Skills in the Digital Era. In *2024 International Conference on IoT, Communication and Automation Technology (ICICAT)* (pp. 360-365). IEEE. DOI: 10.1109/ICICAT62666.2024.10922871

Salhab, R. (2024). AI Literacy across Curriculum Design: Investigating College Instructors' Perspectives. *Online Learning : the Official Journal of the Online Learning Consortium*, *28*(2), n2. DOI: 10.24059/olj.v28i2.4426

Salimova, R., Nurmanbetova, J., Kozhamzharova, M., Manassova, M., & Aubakirova, S. (2025). Philosophy of education in a changing digital environment: An epistemological scope of the problem. *AI & Society*, *40*(3), 1655–1666. DOI: 10.1007/s00146-024-01928-w

Shiri, A. (2024). Artificial intelligence literacy: A proposed faceted taxonomy. *Digital Library Perspectives*, *40*(4), 681–699. DOI: 10.1108/DLP-04-2024-0067

Su, J., Ng, D. T. K., & Chu, S. K. W. (2023). Artificial intelligence (AI) literacy in early childhood education: The challenges and opportunities. *Computers and Education: Artificial Intelligence*, *4*, 100124. DOI: 10.1016/j.caeai.2023.100124

Tran, T. N., & Tran, H. P. (2023, November). Exploring the role of ChatGPT in developing critical digital literacies in language learning: A qualitative study. In *Proceedings of the AsiaCALL International Conference* (Vol. 4, pp. 1-17). DOI: 10.54855/paic.2341

Van Audenhove, L., Vermeire, L., Van den Broeck, W., & Demeulenaere, A. (2024). Data literacy in the new EU DigComp 2.2 framework how DigComp defines competences on artificial intelligence, internet of things and data. *Information and Learning Science*, *125*(5/6), 406–436. DOI: 10.1108/ILS-06-2023-0072

Vijayvergiya, M., Salawa, M., Budiselić, I., Zheng, D., Lamblin, P., Ivanković, M., & Just, R. (2024, July). Ai-assisted assessment of coding practices in modern code review. In *Proceedings of the 1st ACM International Conference on AI-Powered Software* (pp. 85-93). DOI: 10.1145/3664646.3665664

West, P., Lu, X., Dziri, N., Brahman, F., Li, L., Hwang, J. D., . . . Choi, Y. (2023). The Generative AI paradox:" What it can create, it may not understand". *arXiv preprint arXiv:2311.00059*.

Yim, I. H. Y. (2024). A critical review of teaching and learning artificial intelligence (AI) literacy: Developing an intelligence-based AI literacy framework for primary school education. *Computers and Education: Artificial Intelligence*, *7*, 100319. DOI: 10.1016/j.caeai.2024.100319

Chapter 15
Positionality of AI and Human-Machine Collaborations in AI-Augmented Qualitative Research:
Methodological Recalibration in Critical Research

Ashraf Alam
http://orcid.org/0000-0001-6178-1187
Department of Education, National Institute of Technology Calicut, India

ABSTRACT

This chapter interrogates the positionality of artificial intelligence within contemporary qualitative inquiry and re-engineers human-machine collaborations as explicitly epistemic, ethical, and political projects. Mobilising interpretivist, critical, feminist, decolonial, and posthuman frameworks, it contrasts dataist AI epistemologies with hermeneutic, phenomenological, and justice-oriented logics of knowledge production. The chapter theorises human-centered AI assemblages, reframes trustworthiness, reflexivity, and methodological integrity for algorithmically mediated analysis, and systematises justice grammars of data justice, decolonial AI, and rights-based governance. It finally advances a typology of AI integrated qualitative designs and a programme for AI literacy, governance charters, and reporting standards that can anchor plural, reflexive, and equity enhancing qualitative research in an intensively computational era.

DOI: 10.4018/979-8-3373-7640-0.ch015

ht © 2026, IGI Global Scientific Publishing. Copying or distributing in print or electronic forms without written permission of IGI Global Scientific Publishing is prohibited. Use of this chapter to generative artificial intelligence (AI) technologies is expressly prohibited. The publisher reserves all rights to license its use for generative AI training and machine learning model development.

1. INTRODUCTION

1.1 Situating AI Within Contemporary Qualitative Methodology

Artificial intelligence has rapidly become embedded in the infrastructures of contemporary knowledge production, with *agentic AI* systems and large *language models* orchestrating workflows in universities, think tanks, corporations, public agencies, and civil society organisations. In qualitative inquiry, these systems perform concrete operations such as automated transcription, iterative summarisation, pattern surfacing across large text corpora, and generative rewriting of fieldnotes, interview excerpts, or analytic memos (Maj et al., 2025). Such capabilities instantiate a logic of computation grounded in statistical patterning and large-scale optimisation that sits in structural tension with the ontological and epistemological commitments of *interpretivism*, *constructivism*, and *phenomenology*, which prioritise meaning, situatedness, and lived experience. When qualitative researchers route their engagements with participants and texts through AI intermediaries, the tempo, granularity, and trajectory of analysis are subtly recalibrated by algorithmic defaults such as tokenisation, ranking, and probabilistic completion. The result is an emergent research ecology where human interpretation, machine inferences, institutional incentives, and platform infrastructures interlock, raising foundational questions about whose voices are amplified, whose realities are compressed, and how qualitative knowledge is authorised in an AI-saturated era.

1.2 Rationale for a Conceptual-Theoretical Intervention

The rapid incorporation of AI tools into qualitative practice has preceded the development of robust conceptual vocabularies that can guide their use in a principled manner, creating a methodological vacuum filled by marketing rhetoric and instrumental narratives of efficiency. In the absence of a systematic conceptual-theoretical intervention, qualitative methodologists risk a form of *methodological colonisation* in which practices shaped by *dataism* and *algorithmic positivism* silently displace traditions of dialogic interpretation, reflexive self-location, and attention to power-laden context. When AI is treated primarily as a labour-saving infrastructure, decisions about pre-processing, coding, summarisation, and visualisation migrate from human judgement to opaque optimisation routines, while remaining largely invisible in published accounts. A conceptual re-grounding is therefore necessary to specify how AI can serve as *epistemic augmentation* rather than *epistemic outsourcing*, and how interpretive sovereignty can be preserved when collaborating with non-human actants. This chapter addresses that gap by articulating a dense constellation of concepts, theories, and constructs that reposition AI within critical qualitative methodology rather than at its periphery.

1.3 Focus, Objectives, and Guiding Questions

The chapter concentrates on a set of interlocking constructs that are central for AI-augmented qualitative inquiry, including *human-centered AI*, *interpretivism*, *constructivism*, *critical realism*, *reflexivity*, *trustworthiness*, *algorithmic bias*, *positionality of AI*, and *AI literacy*. The primary objective is to develop a rigorous conceptual architecture that clarifies how these constructs can be re-specified in an era when AI systems participate in the generation, transformation, and circulation of qualitative data. This involves asking how different epistemological traditions conceptualise the permissible roles of AI in interpretive work, how the research assemblage is reconfigured when AI becomes a persistent actant, and how classical criteria such as credibility, authenticity, and methodological integrity must be recalibrated in light of algorithmic mediation. A further objective is to elaborate justice-oriented frameworks that foreground *data justice*, *design justice*, *information ethics*, and *decolonial AI*, so that methodological discussions about AI are anchored in questions of power, vulnerability, and epistemic responsibility rather than confined to tool selection or workflow optimisation.

1.4 Scope, Audiences, and Contributions

The scope of the chapter is conceptual and normative, yet it is oriented toward the concrete dilemmas that arise when researchers incorporate AI into real qualitative projects in education, health, labour, governance, culture, and digital life. The intended readership spans disciplinary boundaries and includes qualitative methodologists, doctoral and master's students, public policy analysts, evaluation practitioners, human resources and workforce development professionals, digital learning technologists, and members of ethics committees who must assess AI-mediated protocols. For this heterogeneous audience, the chapter offers three interrelated contributions. First, it provides an epistemological map that situates AI within traditions such as *interpretivism*, *critical theory*, *feminist standpoint epistemology*, and *decolonial thought*, thereby clarifying conceptual compatibilities and tensions. Second, it advances a set of constructs for understanding *human–AI collaboration* in qualitative work, including the *positionality of AI*, multi-layered reflexivity, and typologies of AI roles across the research life cycle. Third, it develops a justice-inflected vocabulary for trustworthiness, integrity, and governance that can inform institutional policies, supervisory conversations, and cross-sector collaborations involving AI-augmented qualitative research.

1.5 Structure of the Chapter

The chapter unfolds in seven sections that build a cumulative argument, while remaining accessible for selective reading by different professional constituencies. Following this introduction, Section 2 elaborates epistemological and ontological foundations that determine how AI can be meaningfully integrated into qualitative inquiry, contrasting logics such as *constructivism*, *critical realism*, *pragmatism*, and *post-structuralism*, and synthesising their implications within a comparative schema that will later be crystallised in Table 1 in that section. Section 3 develops a conceptualisation of *human-centered AI*, reconceiving the qualitative research assemblage as a constellation of human and non-human actants, and articulating a taxonomy of AI roles that underpins the later matrix in Table 2. Section 4 revisits trustworthiness, reflexivity, and methodological integrity in AI-mediated contexts and proposes reconfigured evaluative criteria that are systematically aligned in Table 3. Section 5 turns to algorithmic bias, ethics, and justice-oriented frameworks, integrating concepts such as *data justice*, *care ethics*, and *epistemic reparations*, and distilling their methodological consequences in Table 4. Section 6 then re-imagines qualitative designs that braid computational techniques with interpretive logics, including hybrid, post-coding, and archival re-use methodologies, presenting a synoptic typology in Table 5. Section 7 concludes by synthesising these contributions into a call for a plural, reflexive, and justice-oriented paradigm of AI-augmented qualitative inquiry.

2. EPISTEMOLOGICAL AND ONTOLOGICAL FOUNDATIONS FOR AI IN QUALITATIVE RESEARCH

2.1 Classical Qualitative Epistemologies and Hermeneutic Traditions

Classical qualitative epistemologies such as *interpretivism*, *constructivism*, and *phenomenology* are anchored in the premise that reality is experienced as a plurality of meaning-saturated lifeworlds, co-constituted through language, interaction, and embodiment. In these traditions, qualitative data are not inert text streams but situated articulations embedded in temporality, affect, and social practice. Hermeneutic logics frame understanding as a recursive movement between part and whole, where the analyst's horizon of pre-understanding is continually reconfigured through interpretive engagement with narratives, performances, and artefacts. Approaches such as *symbolic interactionism* and *ethnomethodology* foreground indexicality, conversational sequencing, and the micro-organisation of everyday conduct, insisting that analytic insight emerges from meticulous attention to situated contingencies rather than abstracted regularities (Meylani, 2024). Within such frameworks, any AI intervention that tokenises language, compresses narrative arcs, or generalises across settings risks truncating the very contextual granularity that is epistemically prized. At the same time, these epistemologies create a demanding benchmark for AI, requiring that any computational scaffold must remain subordinate to dialogic, context-sensitive human interpretation.

2.2 Critical, Feminist, and Decolonial Epistemologies for AI-Saturated Inquiry

Critical, feminist, and decolonial epistemologies reposition qualitative inquiry as a site of contestation over power, ideology, and epistemic justice, rather than as a neutral mirror of social reality. *Critical theory* and *critical realism* conceptualise social worlds as stratified, where observable events are shaped by underlying structures, generative mechanisms, and historically sedimented relations of domination. *Feminist standpoint epistemology* and *intersectionality* emphasise the epistemic privilege of marginalised standpoints, arguing that knowledge is always situated within matrices of gender, race, class, disability, caste, and sexuality. Decolonial and Indigenous epistemologies foreground relational ontologies, community custodianship of knowledge, and resistance to *epistemic extractivism* and *data colonialism*. In AI-augmented contexts, these frameworks interrogate whose languages, categories, and narratives are encoded within training corpora, and how algorithmic infrastructures may reproduce colonial hierarchies. They also demand attention to collective rights, Indigenous data sovereignty, and community governed consent. The comparative logics that animate these epistemologies are later crystallised in Table 1, which synthesises how distinct stances construe qualitative data and legitimate AI roles in interpretive work.

2.3 AI Epistemologies, Dataism, and Epistemic Hybridity

AI epistemologies are grounded in *computationalism* and *dataism*, which construe the social world as a vast field of measurable signals that can be decomposed, vectorised, and recombined through probabilistic inference. In this regime, qualitative materials are transmuted into high dimensional embeddings, where proximity in representational space indexes statistical association rather than lived relatedness. Large *language models* operationalise a form of synthetic textual competence, generating plausible continuations of discourse without intentional subjectivity, biographical experience, or situated accountability (Friese, 2025). Such systems instantiate what can be called *algorithmic positivism*, where patterned regularities across corpora are treated as sufficient grounds for knowledge claims, even when underlying processes remain opaque. These logics collide with interpretivist emphases on voice, singularity, and context, yet they also open possibilities for exploratory pattern discovery, lexical diversity mapping, and cross corpus dialogue tracking that would be infeasible through manual inspection alone. The epistemic terrain that emerges is hybrid, where computational patterning must be critically aligned with hermeneutic sensibilities, and where AI outputs are treated as prompts for human inquiry rather than authoritative verdicts, a synthesis that is concisely mapped in Table 1.

Table 1. Epistemological Stances for AI-Linked Qualitative Inquiry

Epistemological Stance	Core Ontological Presuppositions	Conception of Qualitative Textualities	Permissible AI Roles in Inquiry	Principal Epistemic Risks and Potentials
Interpretivism-Constructivism	Plural, intersubjectively constituted realities, context dependent and historically embedded	Situated narratives, thick descriptions, dialogic meaning performances	Assistive pre-processing, suggestive pattern surfacing under strict human adjudication	Decontextualisation, narrative flattening, yet potential for heuristic sensitisation
Critical Realism	Stratified social ontology, causal mechanisms beneath observable events	Accounts that index structures, mechanisms, and emergent properties	Structural cue detection, discourse mapping, support for retroduction when theory led	Reification of surface regularities, neglect of deep causation, but scope for mechanism probing
Feminist-Intersectional Epistemology	Situated knowledges, relational embodiment, power saturated standpoints	Testimonies of oppression, resistance, and care, entangled with identity axes	Bias diagnostics, visibility amplification for marginalised lexicons, participatory analytic aids	Reinscription of hegemonic categories, erasure of nuance, yet leverage for counter narrative surfacing
Decolonial-Indigenous Epistemology	Relational, land linked, community stewarded realities, pluriversal worldviews	Oralities, ceremonial narratives, communal memories, and place-based stories	Community controlled tools for archiving, translation, and intra community reflection	Data colonialism, cultural appropriation, epistemic dispossession, balanced by reparative archival possibilities

continued on following page

Table 1. Continued

Epistemological Stance	Core Ontological Presuppositions	Conception of Qualitative Textualities	Permissible AI Roles in Inquiry	Principal Epistemic Risks and Potentials
Pragmatism and Methodological Pluralism	Reality as processual, consequences oriented, inquiry as problem solving	Texts as resources for action, deliberation, and experimental redesign	Flexible hybrid scaffolds for mixed methods, rapid prototyping of analytic lenses	Instrumentalism, technocratic drift, but strong potential for context tuned innovation
Post-Structural Posthuman Approach	Decentred subjectivity, discursive-material assemblages, human-nonhuman entanglements	Discourses, affects, and artefacts within shifting assemblages of power	Generative simulations, speculative reconfigurations of discourse, assemblage mapping	Hyper abstraction, aestheticisation of inequality, yet capacity for unsettling dominant ontologies

Table 1 demonstrates that there is no epistemologically neutral way to insert AI into qualitative inquiry, since each stance brings distinct ontological assumptions about reality, subjectivity, and causation. For interpretivist and constructivist orientations, AI can at most provide heuristic scaffolding that remains firmly bracketed by human interpretive labour, whereas critical realist and pragmatist positions may legitimise more extensive use of pattern detection, provided that causal reasoning and contextual judgement remain central. Feminist, decolonial, and posthuman approaches highlight how AI systems themselves become actors within constellations of power, embodiment, and materiality, which means that questions of bias, sovereignty, and ontological multiplicity are not peripheral technicalities but constitutive issues. By making these divergences explicit, the table functions as a conceptual decision aid, enabling researchers, policy makers, and institutional gatekeepers to align AI deployments with explicit epistemic commitments rather than with unexamined assumptions about computational neutrality.

2.4 Ontological Commitments and the Status of AI Artifacts

Ontological commitments determine whether AI outputs are treated as traces of reality, speculative artefacts, or heuristic instruments and therefore shape their admissibility within qualitative reasoning. Relativist ontologies emphasise the multiplicity of world disclosures and thus position AI artifacts as one more discursive layer that must be interrogated for whose perspective it encodes. *Critical realism* advances a stratified ontology in which AI generated patterns may correspond to real causal regularities, but only when interpreted through theory driven retroduction that connects them to mechanisms and structures. *Pragmatist* ontologies foreground consequences and therefore evaluate AI artifacts in terms of their capacity to support problem solving, deliberation, or social innovation rather than their representational fidelity (Ozuem et al., 2025). *Post-structural* and *posthuman* ontologies view AI as part of socio-technical assemblages where human and non-human agencies co produce meaning, which implies that AI outputs participate in world making rather than merely reflecting it. In each case, Table 1's distinctions remind researchers that the ontological status assigned to AI productions is never self-evident but flows from prior philosophical commitments.

2.5 Epistemic Compatibility, Tensions, and Hybrid Configurations

The question of whether AI can be meaningfully integrated into qualitative inquiry is not answerable in the abstract, but only with respect to specific epistemic constellations, value commitments, and research problems. Epistemic compatibility requires that AI involvement does not undermine core constructs such as *reflexivity*, *voice*, *thick description*, and *situated accountability*, which are differentially weighted across paradigms. In some hybrid configurations, AI may serve as an *epistemic amplifier*, helping interpretivist analysts locate deviant cases or trace thematic echoes across expansive corpora that would otherwise remain intractable, while still leaving interpretive adjudication with humans (Pattyn, 2025). In others, such as technocratic deployments of sentiment analysis for policy dashboards, AI risks becoming an *epistemic usurper* that replaces dialogic engagement with metricised proxies. Tensions also emerge around scale, since AI thrives on large datasets, while many qualitative projects are intentionally small scale and relationally intensive. The comparative schematisation offered in Table 1 thus becomes a scaffold for designing hybrid arrangements that treat AI as a contingent collaborator within an explicitly articulated epistemic ecology rather than as a universal solution.

3. HUMAN-CENTERED AI AND THE RECONFIGURED QUALITATIVE RESEARCH ASSEMBLAGE

3.1 Foundations of Human-Centered AI

Human-centered AI can be conceptualised as a normative-metatechnical orientation in which computational architectures, algorithmic affordances, and interface designs are subordinated to human flourishing, epistemic dignity, and rights-based accountability. Rather than construing AI as an autonomous optimisation engine, a human-centered framing treats AI as an embedded element within socio-technical systems that must cohere with principles of *value-sensitive design*, *design justice*, and *responsible innovation*. In qualitative inquiry, this entails aligning AI mediated operations such as transcription, clustering, and summarisation with interpretive commitments to *voice*, *thick description*, and *contextual integrity*. Human-centered AI therefore foregrounds transparent configurability, controllable levels of automation, and intelligible feedback channels so that researchers retain final interpretive sovereignty while benefiting from computational scaffolding (Morgan, 2023). It also requires attention to institutional governance and regulatory ecosystems, since procurement decisions, licensing arrangements, and infrastructural dependencies shape which communities, languages, and epistemic traditions are rendered legible to AI systems. The modalities through which AI is positioned within such human-centered assemblages will be systematically compared in Table 2, which later consolidates diverse role metaphors into a compact analytic typology.

3.2 The Positionality of AI within the Research Ensemble

The concept of positionality can be extended from human researchers to AI systems by recognising that models embody genealogies of training data, optimisation objectives, and infrastructural affordances that locate them within global hierarchies of knowledge and power. AI is not an abstract, placeless intelligence but a situated *socio-technical actant* whose responses sediment the linguistic distributions, cultural biases, and epistemic silences of the corpora on which it was trained. Within a qualitative project, AI can occupy different positions such as background utility, gatekeeping filter, interpretive lens, or quasi-interlocutor, and each configuration reconfigures how participants and researchers encounter one another (Finkenstadt, 2025). Metaphors such as AI as assistant, sentinel, editor, or curator are not merely rhetorical tropes, they encode implicit theories of authority, responsibility, and accountability. When researchers conceptualise AI as part of the research ensemble, they must therefore articulate its positionality explicitly and align it with their epistemic commitments, a task that is later operationalised in Table 2 through a structured comparison of AI role modalities and their conceptual implications.

3.3 The Researcher-AI-Participant Triad

As AI mediated interfaces enter qualitative practice, the dyadic relation between researcher and participant increasingly becomes a triadic configuration in which AI functions as an intermediary presence that shapes interactional texture and narrative form. Conversational agents, transcription services, and analytic dashboards modulate pacing, turn taking, and topical salience, thereby influencing what is disclosed, how it is framed, and which utterances acquire analytic prominence. Participants may experience AI as a neutral listener, a quasi-institutional voice, or a surveillant apparatus, depending on cultural context, digital literacy, and prior encounters with automated systems (Malik & Terzidis, 2025). Researchers, in turn, may oscillate between trust in AI efficiency and scepticism toward algorithmic opacity, producing ambivalence that percolates into methodological decision making. This triadic constellation alters power geometries, since AI can amplify some linguistic registers while marginalising others, and can normalise certain interactional scripts at the expense of locally meaningful communicative practices. The modalities through which AI inhabits this triad, such as tool, probe, collaborator, or synthetic participant, form the backbone of the classificatory schema that Table 2 later presents, linking role metaphors to theoretical lenses and reflexive obligations.

3.4 Multi-Layered Reflexivity in AI Contexts

Reflexivity in AI saturated qualitative inquiry must operate simultaneously across multiple strata, encompassing the researcher's subjective dispositions, the technical affordances of models, and the institutional logics that incentivise particular uses of automation. At the level of personal epistemic habitus, researchers need to interrogate their attachments to computational convenience, their affective responses to AI generated texts, and their evolving sense of expertise when working alongside algorithmic systems. Technical reflexivity requires scrutiny of model provenance, training corpora composition, and parameter settings, as well as an understanding of how tokenisation, sampling, and filtering procedures mediate what is analytically visible. Institutional reflexivity involves examining how managerial discourses of innovation, productivity, and risk management frame the adoption of AI tools, and how these discourses intersect with disciplinary norms about rigour and ethical responsibility (Shmueli & Ray, 2024). Feminist, critical, and decolonial traditions reinforce the need for reflexive attention to power asymmetries and structural inequalities that are encoded within algorithmic infrastructures. Each AI role modality listed in Table 2 carries distinct reflexive demands, since treating AI as a collaborator, a synthetic participant, or a methodological lens invokes different responsibilities for disclosure, authorship, and community accountability.

3.5 Modalities of Human-AI Collaboration in Qualitative Work

Human-AI collaboration in qualitative inquiry can be understood as a spectrum of role modalities in which AI systems participate in the research process with varying degrees of agency, visibility, and normative weight. At one end of this spectrum, AI functions as a largely instrumental tool that undertakes low level operations such as transcription or spell checking, operating under strict human supervision and interpretive framing (Nguyen-Trung, 2025). Progressively more involved modalities include AI as diagnostic probe that surfaces latent lexical or thematic patterns, AI as collaborator that proposes alternative codes or memoing prompts, and AI as co-author that participates in articulating analytic narratives and conceptual framings. At the most controversial end, AI may be treated as synthetic participant that generates simulated accounts or counterfactual narratives, or as methodological lens that refracts data through particular representational geometries. Table 2 consolidates these modalities into a dense matrix that links role metaphors with theoretical metaframes, typical processual engagements, authority reconfigurations, and salient reflexive-ethical interrogatives, thereby providing a compact yet rigorous decision scaffold for designing human-AI assemblages in qualitative research.

Table 2. Modalities of AI Participation in Qualitative Assemblages

AI Role Metaphor or Modality	Dominant Theoretical Metaframe	Typical Qualitative Processual Engagements	Reconfigurations of Authority and Voice	Salient Reflexive-Ethical Interrogatives
Instrumental Tool	Technological instrumentalism, procedural rationality	Transcription, de-noising, token alignment, format normalisation	Consolidates researcher primacy, occludes AI agency, stabilises existing hierarchies	How to prevent automation bias while exploiting efficiency gains
Diagnostic Probe	Exploratory dataism, pattern-oriented hermeneutics	Lexical clustering, topic surfacing, anomaly detection, salience mapping	Grants AI preliminary agenda setting power over what appears noteworthy	Which patterns are foregrounded or occluded by algorithmic logics
Analytic Collaborator	Socio-technical co-production, distributed cognition	Code suggestion, memo prompting, rival explanation generation	Negotiates interpretive sovereignty between human judgement and machine proposals	How to document negotiations between human and AI interpretive moves
Textual Co-Author	Posthuman authorship, assemblage theory	Draft generation, reframing of findings, narrative elaboration	Blurs boundaries of authorship, complicates credit and responsibility regimes	Who is accountable for meaning, harm, or misrepresentation in hybrid texts

continued on following page

Table 2. Continued

AI Role Metaphor or Modality	Dominant Theoretical Metaframe	Typical Qualitative Processual Engagements	Reconfigurations of Authority and Voice	Salient Reflexive-Ethical Interrogatives
Synthetic Participant	Simulation ethics, speculative pragmatics	Generation of counterfactual accounts, scenario elaboration, dialogic simulation	Inserts artefactual voices into analytic corpus, risks ontological confusion	How to clearly demarcate synthetic utterances from lived testimonies
Methodological Lens	Critical STS, epistemic infrastructuralism	Constraint of inquiry through model embeddings, representational geometries	Configures entire field of visibility and legibility for analysis	Which realities become intelligible or unintelligible through this lens

Table 2 shows that the choice of AI modality is never a merely technical configuration, but a deeply normative and epistemological act that redistributes authority, visibility, and accountability within qualitative projects. When AI is treated as an instrumental tool, ethical concerns concentrate on data protection and error rates, whereas collaborator and co-author modalities require robust protocols for documenting interpretive negotiations and clarifying credit allocation in publications or policy briefs (Hitch et al., 2025). Synthetic participant and methodological lens modalities are particularly sensitive, since they alter the ontological composition of the corpus and the very horizon of what counts as evidence, which can either enable imaginative exploration or generate epistemic distortions. For practitioners, policy makers, and technologists, the table functions as a heuristic for aligning system configuration, interface design, and governance practices with explicit commitments to *human-centered AI*, *reflexive integrity*, and *justice-oriented inquiry*, thereby anchoring human-AI collaboration within a clearly articulated methodological and ethical architecture.

4. TRUSTWORTHINESS, REFLEXIVITY, AND METHODOLOGICAL INTEGRITY IN AI-MEDIATED QUALITATIVE INQUIRY

4.1 Rearticulation of Trustworthiness within AI-Suffused Inquiry Ecologies

Trustworthiness in qualitative inquiry traditionally orbits around the interlinked constructs of *credibility*, *transferability*, *dependability*, *confirmability*, and *authen-*

ticity, each of which encodes a particular configuration of evidentiary sufficiency, interpretive coherence, and ethical accountability. When AI systems participate in transcription, segmentation, coding, and synthesis, these criteria must be recalibrated rather than merely reiterated (Barrera et al., 2025). Automated pipelines can distort paralinguistic nuance, homogenise linguistic heterogeneity, or filter out so called noise that in fact carries critical contextual signals, thereby destabilising the very conditions that underwrite thick description and interpretive depth. Conversely, computational infrastructures can extend the analytic horizon by enabling cross case pattern recognition, iterative revisiting of large corpora, and rigorous documentation of analytic transformations. The central task is to re-specify trustworthiness as a relational property of human-machine-institutional assemblages rather than as a purely human virtue. Table 3 crystallises this rearticulation by juxtaposing classical dimensions with their AI specific vulnerabilities, remedial strategies, and reporting imperatives, providing a dense schema for practice-oriented reflection.

Table 3. Reframed Trustworthiness Dimensions for AI-Intensive Qualitative Methodology Designs

Trustworthiness Dimension	Classical Qualitative Interpretation	AI-Linked Vulnerabilities and Threats	AI-Informed Remediation and Enhancement Strategies	Key Elements for Transparent Methodological Exposition
Credibility	Plausible, experientially resonant accounts anchored in thick, situated description	Loss of nuance through aggressive compression, misrecognition of idioms, hallucinated paraphrases	Human verification of critical passages, collaborative member sense checking, constrained summarisation	Explicit description of AI stages affecting meaning, criteria for human override decisions
Transferability	Contextually grounded insights enabling analytic generalisation to cognate settings	Over extrapolation from pattern recognition, neglect of boundary conditions, decontextualised clustering	Rich contextual metadata retention, corpus stratification, boundary probing through contrasting cases	Specification of data provenance, contextual granularity retained or discarded by AI tools
Dependability	Coherent, traceable analytic trajectory under conditions of methodological adaptability	Opaque model updates, unstable outputs across runs, undocumented prompt variability	Version control for models and prompts, reproducible workflow scripts, stability probing	Disclosure of model identity, update cadence, and procedural safeguards for output consistency

continued on following page

Table 3. Continued

Trustworthiness Dimension	Classical Qualitative Interpretation	AI-Linked Vulnerabilities and Threats	AI-Informed Remediation and Enhancement Strategies	Key Elements for Transparent Methodological Exposition
Confirmability	Analytic claims grounded in systematic audit trails rather than idiosyncratic bias	Invisible algorithmic value loadings, training data asymmetries, automation bias in researcher judgement	Reflexive AI logbooks, adversarial prompting, counterfactual testing of emergent themes	Articulation of how AI bias was probed, logged, and mitigated across the project
Authenticity	Faithful representation of participant perspectives, affective textures, and marginal voices	Silencing of minoritised dialects, normalisation of dominant registers, sentiment misclassification	Custom vocabularies, accent sensitive pipelines, participatory review of AI derived portrayals	Clarification of how voice preservation was operationalised in AI configuration and checking
Rich Rigor	Adequate quantity, variety, and complexity of data and analytic engagement	Illusory rigor through volumetric processing, superficial breadth without interpretive depth	Multistage close reading, triangulation of AI outputs with manual micro analysis, iterative deep dives	Description of how computational extensiveness was balanced with intensive interpretive labour

Table 3 demonstrates that none of the classical criteria becomes obsolete in AI mediated environments, yet each acquires a new set of fault lines and affordances that demand explicit attention. *Credibility* becomes vulnerable when language models introduce spurious coherence, whereas *transferability* is jeopardised when probabilistic clustering is mistaken for context independent generality. *Dependability* can be destabilised by silent model retraining or undocumented prompt engineering, while *confirmability* is threatened by unexamined algorithmic bias and automation bias in human judgement. *Authenticity* and *rich rigor* risk erosion when minority voices are linguistically normalised or when high volume processing is equated with analytic profundity. By specifying countermeasures and reporting expectations alongside these risks, the table reframes trustworthiness as an outcome of conscious design, disciplined reflexivity, and transparent documentation across the entire human-AI analytic pipeline.

4.2 Reflexive Methodological Transparency for Algorithmically Mediated Workflows

Reflexive methodological transparency in AI intensive qualitative projects involves more than the inclusion of a brief software note, it entails a granular narrative of how computational infrastructures intersected with epistemic intentions and interpretive labour. Researchers need to specify which stages of the workflow were delegated to AI, what constraints governed prompt construction, which outputs were accepted or discarded, and how disagreements between machine suggestions and human intuition were adjudicated (Sun et al., 2025). Such transparency functions as an antidote to *methodological laundering*, where AI involvement is either minimised or aestheticised, thereby obscuring its influence on findings. Structured templates inspired by the architecture of Table 3 can support this effort by aligning descriptions of AI usage with specific trustworthiness dimensions, rather than relegating them to generic technical appendices. For funders, policy makers, and institutional review boards, reflexive transparency provides a basis for evaluating both risk and rigor, enabling them to distinguish between projects that superficially adopt AI tools and those that integrate them within a coherent, ethically attentive methodological design.

4.3 Black-Box Opacity, Algorithmic Explainability, and Interpretive Accountability

The proliferation of complex AI architectures introduces a persistent tension between black-box opacity at the algorithmic level and the demand for interpretive accountability at the methodological level. Tokenisation, multidimensional embeddings, and non-linear transformation layers can generate outputs that are locally coherent but globally untraceable, thereby challenging expectations that analysts should be able to reconstruct the inferential path from data to claim. Traditional qualitative justification relies on practices such as evidentiary exemplification, negative case analysis, and dialogic interrogation of rival explanations, all of which presuppose a humanly intelligible analytic trail. Explainability techniques, such as saliency mapping or feature importance estimates, only partially bridge this gap, as they translate internal model dynamics into yet another level of abstraction rather than into lived context. Consequently, interpretive accountability must be reconceptualised as a layered phenomenon, where researchers may accept algorithmic opacity at micro computational levels while insisting on radical transparency about data flows, task delegations, and interpretive checkpoints. This reconceptualisation resonates with the multidimensional framing of trustworthiness elaborated in Table 3, even though it cannot fully domesticate black-box complexity.

4.4 Reflexivity as an Ethical-Practical Discipline in AI Augmented Inquiry

Reflexivity within AI augmented qualitative research cannot remain a purely introspective stance, it must crystallise into an ethical-practical discipline that structures everyday research decisions and collective deliberations. This discipline includes systematic logging of AI interactions, critical reflection on moments of over trust or unwarranted scepticism toward machine outputs, and explicit discussion of how computational convenience shapes sampling, coding, and representation choices (Landerholm, 2025). Concepts from *virtue ethics* highlight the importance of cultivating intellectual humility, practical wisdom, and courage in refusing seductive yet epistemically corrosive shortcuts. *Care ethics* underscores the relational obligations researchers bear toward participants whose narratives may be filtered or reconfigured by AI systems, especially when these participants occupy precarious social locations. Reflexive routines can also incorporate *critical pedagogy*, where research teams periodically interrogate how macro structures such as platform capitalism, data extractivism, and regulatory regimes constrain or enable their methodological imagination. When aligned with the trustworthiness architecture in Table 3, reflexivity thus becomes a patterned practice that binds epistemic integrity to socio-political responsibility.

4.5 Methodological Integrity and AI-Literate Reporting Architectures

Methodological integrity in AI mediated qualitative inquiry can be defined as a state of coherent alignment between epistemological commitments, research questions, design choices, and the specific AI roles instantiated throughout the project. Integrity is compromised when AI tools are bolted onto interpretivist frameworks without considering how their underlying logics of optimisation, scale, and abstraction interact with constructs such as *voice*, *thick description*, or *emergent theory*. AI literate reporting architectures address this risk by mandating explicit articulation of epistemic stance, justification for each AI delegation, boundaries placed on automation, and criteria for evaluating AI outputs (Jones, 2025). Journals, professional bodies, and regulatory authorities can embed these expectations into submission guidelines, ethics templates, and evaluation rubrics, thereby normalising a culture in which AI usage is neither mystified nor trivialised. In such a culture, Table 3 functions as a conceptual touchstone that links micro level reporting practices to macro level commitments regarding trustworthiness, granting readers, reviewers, and policy makers a structured lens through which to appraise the integrity of AI intensive qualitative studies.

5. ALGORITHMIC BIAS, ETHICS, AND JUSTICE-ORIENTED FRAMEWORKS FOR QUALITATIVE AI

5.1 Typologies and Sources of Algorithmic Bias in Qualitative Pipelines

Algorithmic bias in qualitative AI is not an incidental malfunction but an emergent property of how data assemblages, modelling choices, and deployment environments interact. Within AI augmented qualitative workflows, bias manifests through dataset skew that overrepresents dominant groups, measurement bias that encodes culturally parochial proxies, representation bias that marginalises minoritised lexicons, and deployment bias that repurposes tools outside their originally envisaged sociotechnical context (Brâncoveanu, 2025). Feedback loop bias intensifies these distortions when outputs are recycled into new training corpora, thereby amplifying initial inequities. In transcription, such biases appear as systematic misrecognition of accents, dialects, and code-switching practices, while in topic modelling and clustering they surface as under segmentation of subaltern narratives or over aggregation of heterogeneous experiences. Synthetic text generation can normalise hegemonic idioms, making structurally privileged perspectives appear linguistically natural. Conceptualising these phenomena as expressions of *structural injustice*, *symbolic violence*, and *epistemic exclusion* prepares the ground for the normative frameworks systematised in Table 4, which align ethical theories with justice-oriented methodological responses.

5.2 Ethical Frameworks and Moral Philosophies for AI-Inflected Inquiry

Ethical reflection on AI mediated qualitative research requires more than generic appeals to beneficence or nonmaleficence, it demands the deliberate selection and articulation of coherent moral frameworks that can orient design and practice. *Utilitarian ethics* privileges aggregate welfare and harm minimisation, encouraging efficiency gains, scalable analysis, and cost sensitive decisions, yet struggles to protect vulnerable minorities when their interests do not significantly affect aggregate utility. *Deontological ethics* foregrounds duties, rights, and respect for persons, emphasising non instrumentalisation of participants, robust consent, and constraints on surveillance regardless of purported benefits. *Virtue ethics* focuses on the character of researchers as moral agents, cultivating epistemic humility, courage, and integrity in their engagements with AI tools (Cheah, 2025). *Care ethics* accentuates relational responsibility, attentiveness, and contextual sensitivity, especially in work with traumatised or precarious communities. These frameworks intersect and sometimes conflict, particularly when efficiency-oriented logics support forms of algorithmic intensification that clash with obligations to autonomy or care. Table 4 condenses these ethical grammars alongside justice-oriented constructs into a compact comparative matrix that can guide practical decision making.

5.3 Justice-Oriented AI Frameworks for Critical Qualitative Methodology

Justice oriented frameworks move beyond individual level ethics and interrogate the structural distributions of benefits, risks, and epistemic recognition that AI systems help to configure. *Data justice* examines how data generation, aggregation, and analytics reproduce or contest inequalities in visibility and control, while *information justice* considers who can meaningfully understand, contest, or repurpose outputs. *Decolonial AI* interrogates how inherited colonial logics of extraction, classification, and universalisation are rearticulated through contemporary digital infrastructures, and calls for *epistemic reparations* that foreground pluriversal knowledges and Indigenous data sovereignty (Kumar, 2025). Human rights frameworks introduce legally resonant constructs such as dignity, equality, and non-discrimination, providing levers for regulatory intervention and public accountability. Table 4 synthesises these frameworks with more classical ethical theories, demonstrating how each articulates distinctive central questions, typical uses, and methodological tensions in AI assisted qualitative inquiry.

Table 4. Ethical and Justice Frameworks for AI

Normative Framework	Core Value Commitments	Central Orienting Questions	Typical Qualitative AI Uses	Methodological and Ethical Tensions
Utilitarian Ethics	Aggregate welfare maximisation, efficiency, risk minimisation	What configuration yields greatest net benefits for affected populations	High throughput transcription, rapid sentiment dashboards, scalable pattern scanning	Marginal harms to vulnerable groups, crowding out of deep engagement by throughput logics
Deontological Ethics	Duties, rights, non instrumentalisation of persons	Which uses respect autonomy, consent, and inviolable moral constraints	Strict consent governance, opt out mechanisms, narrow task scoping for sensitive topics	Tension between rigid constraints and organisational demands for flexibility and innovation
Virtue and Care Ethics	Character, relational responsibility, attentiveness to vulnerability	What would a conscientious, caring researcher do in this context	Deliberative configuration of tools, slow checking of narratives, participatory review of outputs	Potential paternalism, unequal distribution of care labour, friction with productivity metrics
Data Justice	Fair distribution of data burdens and benefits, contestability	Who is rendered visible, invisible, or hyper visible through AI infrastructures	Bias audits, visibility mapping for marginalised lexicons, community facing analytic dashboards	Risk of technocratic framing of justice, limited community control over interpretation
Decolonial and Indigenous Epistemologies	Pluriversality, self-determination, Indigenous data sovereignty	How do AI systems reproduce or resist colonial knowledge hierarchies	Community governed archives, language revitalisation aids, Indigenous controlled analytic environments	Cooptation of decolonial rhetoric, infrastructural dependence on non-Indigenous platforms
Human Rights and Public Interest Governance	Dignity, equality, accountability, collective interests	How do uses align with rights, freedoms, and public goods	Impact assessments, rights-based evaluation of use cases, regulatory disclosure regimes	Formal compliance without substantive transformation, uneven enforcement across jurisdictions

Table 4 shows that ethical and justice frameworks do not merely add rhetorical decoration to AI mediated qualitative research, they fundamentally shape how problems are framed, whose interests are prioritised, and which trade-offs are deemed acceptable. A utilitarian orientation may endorse extensive automation when aggregate informational gains seem high, whereas a deontological stance might prohibit identical uses that infringe on consent or privacy. Data justice perspectives compel researchers to consider who bears the brunt of misclassification or misrecognition, while decolonial lenses interrogate the historiographical and geopolitical underpinnings of supposedly neutral tools. Human rights and public interest frameworks connect micro level methodological decisions with macro level governance architectures. For practitioners, the table provides a condensed decision scaffold that can be operationalised in protocol design, ethics review, and institutional policy making.

5.4 Data Privacy, Consent, and Confidentiality in AI Mediated Contexts

Data privacy, consent, and confidentiality acquire new configurations when qualitative materials are routed through AI infrastructures that often operate across jurisdictions, corporate boundaries, and temporal horizons. Privacy can no longer be conceptualised solely as individual control over identifiable information, it also encompasses group privacy, relational privacy, and contextual integrity. Consent shifts from a one-time signature to an ongoing, dialogic process, since future model updates, secondary uses, and cross system integrations are frequently unknowable at the moment of data collection. Confidentiality becomes fragile when small, distinctive narratives are transmitted to external servers, where they may contribute to latent representations that are not easily scrubbed. Concepts such as *differential privacy* or *federated learning* offer partial technical mitigations, yet they do not obviate the need for careful scoping of data flows, rigorous vendor due diligence, and clear communication with participants. Within the landscape mapped by Table 4, data justice and human rights frameworks provide especially strong resources for interrogating the adequacy of privacy and consent practices in AI assisted qualitative inquiry.

5.5 Governance Architectures, Normative Standards, and Institutional Responsibility

The governance of AI mediated qualitative research does not reside solely in individual researcher choices, it is structured by multi-level architectures that span institutional policies, professional standards, regulatory regimes, and platform governance. Institutions decide which vendors to contract, which data protection clauses to negotiate, and which infrastructural safeguards to adopt, thereby preconfiguring what is technologically possible for individual projects. Professional bodies can articulate *methodological charters* that specify acceptable and unacceptable uses of AI in interpretive work, embedding constructs such as *data justice*, *decolonial AI*, and *human-centered design* into disciplinary norms (Eryaman, 2025). Regulatory authorities and oversight bodies can mandate transparency, impact assessment, and rights protecting mechanisms, particularly for uses involving high risk populations or sensitive domains. Nonetheless, governance frameworks risk becoming perfunctory if they are not anchored in the substantive ethical grammars systematised in Table 4. Methodological integrity in AI augmented qualitative research therefore requires that institutions and practitioners co construct governance ecologies where normative frameworks, technical affordances, and lived practices mutually reinforce a justice-oriented vision of human-machine collaboration.

6. METHODOLOGICAL RE-IMAGINATION FOR AI-INTEGRATED QUALITATIVE INQUIRY DESIGNS

6.1 Conceptualising Hybrid and Computational-Qualitative Methodologies

Hybrid and computational-qualitative methodologies can be understood as deliberately engineered assemblages in which *interpretivist*, *critical*, or *phenomenological* logics of inquiry are orchestrated with computational procedures such as vectorisation, topic induction, and sequence modelling. Rather than collapsing qualitative inquiry into a derivative of *dataism*, these designs treat AI systems as contingent mediators that extend the sensory and analytic range of researchers while preserving human-centred hermeneutic sovereignty. Within such architectures, qualitative reasoning remains anchored in *thick description*, *reflexivity*, and *contextual integrity*, whereas AI subsystems undertake tasks such as pattern prefiguration, salience mapping, or high-volume text orchestration (Hitch et al., 2025). The methodological challenge lies in specifying where computation serves as *epistemic prosthesis* and where it risks becoming *epistemic usurper*. Table 5 later provides a condensed typology of such designs, mapping how different hybrids articulate primary logics of inquiry, dominant AI techniques, canonical applications, and characteristic vulnerabilities, thereby functioning as a conceptual cartography for practitioners, policy makers, and technologists who must architect AI-integrated qualitative infrastructures.

6.2 AI-Assisted Thematic Analysis, Grounded Theory, and Narrative Inquiry

AI-assisted thematic analysis, grounded theory, and narrative inquiry exemplify hybrid constellations in which pattern-sensitive computation is interleaved with theory-generative interpretation. In AI-assisted thematic analysis, clustering algorithms and embedding similarities can flag recurrent lexical constellations or anomalous segments, functioning as heuristic prompts for the construction of *candidate themes* that are subsequently refined through close reading and *constant comparison* (Hitch, 2024). In *computationally scaffolded grounded theory*, models may assist in surfacing families of codes across large corpora while the core logic of *theoretical sampling*, *theoretical saturation*, and *axial integration* remains firmly human directed. Narrative inquiry can enlist language models to propose alternative framings or to trace character trajectories across dispersed documents, yet meaning making still rests on interpretive engagement with plot, temporality, and agency. The designs codified in Table 5 treat such configurations not as technical recipes but as epistemic architectures that must be explicitly aligned with constructs such as *trustworthiness*, *authenticity*, and *critical reflexivity* if they are to maintain qualitative integrity.

6.3 Conversational Analysis with AI Augmentation and Computational Ethnography

Conversational analysis with AI augmentation and computational ethnography illustrate how interactional and field-based traditions can be reconfigured within AI intensive ecologies while retaining their analytic distinctiveness. In what can be termed *conversational analysis to the power of AI*, models assist in segmenting turn sequences, quantifying overlaps, tracking lexical entrainment, and mapping repair trajectories across extensive corpora, yet the core commitments of *sequential analysis*, *indexicality*, and *membership categorisation* remain interpretively adjudicated by human analysts. Computational ethnography extends the field site into digital platforms and sensor saturated environments, where AI pipelines ingest chat logs, forum posts, geo-tagged traces, and audiovisual materials, supporting multi scalar patterning of participation, rhythm, and circulation (Meylani, 2024). These designs are synthesised in Table 5, which situates them alongside other AI-integrated qualitative architectures and articulates their characteristic analytics, knowledge claims, and vulnerability profiles in a compact, high-density schema.

Table 5. Typology of AI-Integrated Qualitative Methodological Designs

Design Archetype	Primary Qualitative Logic of Inquiry	Core AI-Computational Techniques	Canonical Applications and Knowledge Claims	Key Epistemic-Ethical Vulnerabilities
AI-Assisted Thematic Analysis	Pattern oriented interpretivism, experiential phenomenology	Embeddings, clustering, keyword extraction, constrained summarisation	Articulation of recurrent meaning constellations across large yet bounded corpora	Decontextualised themes, erasure of subtle experiential textures, overreliance on salience metrics
Computational Grounded Theory	Emergent theory building, constant comparative methodology	Iterative code suggestion, co-occurrence mining, concept network visualisation	Mid-range conceptual models linking categories, properties, and conditions	Premature category crystallisation, importation of latent model ontologies into emergent theory
AI-Augmented Conversational Analysis	Micro interactional sequencing, ethnomethodological scrutiny	Turn segmentation, prosodic contour analysis, n-gram patterning	Fine grained accounts of interactional organisation across extensive datasets	Reduction of embodied nuance, privileging text over gesture, risk of mechanistic interpretation
Computational Ethnography	Field based thick description, ecological holism	Multi modal fusion, trajectory clustering, anomaly detection	Multi scalar mappings of practices, rhythms, and participation in socio-technical ecologies	Surveillance creep, participant trace overexposure, misalignment with community expectations
Post-Coding Meta-Inquiry	Secondary hermeneutics, archival re-reading	Cross project pattern mining, topic evolution tracking, meta summarisation	Higher order syntheses across legacy datasets, identification of latent cross cutting issues	Consent drift, historiographical distortion, re narration without community re-engagement
Hybrid Participatory AI-Infused Design	Participatory action research, design justice praxis	Interactive dashboards, co-creative generative models, explainable interfaces	Co-produced problem framings, scenario elaborations, and intervention blueprints	Unequal epistemic authority over tools, tokenistic participation, hidden infrastructural constraints

Table 5 demonstrates that AI-integrated qualitative designs are not a homogeneous category, but a spectrum of epistemic architectures that differ in their logics of inquiry, dominant computational manoeuvres, and characteristic risk profiles. AI-assisted thematic analysis and computational grounded theory are primarily text analytic, whereas AI-augmented conversational analysis and computational ethnography operate at interactional and ecological levels, and hybrid participatory designs instantiate *design justice* and *participatory action research* commitments within AI mediated environments. Each archetype carries specific vulnerabilities, from decontextualised themes and mechanistic interpretations to surveillance intensification, consent drift, and participatory tokenism. For practitioners and institutional decision makers, the table functions as a compact decision support instrument that clarifies how design choices bear on epistemic robustness, community accountability, and justice-oriented aspirations in AI intensive qualitative work.

6.4 Post-Coding Qualitative Inquiry and Re-Use of Archival Corpora

Post-coding qualitative inquiry and the AI mediated re-use of archival corpora create opportunities for longitudinal, cross project, and meta interpretive analyses that would be infeasible through manual techniques alone, yet they simultaneously introduce complex ethical and historiographical dilemmas. In post-coding inquiry, already analysed datasets are ingested into AI pipelines that scan for latent thematic alignments, discursive shifts, and neglected semantic micro-terrains, enabling the identification of *shadow narratives* and *latent contradictions* that earlier projects may have bracketed. Archival re-use can connect disparate studies across time, region, or institutional setting, underwriting claims about continuity, rupture, or emergent socio-cultural formations. However, such operations risk *consent drift*, where participants' original understandings of how their data would be used no longer match contemporary analytic horizons, and *historiographical distortion*, where presentist categories overwrite the idioms and horizons within which accounts were first articulated. The design archetypes labelled as *post-coding meta-inquiry* in Table 5 crystallise these possibilities and perils, underscoring the need for robust governance, community re-engagement, and reflexive documentation.

6.5 AI Literacy, Capacity-Building, and Pedagogical Infrastructures

AI literacy, capacity-building, and pedagogical infrastructures are preconditions for any ethically robust and epistemically defensible adoption of AI within qualitative research ecosystems. AI literacy for qualitative practitioners is not reducible to technical fluency with interfaces, it involves cognitive understanding of model logics, critical appreciation of *bias* and *representational geometries*, ethical discernment regarding delegation boundaries, and political insight into platform economies and regulatory landscapes (Finkenstadt, 2025). Capacity-building must therefore integrate dimensions of *technological pedagogical content knowledge* and *communities of practice*, where interdisciplinary teams engage in iterative, reflexive experimentation rather than one-off training events. Pedagogical infrastructures such as methods curricula, doctoral seminars, and supervision protocols can embed artefacts like Table 5 as didactic scaffolds, guiding students and mid-career professionals to design AI-integrated projects that are aligned with explicit epistemic and justice commitments. When such infrastructures foreground *human-centered AI*, *data justice*, and *methodological integrity*, they help to ensure that AI becomes a catalyst for imaginative, context-sensitive qualitative inquiry rather than a vehicle for technocratic homogenisation or epistemic erosion.

7. CONCLUSION

7.1 Synthesising Conceptual Architectures for AI-Augmented Qualitative Inquiry

The chapter has constructed an integrated conceptual architecture in which *epistemology*, *ontology*, *socio-technical design*, and *ethico-political governance* are treated as mutually entangled dimensions of AI-augmented qualitative research, rather than as isolated concerns. Epistemological cartographies distinguished *interpretivist*, *critical realist*, *feminist-intersectional*, *decolonial*, *pragmatist*, and *posthuman* orientations, each with distinct thresholds for what counts as legitimate AI involvement in meaning making. Ontological reflections clarified that AI artifacts can be framed as heuristic prompts, speculative simulations, or infrastructural lenses, whose status is always derivative of prior philosophical commitments. The reconfigured research assemblage placed AI alongside researchers, participants, and institutions as a *socio-technical actant* whose positionality must be explicitly theorised. Across this architecture, constructs such as *trustworthiness*, *reflexivity*, *methodological integrity*, *algorithmic bias*, *data justice*, and *human-centered AI* functioned as cross-cutting invariants that coordinate design choices, analytic practices, and governance arrangements within AI-intensive qualitative ecologies.

7.2 Paradigmatic Reorientation and Multilevel Implications for Stakeholders

The conceptual edifice developed across earlier sections calls for a paradigmatic reorientation in which AI is neither romanticised as an omniscient oracle nor trivialised as a neutral utility, but situated within contested fields of power, knowledge, and institutional interest. For academic methodologists, this entails revising canons of qualitative pedagogy so that constructs like *positionality of AI*, *multi-layered reflexivity*, and *socio-technical assemblage* become core rather than peripheral.

Graduate researchers must cultivate AI literacy that integrates technical comprehension, ethical discernment, and political economy awareness, thereby resisting both technophilic capitulation and reflex anti-automation. Policy analysts and industry researchers face the task of reconciling demands for scalability and rapid evidence generation with commitments to *thick description*, *voice preservation*, and *justice-oriented evaluation*. Ethics committees, professional associations, and regulatory bodies need to operationalise the normative grammars of *data justice*, *decolonial AI*, and *human rights* so that AI-intensive qualitative protocols are scrutinised not only for procedural compliance but for their implications for epistemic equity and public interest.

7.3 Future Trajectories and Collective Call for Reflexive, Justice-Oriented Practice

Future trajectories for AI-augmented qualitative inquiry involve the co-development of reporting standards, methodological charters, and pedagogical curricula that encode the constructs articulated throughout this chapter into everyday research infrastructures. Reporting standards must require explicit exposition of epistemic stance, AI role modalities, delegation boundaries, and reflexive logics, so that published studies render their human-machine entanglements inspectable and contestable. Methodological charters can codify red lines around uses that exacerbate surveillance, extraction, or epistemic dispossession, while encouraging *disobedient*, reparative, and community governed deployments of AI. Curricula for qualitative methods in universities, professional schools, and workforce development programmes should integrate hands-on engagement with AI tools under critically reflexive supervision, rather than relegating AI to ancillary technical workshops. Collectively, these trajectories amount to a call for a plural, reflexive, justice-oriented paradigm in which AI is treated as a condition to be governed, interpreted, and re-imagined, rather than as an inevitable fate, thereby preserving the transformative and emancipatory potentials of qualitative inquiry in an increasingly algorithmic world.

REFERENCES

Barrera, B., Feliu, A., Espina, C., Brand, T., Zeeb, H., & Ahmed, F. (2025). Leveraging AI to enhance qualitative research: Experiences and recommendations from case studies in cancer prevention literacy across the European Union. *International Journal of Qualitative Methods, 24*, 16094069251365766. DOI: 10.1177/16094069251365766

Brâncoveanu, C. (2025, August). Augmenting Systematic Literature. In *KI 2025: Advances in Artificial Intelligence: 48th German Conference on AI, Potsdam, Germany, September 16–19, 2025, Proceedings* (p. 3). Springer Nature. DOI: 10.1007/978-3-032-02813-6_1

Cheah, C. W. (2025). AI-Augmented Netnography: Ethical and Methodological Frameworks for Responsible Digital Research. *International Journal of Qualitative Methods, 24*, 16094069251338910. DOI: 10.1177/16094069251338910

Eryaman, M. Y. (2025). The Evolving Landscape of Qualitative Research: Exploring the Potential and Navigating the Challenges of Artificial Intelligence. *International Journal of Qualitative Methods, 24*, 16094069251400178. DOI: 10.1177/16094069251400178

Finkenstadt, D. (2025). AI-Augmented Theory Building: From Theoretical Foundations to Practical Application. *Customer Needs and Solutions, 12*(1), 6. DOI: 10.1007/s40547-025-00155-8

Friese, S. (2025). Generative AI: A catalyst for paradigmatic change in qualitative data analysis. In *Artificial intelligence (AI) in social research* (pp. 71-83). GB: CABI. DOI: 10.1079/9781800626607.0007

Hitch, D. (2024). Artificial intelligence augmented qualitative analysis: The way of the future? *Qualitative Health Research, 34*(7), 595–606. DOI: 10.1177/10497323231217392 PMID: 38064244

Hitch, D., Richards, K., Gupta, A., Thanekar, U., Edwards, J., & Goldingay, S. (2025). The Ethical Implications of Using AI in Qualitative Research. In *Artificial Intelligence (AI) in Social Research* (pp. 137-149). GB: CABI. DOI: 10.1079/9781800626607.0013

Hitch, D., Richards, K., Knight, R., Pepin, G., Thanekar, U., & Goldingay, S. (2025). AI Augmented Qualitative Analysis: Is it the Right Choice for Your Study? In *Artificial Intelligence (AI) in Social Research* (pp. 59-70). GB: CABI. DOI: 10.1079/9781800626607.0006

Jones, K. M. (2025). Generative AI in Qualitative Research and Related Transparency Problems: A Novel Heuristic for Disclosing Uses of AI. *International Journal of Qualitative Methods, 24*, 16094069251404329. DOI: 10.1177/16094069251404329

Kumar, P. (2025). Artificial intelligence (AI)-augmented knowledge management capability and clinical performance: Implications for marketing strategies in health-care sector. *Journal of Knowledge Management, 29*(2), 415–441. DOI: 10.1108/JKM-01-2024-0111

Landerholm, A. (2025). AI in Qualitative Health Research Appraisal: Comparative Study. *JMIR Formative Research, 9*, e72815. DOI: 10.2196/72815 PMID: 40627827

Maj, A., Makowska, M., & Sacharczuk, K. (2025). The content analysis used in nursing research and the possibility of including artificial intelligence support: A methodological review. *Applied Nursing Research, 151919*, 151919. Advance online publication. DOI: 10.1016/j.apnr.2025.151919 PMID: 40086938

Malik, F. S., & Terzidis, O. (2025). A hybrid framework for creating artificial intelligence-augmented systematic literature reviews. *Management Review Quarterly*, 1-27. DOI: 10.1007/s11301-025-00522-8

Meylani, R. (2024). Artificial intelligence in the education of teachers: A qualitative synthesis of the cutting-edge research literature. *Journal of Computer and Education Research, 12*(24), 600–637. DOI: 10.18009/jcer.1477709

Morgan, D. L. (2023). Exploring the use of artificial intelligence for qualitative data analysis: The case of ChatGPT. *International Journal of Qualitative Methods, 22*, 16094069231211248. DOI: 10.1177/16094069231211248

Nguyen-Trung, K. (2025). ChatGPT in Thematic Analysis: Can AI become a research assistant in qualitative research? *Quality & Quantity, 59*(6), 1–34. DOI: 10.1007/s11135-025-02165-z

Ozuem, W., Willis, M., Ranfagni, S., & Omeish, F. (2025). Thematic analysis in an artificial intelligence-driven context: A stage-by-stage process. *International Journal of Qualitative Methods, 24*, 16094069251362982. DOI: 10.1177/16094069251362982

Pattyn, F. (2025). The value of generative AI for qualitative research: A pilot study. *Journal of Data Science and Intelligent Systems, 3*(3), 184–191. DOI: 10.47852/bonviewJDSIS42022964

Shmueli, G., & Ray, S. (2024). Reimagining the Journal Editorial Process: An AI-Augmented Versus an AI-Driven Future. *Journal of the Association for Information Systems, 25*(1), 10. DOI: 10.17705/1jais.00864

Sun, H., Kim, M., Kim, S., & Choi, L. (2025). A methodological exploration of generative artificial intelligence (AI) for efficient qualitative analysis on hotel guests' delightful experiences. *International Journal of Hospitality Management*, *124*, 103974. DOI: 10.1016/j.ijhm.2024.103974

About the Contributors

Caroline M. Crawford, Ed.D., is a Professor of Instructional Design and Technology at the University of Houston-Clear Lake in Houston, Texas, United States of America. She earned her doctoral degree from the University of Houston in Houston, Texas, United States of America, in 1998, with specialization areas in Instructional Technology and Curriculum Theory, and began her tenure at the University of Houston-Clear Lake (UHCL) the same year. At this point in Dr. Crawford's professional career, her main areas of interest focus upon student success, engagement, and belonging, within learning environments. Further, Dr. Crawford's areas of expertise extend into communities of learning and the appropriate and successful integration of technologies into the learning environment; the learning environment may be envisioned as face-to-face, blended and online (virtual or text-driven) environments, as well as microlearning and micro-assist deliverables.

***

Freda Awonakie Akwetey is an educational administrator and consultant with expertise in higher education management, policy, and institutional development. She holds a PhD in Education (Management) from the University of South Africa, as well as an MPhil and MEd in Administration in Higher Education from the University of Cape Coast. She also earned a Bachelor of Education in Economics and Geography from the University of Cape Coast. Dr. Akwetey's professional and research interests focus on quality assurance in higher education, educational leadership and management, public policy in education, and the integration of emerging technologies such as artificial intelligence into research and academic practice. She is actively involved in academic writing, Educational Training and development and a professional teacher

Ashraf Alam is currently serving as an Assistant Professor in the Department of Education at NIT Calicut, Kerala, India. He has been recognised among the *World's Top 2% Scientists* by Stanford University and Elsevier. His academic contributions include more than 75 Scopus-indexed publications and over 100 conference presentations delivered at national and international academic forums. His work has appeared in reputed research publication platforms including Elsevier, Springer, Taylor & Francis, CRC Press, Routledge, Atlantis Press, Springer Nature, Palgrave Macmillan, and IEEE. He earned his Ph.D. from IIT Kharagpur, where his research engaged with foundational concerns in educational policy, curriculum design, educational technology, philosophy of education, innovation in education, educational administration, governance, and pedagogical transformation. His work is grounded in a transdisciplinary orientation that draws from educational sciences, digital pedagogy, curriculum theory, learning analytics, sustainability education, instructional design, artificial intelligence, ethics of technology, and socio-technical studies. Dr. Ashraf's scholarship strives to interrogate and expand the boundaries of educational thought by examining the epistemological, ethical, and social dimensions of digital and post-digital pedagogies. His work critically engages with the changing nature of knowledge production, learner agency, pedagogical mediation, and institutional transformation in technologically saturated educational environments. He is particularly invested in advancing inclusive digital learning, gamified instructional strategies, AI-mediated educational ecosystems, and ethically governed technology integration in education. His research agenda is animated by the pursuit of design-oriented, ethically attuned, and contextually responsive frameworks that can support meaningful educational reform. Through his work, he seeks to contribute to the development of pedagogical models that are not only technologically innovative but also socially just, epistemically inclusive, developmentally sensitive, and aligned with the broader goals of sustainable and human-centred education.

Alda Fildza Amima is a writer from Prawoto, Sukolilo, Pati, with a deep passion for writing. She enjoys creating inspirational stories and academic journal articles as a form of meaningful contribution to the scholarly world. Several of her academic works have been published on SINTA-indexed journals, reflecting her dedication to research and academic growth. In 2025, she graduated cum laude from Raden Mas Said State Islamic University of Surakarta. For Alda, this achievement marks the beginning of a lifelong journey of learning, growth, and sharing valuable insights through writing. She is currently pursuing a Master's degree at Sunan Kalijaga State Islamic University of Yogyakarta.

Francis Arthur, Ph.D, is a researcher and educator in Economics Education with expertise in advanced quantitative research methods, educational technology, and data analytics. He earned his doctorate in Economics Education from the University of Cape Coast and has developed strong methodological competence in Network Analysis, Covariance-Based Structural Equation Modelling (CB-SEM), Partial Least Squares Structural Equation Modelling (PLS-SEM), fuzzy-set Qualitative Comparative Analysis (fsQCA), Latent Profile Analysis, Artificial Neural Networks (ANN), hierarchical and multilevel modelling, and other sophisticated analytical techniques for examining complex relationships in education and economics. His research focuses on artificial intelligence in education, self-regulated learning, academic engagement, teacher self-efficacy, motivational processes, and technology-enhanced pedagogy. He has published extensively in leading international journals and conference proceedings, with studies addressing generative artificial intelligence, digital citizenship, educational metaverse, and innovative teaching and learning in higher education. Dr. Arthur is the recipient of several distinguished honours, including the International Society for Educational Planning Glen Earthman Outstanding Dissertation Award (2025), the Overall Best Presenter Award at the 5th International Conference of Postgraduate Studies at the University of Cape Coast, and the Outstanding Academic Achievement Award as the Overall Best PhD Student in Economics Education with a CGPA of 4.0. Also, Dr. Arthur serves on the editorial boards of international journals and is an active peer reviewer for numerous high-impact, Scopus-indexed, and Web of Science journals, including Journal of Cleaner Production, Interactive Learning Environments, European Journal of Education, Scientific Reports, and PLOS One. Through his scholarship and professional service, Dr. Arthur contributes to the advancement of rigorous, innovative, and policy-relevant research in economics education and the broader field of educational technology.

Nesha Rizky Ashari is from Medan, North Sumatra Province. She earned a Bachelor's degree in Sharia Economics from the State Islamic Institute (IAIN) Curup and a Master's degree in Sharia Economics from the State Islamic University (UIN) Sunan Kalijaga, Yogyakarta. She holds a professional certification as an Associate Wealth Planner (AWP) and a Training Methodology Trainer certification from BNSP. During her education, she was active in various student organizations, both internal and external, and was a recipient of a student exchange scholarship to China at the Wuxi Institute of Technology. Her areas of expertise and research interests include sharia economics, macro and microeconomics, development economics, and social welfare issues. She is currently active as an online tutor at Universitas Terbuka (Open University), a reviewer for the Jurnal Ekonomika (Publisher Bacaan Media), and a practitioner at PT Bank Syariah Indonesia.

Mehdi Bensouda is a professor and researcher at the National School of Business and Management (ENCG) of Fez, part of Sidi Mohamed Ben Abdellah University. He holds a PhD in Business Administration and is a member of the Research Laboratory in Management, Entrepreneurship, and Finance (LAREMEF), he investigates the role of mimetic pressures and financial slack in overcoming risk aversion toward green practices. Dr. Bensouda is a dedicated educator specializing in Advanced Accounting and Management Control, bridging the gap between theoretical institutional theory and practical organizational excellence, while emphasizing corporate sustainability and its implications for organizational performance.

Bartholomeus Dimanche Carl Beth is a computer science enthusiast with a strong interest in research and academic writing. His expertise lies in writing scientific journals and exploring topics related to technology, data systems, and innovation in computing. With a passion for continuous learning, he aims to contribute to the advancement of knowledge in the field of computer science through high-quality research and publications.

Eric Boateng holds a Bachelor degree in Economics Education from the Department of Business and Social Sciences Education, University of Cape Coast. His research interests cover education, economics and health policy.

Sahla Fauziah is a motivated individual with a strong interest in information technology, digital systems, and applied computing. She has participated in various academic and practical activities related to web-based application development, system analysis, and data management. Her work emphasizes efficiency, usability, and structured problem-solving. Sahla demonstrates solid analytical thinking and a collaborative mindset, allowing her to contribute effectively in team-based projects and multidisciplinary environments. She is accustomed to working with structured workflows and adapting to new technological challenges. In addition, Sahla continuously develops her skills by learning emerging technologies and applying them to real-world scenarios. With a strong commitment to growth and innovation, she aspires to support and contribute to the development of effective and innovative digital solutions, both in academic settings and future professional environments.

Dorcas Frempong is currently a second year Master of Philosophy student offering Clinical Health Psychology at the University of Cape Coast. She has a Bachelor of Science degree in Psychology and a certificate in Child and Adolescent Psychotherapy. Her research interests include children and adolescent mental health, geriatric mental health, parenting interventions, and the impact of caregiving on mental health outcomes.

Angga Muhammad Galih is a dedicated individual with interests in information technology, digital systems, and applied computing. He has been involved in academic and practical activities related to web-based application development, system analysis, and data management. His work emphasizes efficiency, reliability, and user-centered design in solving real-world problems. Angga demonstrates strong analytical skills and a collaborative mindset, enabling him to work effectively across technical and multidisciplinary environments. He continues to develop his expertise in emerging technologies and aspires to contribute to digital innovation in academic and professional contexts.

Francis Obeng Gyedu holds a Master of Philosophy degree in Education with focus on Economics from the Department of Business and Social Sciences Education, University of Cape Coast. His research interests cover education, economics of education, and educational psychology.

Harry is a faculty member in the Department of Sociology at Lovely Professional University, India. His expertise lies in migration studies, where he has made significant contributions through research papers and book chapters published in reputed journals and edited volumes. Beyond his academic pursuits, Dr. Harry has actively organized numerous cultural events, fostering student engagement and interdisciplinary dialogue. As an active member of the department, he plays a vital role in promoting academic excellence and cultural awareness.

Chheangkhy Hok is a senior research scholar of the Department of Humanistic Studies, Indian Institute of Technology (BHU), Varanasi, Uttar Pradesh, India. He obtained admission to the PhD programme in 2021 under the Doctoral Fellowship in India for ASEAN, where his research focuses on language and humanities-related studies within a comparative and interdisciplinary framework.

Sovanna Huot, currently enrolled as a PhD candidate in Political Science at the University of Delhi, New Delhi, India, is also a vice principal at a Government Public High School of Cambodia, where he holds the responsibilities of the school's managerial and academic institutional leadership. An accomplished scholar in a vast array of fields, he has undergone a two-year teacher training course from 2010-2012 and has earned bachelor's degrees in English and Educational Administration. Moreover, he holds two master's degrees, which he earned from a highly selective scholarship program, in TEFL and a Master of Education, in School Leadership, Upper Secondary School Principalship, from the National Institute of Education (NIE), Phnom Penh, 2021. His areas of focus are on educational reform and the practice-policy nexus of digit pedagogy, alongside glonacal (global–national–local) education systems. His scholarly work encompasses both political science and education; he has formal training in both fields and has published in both, including Scopus-indexed articles (Author ID: 59316244400).

Mark Inkoom is currently a Master of Philosophy student in Petroleum and Energy Studies at the Institute for Oil and Gas Studies, University of Cape Coast. He holds a Bachelor of Education degree in Social Sciences (Economics) and a certificate in AI Career Essentials from ALX Africa. His research interests are interdisciplinary and include Energy Economics, Resource Governance, Climate Change and Sustainable Development, AI and Machine Learning Applications in Energy and AI in Education.

Aqla Harun Ramadhani Johananda is an artificial intelligence researcher at Esa Unggul University, Jakarta, Indonesia. His research interests focus on machine learning, deep learning, data analytics, and intelligent information systems. He has been actively involved in academic research projects and applied AI development, particularly in designing data-driven models and intelligent applications to solve real-world problems. His work reflects strong analytical thinking, technical competence, and a commitment to continuous innovation. Aqla actively participates in collaborative research initiatives and scientific publications aimed at advancing artificial intelligence applications in digital transformation, education, and industry.

Nikolaos Koutsoupias is a Professor of Applied Informatics & Data Analysis in the Department of International & European Studies of the School of Social Sciences, Humanities & Arts, University of Macedonia. He graduated from the Department of Computer Science, North Carolina State University (NCSU), Raleigh NC, USA and a PhD holder from the Department of Applied Informatics of the University of Macedonia. He worked as a Computing Lab Assistant at the Department of Computer Science and the Faculty of Social Sciences & Humanities, as a Teaching Assistant at the School of Electrical & Computer Engineering (NCSU) and as a Database Programmer of an International Insurance Group (ERGO). He taught IT & Data Analysis courses at the School of NCO Administration of the Hellenic Air Force and at the Departments of General Studies & Marketing of TEI of Thessaloniki. As Lecturer/pd407, Assistant and Associate Professor taught at the Aristotle University and later at the University of W.Macedonia.

Seyha Kun holds a Doctorate in Education from the University of Delhi, India, and a Master's Degree in Educational Planning and Management from the International Institute for Educational Planning (IIEP) in Paris, France (2019). His extensive academic background also includes an MBA from the University of Management and Economics, and undergraduate degrees in English, Management, and Education (NIE). Since joining the Ministry of Education, Youth and Sport

(MoEYS) in 2007, Dr. Kun has transitioned from classroom teaching to a leadership role as Deputy Director of the Department of Planning. He is a key architect behind several major national policy documents, including the Education Strategic Plan (ESP), Budget Strategic Plan (BSP), Annual Operational Plan (AOP), Cambodia SDG4 Education 2030 Roadmap, and Gender Mainstreaming Strategic Plan (GMSP), among others, significantly contributing to capacity development within Cambodia's education sector.

Sovanna Loch is a prominent figure at the Cambodian Monastery Rajgir, serving as a deputy in charge. He is also a PhD research scholar and ICCR Fellow in the Dept. of English at Nava Nalanda Mahavihara (Deemed University), an institution under the Ministry of Culture, Govt. of India. His academic background includes a bachelor's degree in Pāli and a master's in English, both obtained from NNM. His research primarily focuses on Cambodian classical and Buddhist English literature.

Ghada Nakhla is the Undergraduate Mathematics Programme Leader in the Faculty of Arts and Sciences at Edge Hill University. She holds a PhD from Lancaster University and teaches across a wide range of mathematical disciplines, including calculus, linear algebra, statistical inference, statistical modelling, discrete mathematics, and data analytics. In addition to her teaching, she supervises MSc and PhD students, supporting the development of future researchers. Her research portfolio is extensive, and she has served as principal investigator on numerous projects. Her primary interests lie in advanced statistical methods, with a particular focus on the applications of the General Linear Model and Structural Equation Modelling in diverse fields. Her scholarship has been published in peer reviewed journals, featured as chapters in edited volumes, and presented at conferences. Beyond statistical modelling, Ghada's work extends into educational psychology. She has established a strong research agenda examining the role of fear in academic settings, exploring how emotional and psychological dynamics shape student performance. Her studies reveal both the hindering and motivating effects of fear, providing nuanced insights into how emotions influence academic engagement. Through this interdisciplinary approach, Ghada contributes not only to the advancement of quantitative research methods but also to a deeper understanding of the psychological factors that drive teaching, learning, and student success.

Sharon Abam Nortey holds an MPhil in Economics Education from the University of Cape Coast, Ghana. Her research interests include economics education, qualitative research, and artificial intelligence and emerging technologies in education. She also serves as a peer reviewer for Scopus-indexed journals.

Kimyut Phin is a diligent and committed educator and emerging scholar. He is currently pursuing a Ph.D. in English at the University of Delhi, India. He holds bachelor's and master's degrees in English (TEFL and TESOL). He serves as the Director of the Institute of Foreign Languages of the University of Heng Samrin Thbongkhmum, Cambodia. He actively engages with academic management, leadership, innovation, institutional development, curriculum development, instructional design, assessment, projects, and research. He has involved in many Erasmus+ projects, such as the REACT, SICA, BALANCE, IND4.0, and UNICATION projects. He has also had teaching experiences in linguistics, applied linguistics, teaching methodology, second language acquisition, grammar for writing, academic writing skills, global studies, core English, and literature studies for years. Further, he has experience in lesson planning, classroom management, assessment and evaluation, and technology for teaching and learning. His research centers on teaching and learning approaches and methods, English language education, education quality, educational reform, digital pedagogy, literary theory and criticism, and learning motivation in Cambodia and beyond.

Riska Prasasti is from Bone Regency, South Sulawesi Province. She completed her bachelor's and master's degrees in Sharia Economics at IAIN Bone and UIN Sunan Kalijaga Yogyakarta, respectively. During her studies, Riska was active in various internal and external student organizations, both academic and social. Her experience at the local and national levels includes various strategic roles that have shaped her competencies in administrative management, finance, organizational governance, as well as media and scientific publications. Currently, she is active as an online tutor at the Open University, a lecturer assistant at Nahdlatul Ulama University, and an administrator at KOPRI PB PMII, while continuing to develop her interest in writing

Dafelia Qodhari comes from Rejang Lebong Regency, Bengkulu Province. She holds a Bachelor's degree in Islamic Banking from the State Institute of Islamic Studies (IAIN) Curup and a Master's degree in Islamic Economics from Sunan Kalijaga State Islamic University, Yogyakarta. She is also a certified Associate Wealth Planner. During her academic years, she was actively involved in both internal and external organizations. She continues to actively write scholarly works, particularly on topics related to macroeconomics, development, and finance. Currently, she serves as a Master Teacher at Ruang Guru, a teaching assistant at the State Institute of Islamic Studies Curup, and a journal editor at PT Business Finance Analyst.

Silas Afutu Quaye is an MPhil student in Educational Planning at the Institute for Educational Planning and Administration, University of Cape Coast. His research interests cover Economics Education, School Physical Facilities, and the Application of Technology in Teaching and Learning

Emmanuel Quayson is currently a PhD student in Economics Education at the University of Cape Coast, Ghana. He holds a Master of Philosophy degree in Economics Education from the Department of Business and Social Sciences Education, UCC. His research interests span educational technology, applied economics, teacher education, and educational planning, with a current focus on the application of AI in teacher education and higher education.

Mouhcine Rhouiri PhD in Management Sciences from the LAREMEF laboratory at USMBA Fez. Specializing in the role of interactive management control in addressing hidden costs during times of crisis. His primary areas of research include management control, organizational resilience, Organizational Capability, and the application of Artificial Intelligence in Business contexts. In his teaching he juxtaposes theoretical and practical approaches to improve organizational efficiency and adaptability.

Rismawati is a distinguished academic and consultant with a robust focus on sustainability reporting. She currently holds the position of lecturer and Head of the Quality Assurance Institute at Universitas Muhammadiyah Palopo, where she has been pivotal in ensuring the quality of university processes and outputs since February 2020. Additionally, she serves as an Assurer for Sustainability Reporting at the National Center for Sustainability Reporting in Jakarta, specializing in the evaluation of sustainability reports for mining companies across Indonesia. Her academic qualifications are extensive, comprising a Bachelor's degree in Economics and Accounting, a Master's degree in Accounting Science, and a Doctorate in Accounting from Universitas Brawijaya, Malang. Rismawati is also a certified specialist in sustainability reporting, risk management, and tax planning, holding several prestigious certifications including Certified Sustainability Reporting Specialist (CSRS), Certified Risk Management Practitioner (CRMP), Certified Management Accountant (CMA) and Certified Enterprise Risk Analysis (CERA). In her professional capacity, she has provided consultancy services in financial management and tax compliance for companies such as PT. Reski Barito Timur and PT. Helay Raya. Furthermore, she has authored multiple publications on Social Accounting, Sustainability Accounting, Management Accounting and CSR, contributing significantly to the academic community. Her interests in reading, writing, and travel reflect her commitment to lifelong learning and professional growth in her field.

Hicham Saidi is a PhD (c) in management sciences at Sidi Mohamed Ben Abdellah University in Morocco. His research focuses primarily on open innovation in SMEs, examining the role of leadership and the impact of continuity practices in human resource management, particularly in times of crisis and major transformations such as digital transformation. His work draws on a variety of methodologies, combining quantitative and qualitative approaches, as well as systematic, integrative, and bibliometric literature review methodologies.

Gusti Muhamad Sardana is an information technology professional specializing in Data Science, with strong expertise in data analytics, machine learning, and intelligent systems. Professional experience includes developing and implementing data-driven approaches to support decision-making, optimize system performance, and solve complex real-world problems. He is actively involved as a researcher, focusing on emerging technologies and their practical implementation in information systems. Research activities emphasize bridging theoretical frameworks with applied solutions, ensuring academic rigor and real-world relevance. A strong interest in innovation, analytical thinking, and interdisciplinary collaboration underpin his professional and research work, enabling the delivery of impactful, scalable, and sustainable data-driven solutions that align with evolving technological and organizational needs.

Binastya Anggara Sekti has been a lecturer at the Faculty of Computer Science, Esa Unggul University, Jakarta, since 2016. He is actively involved in the Artificial Intelligence and Information Security Study Group. He holds a Bachelor's degree in Informatics Engineering from the Bandung Institute of Technology and a Master's degree in Information Systems Management from Trisakti University, Jakarta. Prior to entering academia, he accumulated over 25 years of professional experience as an IT practitioner in various companies across Jakarta. His work focused on developing data centers for internet service providers and delivering IT solutions for the manufacturing and oil & gas sectors. His research interests include artificial intelligence, cybersecurity, and the development of IT infrastructure.

Kritika Sharma is a Research Scholar in Sociology at Lovely Professional University, India. Her research interests lie in tribal studies, cultural studies, gender studies, and ethnography, with a particular focus on qualitative methods such as thematic analysis. She has published several research papers and book chapters in reputed journals and edited volumes, contributing to the growing discourse on indigenous knowledge systems and social structures. In addition to her scholarly contributions, she actively serves as a reviewer for academic journals, reflecting her commitment to advancing research in her field.

Marina Soltan is a Clinical Lecturer in Medicine at the University of Birmingham with research interests in clinical leadership, educational scholarship, public health policy, and data-driven research. Working at the intersection of healthcare systems and learning systems, she investigates how evidence, education, and leadership can combine to improve health outcomes and advance health equity. Marina holds both a Master's degree and a PhD in Medical Education, with her programmes of study grounded in experimental methods and the use of large-scale national databases such as UKMED to examine educational and workforce inequalities. She is currently undertaking a further PhD in Respiratory Medicine, using national epidemiological datasets to deepen understanding of respiratory disease risk and to inform population-level prevention strategies. Her leadership in translational and data-informed research has been recognised through funding from Health Data Research UK, enabling collaboration with the Helmholtz Institute to extend this work using biobank data. Dr Soltan is also an active public communicator, and her research on health inequality and innovation in healthcare has been featured on BBC News and in a BBC documentary with David Harewood.

Usep Suhud, M.Si., Ph.D. is a Professor of Marketing at the Faculty of Economics, Universitas Negeri Jakarta, Indonesia. He earned his Ph.D. in Marketing from Edith Cowan University, Australia, and his M.Si. in Marketing Communication from Universitas Indonesia. His research expertise spans consumer behavior, digital marketing, tourism marketing, and sustainable consumption. With over 2,400 citations and an h-index of 22, Prof. Suhud's works explore how technology, culture, and emotion shape consumer decision-making in both digital and tourism contexts Aligned with the chapter his scholarship critically examines the intersection of marketing and digital culture—particularly how algorithmic systems influence human perception, online engagement, and market behavior. His intellectual vision emphasizes ethical, human-centered approaches to marketing in AI-mediated societies.

Eko Sukoharsono, S.E., M.Com-Hons., CA, Ph. D. is professor of accounting and sustainability at the Faculty of Economics and Business Universitas Brawijaya, Indonesia. His research interests include environmental and social accountability, sustainability accounting, and digital accounting. Drawing upon postmodern and critical traditions, he investigates how technology, ethics and culture influence accounting and organisation. Introductory chapter Prof. Sukoharsono considers, as part of the book's first chapter, elements embedded in algorithmic and AI systems that are influencing decision making-making by public and private individuals, governance formation and sustainability reporting. He has an h-index of 26 with more than 2,700 citations, and his recent publications focus on digital accountability, AI-driven governance ethics, and Pancasila-based sustainability frameworks. His intellectual gaze connects accounting, tech and culture to catalyze responsible/ human-centered innovation.

Wahyu Wibowo is a graduate student in the Master's Program in Shariah Economics at UIN Sunan Kalijaga Yogyakarta. His academic interests include the Halal industry, Islamic philanthropy, and economic literacy. He received four DIKTI-funded PKM grants (2012–2014), three of which focused on community service—ranging from Jarimatika training to literary-mathematical education for orphans, one of which was selected as a national finalist. He also initiated an entrepreneurship project titled IBU SASTIKA (Innovative Mathematical Literature Book) that combines educational innovation with creative content.

Index

A

B

C

D

E

G

H

I

L

M

N

P

Q

R

S

T

www.ingramcontent.com/pod-product-compliance
Lightning Source LLC
LaVergne TN
LVHW081316300926
849161LV00008B/1779